AF412955

The Handbook of Environmental Chemistry

Volume 3   Anthropogenic Compounds
Part K

O. Hutzinger
Editor-in-Chief

Advisory Board:

P. Fabian · H. Fiedler · H. Frank · M. A. K. Khalil
D. Mackay · H. Neidhard · H. Parlar · S. H. Safe
A. J. B. Zehnder

Springer

Berlin
Heidelberg
New York
Barcelona
Hong Kong
London
Milan
Paris
Singapore
Tokyo

New Types of Persistent Halogenated Compounds

Volume Editor: J. Paasivirta

With contributions by

L. Asplund · Å. Bergman · J. de Boer · K. de Boer
J. P. Boon · E. Jakobsson · N. Kannan
E. Klasson-Wehler · J. Koistinen · R. J. Letcher
D. Muir · M. Oehme · J. Paasivirta · S. Sinkkonen
G. Stern · G. Tomy · W. Vetter

 Springer

Environmental chemistry is a rather young and interdisciplinary field of science. Its aim is a complete description of the environment and of transformations occuring on a local or global scale. Environmental chemistry also gives an account of the impact of man's activities on the natural environment by describing observed changes.

"The Handbook of Environmental Chemistry" provides the compilation of today's knowledge. Contributions are written by leading experts with practical experience in their fields. The Handbook will grow with the increase in our scientific understandig and should provide a valuable source not only for scientists, but also for environmental managers and decision makers.

ISSN 1433-6847
ISBN 3-540-65838-6
Springer-Verlag Berlin Heidelberg New York

Library of Congress Cataloging-in-Publication Data
The Natural environment and the biogeochemical cycles /
with contributions by P. Craig ... [et al.].
v. <A-F > : ill. ; 25 cm. -- (The Handbook of environmental chemistry :
v. 1) Includes bibliographical refereces and indexes.
ISBN 0-387-09688-4 (U.S). -- ISBN 3-540-55255-3 (pt. F : Berlin). --
ISBN 0-387-55255-3 (pt. F : New York)
1. Biogeochemical cycles. 2. Environmental chemistry.
I. Craig. P. J., 1944- . II. Series.
QD31. H335 vol. 1   [QH344]   628.5 s

This work is subject to copyright. All rights are reserved, whether the whole or part of the material is concerned, specifically the right of translation, reprinting, reuse of illustrations, recitation, broadcasting, reproduction on microfilm or in any other way, and storage in date banks. Dublication of this publication or parts thereof is permitted only under the provisions of the German Copyright Law of September 9, 1965, in its current version, and permission for use must always be obtained from Springer-Verlag, Violations are liable for prosecution under the German Copyright Law.

© Springer-Verlag Berlin Heidelberg 2000
Printed in Germany

The use of general descriptive names, registered names, trademark, etc. in this publication does not imply. Even in the absence of a specific statement, that such names are exempt from the relevant protective laws and regulations and therefore free for general use.

Herstellung: ProduServ GmbH Verlagsservice, Berlin
Typesetting: Fotosatz-Service Köhler GmbH, Würzburg
SPIN: 10683800   52/3020 – 5 4 3 2 1 0 – Printed on acid-free paper

Advisory Board

Prof. Dr. P. Fabian
Lehrstuhl für Bioklimatologie
und Immisionsforschung
der Universität München
Hohenbachernstraße 22
D-85354 Freising-Weihenstephan, Germany

Dr. H. Fiedler
Universität Bayreuth
Lehrstuhl für Ökologische Chemie
Postfach 10 12 51
D-95440 Bayreuth, Germany

Prof. Dr. H. Frank
Lehrstuhl für Umwelttechnik
und Ökotoxikologie
Universität Bayreuth
Postfach 10 12 51
D-95440 Bayreuth, Germany

Dr. M. A. K. Khalil
Oregon Graduate Institute
of Science and Technology
19600 N.W. Von Neumann Drive
Beaverton, Oregon 97006-1999, USA

Prof. Dr. D. Mackay
Department of Chemical Engineering
and Applied Chemistry
University of Toronto
Toronto, Ontario, Canada M5S 1A4

Dr. H. Neidhard
Umweltbundesamt
Bismarckplatz 1
D-13585 Berlin, Germany

Prof. Dr. Dr. H. Parlar
Institut für Lebensmitteltechnologie
und Analytische Chemie
Technische Universität München
D-85350 Freising-Weihenstephan, Germany

Prof. Dr. S. H. Safe
Department of Veterinary
Physiology and Pharmacology
College of Veterinary Medicine
Texas A & M University
College Station, TX 77843-4466, USA

Prof. Dr. A. J. B. Zehnder
EAWAG
Ueberlandstrasse 133
CH-8600 Dübendorf
Switzerland

Editor-in-Chief

Prof. Dr. em. Otto Hutzinger
Universität Bayreuth
Postfach 10 12 51
D-95440 Bayreuth, Germany
E-mail: otto.hutzinger@uni-bayreuth.de

Volume Editor

Prof. em. Jaakko Paasivirta
Department of Chemistry
University of Jyväskylä
P.O. Box 35
FIN-40351 Jyväskylä, Finland
E-mail: JPTA@CC.JYU.FI

Preface

Environmental Chemistry is a relatively young science. Interest in this subject, however, is growing very rapidly and, although no agreement has been reached as yet about the exact content and limits of this interdisciplinary discipline, there appears to be increasing interest in seeing environmental topics which are based on chemistry embodied in this subject. One of the first objectives of Environmental Chemistry must be the study of the environment and of natural chemical processes which occur in the environment. A major purpose of this series on Environmental Chemistry, therefore, is to present a reasonably uniform view of various aspects of the chemistry of the environment and chemical reactions occurring in the environment.

The industrial activities of man have given a new dimension to Environmental Chemistry. We have now synthesized and described over five million chemical compounds and chemical industry produces about hundred and fifty million tons of synthetic chemicals annually. We ship billions of tons of oil per year and through mining operations and other geophysical modifications, large quantities of inorganic and organic materials are released from their natural deposits. Cities and metropolitan areas of up to 15 million inhabitants produce large quantities of waste in relatively small and confined areas. Much of the chemical products and waste products of modern society are released into the environment either during production, storage, transport, use or ultimate disposal. These released materials participate in natural cycles and reactions and frequently lead to interference and disturbance of natural systems.

Environmental Chemistry is concerned with reactions in the environment. It is about distribution and equilibria between environmental compartments. It is about reactions, pathways, thermodynamics and kinetics. An important purpose of this Handbook, is to aid understanding of the basic distribution and chemical reaction processes which occur in the environment.

Laws regulating toxic substances in various countries are designed to assess and control risk of chemicals to man and his environment. Science can contribute in two areas to this assessment; firstly in the area of toxicology and secondly in the area of chemical exposure. The available concentration ("environmental exposure concentration") depends on the fate of chemical compounds in the environment and thus their distribution and reaction behaviour in the environment. One very important contribution of Environmental Chemistry to the above mentioned toxic substances laws is to develop laboratory test methods, or mathematical correlations and models that predict the environmental fate of new chemical compounds. The third purpose of this Handbook is to help in the basic understanding and development of such test methods and models.

The last explicit purpose of the Handbook is to present, in concise form, the most important properties relating to environmental chemistry and hazard assessment for the most important series of chemical compounds.

At the moment three volumes of the Handbook are planned. Volume 1 deals with the natural environment and the biogeochemical cycles therein, including some background information such as energetics and ecology. Volume 2 is concerned with reactions and processes in the environment and deals with physical factors such as transport and adsorption, and chemical, photochemical and biochemical reactions in the environment, as well as some aspects of pharmacokinetics and metabolism within organisms. Volume 3 deals with anthropogenic compounds, their chemical backgrounds, production methods and information about their use, their environmental behaviour, analytical methodology and some important aspects of their toxic effects. The material for volume 1, 2 and 3 was each more than could easily be fitted into a single volume, and for this reason, as well as for the purpose of rapid publication of available manuscripts, all three volumes were divided in the parts A and B. Part A of all three volumes is now being published and the second part of each of these volumes should appear about six months thereafter. Publisher and editor hope to keep materials of the volumes one to three up to date and to extend coverage in the subject areas by publishing further parts in the future. Plans also exist for volumes dealing with different subject matter such as analysis, chemical technology and toxicology, and readers are encouraged to offer suggestions and advice as to future editions of "The Handbook of Environmental Chemistry".

Most chapters in the Handbook are written to a fairly advanced level and should be of interest to the graduate student and practising scientist. I also hope that the subject matter treated will be of interest to people outside chemistry and to scientists in industry as well as government and regulatory bodies. It would be very satisfying for me to see the books used as a basis for developing graduate courses in Environmental Chemistry.

Due to the breadth of the subject matter, it was not easy to edit this Handbook. Specialists had to be found in quite different areas of science who were willing to contribute a chapter within the prescribed schedule. It is with great satisfaction that I thank all 52 authors from 8 countries for their understanding and for devoting their time to this effort. Special thanks are due to Dr. F. Boschke of Springer for his advice and discussions throughout all stages of preparation of the Handbook. Mrs. A. Heinrich of Springer has significantly contributed to the technical development of the book through her conscientious and efficient work. Finally I like to thank my family, students and colleagues for being so patient with me during several critical phases of preparation for the Handbook, and to some colleagues and the secretaries for technical help.

I consider it a privilege to see my chosen subject grow. My interest in Environmental Chemistry dates back to my early college days in Vienna. I received significant impulses during my postdoctoral period at the University of California and my interest slowly developed during my time with the National Research

Council of Canada, before I could devote my full time of Environmental Chemistry, here in Amsterdam. I hope this Handbook may help deepen the interest of other scientists in this subject.

Amsterdam, May 1980   *O. Hutzinger*

Seventeen years have now passed since the appearance of the first volumes of the Handbook. Although the basic concept has remained the same some changes and adjustments were necessary.

Some years ago publishers and editor agreed to expand the Handbook by two new open-ended volume series: Air Pollution and Water Pollution. These broad topics could not be fitted easily into the headings of the first three volumes. All five volumes series are integrated through the choice of topics and by a system of cross referencing.

The outline of the Handbook is thus as follows:

1. The Natural Environment and the Biochemical Cycles,
2. Reactions and Processes,
3. Anthropogenic Compounds,
4. Air Pollution,
5. Water Pollution.

Rapid developments in Environmental Chemistry and the increasing breadth of the subject matter covered made it necessary to establish volume-editors. Each subject is not supervised by specialists in their respective fields.

A recent development is the 'Super Index', a subject index covering chapters of all published volumes, which will soon be available via the Springer Homepage http://www.springer.de or http://www.springer-ny.com or http://Link. springer.de.

With books in press and in preparation we have now published well over 30 volumes. Authors, volume-editors and editor-in-chief are rewarded by the broad acceptance of the 'Handbook' in the scientific community.

May 1997   *Otto Hutzinger*

Contents

Foreword

Since the 1960s, persistent halogenated compounds as anthropogenic hazardous contaminants in the environment have been of high public concern and subject to restrictions in their uses and waste emissions. As a result, decreasing amounts of many organohalogen pesticides, PCDDs, PCDFs and PCBs have been observed regionally or even globally. However, substitution of PCBs, DDT etc. by presumably less harmful compounds has caused new contamination problems in some cases.

In addition, some PCB congeners, especially the planar ones, still seem to be on the increase in higher trophic levels due to their extreme persistency. In addition, new ecotoxic effects apparently caused by emissions and wastes have shown up. Breeding damages, sex impairment effects and decreasing fertility are suspected to be caused by persistent chemical pollutants. Some toxic impurities or metabolites of the originally used or emitted organohalogen compounds have been observed to be persistent and bioaccumulating and thus form an environmental threat in the future.

To obtain convincing evidence on chronic ecotoxicity of certain pollutants increasing in environmental matrixes is a complex and time consuming task requiring wide cooperation of scientists having different expertises. This book surveys the present knowledge of this task for a number of persistent halogenated organic substances which have been studied as potential environmental toxicants during the last two decades of the twentieth century in significantly more details than ever before.

Jyväskylä, July 1999   *Jaakko Paasivirta*

CHAPTER 1

Alkylaromatic Chlorohydrocarbons

Jaakko Paasivirta

J. Paasivirta (e-mail: jpta@cc.jyu.fi)
Department of Chemistry, University of Jyväskylä, P.O.Box 35, FIN-40351, Jyväskylä, Finland

Sources, structures, properties, analysis methods, and environmental fate including bio-accumulation and toxicity studies of some groups of alkylaromatic chlorohydrocarbons are reviewed. They are the most abundant chlorohydrocarbons formed in bleaching of pulp with chlorine chemicals. Major substances among them are chlorocymenes and chlorocymenenes (chlorinated methyl isopropyl(ene) benzenes. Other persistent compounds formed are polychlorobibenzyls (stilbene-related compounds), polychloromethylnaphthalenes, poly-chloroalkylphenanthrenes (especially chlororetenes) and polychlorofluorenes. These compounds occur as persistent contaminants in sediments and bioaccumulate in aquatic biota. Acute and chronic toxicity of only few compounds in these groups has been tested.

Isopropyl-PCBs, the main constituents in PCB substitute formulations Chloralkylene-9 and Chloralkylene-12, are more readily degraded in light irradiation and in metabolism than the corresponding non-alkylated PCBs. However, their bioaccumulation in aquatic ecosystem is of concern.

Tetrachlorobenzyltoluenes (TCBTs) as formulation Ugilec 141 are widely used PCB substitutes, especially in mining. There are 96 possible (70 relevant) TCBT isomers. Both Ugilec 141 mixture and nine pure isomers have been studied for their bioaccumulation, metabolism, and effects in aquatic biota and rodents. Based on similarity of toxic effects to PCBs and accumulation in fish and mussels of recipient waters of coal mining areas, TCBTs were not recommended as PCB substitutes.

Keywords. Alkyl-polychlorobibenzyls, Alkyl-polychlorobiphenyls, Alkyl-polychloronaphthalenes, Alkyl-polychlorophenanthrenes, Chlorocymenes, Chlorocymenenes, Chlororetenes, Tetrachloro-benzyltoluenes

The Handbook of Environmental Chemistry Vol. 3 Part K
New Types of Persistent Halogenated Compounds
(ed. by J. Paasivirta)
© Springer-Verlag Berlin Heidelberg 2000

List of Symbols and Abbreviations

AHH   Aryl hydrocarbon hydroxylase
BCF   Bioconcentration factor
CYMD   Polychlorocymenene
CYMS   Polychlorocymene
DDE   2,2-Dichloro-diphenyl-1,1-dichloroethene
DDT   2,2-Dichloro-diphenyl-1,1,1-trichloroethane
ECD   Electron capture detector
EROD   Ethoxyresorufin-*O*-deethylase
FID   Flame ionization detector

fw	fresh weight
GC	Gas chromatography
HRMS	High resolution mass spectrometry
K_{ow}	Octanol-water partition coefficient
LRMS	Low resolution mass spectrometry
lw	Lipid weight
MC	Methyl cholanthrene
m.p.	Melting point
MS	Mass spectrometry
PB	Phenobarbital
PCB	Polychlorobiphenyl
PCDD	Polychlorodibenzo-p-dioxin
PCDE	Polychlorodiphenyl ether
PCDF	Polychlorodibenzofuran
PROD	Pentoxyresorufin-O-deethylase
PLAC	Planar aromatic chlorocompound
RPCBB	Alkyl polychlorobibenzyl
RPCFL	Alkyl polychlorofluorene
RPCN	Alkyl polychloronaphthalene
RPCPH	Alkyl polychlorophenanthrene
RI	Retention index
RT	Retention time
SIM	Selected ion monitoring
TCBT	Tetrachlorobenzyltoluene

1
Introduction

The first organohalogen compounds found in the 1960s to cause ecotoxic effects were insecticide DDT and its metabolites, especially the very persistent DDE [1, 2]. Since then, the environmental impact from polyhalogenated persistent compounds emitted from agricultural and technical uses, industrial processes, combustion etc. has been intensively studied. Most interest in environmental toxicology was focused on dioxin-type of effects through planar ah-receptors from laterally substituted PCDDs and PCDFs, from non- or mono-*ortho*-substituted PCBs, and from other halogenated planar aromatic compounds such as hexachlorobenzene and polychloronaphthalenes [3]. Recent observations on reproductive and developmental damage in chronically exposed biota have indicated a significant potential of alkylated aromatic compounds (among others) as the original cause of this damage by endocrine disruption [4]. Because chlorine substitution makes the substances, in general, more persistent and bioaccumulative, the alkylaromatic chlorohydrocarbon class of substances, which includes the known xenoestrogen or antiandrogen DDT [5], is of increasing concern as potential long-term ecotoxicants.

2
Chlorocymenes and Cymenenes

2.1
Sources, Structures, and Properties

The most abundant chlorohydrocarbons formed in the bleaching of pulp with chlorine chemicals are chlorocymenes (CYMS) and chlorocymenenes (CYMD).

CH$_3$
Cl
Cl
Cl
CH
H$_3$C   CH$_3$
2,3,6-trichloro-
p-cymene

CYMS   CYMD   236CYMS

Emission of chlorocymenes (CYMS) to the environment was first detected by GC/MS analysis of sulfite pulp mill effluents [6–9]. CYMS and chlorocymenenes (CYMD) were found in kraft mill effluents [6]. The latter were more abundant than CYMS [6, 10, 11]. Later, Rantio reported that CYMDs, especially dichlorocymenenes, were formed in sulfite mill bleaching as well [12]. It could be deduced that *o*-, *m*-, and especially *p*-cymenes were formed from wood extractive monoterpenes in cooking of pulp. Then, in chlorobleaching, chlorocymenes were formed. Further, formation of chlorocymenenes could be attributed to HCl loss from cymenes chlorinated to the isopropyl side chain [11].

Kuokkanen has synthesized and characterized 13 CYMS and 7 CYMD model compounds (Table 1) [10, 11, 13, 14]. For chlorocymenes, three known methods were applied: chlorination of *p*-cymene in carbon tetrachloride; alkylation of chlorinated toluenes with acetone in presence of BF$_3$ and P$_2$O$_5$; and HNO$_3$/H$_2$SO$_4$ reaction of *p*-cymene, reduction of the 2,6-dinitro product to 2,6-diamino-*p*-cymene and Sandmeyer reaction of the latter to 2,6-dichloro-*p*-cymene (26CYMS). Chlorocymenenes were synthesized from the appropriate chlorocymenes by bromination of the isopropyl group and alkali treatment to eliminate HBr [10]. Preparative GC was used for separation of the compounds from their mixtures [11, 13, 15]. The products were characterized by mass, ^{1}H NMR, ^{13}C NMR, and IR spectroscopy [11, 14, 15]. Only 1 of 19 compounds prepared, thus far, is solid at room temperature: tetrachloro-*p*-cymene (2356CYMS) has a m.p. of 60 °C [16]. CYMS and CYMD reference compounds used for analyses (including mass spectral data) are listed in Table 1.

Mass spectra of CYMS and CYMD isomers were too similar to allow distinction between their structures. In contrast, structure verification was readily obtained from NMR data. IR spectra as fingerprints for identification of the individual congeners were decisively different from each other [11]. Gas chromatographic separation of congeners was feasible with non-polar capillary columns [11, 17–20].

Table 1. GC retention, CAS numbers, and low resolution mass spectral data of chlorocymenes and chlorocymenenes [11, 13, 17–19]

Compound	Abbr.	RRT[a]	CAS Nr	LRMS peaks m/z (Intensity %) [10]							
2-chloro-*p*-cymene	2CYMS	0.437	4395-79-3	168(33)	153(100)	133(18)	115(68)	105(11)	91(31)	75(17)	58(21)
3-chloro-*p*-cymene	3CYMS	0.457	4395-80-6	168(24)	153(100)	125(10)	115(62)	105(8)	91(39)	77(10)	63(17)
2,3-dichloro-*p*-cymene	23CYMS	0.608	81686-44-4	202(27)	187(100)	167(4)	151(30)	139(7)	115(52)	102(5)	89(12)
2,5-dichloro-*p*-cymene	25CYMS	0.544	86186-41-1	202(35)	187(100)	167(8)	151(31)	139(5)	115(38)	102(3)	89(8)
2,6-dichloro-*p*-cymene	26CYMS	0.566	81686-46-6	202(29)	187(100)	167(19)	151(33)	139(10)	115(60)	102(6)	89(15)
3,5-dichloro-*p*-cymene	35CYMS		81686-45-5	202(29)	187(100)	167(2)	151(32)	139(3)	115(40)	102(3)	89(8)
2,3,5-trichloro-*p*-cymene	235CYMS			236(24)	221(100)	185(24)	173(4)	150(25)	128(8)	115(29)	89(5)
2,3,6-trichloro-*p*-cymene	236CYMS	0.755	81686-42-2	236(57)	221(100)	185(41)	173(11)	150(38)	128(8)	115(44)	89(5)
2,3,5,6-tetrachloro-*p*-cymene	2356CYMS	0.957	81686-43-3	272(37)	257(100)	121(21)	184(21)	149(20)	128(5)	115(5)	92(4)
5-chloro-*o*-cymene	5oCYMS	0.432		168(25)	153(100)	125(6)	115(63)	103(7)	91(27)	77(12)	63(18)
3,5-dichloro-*o*-cymene	35oCYMS			202(26)	187(100)	151(23)	139(2)	125(3)	115(29)	89(5)	75(6)
5,6-dichloro-*o*-cymene	56oCYMS			202(31)	187(100)	151(17)	139(2)	125(3)	115(15)	91(1)	75(2)
2,6-dichloro-*m*-cymene	26mCYMS			202(54)	187(100)	167(8)	151(33)	139(5)	115(31)	89(4)	75(3)
2-chloro-*p*-cymenene	2CYMD	0.500									
3-chloro-*p*-cymenene	3CYMD	0.495		166(100)	151(37)	131(92)	129(21)	115(69)	103(6)	91(48)	64(21)
2,3-dichloro-*p*-cymenene	23CYMD	0.637		200(100)	185(10)	165(86)	160(25)	150(19)	130(45)	115(22)	63(12)
2,5-dichloro-*p*-cymenene	25CYMD	0.564		200(100)	185(22)	165(95)	150(24)	130(75)	115(39)	102(9)	63(67)
2,6-dichloro-*p*-cymenene	26CYMD	0.645	145526-34-7	200(100)	185(10)	165(87)	160(29)	150(19)	130(48)	115(29)	64(22)
3,5-dichloro-*p*-cymenene	35CYMD	0.583		200(73)	185(19)	165(100)	150(27)	130(80)	115(36)	102(13)	63(26)
2,3,5-trichloro-*p*-cymenene	235CYMD	0.731									
2,3,6-trichloro-*p*-cymenene	236CYMD	0.743	118795-95-2	234(93)	219(11)	199(100)	184(11)	164(78)	149(25)	129(52)	63(31)

[a] Retention times in GC on non-polar (SE-54) column relative to 2,4,6-trichlorobiphenyl from five different runs: means of 1–5 values.

2.2
Analyses

2.2.1
Extraction

CYMS and CYMD have been analyzed from water, effluent, biosludge, sediment, and biota samples [17]. Before extraction, samples were spiked with internal standards for quantitation. Water and effluent samples were filtered using pre-weighed glass microfibre filters. Filtrates were made alkaline (pH > 9) and extracted twice with hexane in a separating funnel. Particles on filter, sludge, and sediment samples were air-dried, weighed, and extracted in a Soxhlet with toluene for 48 h [12, 17, 20, 21]. Biological samples were dried either by mixing with anhydrous sodium sulfate or freeze-dried and extracted in Soxhlet with a mixture of petroleum ether (40–60 °C):acetone:hexane:diethyl ether 18:11:5:2 by volume for 6 h [12, 16, 22].

The extracts were evaporated with Rotavapor and finally with a cold nitrogen gas stream and, in case of biota samples, the residue weighed to give lipid contents. Then the residue was dissolved in hexane for cleanup [12, 17, 21–23].

2.2.2
Cleanup

Earlier cleanup for CYMS and CYMD determination was performed by shaking with concentrated sulfuric acid [12, 21, 24]. However, later it was found that 23 CYMD and 25 CYMD degraded 100 and 50 %, respectively, with sulfuric acid treatment [17]. Therefore, in later studies the extract was purified in a Pasteur pipette column packed with 2 g of neutral aluminum oxide (Merck, activity I), which was first activated at 800 °C and then deactivated with water. About 0.5 ml of a sample in hexane was added into the dry column, and the analytes were eluted out with 15 ml of hexane [17, 22, 23].

2.2.3
Final Determination

Temperature programmed GC on capillary columns has been mostly used for determination of CYMS and CYMD congeners. Mean relative retention times from non-polar column are presented in Table 1. Detectors used have been ECD [10, 21, 24], FID [13], and MS [11, 17–20]. Generally, the most sensitive and selective determination was achieved by selected ion monitoring (SIM) mass spectrometry [17].

Table 2. Calculated, measured, [16] and estimated [17, 25, 26] properties at 25 °C

Property	23 CYMS	26 mCYMS	25 CYMS	236 CYMS	2356 CYMS	236 CYMD
M g/mol^{-1}	203.1	203.1	203.1	237.6	272.0	235.5
Log K$_{ow}$	5.5[a]	5.55[b]	5.6[a]	6.2[a]	6.8[a]	5.42[b]
Vapor pressure Pa	4.26[b]	4.28[b]	5.36[b]	1.33[b]	0.00439[b,c]	0.450[b]
Solubility mol m^{-3}	0.029[a]	0.0172[b]	0.024[a]	0.004[a]	0.0012[a,c]	0.023[b]
Half-life h (19 °C)						
Sediment	10,000[b]	10,000[b]	10,000	10,000	10,000	10,000
Water	100[b]	100[b]	100[b]	100[b]	100[b]	100[b]
Fish	1500[b]	1500[b]	1500[b]	1600[b]	1700[b]	1500[b]

[a] Measured properties.
[b] Estimated properties.
[c] Value for subcooled liquid (of 2346 CYMS, m. p. 60 °C).

2.3
Environmental Fate

2.3.1
Fate-Related Properties and Toxicity

CYMS and CYMD are major chlorohydrocarbons formed in chlorobleaching of pulp and in chlorodisinfection of waters containing dissolved organic materials. Therefore, their fate in recipient aquatic ecosystem is of concern. Their persistency was found to be high in laboratory tests against concentrated sulfuric acid [11, 17], and also in secondary biological treatment of pulp bleaching effluents [18]. For the first environmental fate predictions, some properties (listed in Table 2) of five CYMS and one CYMD have been measured [16] or estimated [25, 26]. Toxicity of CYMS and CYMD has not been reported except one series of tests with *Daphnia magna*, which gave for 2356 CYMS LC$_{50}$ of 42 µg l^{-1} and 370 µg l^{-1} in pure and humic water, respectively [27].

2.3.2
Occurrence in Sediments and Biosludges

In a study of dated sediment layers of Central Finland 236 CYMS and 236 CYMD were measured and found most abundant in the near recipients of bleaching pulp mills. They were measured as significant contaminants with decreasing gradient in bottoms of the recipient watercourse lakes tens of kilometers downstream from the mill discharge points [24]. An example of the contents is shown in Fig. 1.

According to a further studies five CYMS congeners (2 CYMS, 23 CYMS, 25 CYMS, 26 mCYMS, and 236 CYMS), but no CYMDs could be measured in three other pulp mill recipients of Finland in amounts of tens of nanograms per gram of dry sediment [17, 23].

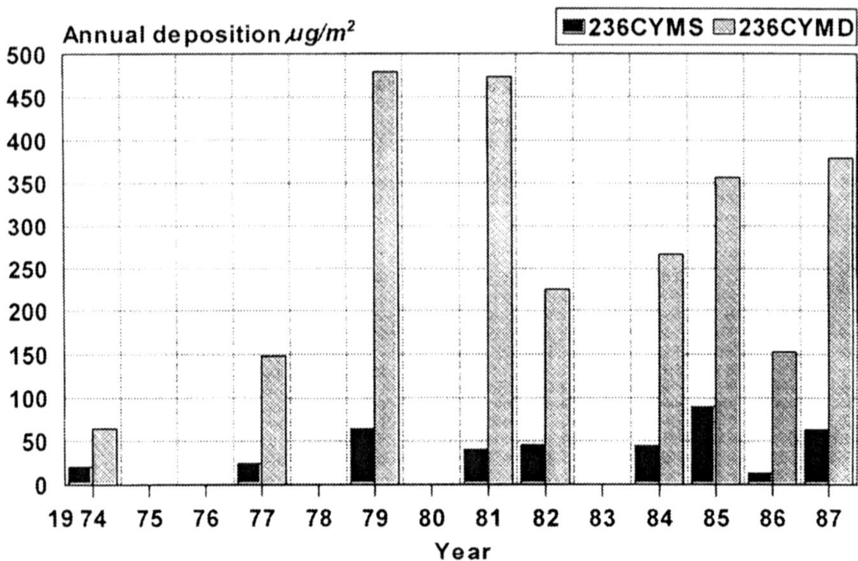

Fig. 1. Analysis results of dated sediment layers at Lake Kuhnamo 1 km downstream from pulp mill discharge at Äänekoski, Central Finland [24]

Contents of surface sediments showed a slight decreasing gradient in a marine recipient at the west coast of Finland but increasing gradient in a fast-flowing recipient river in South-East Finland (Fig. 2). The latter virtual paradox was explained by deposition of pure particles from upstream [17, 23].

23CYMS and 25CYMS, major components in bleaching mill discharges [12, 17, 22, 23], were found in bottom solids of Lake Baikal near pulp mill discharge point [17] and in dated sediment layers of Lake Ladoga, in accordance with the history of closed and operating pulp mills [28]. CYMS and CYMD compounds were also found to accumulate significantly to the return and waste biosludges of the activated sludge treatment plants of bleaching pulp mills [12, 17, 29]. These results indicate a high environmental persistency of CYMS and CYMD compounds. Biodegradation rates seem to vary very much depending on conditions. Activated sludge treatment reduced chlorocymenes from effluent to 30% in a Swiss mill [30] and to 3% in a sulfite mill in Finland [12]. In contrast, treatment in a Finnish kraft mill showed no reduction at all [12, 18]. Levels of CYMS and CYMD were abundant both in kraft mill biosludge (800–1500 ng g^{-1} dw) [29] and in effluent after the activated sludge treatment [18].

Contents of SCYMS (23CYMS + 26mCYMS + 25CYMS + 236CYMS + 2356CYMS) in surface bottom sediments of Kymijoki River, Finland, in spring 1993 were observed 8 km and 16 km downstream of kraft mill discharge point and found to be 41.4 ng g^{-1} and 83.2 ng g^{-1} dw [17, 25]. Fate modeling by program PPEF [31], based on measured contents in mill effluent in March 1993, predicted concentrations of discharge at 4 km and 26 km to be 12.3 ng g^{-1} and 7.3 ng g^{-1} dw [17, 25]. While emissions to Kymijoki River decreased steeply

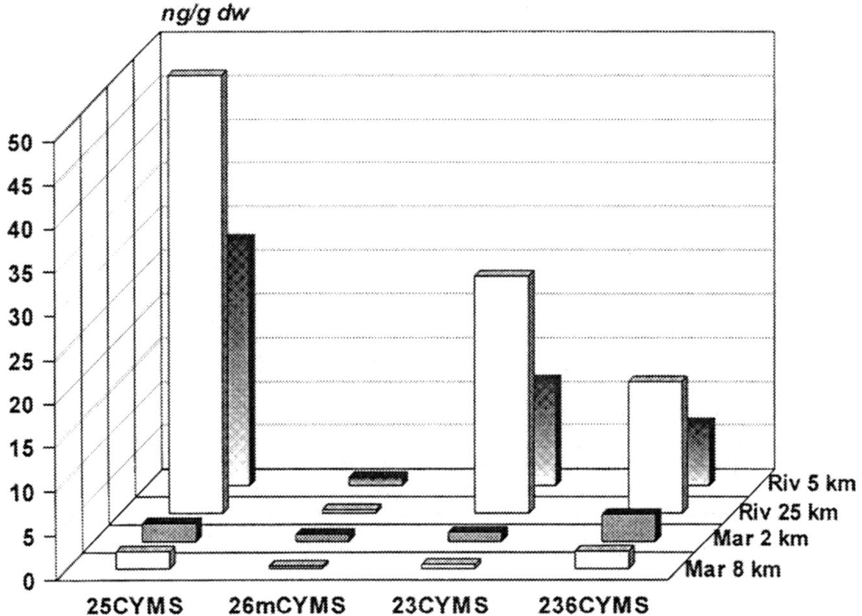

Fig. 2. Contents in surface sediments of marine coastal environment (Mar) of Pietarsaari pulp mill at West-Coast of Finland and of bottoms of the Kymijoki River (Riv) downstream of the Kuusankoski pulp mill at two distances from the discharge point [17, 23]

during 1990–1993, the significantly higher levels observed compared to modeled concentrations are due to earlier discharges and indicate high persistency of CYMS. Modeling with an interim version of program WATER [32] gave the same results as PPEF at Kymijoki [23] and a similar result for 236CYMS at Äänekoski watercourse [33].

2.3.3
Contents in Biota

First detection of chlorocymenes and chlorocymenenes in fish took place in a food chain study of contaminants in clean and polluted lakes of Finland [34]. Eleven roach and four pike specimens from lake Vatia 15 km downstream from a pulp mill were analyzed for persistent chlorohydrocarbons. 236CYMS and 236CYMD were identified with authentic model substances by GC/ECD at levels near to the detection limit of 0.2 ng g^{-1}. 236CYMS was detected in all fish samples and 236CYMD in three roach and one pike [34]. However, three pike from a more remote recipient and three pike from a pure reference lake also contained a trace of 236CYMS, which indicated partial air-transport [34].

Further analyses of chlorocymenes and cymenenes in kraft mill recipients were carried out on Kymijoki River perch and pike and Pietarsaari pike in 1993 [17, 23]. Perch contained 31 ng g^{-1} and 32 ng g^{-1} lw of 25CYMS and 2356CYMS,

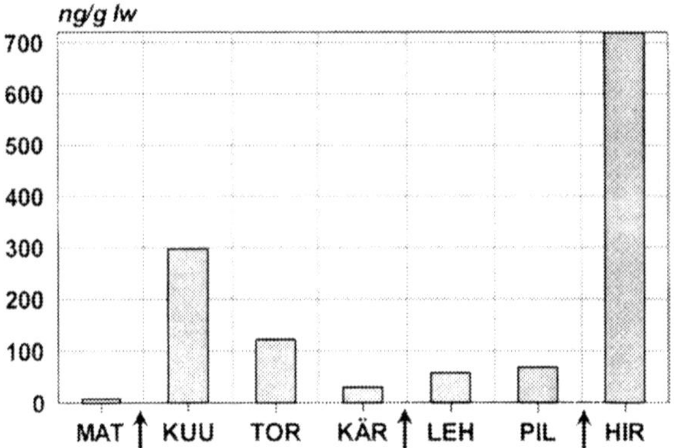

Fig. 3. Mean concentrations of 236 CYMS in mussels incubated in August 1986 at seven stations of the Kymijoki basin, Finland [32]. MAT is reference station 40 km upstream of the first bleaching pulp mill in Äänekoski. Stations KUU, TOR, and KÄR are 18, 40, and 75 km downstream. LEH is 15 km downstream from another mill, where bleaching was stopped in 1981. PIL and HIR are stations of the Kymijoki River (outflow from the Lake Päijänne) upstream and downstream from additional large pulp mill. Places of discharge are indicated as *arrows*

respectively. Kymijoki pike (four in number) contained 32 ng g^{-1}, 84 ng g^{-1}, and 220 ng g^{-1} lw of 25 CYMS, 236 CYMS, and 236 CYMD. Pietarsaari pike (six in number) contained on average 42 ng g^{-1}, 53 ng g^{-1}, and 56 ng g^{-1} lw 25 CYMS, 23 CYMS, and 236 CYMS. In addition, one Pietarsaari pike was found to contain about 40 ng g^{-1} lw of an unknown tetrachlorocymene (*o*- or *m*-) [17, 23].

Mussels (*Anodonta piscinalis*) incubated for regular monitoring each August in Finnish pulp mill recipient watercourses were, since 1985, found to accumulate significant amounts of chlorocymenes and chlorocymenenes [21, 35, 36]. Steeply degreasing gradient of contents at incubation stations downstream of pulp mill discharges (example in Fig. 3) clearly showed these chlorohydrocarbons to originate from bleaching.

In the 1980s, process improvements in pulp mills, e.g., continued cooking, transfer from free chlorine to chlorine dioxide in bleaching and including secondary wastewater treatment before discharge, caused a significant decreasing time trend of organochlorine emission from pulp mills to recipient watercourses in Finland. Such a trend is demonstrated in decrease of mean 236 CYMS and 236 CYMD contents in incubated mussels (Fig. 4) [21, 35, 37].

Chlorocymenes and chlorocymenenes have shown their high bioaccumulation rates in fish [17, 23] and incubated mussels [21, 35, 37]. Passive accumulation from watercourse to SPMD devices (triolein-filled polyethylene layfat tubings) in use for mussel incubation also took place at a significantly high rate [37]. Accumulation of CYMS and CYMD from effluent water to biosludge was also high [23]. Overall bioaccumulation rates of CYMS were slightly higher than those of CYMD [38]. While CYMS also have high environmental persistency and presumably potency to cause toxic effects, they must be considered as possible long-term environmental hazards from chlorination in bleaching and water disinfection processes.

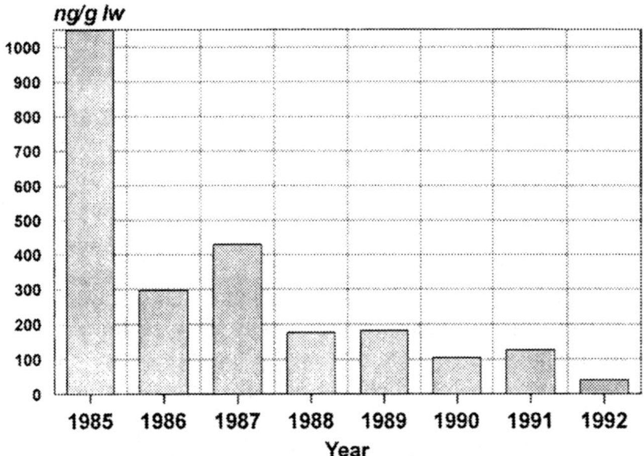

Fig. 4. Trend of the mean contents of 236CYMS in mussels incubated at station KUU in successive years 1985–1992 [33, 35]

3
Other Alkylaromatic Chlorohydrocarbons from Bleaching

3.1
Structures, Syntheses, and Origin

In addition to chlorocymenes and chlorocymenenes, four groups of persistent alkylaromatic chlorohydrocarbons have been detected and characterized from chlorobleaching of pulp. They were assumed to be planar aromatic chlorocompounds (PLAC) because they were found in cleanup fractions of PCDDs and PCDFs after reversed elution with toluene from a carbon column [39]. PLAC groups were alkyl-polychloronaphthalenes (RPCN), alkyl-polychlorobibenzyls (RPCBB), alkyl-polychlorofluorenes (RPCFL), and alkyl-polychlorophenanthrenes (RPCPH). Among the latter, chlororetenes are most abundant [40].

Occurrence of alkyl polychloronaphthalenes (RPCN) in spent bleach liquors was first time reported in 1979 [8]. Their structure was deduced to contain one methyl and one isopropyl group [8] which indicates origin from wood. Later analyses indicated that most common RPCNs in kraft pulp mill effluents and sludges were mono- to tetramethyl polychloro naphthalenes (C1-4PCN) [41–45]. In sludges their total contents were 28–72 ng g^{-1} dw and in effluents only trace levels of a few ng l^{-1} [42, 45]. Their origin was suggested to be chlorination of alkyl naphthalenes abundant in oil-based defoamers used in pulp mills [42].

No source of environmental RPCNs other than chlorination in bleaching-related processes is known. Analysis of urban air of Bordeaux showed abundant occurrence of methyl, dimethyl, and trimethyl naphthalenes (obvious oil components) and also mono- and dichloro naphthalenes (combustion products), but no traces of RPCNs [46].

C5-PCBBs were first detected as "TeCDD imposters" in analyses of coffee filter paper and pulp mill effluent by monitoring tetrachlorodibenzo-p-dioxins in GC/LRMS/SIM with ions m/z = 320 and 322. Their large peaks at the TeCDD window had intensity ratio peak 320:peak 322 = 100:65 which corresponded to two chlorines in the molecule instead of the four of TeCDD.

The compounds were proposed to be dichloro C6-polychlorodibenzo-furans [46]. Similar observations from pulp mill samples were done by Kuehl et al. [47]. They suggested the structures of these compounds to be chlorinated xanthenes and xanthones. Buser et al. reported the occurrence of methyl-, poly-methyl-, and alkyldibenzofurans in pulp mill sludge and sediments [48].
Later, C5-PCBBs were detected as three dichloro and four trichloro congeners in pulp mill effluents, recipient sediment, and mussels incubated in recipient water [49]. Mass spectra of these congeners ruled out the structure of alkyl polychlorodibenzofurans and supported chlorinated alkyl bibenzyls instead, which was verified by model compound syntheses of the both structure types [50–52].

The first synthetic proof of the RPCBB structures was obtained by treating alkylbenzenes with 1,2-dichloroethane in presence of AlCl$_3$ at 70 °C for 1–2 h and chlorinating the products with elemental chlorine in CCl$_4$ in the presence of FeCl$_3$ [50, 51]. A further group of alkyl-substituted diarylethanes including RPCBBs (from chloro-substituted precursors) was obtained by interaction of substituted benzaldehydes or acetophenones with substituted benzylmagnesium chlorides. Then the stilbene products were hydrogenated over Pd-C in 94% ethanol at atmospheric pressure to give 1,2-diarylethanes [52]. The origin of C5-PCBB formation in bleaching is possibly from wood raw material of pulping; possible dimerization of cymenes or cymenenes was suggested [51]. However, abundant stilbene derivatives in wood could also react in bleaching to produce RPCBBs. Other possibilities are an origin from oil-based defoamer and from alkyl 1,2-diphenylethanes patented as PCB replacement [52, 53].

The levels of RPCBBs were estimated to be 10–120 ng l^{-1} in effluents and 5–33 ng g^{-1} dw in biosludges from the pulp mills [43]. However, because the analysis method used was not optimized for diarylethane structures the actual contents might have been significantly higher (see below) [45].

Alkyl-polychlorofluorenes (RPCFL) were detected at sub-ng l^{-1} levels in effluents, at a few ng g^{-1} dw levels in biosludge and < ng g^{-1} dw concentrations in softwood pulp product of a kraft mill. The structures were deduced from high resolution mass spectra and verified by comparison to GC/HRMS/SIM of model substance mixtures from chlorination of methyl- and dimethylfluorenes [54]. While alkylfluorenes are abundant and persistent crude oil components [55], the origin of RPCFLs in bleach liquor could be the use of oil-based defoamer in the mill.

Alkyl-chlorophenanthrenes (RPCPH) were observed by GC/MS in pulp mill originated samples in amounts comparable to those of RPCNs [42, 44, 45]. Contents of C4-PCPHs (methyl isopropyl chlorophenanthrenes), mainly chlororetenes, were 2–56 ng l^{-1} in effluent, 0.9–8.7 ng g^{-1} dw in pulp, and 61–72 ng g^{-1} in biosludge of two mills [44]. Synthetic verification of the major RPCPH in bleach liquors was obtained by chlorination of retene with chlorine gas in CCl$_4$ solution [52]. The product, 9-chlororetene, was identified by HRGC/HRMS as a major component of the RPCPHs in mill effluent, sludge and pulp [42]. Nevalainen [52] suggested that abietic acid could be the precursor to chlorinated retenes (C4-PCPHs) and possibly also to alkyl chlorobibenzyls (RPCBBs).

3.2
Analyses

3.2.1
Extraction

Extractions of RPCNs, RPCBBs, RPCFLs, and RPCPHs from various samples have been applied with the same procedures as for CYMS and CYMD (see above) [43].

3.2.2
Cleanup

While all planar aromatic PLACs and RPCBBs of interest were persistent against sulfuric acid, the cleanup of the evaporated extract in hexane was started by sulfuric acid shaking or with sulfuric acid impregnated on a silica column [43, 56]. For sediment samples, elemental sulfur interference was eliminated using copper activated with HCl [43, 57]. Further cleanup of PLACs was done according to Scheme 1 [43].

The procedure gave good recovery of PLACs (RPCN, RPCFL, and RPCPH) [42, 43, 54]. RPCBBs, which were found to have non-rigid 1,2-diphenylethane structures in 1990, were later observed to escape in cleanup in great part to Fr I (as marked in Scheme 1) [42]. No optimization of RPCBB cleanup has been reported. Therefore, the concentrations of RPCBBs reported in effluents, products, wastes, and the environment, thus far, must be underestimates.

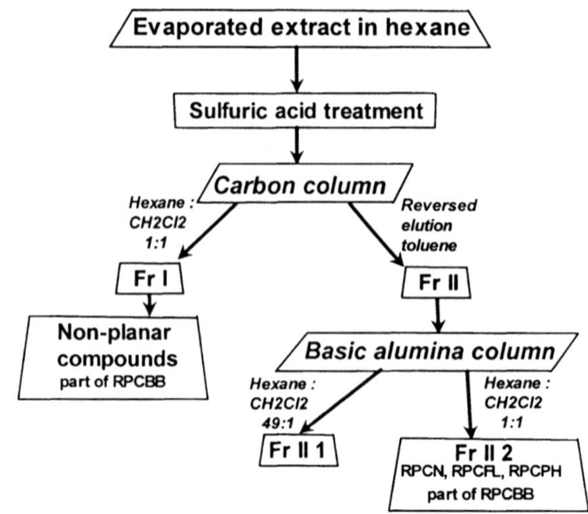

Scheme 1

3.2.3
Final Determination

PLAC fractions (Fr II 2, Scheme 1) were analyzed by temperature-programmed GC with fused silica capillary columns coated with SE-54 (HP5 0.2 mm inner diameter, 0.11 cm film thickness or U2 0.2 mm inner diameter, 0.33 μm film thickness). Length of the columns was 25 or 50 m. ^{13}C-labeled PCDD/Fs added before extraction and after cleanup were used as internal standards for quantitative results and recovery, respectively. Mass selective detector (HP 5970) or magnetic sector mass spectrometer (mostly VG AutoSpec) was used for electron impact LRMS or HRMS, respectively. Mass spectral data of the analyzed substance groups are collected in Table 3. Exact quantification was achieved only in a few cases when authentic standard compound was available. In most

Table 3. Formula and mass spectral data (m/z) of RPCNs, RPCBBs, RPCPHs, and RPCFLs

Compound	Mol. formula	$M^{+\bullet}$	$(M+2)^{+\bullet}$	Abundant fragment ions					Ref.
C2-MonoCN	$C_{12}H_{11}Cl$	190.0549	192.0520						
C3-MonoCN	$C_{13}H_{13}Cl$	204.0706	206.0677	189	169	152	139		[42]
C4-MonoCN	$C_{14}H_{15}Cl$	218.0862	220.0833						
C2-DiCN	$C_{12}H_{10}Cl_2$	224.0160	226.0131						
C3-DiCN	$C_{13}H_{12}Cl_2$	238.0316	240.0287	203	197	165	152	139	[42]
C4-DiCN	$C_{14}H_{14}Cl_2$	252.0473	254.0444						
C2-TriCN	$C_{12}H_9Cl_3$	257.9770	259.9741[a]						
C3-TriCN	$C_{13}H_{11}Cl_3$	272.9926	274.9897						
C4-TriCN	$C_{14}H_{13}Cl_3$	286.0083	288.0054						
C2-TetraCN	$C_{12}H_8Cl_4$	291.9380	293.9351						
C3-TetraCN	$C_{13}H_{10}Cl_4$	306.9615	308.9586						
C4-TetraCN	$C_{14}H_{12}Cl_4$	319.9693	321.9664						

Table 3 (continued)

Compound	Mol. formula	$M^{+\cdot}$	$(M+2)^{+\cdot}$	Abundant fragment ions					Ref.
C5-MonoCBB	$C_{19}H_{23}Cl$	286.1488	288.1459	153	133	119			[43]
C5-DiCBB	$C_{19}H_{22}Cl_2$	320.1099	322.1970	167	153	119			[43]
C5-TriCBB	$C_{19}H_{21}Cl_3$	354.0709	356.0680	187	167	153			[43]
C5-TetraCBB	$C_{19}H_{20}Cl_4$	388.0319	390.0290	201	187	153			[43]
C4-MonoCPH	$C_{18}H_{17}Cl$	268.1019	270.0990	253	255	218	202	189	[52]
C4-DiCPH	$C_{18}H_{16}Cl_2$	302.0629	304.0600						
C4-TriCPH	$C_{18}H_{15}Cl_3$	336.0239	338.0210						
C4-TetraCPH	$C_{18}H_{14}Cl_4$	369.9850	371.9821						
C1-MonoCFL	$C_{14}H_{11}Cl$	214.0549	216.0520						
C1-DiCFL	$C_{14}H_{10}Cl_2$	248.0160	250.0131						
C1-TriCFL	$C_{14}H_9Cl_3$	281.9770	283.9741						
C1-TetraCFL	$C_{14}H_8Cl_4$	315.9380	317.9351						
C1-PentaCFL	$C_{14}H_7Cl_5$	349.8990	351.8961[b]						
C2-MonoCFL	$C_{15}H_{13}Cl$	228.0706	230.0677						
C2-DiCFL	$C_{15}H_{12}Cl_2$	262.0316	230.0677						
C2-TriCFL	$C_{15}H_{11}Cl_3$	295.9926	264.0287						
C2-TetraCFL	$C_{15}H_{10}Cl_4$	329.9537	331.9508						
C2-PentaCFL	$C_{15}H_9Cl_5$	363.9147	365.9118[b]						

[a] In GC/HRMS/SIM $(M+4)^{+\cdot}$, m/z = 261.9712, was used instead of $(M+2)^{+\cdot}$ due to interference [44].

[b] $(M+2)^{+\cdot}$ and $(M+4)^{+\cdot}$, m/z = 353.8932 and 367.9089 for C1 and C2-PentaCFL, respectively, were used in GC/HRMS/SIM for better sensitivity [44].

cases the results were relative assuming 1:1 response ratio with an internal standard peak in the same SIM window [39, 40, 42–45, 50–52, 54, 58].

3.3
Environmental Fate

3.3.1
Occurrence in Sediments

The possible importance of alkyl polychlorobibenzyls (RPCBBs) as persistent pollutants became obvious, when their congeners were found in sediments of pulp mill recipient watercourses of Central Finland (Fig. 5) at high ng g^{-1} dw levels [58]. In pulp mill recipient sediment of the Lake Ladoga C4-mono- and di-CPHs (chlororetenes) occurred at 5–110 ng g^{-1} dw contents [28].

3.3.2
Bioaccumulation and (Eco)Toxic Potential

Mussels incubated in pulp mill recipient watercourses of the Kymijoki River Basin in 1986–1990 collected 8.5–34.8 ng g^{-1} lw C5-PCBB congeners [43], and 20–89 ng g^{-1} lw chlororetenes (C4-PCPHs) [38]. In Kymijoki River recipient fish, an average C5-PCBBs content of 1.33 ng g^{-1} lw has been measured [38].

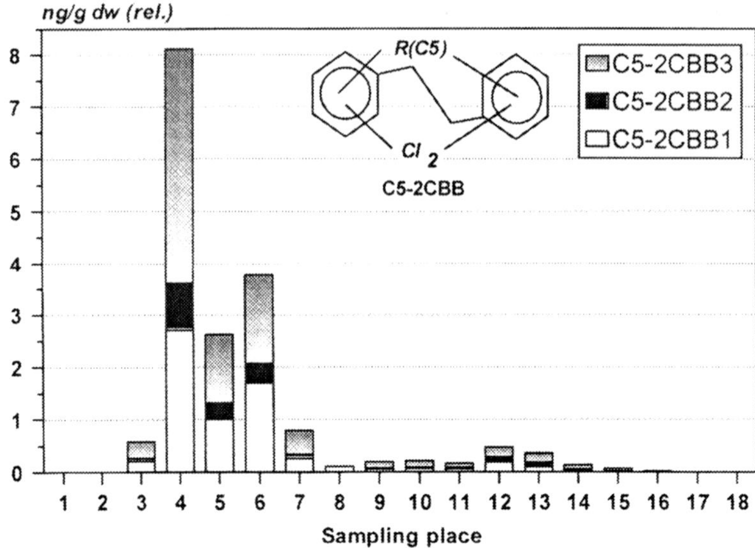

Fig. 5. Relative contents of three C5–2CBB isomers in surface sediments of Central Finland in 1988–1989 [58]. The places 1, 2, 17, and 18 were non-recipient areas. The place 3 was near a pulp mill closed in 1985. Places 4–11 were downstream from an operating large mill and 12–16 from mill closed in 1981

While very few fish and mussel data were available, estimation of bioaccumulation power of alkylaromatic chlorohydrocarbons was also made from accumulation from effluent to biosludge. The rank order of highest to lowest accumulation factor (AF, mean value) obtained was [38, 54] RPCN (4120) > RPCPH (2680) ~ RPCFL (2500) > CYMS (1227) > RPCBB (375) > CYMD (158).

While RPCNs, RPCPHs, RPCFLs, and RPCBBs are not readily biodegradable, they must be classified as persistent bioaccumulative chemicals. Their (eco)toxicity has not been tested.

The non-chlorinated skeleton compound of C4-PCPHs, retene, occurs at high levels in pulp mill effluents and in recipient sediments up to 2 mg g^{-1} dw. Retene causes MFO induction in fish and is toxic to early life-stages of zebrafish at LOEL of 16 µg l^{-1}. TEF of egg mortality related to 2,3,7,8-tetrachlorodibenzo-*p*-dioxin is on average about 0.007 [59]. Accordingly, chlorinated retenes are of concern as possible dioxin-type of persistent xenobiotics.

4
Isopropyl-PCBs

4.1
Production, Structures, and Uses

Alkylchlorobiphenyls (Chloralkylenes) were suggested as a PCB substitute for applications in capacitors, transformers, and similar applications where favor-

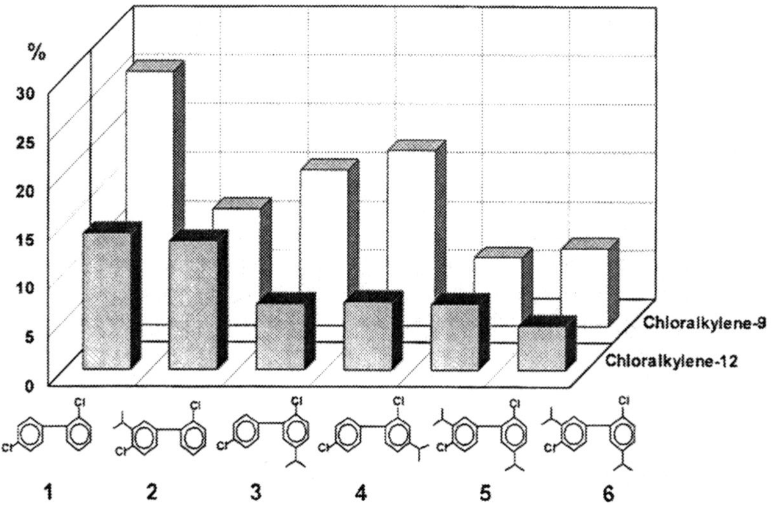

Fig. 6. Amounts and structures of the major components of Chloralkylene-12 [62] and Chloralkylene-9 [63]

able electrical properties comparable to PCBs are required. Their technical production included alkylation of PCB or PCT (polychlorinated terphenyl) mixtures containing 1–3 chlorine atoms per molecule [60]. The product had good electrical properties, but thermal stability and resistance to oxidation were lower than those of PCBs [61]. Trade names of the isopropyl-PCBs with slightly different degree of alkylation were Chloralkylene-9 and Chloralkylene-12. Structures and amounts of their six main components [62, 63] are shown in Fig. 6.

4.2
Analyses

4.2.1
Fractionation of Chloralkylene Products

Distillation in reduced pressure through 1 m Vigreux column or liquid chromatography (LC) on reversed phase produced three main fractions (A–C) from Chloralkylene-9. Fraction A contained only compound 1 (2,4′-dichlorobiphenyl; Fig. 6), starting material in production of the alkyl derivative mixture. Fraction B could be divided by thin-layer chromatography (TLC) on alumina to Bb and compound 2 (2,4′-dichloro-3′-monoisopropylbiphenyl). Subfraction Bb was divided further with TLC on silica to 3 (2,4′-dichloro-5-isopropylbiphenyl) and to 4 (2,4′-dichloro-4-isopropylbiphenyl). HPLC with *n*-hexane:perfluorocyclohexane 4:1 on reversed phase column isolated from fraction C compound 5 (2,4′-dichloro-3′,5-di-isopropylbiphenyl) and compound 6 (2,4′-dichloro-3′,4-di-isopropylbiphenyl) from each other. The structures of 1–6 were deduced from their mass and infrared spectra and identified by GC retention times (RT) on two different columns compared with RTs of authentic samples [63].

4.2.2
Extraction and Cleanup from Environmental Samples

Extraction procedures for persistent organic pollutants like PCB and DDT residues can be applied with success for Chloralkylene components. Hexane-acetone or similar solvent mixtures are adequate [62]. Sulfuric acid treatment in cleanup leaves the isopropyl-PCBs unaffected. However, oxidation, e.g., with chromic acid degrades the alkyl-PCBs [62].

4.2.3
Final Determination

Gas chromatography on capillary columns coated with Apiezon L has been demonstrated to be very suitable for determination of alkyl-PCBs. Pure model standard substances are missing in most cases. However, rather good semi-quantitative data are obtained with a specific flame ionization detector method with anthracene as internal standard [64]. For environmental levels, mass spectrometric detection methods are necessary.

Chromatographic properties and EI mass spectra data of main Chloralkylene components are well known [62, 63, 65]. Useful property data are collected in Table 4.

4.3
Environmental Fate

4.3.1
Photodegradation

UV irradiation at 260–320 nm in a Rayonette reactor in methanol of Chloralkylene-12 led to formation of reductive dechlorination products at a significant rate. This result indicated a greater overall ease of photodegradation of the alkyl substituted PCBs relative to the parent compound [66]. By irradiation at over 290 nm wavelength UV both in water and on a wet silica gel column in air flow, Chloralkylene-9 was transferred mainly to hydroxylation and reductive dechlorination products by similar mechanisms to PCBs [67].

4.3.2
Bioaccumulation and Metabolism

The log K_{ow} values of Chloralkylene-12 components determined by reversed phase thin layer chromatography ranged from 4.56 to 9.53 with increasing chlorine and alkyl substitution degree [68]. Accordingly, isopropyl-PCBs could have a high bioconcentration potential. However, the first accumulation and excretion study of Chloralkylene-9-^{14}C showed rapid metabolism in rats [69].

Table 4. Formula and mass spectral data (m/z) of main components of Chloralkylenes 9 and 12

Compound[a]	Nr	RT[b]	Mol. formula	$M^{+\bullet}$	$(M+2)^{+\bullet}$	Abundant fragment ions m/z (I%)					Ref.
2,4'-DiCB	1[c]	5.62	$C_{12}H_8Cl_2$	222.0003	223.9974						
2,4'-DiC-3'-IpB	2[c]	8.83	$C_{15}H_{14}Cl_2$	264.0473	266.0444	249(100)	229(11)	222(26)			[65]
2,4'-DiC-5-IpB	3[c]	9.76	$C_{15}H_{14}Cl_2$	264.0473	266.0444	249(100)	229(15)	222(12)			[65]
2,4'-DiC-4-IpB	4[c]	10.26	$C_{15}H_{14}Cl_2$	264.0473	266.0444	249(100)	229(15)	222(9)			[65]
2,4'-DiC-3',5-DiIpB	5[c]	12.83	$C_{18}H_{20}Cl_2$	306.0942	308.0913	291(100)	271(10)	264(23)	249(39)	229(10)	[65]
2,4'-DiC-3',4-DiIpB	6[c]	13.08	$C_{18}H_{20}Cl_2$	306.0942	308.0913	291(100)	271(12)	264(12)	249(24)	229(9)	[65]
2,4'-DiC-TriIpB	8[d]	15.80	$C_{21}H_{26}Cl_2$	348.1412	350.1383	333(91)	306(59)	291(59)	249(44)	222(13)	[65]
2,4'-DiC-TriIpB	9[d]	16.47	$C_{21}H_{26}Cl_2$	348.1412	350.1383	333(75)	306(47)	291(49)	249(43)	222(13)	[65]
2,4'-DiC-TetraIpB	e		$C_{24}H_{32}Cl_2$	390.1881	392.1852						

[a] C = chloro; Ip = isopropyl; B = biphenyl.
[b] RT = retention time on glass column packed with 0.2% Carbowax 20 M on 100–200 mesh Chromosob W [62].
[c] Compound number from text (Fractionation of Chloralkylene products) and Fig. 6.
[d] GC peak Nr from [64]; intensity of $M^{+\bullet}$ ion 100% and of $(M+2)^{+\bullet}$ 67.4%.
[e] Two small peaks (RT ~16–18) detected from Chloralkylene 12 [62].

Two Chloralkylene model compounds in the rat were shown to have two major routes: (1) stepwise oxidation of the isopropyl group to carboxylic acid via intermediate keto- and carboxylic acid derivative, and (2) hydroxylation of the chlorine-substituted phenyl ring [70]. The same model compounds 4-chloro-4'-isopropylbiphenyl and 2,5-dichloro-4'-isopropylbiphenyl showed a very limited metabolism in fish and frogs, but in a pure strain of fungus readily metabolism via the same rotes as in rats (see above) took place. Aerobic bacteria from activated sludge degraded 4-chloro-4'-isopropylbiphenyl extensively to p-chlorobenzoic acid [71]. However, Chloralkylene-9-^{14}C and two reference PCBs were almost chemically unchanged in activated sludge [72]. In marsh plant *Veronica beccabunga* Chloralkylene-9-^{14}C metabolized more rapidly than the corresponding nonalkylated PCBs to phenolic and alcoholic substances [73]. Rapid metabolism and excretion of isopropyl-PCBs in mammals was supported by study of the fate of Chloralkylene-9-^{14}C in the rhesus monkey after single oral administration [74].

4.3.3
Fate in Environment

Administered into the upper layer (0–10 cm) of agricultural soil under outdoor conditions, followed by sowing of carrots, 80% of Chloralkylene-9-^{14}C volatilized within one vegetation period. Most of the remaining radioactivity was still in the upper layer, 0.8% dispersed to a depth of 40 cm, and 3.3% taken up with carrot beet. The upper soil layer residues were 41% unchanged Chloralkylene-9-^{14}C, 19% soluble metabolites, and 40% unextractable metabolites. In the second year (sugar beet grown), the amount of unextractable residues rose to 68%. From the soil extract the following metabolites were detected: one dichlorobiphenyl-OH, one dichlorobiphenyl-OCH$_3$, and two isomers of isopropyl-dichlorobiphenyl-OH. From the carrot root extract, dichlorobiphenyl-OH was detected. Acid hydrolysis of the soil conjugates yielded dichlorobiphenyl-OH among other compounds [75]. PCB replacement compounds have been detected by GC/MS in fish of the North American Great Lakes coastal zones [76, 77]. Nonchlorinated isopropylbiphenyls were also analyzed in fish [77]. As conclusion, isopropyl-PCBs are persistent and bioaccumulative environmental contaminants and might be hazardous, if not to mammals at least to aquatic biota.

5
Tetrachlorobenzyltoluenes

5.1
Production, Uses, and Structures

Tetrachlorobenzyltoluenes (TCBT) manufactured by Prodelec, France have been patented as substitutes for PCBs in dielectric and cooling materials of capacitors and transformers and in hydraulic fluids [78–82]. An important application of 100% TCBT mixture (Ugilec 141) was for hydraulic linkages in machinery used in underground mining [83]. Formulation Ugilec T is a mixture of 60% TCBT and 40% trichlorobenzene. Technical synthesis of TCBT is done in three steps. First, toluene is chlorinated with $FeCl_3$ as catalyst producing 2,3-, 2,4-, 2,5-, 2,6-, and 3,4-dichlorotoluenes (I) in the ratio 8:43:20:18:11. Second, an aliquot of the product is chlorinated photochemically to dichlorobenzylchlorides (II). Third, Friedel-Crafts alkylation of I with II, $FeCl_3$ as catalyst, leads to a mixture of TCBTs [84].

There are 96 possible isomers (congeners) of TCBTs. Based on analogy to PCBs, Ehmann and Ballschmiter have developed a systematic numbering for the isomers [84]. Structures and numbers of the six major components of Ugileg 141 are shown below.

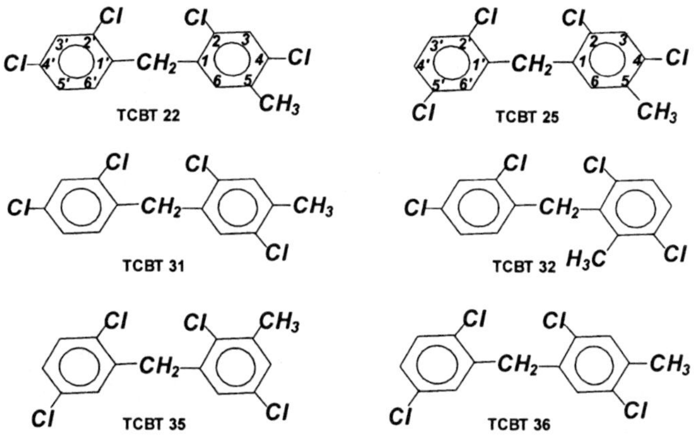

TCBT 22 TCBT 25 TCBT 31 TCBT 32 TCBT 35 TCBT 36

Because 3,5-dichlorotoluene was formed only in very small amounts in the first step of technical TCBT synthesis, the number of relevant isomers in the final Ugilec 141 product was 70. Structures of these congeners were identified by syntheses of defined TCBT reference mixtures and using GC with three phases of differing polarity [84].

5.2
Pyrolysis

TCBTs are relatively persistent and non-flammable (and therefore useful in hydraulic oils). Pyrolysis in a closed system for two days in the presence of hydrogen peroxide at 300 °C produced as main components chlorinated benzophenones, fluorenes, fluorenones, xanthenes, and xanthones. Polychlorodibenzofurans (PCDF) and polychlorodibenzo-p-dioxins (PCDD) were formed in much lower concentrations than from PCB product Pyralene T1 under identical conditions [84]. In pyrolysis at 450–700 °C with excess oxygen TCBTs produce more PCDFs and PCDDs. In parallel with the behavior of PCBs, PCDFs are formed in significantly higher amounts than PCDDs from PCBTs [86].

5.3
Analyses

5.3.1
Extraction

TCBTs are coextracted with PCBs and other lipophilic substances in standard methods of analyses from different environmental matrixes. Waste oil samples were homogenized by ultrasonic shaking, water removed, sample dried by filtration over anhydrous sodium sulfate, and diluted with *n*-hexane [85]. Water samples (1000 ml) were mixed with 50 ml hexane by Turrax (11,000 rpm). The hexane phase was then dried with anhydrous sodium sulfate [86]. Laboratory water sample (1000 ml) was extracted with three portions of hexane (100 ml, 60 ml, and 60 ml) [87]. Another water sample (50 ml) from laboratory exposure experiments was liquid-liquid extracted with 9 ml of hexane [88].

Solid samples (e.g., sediments) were dried to constant weight at 105 °C and extracted in Soxhlet with hexane for about 20 h [86]. Biosamples can be extracted according to PCB analysis procedures, e.g., fish muscle tissue by column extraction [86]. Soft tissue of laboratory-exposed zebra mussels was Soxhlet extracted in 250 ml flasks with 100 ml of toluene and some boiling stones for more than 24 h [88]. In an environmental survey, fish muscle (500 g per sample) was cut into small pieces and freeze-dried and extracted with light petroleum (b.p. 40–60 °C) [89]. Another procedure was to mix the dried fish muscle pieces with dry sodium sulfate and extract in a thimble in a Twisselman-device with petroleum ether for 4 h. The extract was dried with anhydrous sodium sulfate for at least 30 min, filtered, and evaporated in Rotavapor, and the rest of the solvent removed with a gentle nitrogen gas stream [90]. Pooled eel samples were homogenized with a Waring blender, ground with Na_2SO_4, and Soxhlet extracted for 6 h with *n*-pentane/dichloromethane (1/1 v/v) [91]. Rat tissue was extracted in a similar way [87, 92].

5.3.2
Cleanup

In standard cleanup procedures TCBTs follow PCBs to the same analytical fractions. In determination from waste oil, hexane solution was first spiked with $^{13}C_6$-labeled TCBT 80 (2′,3,4,6′-C_{14}-6-Me). Unlabelled TCBT 80 can also be used, because this congener is present only at low levels in Ugilec 141. The sample (1 ml) was then applied to a carbon column (Carbosphere 0.6 g), washed with *n*-hexane (50 ml), and eluted with toluene (25 ml). Evaporated (Rotavapor, then a gentle nitrogen stream at 60 °C) eluate was redissolved in *n*-hexane (1 ml) and applied to a column containing 5 g basic alumina and some sodium sulfate. The column was washed with *n*-hexane (18 ml) and a 95:5 (v/v) mixture of *n*-hexane-dichloromethane, then eluted with 15 ml of the same solvent mixture. The eluate was evaporated at 60 °C under a gentle nitrogen stream, redissolved in *n*-hexane (1 ml), and applied to a column of deactivated silica gel (1 g) and some sodium sulfate. The silica column was rinsed with *n*-hexane

(2×0.5 ml) and eluted with 5 ml of *n*-hexane. The evaporated (as above) eluate was redissolved in *n*-hexane (1 ml) and carbon column chromatography was performed again as described above [85].

In water analyses, dried hexane extract (Turrax, 1 ml) was shaken with reagent DIN 51257 and isopropanol (1 ml), then treated with 0.5 g lots of sodium sulfite to dissolve the possible precipitation. After shaking with 5 ml of water the hexane phase was dried with sodium sulfate. The solution was transferred to an activated ($600\,°C$, 2 h) Florisil ($100 - 200$ mesh) column (15×0.5 cm) washed before use with 20 ml of iso-octane. Eluate collected with 7 ml of iso-octane/toluene (95/5 v/v) was evaporated to 2 ml volume [86].

For water, rat tissue, and fish samples, cleanup with deactivated Florisil (magnesium silicate containing 5% water) was performed. Column length was 300 mm and inner diameter 30 mm. A maximum of 0.6 g of lipidic sample per 20 g of Florisil was applied. The upper portion of column was packed with anhydrous sodium sulfate. TCBTs were eluted with 150 ml of hexane. A second elution with 100 ml of hexane/dichloromethane (85/15) was performed to determine whether any TCBT had been retained in the column. Both eluates were concentrated prior to chromatographic analyses [87].

Another cleanup procedure of lipid extract (from zebra mussels) included first elution with 35 ml of hexane through a column packed with sodium sulfate, 1 g alumina and 2 g silica. Second elution was done with 15 ml of hexane through a 0.7 g silica column [88].

In analyses of extract from freeze-dried fish [89] gel chromatography on Bio-Beads S-X3 with cyclohexane/ethyl acetate (1/1) was used for removal of neutral fat. Then adsorption column chromatography on silica gel (1.5% water) according to the multi-residue procedure of Specht and Tillkes [93] was applied. Elution with *n*-hexane separated TCBTs and PCBs from the main portion of organochlorine pesticides and other more polar compounds. The hexane extract was evaporated to 0.5 ml for GC-MS determination [89, 90].

Lipid was removed from eel extract by a 15 g alumina ($Al_2O_3 \cdot 6H_2O$) column and the concentrated sample was further cleaned up by means of a 1.8 g silica ($SiO_2 \cdot 1.5\% H_2O$) column [91] Lipid removal from rat tissue extracts by shaking with 10 ml hexane solution with conc. sulfuric acid gave excellent results, as well [92].

5.3.3
Final Determination

Capillary gas chromatography must be used to achieve sufficient separation for TCBTs in product, waste, and environmental samples. Coatings DB-1701 (85% dimethyl, 15% cyanopropylsilyl) and CP-Sil 88 (100% cyanopropyl) gave better separation than DB-5 (95% dimethyl, 5% phenyl) [84]. Retention indices of twelve TCBT isomers are listed in Table 5 together with melting points, solubilities in water, and log K_{ow}s of nine pure compounds.

EC-detection was useful for analysis of TCBT formulation or samples from laboratory bioaccumulation experiments [87, 88, 92–94]. However, ECD appeared to be unsuitable for environmental analyses because its responses varied greatly [84], interference with PCBs was difficult to avoid, and sensitivity of

Table 5. Properties of important TCBT isomers (M = 320.026 g mol^{-1})

Nr	Status	Structure [84]	m.p. °C [94]	S^c (mol m^{-3})	log K$_{ow}$ [95]	RI^d DB-5	RI^d DB-1701
28	b	2,2′,4,6′-Cl$_4$-5-Me	114	4.34 E-6	6.73	2320	2235
25	ab	2,2′,4,5′-Cl$_4$-5-Me	78.7–80.5	8.78 E-6	7.54	2320	2243
35	a	2,2′,5,5′-Cl$_4$-3-Me				2327	2252
36	ab	2,2′,5,5′-Cl$_4$-4-Me	74.3–75.6	9.44 E-6	7.48	2330	2252
32	a	2,2′,4′,5-Cl$_4$-6-Me				2339	2252
22	ab	2,2′,4,4′-Cl$_4$-5-Me	61.9–63.7	3.25 E-5	7.43	2345	2262
27	b	2,2′,4,6′-Cl$_4$-3-Me	87	1.02 E-5	7.20	2354	2263
31	a	2,2′,4′,5-Cl$_4$-4-Me				2354	2274
80	b	2′,3,4,6′-Cl$_4$-6-Me	107	5.72 E-5	7.15	2372	2320
21	b	2,2′,4,4′-Cl$_4$-3-Me	76	4.01 E-5	7.20	2385	2300
52	b	2,3′,4,4′-Cl$_4$-5-Me	75	6.59 E-6	7.26	2405	2355
74	b	2′,3,4,4′-Cl$_4$-5-Me	83	3.62 E-5	7.41	2412	2364

[a] Main component of Ugilec 141 [84].
[b] Commercially available [88].
[c] S = Solubility in water at 25 °C [94].
[d] RI = Retention Index (ATA) [84].

ECD to TCBT isomers was a factor of two lower than for the corresponding tetra- and higher chlorinated PCB congeners [85]. Therefore, most practically successful TCBT analyses were done by GC-MS, especially with SIM mode [84–86, 89–91]. The most useful EI-MS ions of TCBT are those of (high resolution) m/z (Int%) values 319.9508(46.5), 317.9537(60.0), 284.9819(97.1), and 282.9848(100), two M$^{+·}$ and two (M-Cl)$^+$ peaks, respectively. Model electron impact mass spectrum (upper part) of a TCBT isomer is presented in Fig. 7.

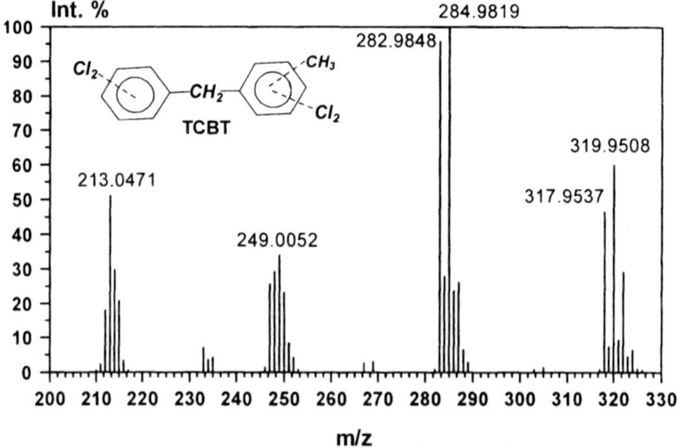

Fig. 7. Simulated (high resolution m/z values calculated, intensity ratios of the highest peaks taken from [89]) upper part of mass spectrum of a TCBT isomer

There the intensity ratios between ion clusters were taken from experimental measurement of low resolution MS [89], but intensity ratios within each isotope cluster was calculated as the sum of binary distributions of the natural abundance of each isotope involved. High resolution m/z values of major peaks (for HRMS) were calculated as sums of exact masses of isotopes. This simulation method (ISOCLUST) developed by the present author (J. Paasivirta) is operable with a desk computer. Low resolution electron impact MS is suitable for TCBT determinations, especially by selected ion monitoring (SIM). In this mode, focusing to four ions, m/z 213.0, 283.0, 285.0, and 320.0, which are not interfered with by PCBs is recommended. In a TCBT study, quantitative results were based on their sum intensity for better signal to noise ratio [91].

5.4
Environmental Fate

5.4.1
Bioaccumulation and Metabolism

The physical properties of TCBTs (Table 5) are typical for substances highly bioaccumulative from water to aquatic biota. Their extensive use in hydraulic fluids in mining has led to significant leakage to the environment. In the coal mining area of Germany, river Lippe contained in spring 1987 Ugilec 141 up to 3.5 mg kg^{-1} in dry sediment and 0.1–4.7 mg kg^{-1} in fresh fish (14–155 mg kg^{-1} in lipid weight) [86]. In another survey of German rivers the fresh fish muscle concentrations were at background areas (upper Ruhr, Lippe, Rhein, and Weser) non-detectable (<0.02 mg kg^{-1}), and in coal mining exposed parts of Ruhr and Lippe 0.05–24 mg kg^{-1} [90]. Downstream in the Netherlands rivers yellow eels were measured to contain up to 8.4 mg kg^{-1} TCBT. The concentrations in the late 1980s showed an increasing trend, while PCB contents were decreasing [91]. Consequently, the Dutch government banned TCBTs in 1988 [91]. In more recent analyses of zebra mussels and eels from Rhine and Meuse rivers in the Netherlands, the measured concentrations of TCBTs were comparable to other persistent organic zenobiotic compounds, except PCBs and PAHs which occurred at significantly higher levels [95].

A laboratory study of bioaccumulation of TCBT mixtures from water to fish (*Brachydanio rerio*) gave bioconcentration factors (BCF) for 3, 7, 15, and 30 days values of 325, 420, 1065, and 988, respectively. In reference experiments with PCB-5 Cl, BCF values were 3950, 4000, 6950, and 6140. Depuration times of TCBTs in these experiments were 26 and 115 days for 50% and 95% depuration, respectively [87]. In feeding experiments to Wistar albino rats [92], TCBTs accumulation to tissue was compared to that of PCBs: BCF values were 25–30 times lower for the brain, 65–123 times lower for the liver, and 21–33 times lower for fatty tissue. The results indicated a significantly faster metabolism of TCBTs than of PCBs. Characteristically, both PCBs and TCBTs deposited in larger amounts in abdominal fat than in the brains and livers of rats [92]. In other rat feeding study, TCBT mixtures were found to accumulate only weakly in rats and eliminate readily, in contrast to the strong accumulation and slow elimination of PCB-5 Cl. Both mixtures induced liver-toxic effects, but the liver lesions from PCBs were more severe [87].

The measured $\log K_{ow}$ values of individual TCBTs were as high as 6.7–7.3 [96]. Bioaccumulation study of eight individual TCBT isomers in the zebra mussel (*Dreissena polymorpha*) gave log BCF values from 4.43 to 5.19, lower than expected from a linear relationship, but higher than calculated from the polynomial relationship between $\log K_{ow}$ and log BCF [88]. The log BCF values measured for guppy (*Poecilia reticulata*) were significantly lower ranging from 1.67 to 2.68. This indicated that bioconcentration potential of TCBTs is species dependent [97, 98].

5.4.2
Toxicity and Environmental Hazard

In addition to their high bioaccumulation potential in aquatic environment (see above), TCBTs have been observed to be quite persistent against microbial degradation. Experiments with activated sludge and with PCB-degrading micro-organism strain *Alcaligenes* sp. JP1 indicated no significant degradation of TCBTs [99]. Accordingly, experiments on acute and chronic toxicity are relevant for estimating their environmental hazard.

Standard acute toxicity tests of Ugilec 141 with bacteria and algae showed no measurable effects. In contrast, *Daphnia* toxicity was comparable to that of a PCB formulation Clophen A60. The EC100, EC50, and EC0 values were 520 µg l^{-1}, 180 µg l^{-1}, and 60 µg l^{-1} for Ugilec 141, and 420 µg l^{-1}, 190 µg l^{-1}, and 53 µg l^{-1} for Clophen A60, respectively [100].

Immunotoxicity of TCBTs to mammals has been established. The tetanus antibody response was reduced in male rats after six weeks feeding with 200 ppm of TCBT, and in both males and females fed with 1000 ppm [92].

Enzyme induction experiments in male and female Wistar rats with administration of TCBT (Ugilec 141) produced a weak increase of total Cytochrome P-450 (<1.9-fold). Benzhepthamine-*N*-demethylase rose up to 3.5-fold, but benzo(*a*)pyrene hydroxylase (AHH) and ethoxyresorufin-*O*-deethylase (EROD) inductions were very low. Instead, the latter two inductions were markedly high with TCBT pyrolysate. TCBTs were concluded to be the phenobarbital (PB) type of inducer, while pyrolysis products contained 3-methylcolanthrene (3-MC) inducers of the dioxin type. The authors recommended use of TCBTs only in sealed systems [101].

Effects of single IP doses of Ugilec-141 (200 mg kg^{-1}), Arochlor 1254 (200 mg kg^{-1}), and PCB-77 (coplanar, 50 mg kg^{-1}) in mice were similar in liver weight and total Cytochrome P-450 increase, as well as in plasma TT4 and plasma retinol reductions. However, with Ugilec 141 EROD induction was significantly lower and PROD induction insignificant compared to the effects of Arochlor 1254 and PCB-77. A mild hypertrophy of thyroid follicle cells and a slight decrease of hepatic retinol was observed with all three substrates. It was concluded that TCBTs possess a weak B-MC type of induction activity. Low EROD of Ugilec-141 was suggested to correspond to TCBT molecule deviation from planarity. TCBTs were concluded to have no apparent differences in toxicity from PCBs, which turned Ugilec 141 and related TCBTs into very bad alternatives for PCBs [102].

Semi-empirical molecular orbital calculations of rotational energy barriers (planarity) and estimation of ah-receptor binding activity (at least three chlorines in lateral positions of the 3×10 Å rectangle) were done for all 96 individual TCBT isomers and compared to the results of the same procedure for 16 PCBs and 22 polychloro diphenyl ethers (PCDEs). According to the results, two isomers, 3,3′,4,4′-Cl4-2Me (TCBT 87) and 3,3′,4,4′-Cl4-5Me (TCBT 88) were suggested as potential compounds which may exhibit a dioxin type of toxicity [103, 104].

References

1. Ratcliffe DA (1967) Nature 215:208
2. Hickey JJ, Anderson DW (1968) Science 162:271
3. Kimbrough RD, Jensen AA (eds) (1989) Halogenated biphenyls, terphenyls, naphthalenes, dibenzodioxins and related products, 2nd edn. Elsevier, Amsterdam New York Oxford
4. Basler A, Lebsanft J (1999) Environ Sci & Pollut Res 6:44
5. Kelee WR, Gray LE, Wilson EM (1998) Reprod Fertil Dev 10:105
6. Björseth A, Lunde G, Gjös N (1977) Acta Chem Scand B 31:797
7. Eklund G, Josefsson B, Björseth A (1978) J Chromatogr 150:161
8. Björseth A, Carlberg CE, Möller M (1979) Sci Tot Environment 11:197
9. Gjös N, Carlberg CE (1980) Adv Mass Spectrom 88:1468
10. Kuokkanen T (1981) NORDFORSK, Miljövårdsserien 1:89
11. Kuokkanen T (1989) Doctoral Thesis, Department of Chemistry, University of Jyväskylä, Finland, Res Rep 32
12. Rantio T (1992) Chemosphere 25:505
13. Kuokkanen T, Autio P (1989) Chemosphere 18:1921
14. Kuokkanen T (1989) Chemosphere 19:1243
15. Kuokkanen T (1989) Chemosphere 19:1349
16. Lun R, Varhanickova D, Shiu W-Y, Mackay D (1997) J Chem Eng Data 42:951
17. Rantio T (1997) Doctoral Thesis, Department of Chemistry, University of Jyväskylä, Finland, Res Rep 57
18. Kuokkanen T, Koistinen J (1989) Chemosphere 19:272
19. Rantalainen, A-L (1998) Unpublished report
20. Koistinen J (1992) Chemosphere 24:559
21. Herve S, Heinonen P, Paukku R, Knuutila M, Koistinen J, Paasivirta J (1988) Chemosphere 17:1945
22. Rantio T (1995) Chemosphere 31:3143
23. Rantio T (1996) Chemosphere 32:239
24. Paasivirta J, Hakala H, Knuutinen J, Otollinen T, Särkkä J, Welling l, Paukku R, Lammi R (1990) Chemosphere 21:1355
25. Rantio T, Paasivirta J (1996) Chemosphere 33:453
26. Lyman WJ, Reehl WF, Rosenblatt DH (eds) (1990) Handbook of chemical property estimation methods. ACS, Washington, DC
27. Virtanen V, Kukkonen J, Oikari A (1989) University of Joensuu, Faculty of Mathematics and Natural Sciences Rep Series 29:84
28. Särkkä J, Paasivirta J, Häsänen E, Koistinen J, Manninen P, Mäntykoski K, Rantio T, Welling l (1993) Chemosphere 26:2147
29. Paasivirta J, Koistinen J, Kuokkanen T, Maatela P, Mäntykoski K, Paukku R, Rantalainen A-L, Rantio T, Sinkkonen S, Welling l (1993) Chemosphere 27:447

30. Leuenberg C, Giger W, Coney R, Graydon JW, Molnar-Kubica E (1985) Water Res 19:885
31. Mackay D, Southwood JM (1992) Modeling the fate of organochlorine chemicals in pulp mill effluent. The Pulp and Paper Centre of the University of Toronto
32. Trapp S, Matthies M (1998) Chemodynamics and environmental modeling. An introduction. Springer, Berlin Heidelberg New York
33. Trapp S, Rantio T, Paasivirta J (1994) Environ Sci Pollut Res 1:246
34. Paasivirta, J, Särkkä J, Surma-Aho K, Humppi T, Kuokkanen T, Marttinen M (1983) Chemosphere 12:239
35. Herve S (1991) Doctoral Thesis. Department of Chemistry, University of Jyväskylä, Res Rep 36
36. Paasivirta J, Rantalainen A-L, Welling l, Herve S, Heinonen P (1992) Water Sci Tech 25:105
37. Herve S, Prest HF, Heinonen P, Hyötyläinen T, Koistinen J, Paasivirta J (1995) Environ Sci Pollut Res 2:24
38. Rantio T, Koistinen J, Paasivirta J (1996) In: Munkittrick KR, Carey JH, Van der Kraak GJ (eds) Environmental fate and effects of pulp and paper mill effluents. St Lucie Press, Delray Beach, FL, p 341
39. Paasivirta J, Mäntykoski K, Koistinen J, Kuokkanen T, Mannila E, Rissanen K (1989) Chemosphere 19:149
40. Paasivirta J (1992) Organochlorines from pulp mills and other sources. Department of Chemistry, University of Jyväskylä, Res Rep 38
41. Koistinen J, Paasivirta J, Lahtiperä M (1993) Chemosphere 27:149
42. Koistinen J, Nevalainen T, Tarhanen J (1992) Envir Sci Technol 26:2499
43. Koistinen J (1992) Chemosphere 24:559
44. Koistinen J, Paasivirta J, Nevalainen T, Lahtiperä M (1994) Chemosphere 28:1261
45. Koistinen J (1993) Doctoral Thesis. Department of Chemistry, University of Jyväskylä, Res Rep 42
46. Rosell A, Gomez-Belinchon JI, Grimalt JO (1991) J Chromatogr 562:493
47. Kuehl DW, Butterworth BC, DeVita WM, Sauer CP (1987) Biomed Environm Mass Spectr 14:443
48. Buser H-R, Kjeller L-O, Swanson SE, Rappe C (1989) Environ Sci Technol 23:1130
49. Koistinen J, Herzschuh R, Stach J (1990) Organohalogen Compounds 3:259
50. Nevalainen T, Koistinen J (1990) Organohalogen Compounds 4:361
51. Nevalainen T, Koistinen J (1991) Chemosphere 23:1581
52. Nevalainen T (1995) Chemosphere 30:847
53. Okada Y, Okubo K, Igarashi Y (1981) Ger Offen DE 8024020, March 26
54. Koistinen J, Paasivirta J, Nevalainen T, Lahtiperä M (1994) Chemosphere 28:2139
55. Sugiura K, Ishihara M, Shimauchi T, Harayama S (1997) Environ Sci Technol 31:45
56. Smith LM, Stalling DL, Johnson JL (1984) Anal Chem 56:1830
57. Paasivirta J, Mäntykoski K, Paukku R, Piilola T, Vihonen H, Särkkä J, Granberg L (1986) Aqua Fenn 16:17
58. Koistinen J, Paasivirta J, Särkkä J (1990) Chemosphere 21:1371
59. Billiard SM, Querbach K, Hodson PV (1997) 3rd International Conference on Environmental Fate and Effects of Pulp and Paper Mill Effluents, Rotorua, New Zealand, 10–13 November 1997. Conference Preprints: 362 (Post Conference Book in press)
60. Progil SA (1971) French Pat 1,603,289, May 7 1971
61. Prodelec Co (1972) Substitute impregnants for capacitor Askarel, the Chloralkylene 12 Information Brochure, Paris, France
62. Sundström G, Hutzinger O, Karasek FW, Michnowicz J (1976) JAOAC 59:982
63. Klesse C, Parlar H, Hustert K, Korte F (1981) Chemosphere 10:1293
64. Krupcik J, Leclerq PA, Garaj J, Simova A (1980) J Chromatogr 191:207
65. Lay JP, Kotzias D, Klein W, Korte F (1976) Chemosphere 5:45
66. Ruzo, LO, Sundström G, Hutzinger O, Safe S (1976) Chemosphere 5:71
67. Kotzias D, Klein W, Korte F (1977) Chemosphere 6:99
68. Renberg L, Sundström G, Sundh-Nygård K (1980) Chemosphere 9:683

69. Begum S, Lay JP, Klein W, Korte F (1975) Chemosphere 4:241
70. Tulp MTh, Sundström G, de Graaff JC, Hutzinger O (1977) Chemosphere 6:109
71. Tulp MTh, Tulp M, Tillmanns GM, Hutzinger O (1977) Chemosphere 6:223
72. Herbst E, Scheunert I, Klein W, Korte F (1977) Chemosphere 6:725
73. Moza PN, Scheunert I, Klein W, Korte F (1977) Chemosphere 6:575
74. Müller WF, McMurray T, Coulston F, Korte F (1979) Chemosphere 8:21
75. Moza PN, Scheunert I, Korte F (1979) Arch Envrionm Contam Toxicol 8:183
76. Pariso ME, St.Amant JR, Sheffy TB (1984) In: Nriagu JO, Simmons MS (eds) Toxic contaminants in the Great Lakes. Wiley, New York Chichester Brisbane Toronto Singapore, p 265
77. Peterman PH, Delfino JJ (1990) Biomed Envir Mass Spectrom 19:755
78. Mathais H, Commandeur R, Pontoglio A, Nebel S (1978) Eur Pat Appl EP 8251 19, 800, 220, CA 94:46, 935
79. Casale L, Nebel S, Pontoglio A (1982) Belg BE 893,387 A1 19,821,202, CA 98:145, 437
80. Bernard MPC, De Saint-Mathais HH, Gagnieur AR (1981) Eur Pat Appl EP 70,218 A1 830,914,19, CA 98:199, 263
81. Gervason P, Commandeur R, Curtner BJR (1982) Eur Pat Appl EP 88, 650 A1 19, 830, 914, CA 99:215, 498
82. Kervennal J, Commandeur R, Huet JM (1982) Eur Pat Appl EP 88,667 A1 19,830, 914, CA 100:52, 475
83. Dobratz G, Flucke W (1986) GER Offen DE 3,419,887, CA 104:71,569u
84. Ehmann J, Ballschmiter K (1989) Fresenius Z Anal Chem 332:904
85. van der Velde EG, Linders SHMA, Wammes HIJ, Meiring HD, Liem AKD (1995) J High Resol Chromatogr 18:647
86. Poppe A, Alberti J, Friege H, Rönnefahrt B (1988) Vom Wasser 70:33
87. Bouraly M, Millischer RJ (1989) Chemosphere 18:2051
88. van Haelst AG, Zhao Q, van der Wielen FWM, Govers HAJ, de Voogt P (1996) Ecotoxicol Environm Safety 34:35
89. Fürst P, Krüger C, Meemken H-A, Groebel W (1987) J Chromatogr 405:311
90. Fürst P, Krüger C, Meemken H-A, Groebel W (1987) Z Lebensm Unters Forsch 185:394
91. Wester PG, van der Valk F (1990) Bull Environ Contamin Toxicol 45:69
92. Leoni V, Mastroeni I, Vescia N, Fabiani L, Giuliani AR (1988) Bull Environ Contamin Toxicol 41:523
93. Specht W, Tillkes M (1985) Fresenius Z Anal Chem 322:443
94. van Haelst AG, Zhao Q, van der Wielen FWM, Govers HAJ (1996) Chemosphere 33:257
95. Hendriks AJ, Pieters H, DeBoer J (1998) Environ Toxicol Chem 17:1885
96. van Haelst AG, Heesen PF, van der Wielen FWM, Govers HAJ (1994) Chemosphere 29:1651
97. van Haelst AG, Loonen HEVM, van der Wielen FWM, Govers HAJ (1993) Polycyclic Aromat Compd 3 (Suppl):1207
98. van Haelst AG, Loonen HEVM, van der Wielen FWM, Govers HAJ (1996) Chemosphere 32:1117
99. van Haelst AG, Bakboord J, Parsons JR, Govers HAJ (1995) Chemosphere 31:2799
100. Friege H, Poppe A (1991) Umwelt 21:428
101. von Meyerinck L, Hufnagel B, Schmoldt A, Benthe HF (1990) Toxicol Lett 51:163
102. Murk AJ, van der Berg JHJ, Koeman JH, Brouwer A (1990) Organohalogen Compounds 1:199
103. van Haelst AG, Tromp PCB, de Voogt P, Govers HAJ (1993) Organohalogen Compounds 14:219
104. van Haelst AG, Tromp PCB, Govers HAJ, de Voogt P (1997) Quant Struct Act Relat 16:214

CHAPTER 2

Tris(4-Chlorophenyl)Methanol and Tris(4-Chlorophenyl)Methane

Jacob de Boer

J. de Boer (e-mail: j.deboer@rivo.dlo.nl)
DLO-Netherlands Institute for Fisheries Research, P. O. Box 68, 1970 AB IJmuiden,
The Netherlands

Tris(4-chlorophenyl)methanol (TCPM) and tris(4-chlorophenyl)methane (TCPMe) are globally widespread contaminants. However, there is a lack of knowledge on production figures, origin, and application of these compounds. One source is technical DDT in which these compounds are present as impurities, but it is assumed that there are other sources as well. Analytical methods for the determination of TCPM and TCPMe comprise extraction, cleanup over florisil columns or by gel permeation chromatography, and gas chromatographic analysis with electron capture or mass spectrometric detection. TCPM and TCPMe have been found in fish, birds, and marine mammals from various parts of the world. TCPM concentrations in marine mammals from the North Sea are around 1–2 mg/kg on a lipid weight basis. TCPM and TCPMe are highly bioaccumulative and a 10–100-fold biomagnification from fish to marine mammals is suggested. There is little information on their toxicity. TCPM is a phenobarbital and a 3-methylcholanthrene inducer producing hepatic effects, splenomegaly, and increased white blood cell and lymphocyte counts in rats after short-term dietary exposure. There are indications of a possible carcinogenic character.

Keywords. Tris(4-chlorophenyl)methanol, Tris(4-chlorophenyl)methane, Origin, Analytical methods, Environmental distribution, Toxicity

1
Introduction

Tris(4-chlorophenyl)methanol and tris(4-chlorophenyl)methane have been detected as micro-contaminants in the marine environment since 1989 [1]. Several abbreviations are used for these compounds, such as 4,4',4"-TCP, TCP, TCPM, and TCPM-OH for tris(4-chlorophenyl)methanol and 4,4',4"-TCPMe,

The Handbook of Environmental Chemistry Vol. 3 Part K
New Types of Persistent Halogenated Compounds
(ed. by J. Paasivirta)
© Springer-Verlag Berlin Heidelberg 2000

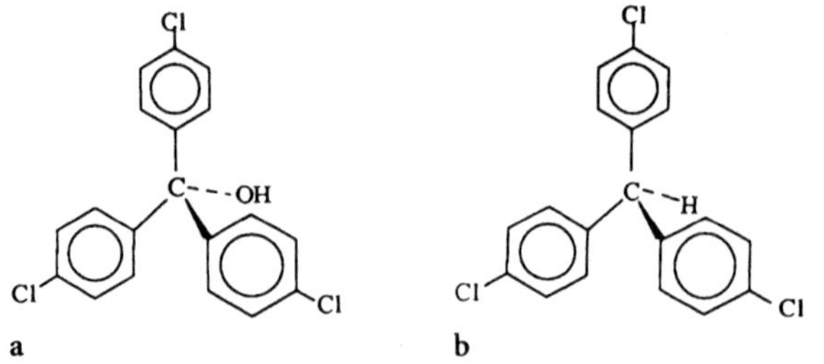

Fig. 1a, b. Structures of: **a** TCPM; **b** TCPMe

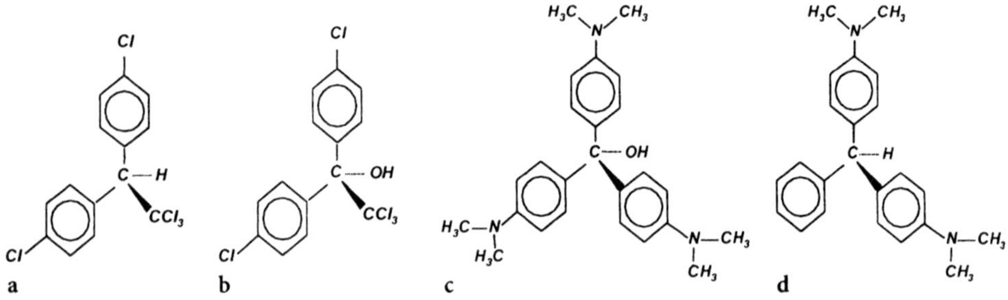

Fig. 2a–d. Structures of: **a** *p,p'*-DDT; **b** dicofol; **c** Basic Violet 3; **d** Malachite Green

TCPMe, and 4,4',4''-TCPM for tris(4-chlorophenyl)methane. In this chapter the abbreviations TCPM and TCPMe will be used. TCPM and TCPMe (Fig. 1) are structurally related to DDT and dicofol (1,1,1-trichloro-2,2-bis(4-chloro-phenyl)methanol) (Fig. 2).

2
Origin, Production, and Use

There is little information on sources of TCPM and TCPMe. From citations in the patent literature, the use of these compounds in synthetic (optically active) high polymers and lightfast dyes for acrylic fibers was suggested as a source for the environmental presence of TCPM and TCPMe [2–4]. TCPM is possibly a metabolite of TCPMe or tris(4-chlorophenyl)methylchloride (TCPC), which are used in the production of dyes [5]. However, some of the patent literature is rather new, and it is questionable whether that use would completely account for the apparently long-time presence of TCPM and TCPMe in the environment.

Another source of TCPM and TCPMe may be the pesticide DDT itself. Buser [6] showed that under conditions such as those used in the technical synthesis of DDT, 4,4',4''-TCPMe and two additional isomers 2,2,'4''- and 2,4,'4''-TCPMe were formed in small amounts. In two technical DDT mixtures Buser [6] determined the same isomers in total amounts of 150–180 mg/kg, of which about one third was 4,4',4"-TCPMe. These impurities in commercial DDT mixtures may have contributed to the environmental levels of TCPM and TCPMe, but probably do not completely explain these levels. It is likely that the presence of TCPM and TCPMe in the marine environment originates from various sources. Similar to DDT, formulations of dicofol may also contain impurities of TCPM [4].

Other citations from an American Chemical Society Chemical Abstracts Search (CAS) mention TCPM as a starting product for the manufacture of anthelmintic drugs [4] and anti-ecdysone activity of TCPM, similar to the agrochemical Triarimol [4].

Obviously, there is an urgent need for more information on the production and use of TCPM and TCPMe.

3
Analytical Methods

Extraction of TCPM and TCPMe from biological materials or sediments is generally carried out by Soxhlet extraction with dichloromethane [1, 7] or mixtures of dichloromethane with n-hexane [4] or n-pentane [8]. Zook et al. [9] used a cold column extraction with dichloromethane/n-hexane (50:50, v/v).

Florisil columns are used for the separation of lipids and the target compounds [1, 4, 7]. Zook et al. [9] used a dialysis technique with a polyethylene film for the removal of lipids, followed by gel permeation chromatography (GPC), using S-X3 Bio beads with dichloromethane/n-hexane (50:50, v/v), carbon column chromatography, and florisil columns. De Boer et al. [8] have tried to avoid the use of florisil, because of the extensive pre-treatment and its relative instability. They used GPC, Bio beads S-X3 with dichloromethane/n-hexane (50:50, v/v) for the separation of lipids and TCPM and TCPMe. The GPC elution was carried out twice and was followed by a silica gel column fractionation to separate TCPM and TCPMe from the PCBs. The recovery of a TCPM spike in a seal blubber extract in this method was 90%.

The same authors reported that TCPM was unstable during a sulfuric acid treatment and that TCPM could not be separated from the lipids by alumina column chromatography [8].

Gas chromatography is the method of choice for the determination of TCPM and TCPMe. Non-polar or medium-polar columns, such as DB5 and CP Sil 8 (5% phenyl 95% dimethylpolysiloxane) [1, 4, 8], DB17 (50% phenyl 50% dimethylpolysiloxane) [4, 10], and CP Sil 12 (58% CP Sil 8 and 42% CP Sil 19 (85% methyl 7% phenyl 7% cyanopropyl 1% vinyl polysiloxane)) [8] can be used. TCPM and TCPMe elute relatively late in the chromatograms, normally just before octachloronaphthalene, with GC oven temperatures around 270–300 °C. Splitless injection can be used [8].

Three different detection methods have been used until now: GC/ECD[1, 4, 9], GC/EI (electron impact)-MS [1, 4, 6–9], and GC/ECNI (electron capture negative ion)-MS [8, 9]. The ions used for identification and quantification in GC/MS are 111, 139, 141, 251, 362, 364 (TCPM, EI-MS), 362, 346, 348 (TCPM,

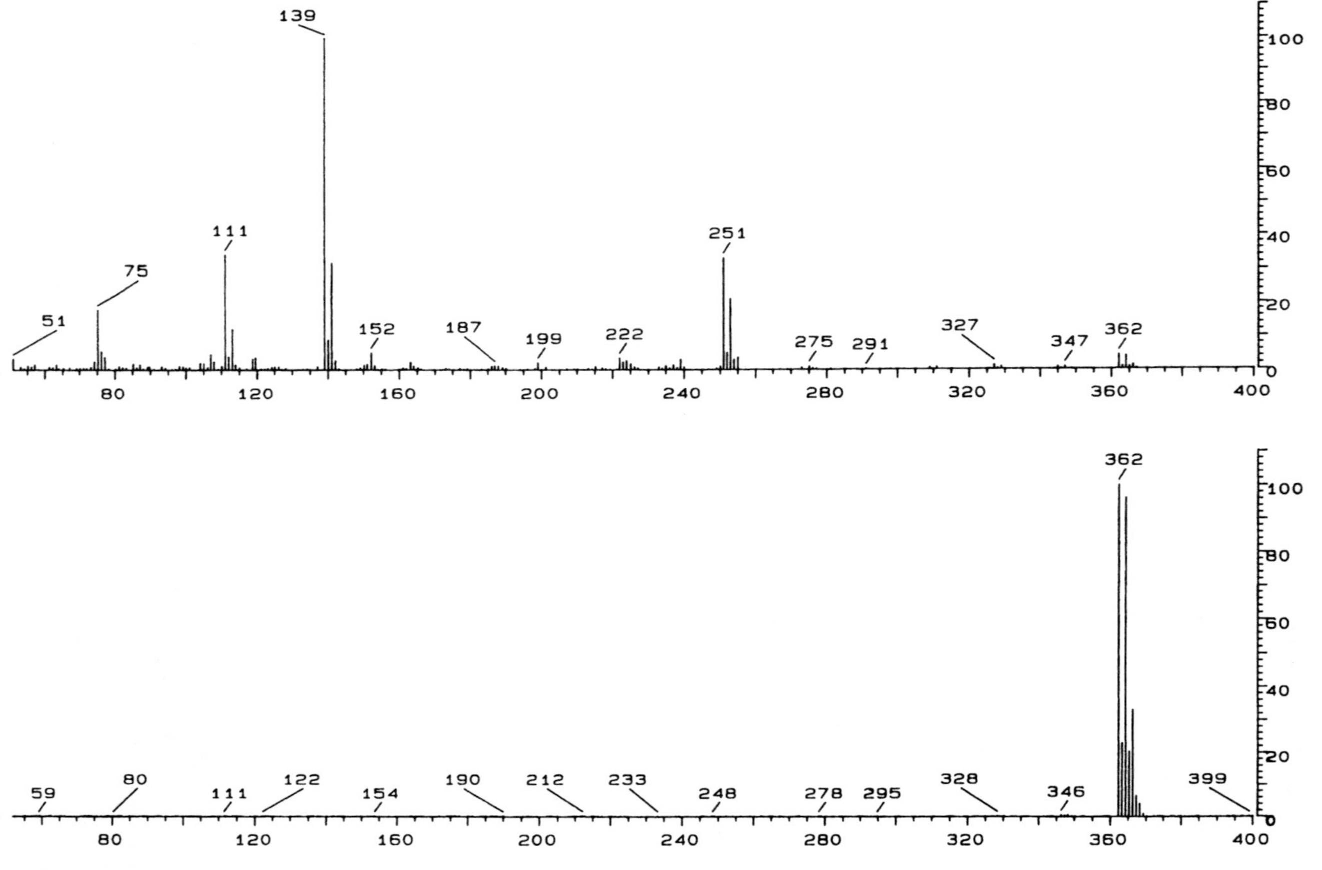

Fig. 3a–d. Mass spectra of: a TCPM in the EI mode; b TCPM in the ECNI mode

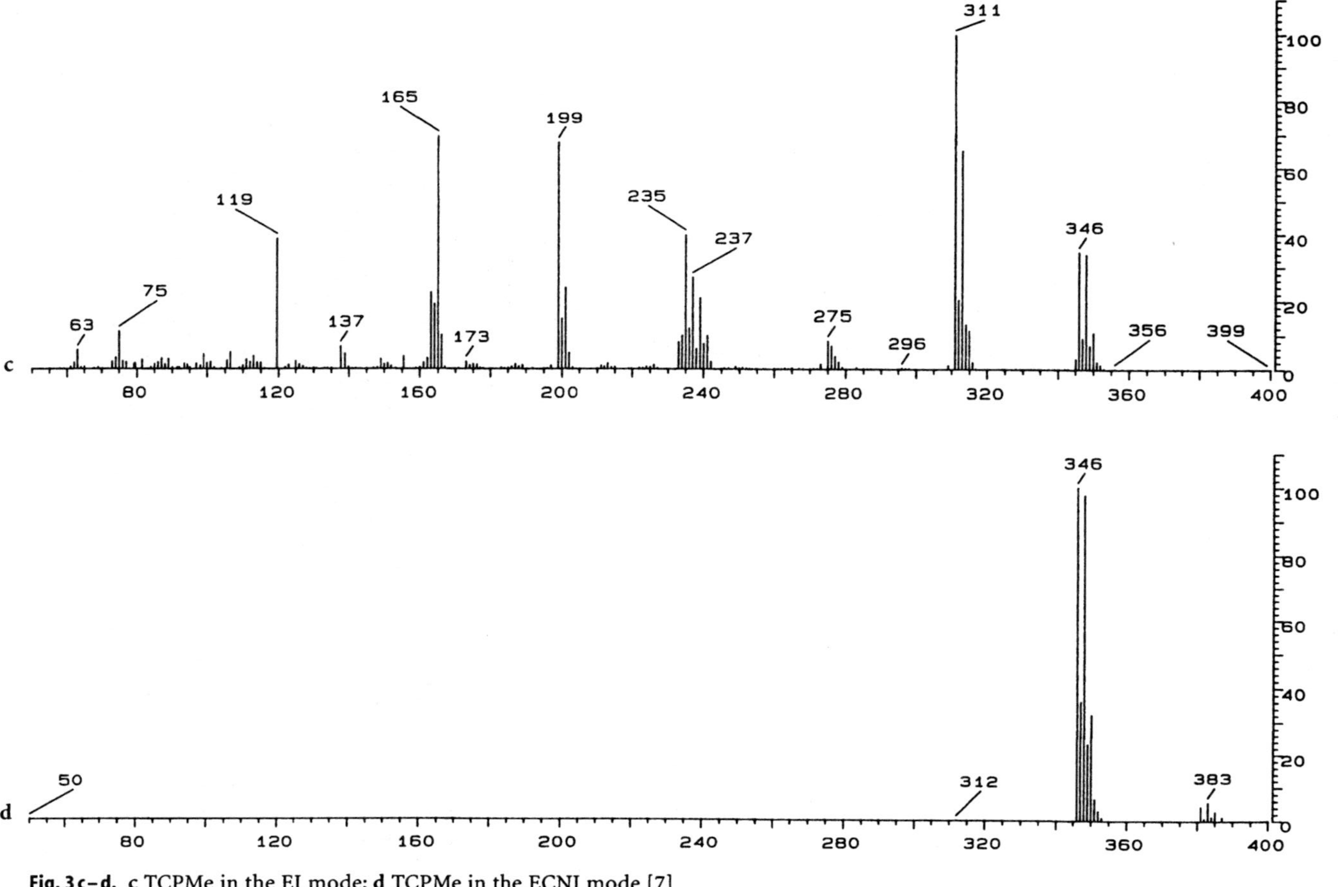

Fig. 3 c–d. c TCPMe in the EI mode; d TCPMe in the ECNI mode [7]

Table 1. Detection limits of TCPM and TCPMe [8]

Method	GC/ECD *CP Sil 8*		GC/EI/MS *CP Sil 8/12*		GC/ECNI-MS *CP Sil 8/12*	
	TCPM	TCPMe	TCPM	TCPMe	TCPM	TCPMe
Detection limit (pg)	10	10	30	10	3	100
Detection limit (µg/kg)	0.1	0.1	0.2	0.07	0.02	0.7

ECNI-MS), 311, 313, 346, 348 (TCPMe, EI-MS), and 346, 348 (TCPMe, ECNI-MS). Much more fragmentation is observed in the EI spectra when compared to the ECNI spectra (Fig. 3). As is to be expected, the most prominent peaks in the ECNI spectra are due to the molecular ions.

De Boer et al. [8] have compared the sensitivity of the three detection techniques for TCPM and TCPMe (Table 1). Generally, compounds with less than four chlorine atoms are rather difficult to determine in environmental samples by means of GC/ECNI-MS because of too low a sensitivity [10]. However, for TCPM the electronegativity of the oxygen apparently helps to increase sensitivity and thus to create good analyte detectability for this compound. GC/ECNI-MS clearly is not the method of choice for TCPMe because of low sensitivity (Table 1). This confirms the role of the oxygen atom observed with TCPM. Both GC/EI-MS and GC/ECD are suitable alternatives, but MS-based detection is preferred because of its higher selectivity. GC/EI-MS can in principle also be used for the determination of TCPM, with m/z 139 to be used for quantification. However, quantification on such a low mass is easily disturbed by interferences of mass fragments of other compounds. This was clearly demonstrated by comparing EI and ECNI results for TCPM in a variety of samples [8]. Depending on the sample, up to ten-fold too high results were obtained by GC/EI-MS. Such problems are normally not encountered in the GC/EI-MS determination of TCPMe for which the relatively high m/z value of 311 is used. The different sensitivity of TCPM and TCPMe for the two MS techniques may necessitate replicate injections of the extract containing TCPM and TCPMe, but with more recent GC/MS systems a switch from EI to ECNI during the run is possible. Consequently, a sensitive and selective analysis of both compounds is possible.

The use of ECD detection causes a small loss in sensitivity (Table 1), but a considerable loss in selectivity. Therefore, ECD can be used for screening of TCPM and TCPMe, but confirmation by GC/ECNI-MS or GC/EI-MS, respectively, will often be essential.

4
Environmental Distribution

TCPM and TCPMe are present in fish, birds, and marine mammal samples on a global scale. Walker et al. [1] reported the presence of TCPM in seals from Puget

Sound, North West USA, after synthesis of TCPM as an analytical standard. Jarman et al. [4] reported the presence of TCPM and TCPMe in various birds and marine mammals from the Arctic and Antarctica, Australia, and the USA. The presence of TCPM in a variety of environmental tissues from different areas in Canada was semi-quantitatively determined by Norstrom and Simon [11]. The presence of TCPM and TCPMe in adipose tissue and livers of seals from the Baltic was reported by Zook et al. [9]. De Boer et al. [8] reported the presence of TCPM and TCPMe in several marine mammals, cod liver, and mussels from the North Sea and Dutch Wadden Sea. TCPMe was synthesized by these authors and its presence in North Sea and Wadden Sea samples was confirmed. They also reported the presence of TCPM and TCPMe in eel from the river Rhine delta and several other Dutch rivers and in sediment from the river Rhine delta. Rahman et al. [7] reported the presence of TCPM in cod liver oil and mackerel oil, originating from the southern North Sea. The same authors also reported the presence of TCPM and TCPMe in Italian human milk at a level of 2.5 µg/kg and 1.6 µg/kg, respectively. An overview of the levels of TCPM and TCPMe reported in marine environmental samples is given in Table 2.

Table 2. Concentrations of TCPM and TCPMe in various samples from the aquatic environment

Sample	Location	Year	TCPM µg/kg[a]	TCPMe	Ref.
Harbor seal blubber	Puget Sound, USA	1972–1982	23–750		[1]
Northern fur seal blubber	St. Lawrence River, Canada	1981	5–29		[4]
Harp seal fat	St. Lawrence River, Canada	1952	2		[4]
Ringed seal liver	Baltic Sea	1980–1987	3000		[9]
Ringed seal adipose tissue	Baltic Sea	1980–1987	900		[9]
Ringed seal blubber	Baltic Sea	<1998	264[b]	211[b]	[20]
Ringed seal liver	Baltic Sea	<1998	13[b]	23[b]	[20]
Common seal blubber	Dutch Wadden Sea	1990–1992	750–2000		[8]
Antarctic fur seal milk	South Georgia Isl., Antarctica	1984–1985	<2–2.7		[4]
Australian sea lion milk	Kangaroo Island, Australia	1987	<7		[4]
California sea lion milk	Central coast, California, USA	1988	19		[4]
Beluga whale fat	St. Lawrence River, Canada	1984–1986	1–41		[4]
Beluga whale fat	St. Lawrence River, Canada	1952	4		[4]

Table 2 (continued)

Sample	Location	Year	TCPM µg/kg[a]	TCPMe µg/kg[a]	Ref.
Beluga whale blubber	St. Lawrence River, Canada	1987–1990	52–333[c]	42–1450[c]	[21]
Whitebeaked dolphin blubber	Newfoundland, Canada	1982	184		[21]
Whitebeaked dolphin blubber	Dutch Wadden Sea	1992	570		[8]
Whitebeaked dolphin blubber	Southern North Sea	1990	1400		[8]
Common dolphin blubber	Dutch Wadden Sea	1992	220		[8]
Harbor porpoise blubber	Southern North Sea	1990	1000		[8]
Polar bear fat	Arctic, Canada	1984	35		[4]
Polar bear liver	Arctic, Canada	1983	4000–6800		[4]
Herring gull egg	Lake Ontario, Canada	1971–1985	94–3300		[4]
Peregrine falcon egg	Croften, centr. coast, BC, Canada	1989	250–5,300		[4]
Great blue heron egg	Croften, centr. coast, BC, Canada	1986	36		[4]
White-tailed sea eagle	Poland	1991–1996	14–54,000	9–33,000	[22]
Cod liver	Northern North Sea	1992	6		[8]
Cod liver	Southern North Sea	1993	40		[8]
Cod liver oil	Southern North Sea	1987	51[d]	3[d]	[7]
Mackerel oil	Southern North Sea	1987	35[d]	0.6[d]	[7]
Mussels	Dutch Wadden Sea	1994	13	<6	[8]
Tuna, perch, angler, mullet	Mediterranean Sea	1992	<1[d]	<1[d]	[8]
Yellow eel	Rhine delta, Netherlands	1994	180–360	37	[8]
Sediment[e]	Rhine delta, Netherlands	1992	26–160	4–13	[8]

[a] µg/kg lipid weight.
[b] Way of determination unknown.
[c] Calculated using a relative response factor to *p,p′*-DDT.
[d] Calculated using a relative response factor to CB-169.
[e] Concentration in sediment in µg/kg total organic carbon.

A considerable data set is available for TCPM, but the number of TCPMe data is limited. The results vary from < 1 µg/kg on a lipid weight basis in some fish samples to 54 mg/kg in white-tailed sea eagles, which shows that TCPM and TCPMe are highly bioaccumulative. A considerable bioconcentration of TCPM and TCPMe is also expected on the basis of the octanol-water partition coefficients calculated by the C log P method [12]. Log K_{ow} for TCPM is 6.0, and for TCPMe 6.5.

Most data originate from marine samples. Only from the North Sea have fish (cod liver and mackerel), shellfish (mussels), and marine mammals been analyzed. The TCPM concentrations in these samples suggest a strong biomagnification of TCPM from fish to seals, porpoises, and dolphins in the order of magnitude of 10-fold to 100-fold. The total number of samples analyzed from the different areas is relatively small. Consequently, it is difficult to say something about the natural variation of TCPM concentrations in marine mammals. However, data of other halogenated micro-contaminants such as PCBs in marine mammals show that the natural variation can be relatively large [13]. The range of TCPM concentrations found in the marine mammals therefore indicates that significant differences in biomagnification between the different species are unlikely.

TCPM concentrations in cod liver are comparable to those reported for HCB and *p,p'*-DDT in the same matrix [10]. Rahman et al. [7] reported that TCPM and TCPMe concentrations in Italian human milk were 2–3 orders of magnitude lower than the detected levels of other organochlorine compounds such as DDT. This could suggest a metabolism of TCPM and TCPMe in the human body. However, the calculation method used in that study may have resulted in less accurate TCPM and TCPMe data because analytical standards were not available and the calculation was based on the assumption of equimolar responses of TCPM, TCPMe, and CB169 in GC/EI-MS.

TCPM was found in two samples originating from 1952 (Table 2). These data show that TCPM has apparently been present in the environment for 45 years. More recent data on TCPM in organisms from the St. Lawrence River show that the current TCPM levels are 10- to 100-fold higher than in 1952. Some data should be interpreted with care because analytical standards were not always available and errors could have been made by using relative response factors to other compounds.

TCPM is reported in samples from areas from all over the world. A reason for this ubiquitous occurrence may be the presence of the 4-chlorophenyl rings in the compound, which are also found in *p,p'*-DDT and its metabolites *p,p'*-DDD and *p,p'*-DDE, which are also very persistent contaminants. This structure is also found in bis(4-chlorophenyl) sulfone, which was recently reported as a new persistent contaminant [14]. The 4-chlorophenyl structure is apparently very resistant to transformation in the environment. Only in Mediterranean fish samples were TCPM and TCPMe not detected at a level of 1 µg/kg lipid weight. TCPM concentrations on a lipid weight basis decrease from 180 µg/kg to 360 µg/kg in eel from the river Rhine delta, to 40 µg/kg in cod liver from the southern North Sea, and to 6 µg/kg in cod liver from the northern North Sea (Table 2). This suggests a relationship between high TCPM concentrations and densely populated, industrialized areas. High TCPM concentrations in samples from the Baltic Sea and Lake Ontario and low TCPM concentrations in samples from Antarctica confirm this hypothesis.

Contamination of biota and sediment in the Rhine estuary with, for example, PCBs and organochlorine pesticides has remained essentially constant over the last three years [10]. Although this may be different with regard to TCPM and TCPMe, a comparison of the sediment samples from 1992 with the eel samples from 1994 from the same area may at least give an indication of the distribution of TCPM and TCPMe in the biota and sediment. When comparing the concentrations in fish expressed on a lipid weight basis with those in sediment, expressed on an organic carbon basis, the fish/sediment ratios are in the range 1.1–14 for TCPM, and 2.8–13 for TCPMe. These ratios roughly correspond with those found for PCBs in the same area (ratios 1–5) [10]. They are much higher than those for polychlorinated terphenyls (PCTs) (ratios 0.2–1) in the same area [15]. These data suggest that the adsorption characteristics of TCPM and TCPMe to sediments are comparable to those of PCBs.

5
Toxicity

Data on the toxicity of TCPM and TCPMe are very scarce. Michaels and Lewis [2] have reported a toxicity study on five azo and triphenylmethane dyes. In that study Basic Violet 3, with a structure in which the basic structure of TCPM can be recognized (Fig. 2), shows the highest toxicity: a survival rate of 20.7 ± 6.6 % at a dye concentration of 5.0 mg/l of microbiota in a plating medium incorporated with dye.

The structure of TCPM and TCPMe also closely resembles that of malachite green (Fig. 2), a well known carcinogenic dye which easily forms radicals. The presence of a compound resembling such a carcinogenic substance in fish could have serious consequences.

Hoogenboom [16] carried out an Ames test with TCPM and TCPMe. For determining whether a result can be regarded positive in this test, two criteria are used. The number of revertants should be dose-related and the positive response should be reproducible. In the study of Hoogenboom no dose-related increase in the number of revertants was observed with either TCPM and TCPMe. However, an increased number of revertants was observed at the highest dose of 5 mg TCPM and 5 mg TCPMe. Due to precipitation of both compounds at this dose-level it was difficult to evaluate the toxicity of the compounds based on the presence of microcolonies. The authors concluded that the positive results require confirmation in an independent trial and, until then, no definite conclusion can be drawn with respect to the mutagenic potential of TCPM and TCPMe. Poon et al. [17] reported hepatic effects, splenomegalmy, and increased white blood cell and lymphocyte counts after dietary exposure of rats to TCPM at concentrations of 10 mg/kg and 100 mg/kg, equivalent to daily intakes of 1.2 mg/kg and 12.4 mg/kg in males and 1.2 mg/kg and 10.9 mg/kg in females. Similar effects have been observed for DDT but at higher levels ranging from 15 mg/kg to 40 mg/kg [18, 19]. TCPM differs from DDT by being both a phenobarbital and a 3-methylcholanthrene type of AHH inducer whereas DDT is only a phenobarbital type of inducer [17].

References

1. Walker W, Riseborough RW, Jarman WM, Lappe BW, Lappe JA, Tefft JA, Long RL de (1989) Chemosphere 18:1799
2. Michaels GB, Lewis DL (1985) Environ Toxicol Chem 4:45
3. Michaels GB, Lewis DL (1986) Environ Toxicol Chem 5:161
4. Jarman WM, Simon M, Norstrom RJ, Bruns SA, Bacon CA, Simaret BRT, Riseborough RW (1992) Environ Sci Technol 26:1770
5. Boer J de (1997) Rev Environ Contam Toxicol 150:95
6. Buser HR (1995) Environ Sci Technol 29:2133
7. Rahman MS, Montanarella L, Johansson B, Larsen B (1993) Chemosphere 27:1487
8. Boer J de, Wester PG, Evers EHG, Brinkman UAT (1996) Environ Pollut 93:39
9. Zook DR, Buser HR, Bergqvist PA, Rappe C, Olsson M (1992) Ambio 21:557
10. Boer J de (1995) Analysis and biomonitoring of complex mixtures of persistent halogenated micro-contaminants. PhD Thesis, Free University, Amsterdam, The Netherlands
11. Norstrom R, Simon M (1988) Presence of tris(chlorophenyl)methanol in Canadian environmental tissue samples. Rep-CRD-88-7, Canadian Wildlife Service, Chemistry Research Division, Hull, Québec, Canada
12. Hansch C, Leo AJ (1979) Substituent constants for correlation analysis in chemistry and biology. Wiley, New York, USA
13. Duinker JC, Hillebrand MT, Zeinstra T, Boon JP (1989) Aquatic Mammals 15:95
14. Olsson A, Bergman Å (1995) Ambio 24:119
15. Wester PG, Boer J de, Brinkman UATh (1996) Environ Sci Technol 30:473
16. Hoogenboom LAP (1996) *Salmonella*/microsome test with tris(4-chlorophenyl) methanol and tris(4-chlorophenyl)methane. Rep LHO.RIK.95.04, RIKILT-DLO, Wageningen, The Netherlands
17. Poon R, Lecavalier P, Bergman Å, Yagminas A, Chu I, Valli VE (1997) Chemosphere 34:1
18. Gupta PH, Mehta S, Mehta SK (1989) Biochem Int 19:247
19. Hart LG, Fouts JR (1965) Naunyn Schmiedebergs Arch Exp Pathol Pharmakol 249:486
20. Strandberg B, Bergqvist PA, Rappe C (1998) Anal Chem 70:526
21. Muir DCG, Ford CA, Rosenberg B, Norstrom RJ, Simon M, Béland P (1996) Environ Pollut 93:219
22. Strandberg L, Strandberg B, Bergqvist PA, Rappe C, Falandysz J (1998) Organohalogen Compounds 39:453

CHAPTER 3

Polychlorinated Terphenyls

Jacob de Boer

J. de Boer (e-mail: j.deboer@rivo.dlo.nl)
DLO-Netherlands Institute for Fisheries Research, P.O. Box 68, 1970 AB IJmuiden,
Netherlands

Polychlorinated terphenyls (PCTs) have been produced over the same period as polychlorinated biphenyls (PCBs), but in 15–20-fold lower quantities (ca. 60,000 tons in total). The
production of PCTs has been terminated in most countries in the mid-1970s. PCTs have
similar properties to PCBs and have been used for the same purposes such as in electronic
equipment, lubricants, sealants, etc. They are also found as global contaminants and the ratios
of environmental levels of PCBs and PCTs approximately reflect the production figures.
Because of the complexity of PCTs the analysis is difficult. Gas chromatography (GC) with
electron capture detection or mass spectrometric detection has been applied, but the best
possible quantification at the moment is still a semi-quantitative total PCT analysis. More
advanced techniques, such as comprehensive multi-dimensional GC, are required to enable a
congener-specific PCT analysis. The lack of commercially available PCT congeners has also
hindered a congener-specific analysis. Relatively high PCT concentrations, more than
10 mg/kg on a lipid weight basis in marine mammals and birds, show the bioaccumulative
properties of PCTs. The persistent character of PCTs was confirmed in an in vitro study in
hepatic microsomes of marine mammals. The toxicity of PCTs is generally considered to be
equivalent to that of PCBs. A difficulty in toxicological studies of PCTs is the contamination
of PCT mixtures with PCBs.

Keywords. Polychlorinated terphenyls (PCTs), Production, Analysis, Environmental distribution, Toxicity

The Handbook of Environmental Chemistry Vol. 3 Part K
New Types of Persistent Halogenated Compounds
(ed. by J. Paasivirta)
© Springer-Verlag Berlin Heidelberg 2000

1
Introduction

Polychlorinated terphenyls (PCTs) are chemical compounds which may be regarded as diphenylbenzenes. They occur in three possible arrangements: ortho-, meta-, and para-terphenyls (Fig. 1). Any hydrogen atom in a terphenyl may be replaced by a chlorine atom to give a (poly) chlorinated terphenyl [1]. The maximum number of different congeners possible is 8149 [2]. The production of both PCTs and polychlorinated biphenyls (PCBs) was started in 1929 in the US. Their chemical stability, high heat capacity, and excellent electric properties made them highly desirable for a number of industrial uses, in most cases similar to the use of PCBs. Examples are the use in plasticizers in synthetic resins, lubricants, paper coatings, printing inks, sealants, fire retardants, and waxes.

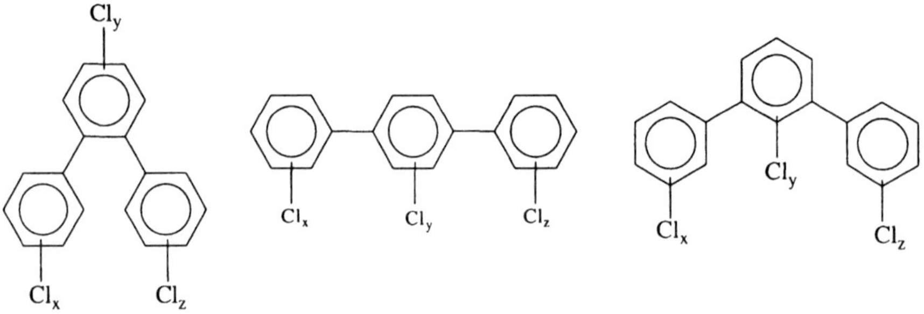

Fig. 1. Structure of PCTs

As highly chlorinated aromatics, PCTs might be expected to have a high resistance to biodegradation and photodegradation. Despite the obvious similarities in environmental properties with PCBs, the attention devoted to the environmental presence of PCTs has always been very limited compared to PCBs [3–7]. Nevertheless, PCT residues have been detected in many environmental samples and environmental concentrations of PCTs have sometimes shown to be up to 10% of PCB concentrations in the same samples [8]. PCTs have been detected in fish, oysters, birds, wild dogs, cats, snakes, cow livers, pigeon fat, mother's milk, human fat/blood, soil, sediment, garbage, and river water [9]. Therefore, the environmental contamination/distribution of PCTs may well be wider than is generally assumed.

2
Chemical and Physical Properties

PCTs have chemical and physical features closely resembling those of PCBs [1, 10]. They are non-oxidizing, inert, permanently thermoplastic, of low volatility, non-corrosive to metals, not hydrolysed by water, and resistant to aggressive and corrosive chemicals like alkalies and strong acids. They impart fire retardant properties to other materials [1]. Most PCTs were produced as technical mixtures under the name of Aroclor by Monsanto, St. Louis, Missouri, USA. The PCTs mixture was given a four digit code, starting with two digits denoting a particular series of Aroclors, while the last two digits indicate the chlorine content of the material, but sometimes the percentage of PCT in a mixture [1]. Some Aroclor mixtures were mixtures of PCBs and PCTs. The Aroclor 25 series was a 75:25 blend of PCBs and PCTs and the Aroclor 44 series was a 60:40 blend of PCBs and PCTs. Thus Aroclor 2565 contained 75% PCB, 25% PCT, and 65% chlorine. The (distilled) Aroclor 54 series only contained PCTs. Members of this series were the Aroclors 5432, 5442, and 5460. These products were colorless to light yellow mixtures, whereas the undistilled PCTs, available as the A50 series, were somewhat more highly in color [1]. The A54 PCT mixtures can contain 0.5–10% of PCBs [1]. Chlorinated dibenzofurans are also found in commercial PCT mixtures [11]. Aroclors 60 (6037, 6040, 6050, 6062, 6070, and 6090) were blends of Aroclor 5460 and 1221, a monochlorinated biphenyl with 21% chlorine. Other products produced by Monsanto and containing PCTs were the hydraulics/lubricants Pydrauls [1].

PCTs are produced by a controlled chlorination of hydrocarbons [12]. The fully chlorinated perchloro-p-terphenyl is obtained from p-terphenyl by means

Table 1. Some properties of four PCT-containing Aroclor mixtures [1]

	Aroclor 2565	Aroclor 4465	Aroclor 5442	Aroclor 5460
Appearance	Black, opaque, brittle resin	Clear, light yellow resin	Clear yellow sticky resin	Clear yellow-to-amber, brittle resin or flakes
Specific Gravity (25 °C)	1.734	1.670	1.470	1.670
Distillation range, °C, corrected (ASTM D-20, modified)	–	230–320 (4 Torr)	215–300 (4 Torr)	280–335 (5 Torr)
Flash Point, °C (Cleveland Open Cup)	–	None to boiling point	247	None to boiling point
Fire Point, °C (Cleveland Open Cup)	–	None to boiling point	>350	None to boiling point
Pour Point, °C (ASTM D-97)	–	–	46	–
Softening Point, °C (ASTM E-28)	66–72	60–66	46–52	98–105.5
Refractive Index n 20 D	–	1.664–1.667	–	1.660–1.665
Viscosity, sec Saybolt Universal (ASTM D-88)	–	90–150 (130 °C)	–	300–400 (98.9 °C)

of SbCl$_5$ at 360 °C [13]. The liquid and resinous Aroclor compounds are soluble in most organic solvents. Crystalline Aroclor compounds are described as relatively insoluble [1]. All Aroclors are insoluble in water and glycerine. Properties of some Aroclor mixtures are given in Table 1. Hutzinger et al. [14] reported melting points of decachlorinated PCTs of 318–360 °C. Chittim et al. [15] synthesized 22 individual PCT congeners by a diazo coupling of biphenylamine with an excess of a symmetrical chlorobenzene. A similar coupling with unsymmetrical chlorobenzenes gave mixtures which could be purified in some cases. Melting points and spectroscopic properties of these 22 PCTs were reported [15]. The chemical abstract number of PCTs is 61788–33–8.

3
Production and Use

The production of both PCTs and PCBs was started in 1929 in the US by Monsanto Chemical Company, which has remained the sole US producer. Since then PCTs have also been produced in France, Italy, Germany, and Japan [12]. Information about production volumes is incomplete, but during the period 1955–1980 about 60,000 tonnes were produced world-wide [9, 12], which is 15–20-fold lower than the total PCB production during that period [10] or the cumulative world-wide PCB production until 1984. Table 2 contains information about producers, countries, and brand names [9].

Monsanto USA voluntarily terminated production in 1972. The supply of Aroclors was discontinued to those users who could not control release to the environment [1]. In Germany the production of PCTs was terminated in 1974, in Italy in 1975, and in France in 1980 [12]. No information is available about the cessation of production of PCTs in Japan.

PCTs have been used in plasticizers in synthetic resins, lubricants, paper coatings, printing inks, sealants, caulking compounds, fire retardants, and waxes. One of the main applications was the use in the investment casting industry, in which PCTs are used in the process of metal-melting. The investment casting industry produces precision cast metal parts and shapes for the aircraft and other machinery manufacturing industries [1]. In the early 1980s 10% of the PCTs were used in electronic equipment [12].

Table 2. Producers and brand-names of PCTs [9]

Producer	Country	Brand-name
Monsanto	USA	Aroclor 5432, 5442 and 5460
Kanegafuchi	Japan	Kanechlor C
Mitsubishi-Monsanto	Japan	Aroclor series
Bayer	Germany	Leromoll 112–90 and 141 Clophen-Harz W
Caffaro	Italy	Cloresil A, B and 100
Prodelec, Produits Chimique Ugine Kuhlman	France	Phenoclor, Electrophenyl T-60

4
Analysis

Although PCTs have a relatively low volatility, gas chromatography is the preferred analytical method for the determination of PCTs. The retention times are evidently longer for PCTs than for PCBs. Often higher oven temperatures are necessary. Despite the relatively long period of ca. 30 years in which PCTs have been analyzed in environmental and industrial samples, not much progress has been made in the analytical methodology. This is mainly due to the high degree of complexity of PCT mixtures, which is about 40-fold higher than that of PCBs. Actually, this complexity may have deterred analysts to some extent. Also, individual PCT congeners have not become commercially available which has hindered the development of a congener-specific analysis. Apart from that, other difficulties are encountered in the development of a congener-specific analysis. A very good capillary column is not able to separate more than ca. 100–200 different congeners. It is known that it is very difficult to separate all PCBs in one run on a single capillary column, although good attempts have been made. However, there are only 209 PCB congeners of which maybe ca. 100–150 can occur in environmental samples. The number of PCT congeners in environmental samples may be estimated at several thousands, given the theoretical total number of 8149. This means that single-column GC analysis by no means can provide sufficient resolution to obtain a complete separation of the PCT congeners. Although multi-dimensional GC techniques have not been applied to the analysis of PCTs, it may even be discussed if heart-cut techniques would be able to provide sufficient resolution. Comprehensive two-dimensional GC may offer a final solution, but this technique is still under development.

4.1
Extraction and Clean-Up

The methods used for extraction and clean-up of PCTs are very similar to those used for PCBs. Both non-destructive and destructive methods are being used to separate the PCTs from the lipids. Soxhlet extraction is used by various authors with dichloromethane/methanol (2:1, v/v) [16], dichloromethane [17], or dichloromethane/pentane (1:1, v/v) [8]. Canton and Grimalt hydrolyzed the extract by a 6% KOH solution in methanol [16]. Jan and Malnersic [18] tested several methods for extraction and fat destruction. They report that destructive methods could degrade some of the PCT congeners. Wester et al. [8] did not find degradation of PCTs after application of sulphuric acid to the Soxhlet extracts of fish samples. Van der Valk and Dao [19] have reported that alkaline hydrolysis may lead to degradation of higher chlorinated PCBs. It is expected that alkaline hydrolysis will also degrade some of the higher chlorinated PCTs. Alumina and silica columns were used by Canton and Grimalt [16] and Wester et al. [8]. Florisil columns were used by Addison et al. [20] and Hale et al. [17]. Hale et al. [17, 21] used gel permeation chromatography (GPC) to separate the PCTs from the lipids and also from the PCBs. It was found that the PCTs eluted before most PCBs both on an S-X3 and S-X8 Bio beads column (100 g,

200–400 mesh). The use of dichloromethane as the sole solvent was preferred. S-X3 was considered to give a better separation between PCTs and lipids than S-X8. Onuska et al. [22] tested microbore and open tubular capillary columns in supercritical fluid chromatography (SFC). The UV detection used limited the sensitivity of the methods to detection of mg/kg levels. The quantitative performance of this method was considered to be comparable to analyses based on perchlorination.

4.2
Gas Chromatography

In the 1970s PCT analyses were mainly carried out on packed GC columns. Because of the enormous difference between the required and the obtained resolution, it was often decided to perchlorinate the PCTs to tetradecachloroterphenyls [9, 14, 23–26]. Obviously, this technique implies a number of disadvantages. Information on PCT patterns, possibly containing information on the distribution in the environment, and on the chlorine degree are lost. Information on individual PCT congeners, of which some will be more toxic than others, cannot be obtained. False positive and false negative results can easily occur due to the chlorination of interferences or insufficient chlorination. The stationary phases used were, e.g., OV-17, OV-210, Dexsil 300, and SE-30. It was essential that these phases were stable at temperatures of 250–300 °C. At lower maximum allowable oven temperatures not all PCTs and certainly not the tetradeca congeners would elute from the columns.

Later, packed columns were replaced by capillary columns and perchlorination techniques were used less frequently. However, a congener-specific determination of PCTs was still not possible due to the high number of PCTs present in environmental and technical samples. An overview of columns and conditions used in more recent studies is given in Table 3. The stationary phases used are generally non-polar and semi-polar. As with packed columns a condition is a sufficiently high maximum allowable temperature to enable elution of all PCTs.

Care should be taken to avoid co-elution with other contaminants. Figure 2 shows that peak-overlap is possible between higher chlorinated PCBs and lower chlorinated PCTs (< 5 chlorine atoms) [8]. Clearly longer columns in combination with smaller internal diameters will provide a better separation. It has been shown for the PCB analysis that these parameters are limited by practical difficulties. Longer columns require a higher pressure and result in very long analysis times, while the gain in resolution is only square root proportional to the additional length. A smaller internal diameter also requires a higher pressure and will finally result in leakages at the septum or at connections. In practice a length of 50–60 m combined with an internal diameter of 0.15 mm gives a maximum resolution without causing practical problems [27]. Multi-dimensional (MD) GC based on heart-cut techniques would offer a better resolution [28, 29]. It is still questionable if such a technique would be able to offer a separation of all relevant PCTs, but it would certainly be a step in that direction. However, until now MDGC has not been applied to PCT analysis. Recently a technique called comprehensive (C) MDGC was introduced [30–32]. This tech-

Table 3. GC columns and conditions for the determination of PCTs

Reference	Column	Maximum oven temp. (°C)	Carrier gas	Detection
[59]	SCOT SE-30 1.3 m × 5 mm	ca. 200	He	ECD
[9]	CP Sil 5 25 m × 0.25 mm	325	N_2	ECD
[35]	DB 5 50 m × 0.25 mm	310	Unknown	ELCD[a]
[43]	DB 17 30 m × 0.32 mm	280	He	EI-MS
[21]	DB 5 30 m × 0.32 mm	300	He	ELCD[a]
[17]	DB 5 30 m × 0.32 mm	310	He	ELCD[a] and EI-MS, NCI-MS
[8]	CP Sil 12 45 m × 0.25 mm	280	He	NCI-MS
[34]	DB 5 60 m × 0.25 mm	310	He	HR EI-MS
[16]	SPB 5 30 × 0.25 mm	300	He H_2	ECD NCI-MS

[a] ELCD = electrolytic conductivity detector at 950 °C with *n*-propanol as electrolyte.

nique offers a complete three-dimensional chromatogram of complex mixtures and again a much better resolution. The separation space obtained by CMDGC is the product of the separation spaces of the two GC columns used, whereas this is the sum of the separation spaces of the two columns used in heart-cut MDGC. CMDGC is based on a modulation technique, in which all peaks from the first column are trapped on a retention gap and very rapidly transferred to the second column [33]. This sequential process finally results in a true three-dimensional chromatogram. Because of the speed of the process, extremely narrow peaks are being formed which enhances the sensitivity to a considerable extent. However, a disadvantage is that some detectors have difficulties in responding fast enough. Flame ionization detectors can successfully be used in CMDGC [32], but combinations of CMDGC with MS and ECD detection result in more difficulties and are still under study. A smaller cell volume such as in the recently introduced HP micro-ECD may offer a solution.

4.3
Mass Spectrometry

The determination of total PCT by MS is a difficult procedure in which errors can easily be made. Chemical transformations in the environment, biological availability, uptake kinetics, metabolism, and elimination from organisms will

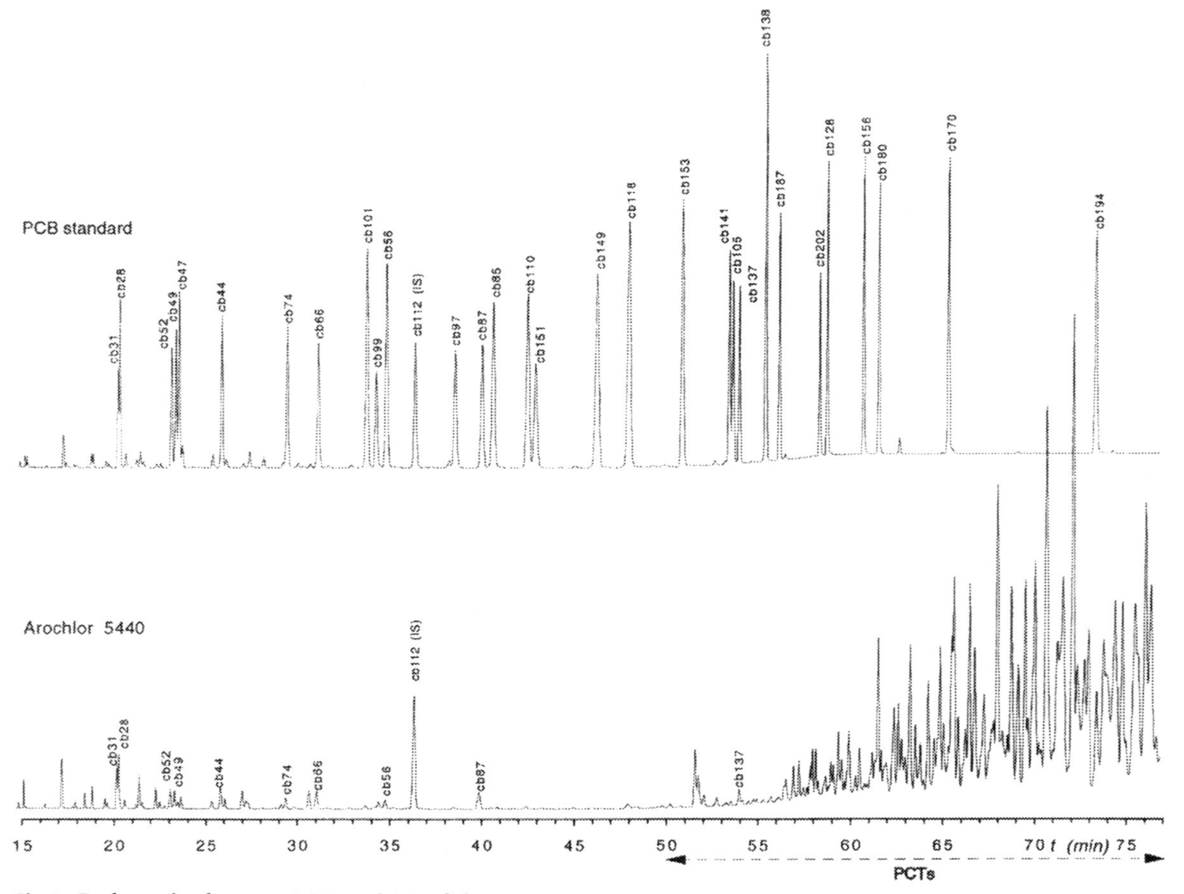

Fig. 2. Peak overlap between PCBs and PCTs [8]

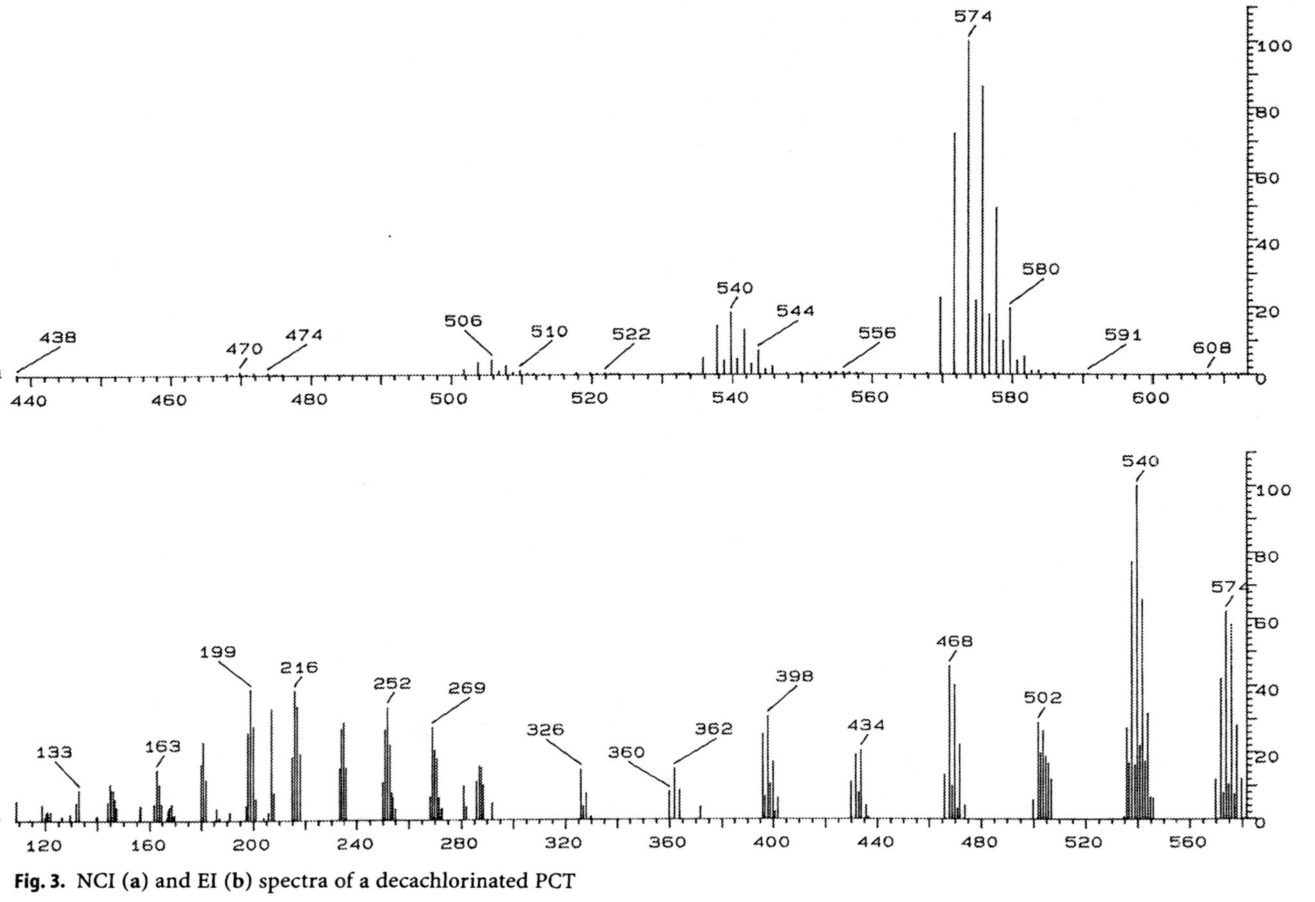

Fig. 3. NCI (a) and EI (b) spectra of a decachlorinated PCT

influence the PCT pattern in samples and make it rather different from that in the technical PCT mixtures. Combined with the lack of commercial availability of most individual PCT congeners this makes accurate quantification of PCTs virtually impossible. The analytical methods used at the moment should therefore be considered as semi-quantitative methods.

Both electron impact (EI) and negative chemical ionization (NCI) techniques can be used for MS detection of PCTs. The use of EI results in a more fragmented spectrum, which may provide more structural information [34]. NCI provides a higher sensitivity [8]. Only lower chlorinated congeners (<4 chlorine atoms) show a lower sensitivity with NCI. However, lower chlorinated PCTs are hardly present in Aroclor 5442, and even less so in Aroclor 5460. Hale et al. [17] reported the presence of lower chlorinated PCTs in sediments and shellfish from the Chesapeake Bay due to a specific discharge of Aroclor 5432. Monsanto has, however, produced substantially less Aroclor 5432 than Aroclor 5460 [8]. EI and NCI spectra of a deca PCT are shown in Fig. 3.

Several errors can be made in the determination of PCTs by GC/MS. Caixach et al. describe the need for high resolution (HR) EI-MS for a reliable determination of PCTs [34]. Due to the interference of $[M-2Cl]^+$ fragments from higher chlorinated PCTs with $[M]^+$ fragments of lower chlorinated PCTs substantial errors can be made in the quantification of homologues in technical PCT mixtures (Fig. 4). They recommended a resolution of 35,000 to avoid calculation errors due to interfering fragments [34]. Wester et al. [8] calculated total PCT concentrations after determination by GC/NCI-MS using different technical PCT standards and different selected ions (Table 4). They showed that

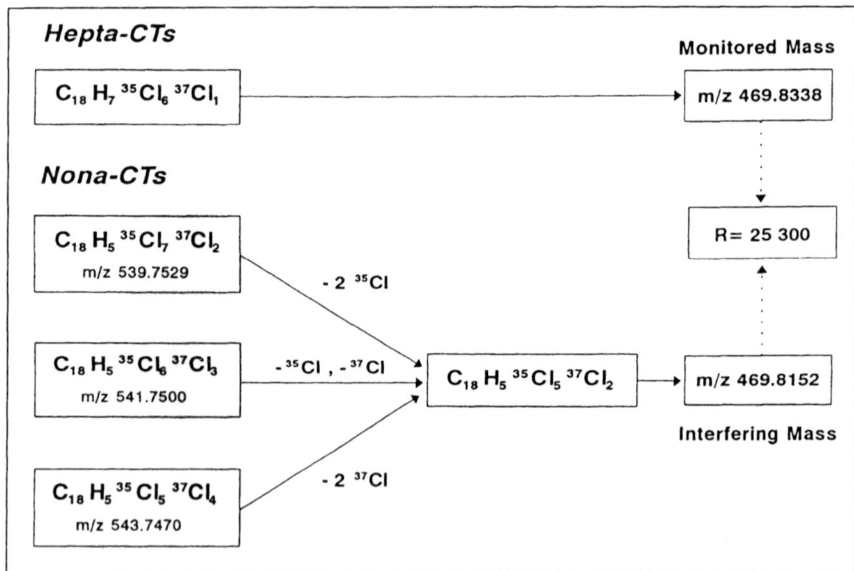

Fig. 4. Possible interferences of higher chlorinated PCT in the mass spectrometric determination of lower chlorinated PCTs [34]

Table 4. PCT concentrations in µg/kg wet weight corrected for recovery and calculated by different methods [8]

Sample	TIC		SIC m/z = 436	SIC m/z = 470		SIC m/z = 436	
	A5442	A5460	A5442	A5442	A5460	A5442	A5460
r, calibration curve	0.9998	0.9998	0.9998	0.9998	0.9999	0.9996	0.9998
Recovery (%)	94	91	93	93	89	89	88
Cod liver, North Sea	410	300	490	2200	440	1300	26
Tufted duck egg, NL[a]	35	25	32	140	26	110	2.0
Dolphin[b], North Sea	7400	5500	9600	42000	8600	19000	370
Eel fillet, Rhine, NL	200	150	250	1300	260	1100	23
Sediment, Waal, NL	100	17	46	710	74	1400	33

[a] NL = Netherlands.
[b] White-beaked dolphin.

the use of a technical PCT mixture which as much as possible resembles the peak pattern in the environmental sample is very important and errors between 30% for biota samples and 80% for sediment samples can be made. In addition they recommended that the use of single ion monitoring should be avoided due to the introduction of relatively large errors.

5
Environmental Distribution

The available dataset on PCT concentrations in the environment is small compared to that on PCBs. In addition, literature data on PCTs must be treated with care because of the large variation in analytical procedures and calculation methods used. Because of the lack of quality control data on PCTs and considering the well-known poor inter comparability of total-PCB determinations in the 1970s, literature data on PCTs probably show at best only the order of magnitude of the PCT contamination present. An overview of PCT concentrations in environmental matrices is given in Table 5.

Extremely high PCT concentrations have been reported by Hale et al. [17, 21] and Gallacher et al. [35] in sediments and biota from the James river and Chesapeake Bay (Tabbs Creek) in the USA (Table 5). The PCT patterns of those samples closely resembled those of A5432, with penta-CTs being the main constituents. Contamination was probably caused by recent, local discharges of PCTs, because PCTs were only found in the upper layer of the sediments. The other data presented in Table 5 are presumably all caused by more diffuse sources. The bioconcentration potential of PCTs is shown by the high PCT concentrations in white-tailed eagles from Sweden [23], eider duck livers from the Dutch Wadden Sea [36], and in white-beaked dolphins and harbor porpoises from the North Sea [8, 36]. The presence of PCTs in sperm whales shows that these compounds have reached deep ocean waters [36, 37]. Unfortunately there is insufficient information to establish temporal trends. PCT concentrations in

Table 5. Selected literature data on total-PCT concentrations in environmental samples

Sample and location	Concentration (μg/kg)	Basis of calculation[a]	Standard	Ref.
Oyster, Eastern Scheldt, '71, Netherlands	80–150	lw	Clophen Harz-W	[38]
Eel, IJssel Lake, '71, Netherlands	200–500	lw	Cloph. H. W	[38]
Herring eggs, Bay of Fundy, '72, Canada	100	ww	A5460	[25]
Herring fat, Bay of Fundy, '72, Canada	1400	lw	A5460	[25]
Cormorant eggs, B. of Fundy, '72, Canada	not detected	ww	A5460	[25]
Sparrowhawks, herons, '76?, England	50–1200	ww	A5460	[60]
Grey seal, '76, Sweden	500–1000	dw	perchlorination	[23]
White-tailed eagle, '76, Sweden	2800–17,200	lw	perchlorination	[23]
Trout Soca river, '77?, Yugoslavia	3–8	ww	A5460	[18]
Seagull fat, '77?, Livorno, Italy	610–10,510	ww		[61]
Black-h. gull liver, '75/76, Polish coast	1600	ww	A5460	[62]
Sediment, B. Biscay coast '86/87, Spain	0–0.4	ww	Leromoll 141	[16]
Oysters, Chesapeake Bay '88/'89, USA	<400–35,000	dw	A5432	[17]
Sediment, Chesapeake Bay '88/'89, USA	<5–250,000	dw	A5432	[17]
Sediment, James River '89, USA	26,000	dw	A5460	[21]
Sediment, Chesapeake Bay '89/90, USA	<170–43,300	dw	A5432	[35]
Oysters, Chesapeake Bay, '89/90, USA	1900–18,300	dw	A5432	[25]
Cordgrass, Chesapeake Bay, '89/90, USA	<100–5080	dw	A5432	[35]
Crab, Chesapeake Bay, '89/90, USA	<101–7030	dw	A5432	[35]
Cormorant liver, Ketelmeer, '90, Netherlands	90–120	ww	A5442	[8]
Harbor porpoise blubber, '90, North Sea	1900	ww	A5442	[8]
Whiteb. Dolphin blubber, '91, North Sea	7400	ww	A5442	[8]
Gentoo penguin liver, '91, Falklands	2	ww	A5442	[8]
Eel fillet, Rhine, '92, Netherlands	60–200	ww	A5442	[8]
Eel fillet, Meuse, '92, Netherlands	60–130	ww	A5442	[8]
Sediment, Rhine, '92, Netherlands	0.5–4	dw	A5442	[8]
Sediment, Meuse, '92, Netherlands	3	dw	A5442	[8]
Cod liver, '92, Dutch coast	410	ww	A5442	[8]
Cod liver, '92, northern North Sea	41	ww	A5442	[8]
Eider Duck liver, Wadden Sea, '94	180–860	ww	A5442	[36]
Minke whale blubber, '94, North Sea	270	ww	A5442	[36]
Sperm whale blubber, '95, North Sea	90–220	ww	A5442	[36]
Whiteb. Dolphin blubber, '95, North Sea	2800	ww	A5442	[36]
Harbor seal blubber, '96, North Sea	<130–2700	ww	A5442	[36]
Bonitos, salmon, '97?, Spain	21–260	ww	A5460	[41]
Mussels, clams, '97?, Spain	1.2–24	ww	A5460	[41]

[a] lw = lipid weight; dw = dry weight; ww = wet weight basis.

eel from the IJssel Lake in the Netherlands were about constant between 1971 and 1992 [8, 38], although the analytical methods were different (Table 5). The relatively high PCT level in the white-beaked dolphin from the North Sea, found twenty years after the majors bans on PCT production, is an indication of the extremely persistent character of the PCTs [8]. Detection of PCTs in the sub-Antarctic area (Falkland Islands) shows the global character of the PCT contamination [8]. Doguchi reports only low levels of PCTs in Japanese environmental samples along with many results below the detection limits [39]. However, PCT levels in human fat and blood samples were almost equivalent to PCB concentrations, while PCT production in Japan was, at 2700 tonnes, only 5% of PCB production.

The PCT level in cod liver from the Dutch coast (0.5 mg/kg wet weight) is substantially higher than the level of organochlorine pesticides as DDT, DDE, HCB, and dieldrin in similar samples (0.01–0.1 mg/kg wet weight) [39]. The same is true for the PCT level in eel from the river Rhine (0.2 mg/kg wet weight) compared to organochlorine pesticide levels (0.001–0.1 mg/kg wet weight) [7]. In spite of these ratios, organochlorine pesticides have been included in various monitoring programs for food control and environmental purposes [40], whereas PCTs have been given little attention.

PCTs were qualitatively determined in fish and shellfish for the river Ebro delta in Catelonia, Spain. PCT concentrations in market samples of marine origin from Spain [41] were comparable with those found in Dutch fish (Table 5). PCTs were also determined in paperboard and food packaging materials [26, 42]. Villeneuve et al. [42] found PCTs present in 5.5% of 73 samples of food packaging materials at concentrations of 0.01–0.05 mg/kg total weight. It may be assumed that, given the bans on PCT production, PCT concentrations in food packaging materials have gone down over the last decade.

Recently, PCTs were identified as indoor environmental contaminants in dust of a classroom of a junior high school [43]. The authors failed to identify the emission source of the PCTs though sealants and electrical appliances were scrutinized for PCT contamination. Further research on PCTs in the indoor environment was recommended.

Quantitative data on PCT concentrations in human milk, blood, and adipose tissue are only known for Japan and Netherlands [38, 39, 44]. All data are from the 1970s. Due to the analytical problems it is difficult to make any assessment. PCT concentrations in Dutch human milk varied between 0.5 and 0.8 mg/kg on a lipid weight basis. PCT concentrations in Japan varied between 0 mg/kg and 0.22 mg/kg on a lipid weight basis, while PCT concentrations in other tissues (blood, liver, adipose tissue, kidney, brain) varied between 0.04 mg/kg and 9.2 mg/kg on a lipid weight basis. PCTs were qualitatively determined in human adipose tissue from the USA [45].

5.1
Correlation of PCT and PCB Concentrations

Wester et al. calculated PCB/PCT ratios in sediment and biota samples [8]. A correlation between the total-PCT and the total-PCB concentrations was found for biota (r = 0.924, n = 25) and for sediment (r = 0.815, n = 5). Figure 5 shows the correlation between total-PCT and total-PCB concentrations in biota on a lipid weight basis. The use of pattern comparison may have caused a bias in the PCT data, that is the slope of the curve in Fig. 5 may be influenced by the standard used for the PCT analysis. The precision of the PCT analysis, with a standard deviation of 15%, was similar to that obtained in PCB analysis [46]. Total-PCT concentrations in biota expressed as A5442 are in general 1–10% of the total-PCB concentration, with a mean of 4% (n = 25). These percentages are higher than those reported by Renberg et al. [23], who found PCT levels of 0.3–1.6% of the PCB levels in white-tailed eagles and grey seals from Sweden. Actually, the ratio in biota roughly reflects the production volumes of PCBs and PCTs. Higher PCT/PCB ratios were found in the five sediment samples analyzed: 10–24% with a mean of 16% (n = 5). A possible explanation for the high PCT/PCB ratios is a better adsorption of PCTs than PCBs to sediments. From four of these five sediment sampling sites, eel fillets were also analyzed. The fish/sediment ratio for PCTs expressed as µg/kg lipid weight in fish relative to µg/kg total organic carbon in sediment ranges from 0.2 to 1, whereas for PCBs this ratio ranges from 2 to 4. This confirms the higher adsorption of PCTs to sediments.

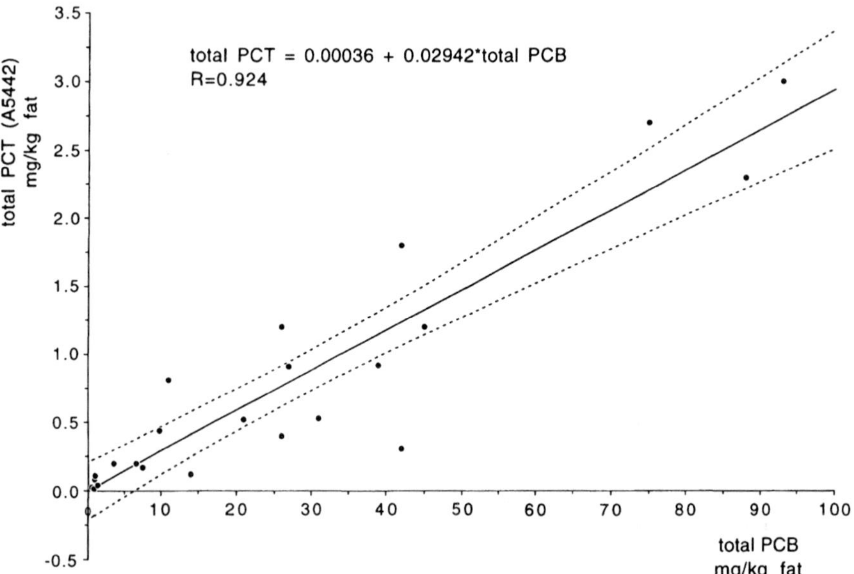

Fig. 5. Correlation between total PCT and total PCB concentrations on a lipid weight basis in biota; *dotted lines* show 95% confidence interval [8]

6
Toxicity

The toxicological properties of PCTs have not been extensively investigated. In general the toxicity of PCTs is considered to be equivalent to that of PCBs, the long-term toxicity being most important [12]. A general difficulty in toxicological studies of PCTs is the contamination of the PCT mixtures with PCBs. Possible effects found can be caused by the PCB contamination instead of by the PCTs themselves.

The acute toxicity of lower chlorinated PCTs is higher than that of higher chlorinated PCT mixtures [12]. The LD_{50} for rats is 4–11 g/kg, which is higher than for PCBs, presumably due to a lower solubility and adsorption in the body [12, 47]. Mice and rats gained less body weight when exposed to PCTs (550 mg/kg for 24 weeks and 10 g/kg for 3 weeks, respectively). Several macroscopic effects such as hair loss, swollen eyelids and lips, progressive generalized subcutaneous edema, and isolated acneform lesions were observed with monkeys after a 5 g/kg dose of Aroclor 5460. The effects were considered less damaging within 6 weeks than those observed for PCBs [48]. PCTs may cause a dose-dependent increase in relative liver weight and a proliferation of the endoplasmatic reticulum [48–50].

As regards efects on enzyme systems, PCTs have been reported to induce cytochrome P-450, aniline hydroxylase, aminoprine N-demethylase, esterase, and nitroreductase, while depressed levels of glycose-6-phosphatase and aromatic hydrolase have been reported [12, 49–53]. PCTs were found to be less effective in some cases than PCBs, which may possibly be related to the absence of a planar configuration such as in case of some PCBs. PCTs induced both phenobarbital and 3-methylcholanthrene inducible forms of cytochrome P-450 [53, 54]. Although 20% of the induction was due to PCB contamination in the PCT mixtures, up to 0.75%, and PCBs were generally a more potent inducer, there is sufficient evidence that PCTs showed P-450 induction themselves [54]. Both PCBs and PCTs caused a 30–40% inhibition of Na^+-K^+ ATPase in brain and kidney preparations of blue gills (*Lepomis machrochirus*) [55].

Aroclor 5442 was eight times more estrogenically active in rats than some low chlorinated PCBs, whereas highly chlorinated PCBs and PCTs were inactive [51]. The reproductive effect of PCTs has only been studied in the laboratory. A greater number of abnormal white leghorn chicken embryos was noted after a 20 mg/kg diet of Aroclor 5460 [56]. The hatchability was not effected, even after a 20 mg/kg diet of Aroclor 5442 [57].

The carcinogenicity of PCTs was demonstrated in a study with mice which were treated with a 250 mg/kg and 500 mg/kg PCT diet for 24 weeks [58]. A significantly higher incidence of nodular hyperplasia of the liver was found in the PCT exposed mice. Also the exposed group had a higher incidence of hepatocellular carcinomas. A combination of HCB and PCTs in the diet of the mice considerably enhanced the occurrence of nodules and carcinomas.

Boon et al. studied the in vitro biotransformation capacity of hepatic microsomes of a sperm whale (*Physeter macrocephalus*), a white-beaked dolphin (*Lagenorhynchus albirostris*), a harbor seal (*Phoca vitulina*), and an eider duck

(*Somateria mollissima*) for PCTs [36]. The biotransformation rates of PCTs were generally low. Eider duck microsomes metabolized more of the lower chlorinated PCTs, whereas the harbor seal microsomes metabolized a wider range of PCTs. The microsomes of the sperm whale and the white-beaked dolphin were almost incapable of metabolizing PCTs. PCTs were considered to be more bioaccumulative than the PCBs but the larger molecular size of PCTs would possibly inhibit bioaccumulation to some extent. In a mutatox assay Aroclor 5442 was not genotoxic [36].

References

1. Jamieson JWS (1977) Report EPS-3-EC-77–22. Environ Protec Service, Ottawa, Canada
2. Reutergårdh L (1988) PhD Thesis, University of Stockholm, Sweden
3. Huschenbeth E (1977) Arch Fish Wiss 28:173
4. Ballschmiter K, Zell M (1980) Intern J Environ Anal Chem 8:15
5. Tanabe S (1988) Environ Pollut 50:5
6. Duinker J, Hillebrand MTJ, Zeinstra T, Boon JP (1989) Aquatic Mammals 15:95
7. Boer J de, Hagel P (1994) Sci Total Environ 141:155
8. Wester PG, Boer J de, Brinkman UATH (1996) Environ Sci Technol 30:473
9. Kok A de, Geerdink B, Vries G de, Brinkman UATh (1982) Intern J Environ Anal Chem 12:99
10. Voogt P de, Brinkman UAT (1989) In: Kimbrough RD, Jensen AA (eds) Production, properties and usage of polychlorinated biphenyls. Halogenated biphenyls, terphenyls, naphthalenes, dibenzodioxins and related products. Elsevier, Amsterdam, p 3
11. Madge DS (1978) Gen Pharmacol 9:361
12. Jensen AA, Jørgensen KF (1983) Sci Total Environ 27:231
13. Suschitzky H (1974) Polychloroaromatic compounds. Plenum Press, New York
14. Hutzinger O, Safe S, Zitko V (1973) Intern J Environ Anal Chem 2:95
15. Chittim B, Safe S, Ruso LO, Hutzinger O, Zitko V (1977) J Agric Food Chem 25:323
16. Canton L, Grimalt JO (1991) Chemosphere 23:327
17. Hale RC, Graeves J, Gallagher K, Vadas GG (1990) Environ Sci Technol 24:1727
18. Jan J, Malnersic S (1978) Bull Environ Contam Toxicol 19:772
19. Valk F van der, Dao QT (1988) Chemosphere 17:1735
20. Addison RF, Fletcher GL, Ray S, Doane J (1972) Bull Environ Contam Toxicol 8:52
21. Hale RC, Gallagher K, Gundersun JL, Mothershead RF (1991) J Chromatogr 539:149
22. Onuska FI, Terry KA, Rokushika S, Hatano H (1990) J High Resolut Chromatogr 13:317
23. Renberg L, Sundström G, Reutergårdh L (1978) Chemosphere 6:477
24. Fries GF, Marrow GS (1973) J Assoc Off Anal Chem 56:1002
25. Zitko V, Hutzinger O, Jamieson WD, Choi PMK (1972) Bull Environ Contam Toxicol 7:200
26. Thomas GH, Reynolds LM (1973) Bull Environ Contam Toxicol 10:37
27. Boer J de, Dao QT (1989) J High Resolut Chromatogr 12:755
28. Boer J de, Geus H-J de, Brinkman UATh (1997) Environ Sci Technol 31:873
29. Boer J de, Dao QT, Wester PG, Bøwadt S, Brinkman UATh (1995) Anal Chim Acta 300:155
30. Geus H-J de, Boer J de, Brinkman UATh (1996) Trends Anal Chem 15:398
31. Bushley MM, Jorgenson JW (1990) Anal Chem 62:61
32. Phillips JB, Xu J (1995) J Chromatogr A703:327
33. Liu Z, Phillips JB (1991) J Chromatogr 29:227
34. Caixach J, Rivera J, Galceran MT, Santos FJ (1994) J Chromatogr A. 675:205
35. Gallacher K, Hale RC, Greaves J, Bush EO, Stilwell DA (1993) Ecotoxicol Environ Safety 26:302
36. Boon JP, Smith DEC, Lewis WE, Klamer HJC, Pastor D, Wester PG, Boer J de (1998) BEON Rep 98–1, Rijkswaterstaat, The Hague, Netherlands

37. Boer J de, Wester PG, Klamer HJC, Lewis WE, Boon JP (1998) Nature 394:28
38. Freudenthal J, Greve PA (1973) Bull Environ Contam Toxicol 10:108
39. Doguchi M (1977) Ecotoxicol Environ Safety 1:239
40. Boer J de (1988) Chemosphere 17:1811
41. Fernandez MA, Hernandez LM, Gonzalez MJ, Eljarrat E, Caixach J, Rivera J (1998) Chemosphere 36:2941
42. Villeneuve DC, Reynolds LM, Thomas GH, Phillips WEJ (1973) J Assoc Off Anal Chem 56:999
43. Seidel U, Schweizer E, Wodarz R, Rettenmaier AW (1996) Environ Health Perspec 104:1172
44. Fuhano S, Doguchi M (1977) Bull Environ Contam Toxicol 17:613
45. Wright LH, Lewis RG, Crist HL, Sovocool GW, Simpson JM (1978) J Anal Toxicol 2:76
46. Boer J de, van der Meer J, Brinkman UATh (1996) J Assoc Off Anal Chem 79:83
47. Fishbein L (1974) Ann Rev Pharmacol 14:139
48. Allen SR, Norback PH (1973) Science 179:498
49. Norback DH, Allen JR (1972) Environ Health Perspect 1:137
50. Sosa-Lucero JC, De la Iglesia FA, Thomas GH (1973) Bull Environ Contam Toxicol 10:248
51. Bitman J, Cecil HC, Harris SJ (1972) Environ Health Perspect 1:145
52. Cecil HC, Harris SJ, Bitman J (1975) Arch Environ Contam Toxicol 3:183
53. Toftgårdh R, Nilsen OG, Glanmann H (1980) Polychlorinated terphenyls and mixed types of inducers of rat liver microsomal cytochrome P-450. In: Gustafsson JA (ed) Biochemistry, biophysics and regulation of cytochrome P-450. Elsevier/North Holland Biomedical Press, Amsterdam, Netherlands
54. Nilsen OG, Toftgårdh R (1981) Arch Toxicol 47:1
55. Yap HH, Desaiah D, Cutkomp LK, Koch RB (1971) Nature 233:61
56. Cecil HC, Bitman J, Lillie RJ, Verret J (1974) Bull Environ Contam Toxicol 11:489
57. Lillie RJ, Cecil HC, Bitman J, Fries GF (1974) Poultry Sci 53:726
58. Shirai T, Miyata Y, Nakanishi K, Murasaki G, Ito N (1978) Cancer Lett 4:271
59. Stalling DL, Huckins JN (1971) J Assoc Off Anal Chem 54:801
60. Hassell KD, Holmes DC (1977) Bull Environ Contam Toxicol 17:618
61. Vanucchi C, Reynolds LM, Phillips WEJ (1978) Chemosphere 6:483
62. Falandysz J (1980) Mar Pollut Bull 11:75

CHAPTER 4

Polybrominated Biphenyls and Diphenylethers

Jacob de Boer · Karin de Boer · Jan P. Boon

J. de Boer (e-mail: j.deboer@rivo.dlo.nl)
DLO-Netherlands Institute for Fisheries Research, P. O. Box 68, 1970 AB IJmuiden,
The Netherlands
K. de Boer, J.P. Boon (e-mail: boon@nioz.nl)
Netherlands Institute for Sea Research, P. O. Box 59, 1790 AB Den Burg, Texel, The Netherlands

Polybrominated biphenyls (PBBs) and polybrominated diphenylethers (PBDEs), also called brominated diphenyl oxides, are produced and used as flame retardants. There are theoretically 209 different congeners of both PBBs and PBDEs. These congeners have specific chemical and physical properties, which lead to different biological and toxicological effects. Most studies have been based on commercial mixtures of brominated flame retardants, which complicates the pursuit of unambiguous data and insights. Only adequate quantification of individual congeners will allow comparative environmental and toxicological studies.

Most of the PBB and PBDE congeners found in commercial flame retardants are persistent, lipophilic and bioaccumulating, which represents a potential threat to both human and environmental life. The PBDE concentrations in environmental samples are generally higher than those of the PBBs, which seems to be related to the production figures of these compounds. The PBB production was strongly reduced after an accident in which PBBs were mixed with cattle feed. In contrast with that the PBDE production is increasing. Nowadays, most of the PBDE production is based on decaBDE. However, environmental patterns mainly consist of tetra and penta BDEs. Possibly, lower brominated mixtures have been used in higher quantities than is known or degradation of the relatively instable deca BDE may take place, but other explanations cannot be excluded. The acute toxicity of PBBs and PBDEs is relatively low, but long term effects on the balance of endocrine systems of animals and humans seem to pose the most serious risk. Because of the potential toxic effects found, a substitution of these flame retardants by environmentally friendly alternatives may be considered.

Keywords. Polybrominated biphenyl, Polybrominated diphenylether, Production, Analytical methods, Environmental distribution, Toxicity

The Handbook of Environmental Chemistry Vol. 3 Part K
New Types of Persistent Halogenated Compounds
(ed. by J. Paasivirta)
© Springer-Verlag Berlin Heidelberg 2000

1
Introduction

Polybrominated biphenyls (PBBs) and polybrominated diphenylethers (PBDEs)
are aromatic hydrocarbons, which are used as flame retardants. Flame retardants are chemicals that are added to polymers which are used in different

materials such as electrical and electronical equipments, paint, textiles (particularly in office buildings) and in cars and aircrafts to prevent them from catching fire [1]. The use of flame retardants has increased due to stricter fire regulations in many countries and an increased use of plastic materials and synthetic fibres [1]. PBBs are formed by substituting hydrogen by bromine [2]. Instead of biphenyl, diphenylether is used in the bromination to PBDEs [3].

The general chemical formulas of PBBs and PBDEs (Fig. 1) show that PBBs and PBDEs have a large number of possible congeners, depending on the number and position of the bromine atoms on the two phenyl rings. Theoretically, 209 congeners of each chemical are possible. A systematic numbering system is developed by Ballschmiter et al. [4] for polychlorinated biphenyl (PCB) congeners which has been adopted for the corresponding PBB and PBDE congeners [5].

PBBs manufactured in the early 1970s for commercial use, consist mainly of hexa-, octa-, nona-, and decabromobiphenyl. They were developed as flame retardants due to their ability to meet flame resistance performance requirements, economical feasibility, and they have little effect on the flexibility of the base compounds. PBBs came to the attention of the public in 1974, when it was discovered that about 1000 pounds had been accidentally substituted for magnesium oxide as an additive in cattle feed in Michigan in 1973. After this, the production of PBBs has decreased [2]. However, decabromobiphenyl (DeBB) and possibly some other PBB mixtures are still produced commercially but alternative chemicals have been introduced to replace them as flame retardants, in particular PBDEs. Products based on penta-, octa-, and decabromodiphenylether are the only commercially interesting PBDEs [3]. The production of PBDEs increased since the end of the 1970s [3]. They are currently being used in the housing and back covers of colour televisions and personal computers, in electronic parts of the same instruments, in seats of cars and busses and in textiles [6]. PentaBDEs are mainly being used in textiles and polyurethane foams, whereas decaBDE is used both in textiles and in many other kinds of synthetic plastics such as polyester used for elctronic circuit boards [1].

Like other organohalogen compounds such as PCBs and DDT, PBBs and PBDEs are lipophilic, and persistent [2, 3]. The high resistance towards acids, bases, heat, light, reduction and oxidation is disadvantageous when these compounds are discharged into the environment, where they persist for a long time. In addition, toxic compounds, polybrominated dibenzofurans (PBDFs) and dibenzodioxins (PBDDs), may be formed when these flame retardants are heated [5].

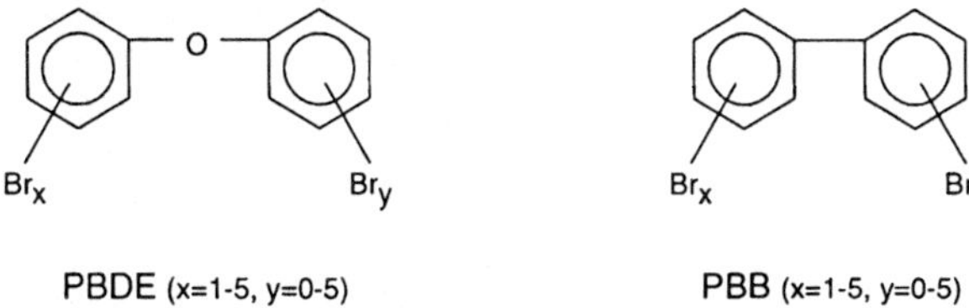

Fig. 1. Basic formulas of PBDEs and PBBs

Chlorinated compounds such as PCBs (in dielectric fluids) and DDT (used as a pesticide) were found in high concentrations in living organisms in the late 1960s. These chemicals were shown to be a potential risk for different organisms. Since then many countries have banned or restricted the use of these chemicals, and the environmental levels have decreased [11]. Whereas these organochlorine compounds were banned, PBBs and PBDEs were ignored. Apart from an European ban on hexaBB production and a voluntary reduction of the PBB production in the USA, no ban has been enacted, while the production and use of brominated flame retardants increased [7]. Taking into account the large worldwide production and application of PBBs and PBDEs and their persistence, it is envisaged that a large part of the total production will eventually reach the environment, including the marine environment [8]. There, the PBBs and PBDEs will bioaccumulate because of their lipophilicity and their resistance to degradative processes [5]. PBDEs and PBBs are considered to be a potential threat for human health, particularly through fish consumption [9].

In this chapter the chemical and physical properties, production and use, analytical methods, environmental fate and occurrence and the toxicology of PBBs and PBDEs, are discussed.

2
Chemical and Physical Properties

2.1
PBBs

PBBs are manufactured using a Friedel-Crafts type reaction in which biphenyl reacts with bromine in (or without) an organic solvent, using aluminium chloride, aluminium bromide, or iron as a catalyst [10]. During the production of technical-grade DeBB (Adine 0102), biphenyl is directly brominated in a large excess of bromine, used as reactant and solvent in the presence of a Lewis acid catalyst (aluminium type). DeBB is further purified by destillating the excess bromine in the presence of a brominated solvent [2]. The composition of the manufactured PBBs is given in Table 1. PBBs are not known as natural products. Of the 209 possible congeners, 101 individual PBB congeners are listed in the Chemical Abstracts Service (CAS) register at this moment [2].

In general, PBBs show an unusual chemical stability and resistance to acids, bases, heat, reduction and oxidation. PBBs are chemically comparable to the PCBs. However, chlorine atoms have a stronger association to polybiphenyl than bromine atoms [2]. Unlike PCBs, the reactivity of PBBs has not been well studied and documented in the literature [11]. Like PCBs, their chemical stability is dependent, in part, on the degree of bromination and the specific substitution patterns [12].

Some chemical and physical data of commercial PBB mixtures are given in Table 2A. The chemical and physical properties depend on the PBB compound, and differ between each congener [2].

Table 1. Composition of commercial PBB mixtures [2]

PBB mixture (manufacturer)	Weight of bromine (%)	Weight of different homologus groups						
		Br_{10}	Br_9	Br_8	Br_7	Br_6	Br_5	Br_4
Hexabromobiphenyl								
FM BP-6 (Michigan Chemical)	75				13.8	62.8	10.6	2
FM [lot RP-158 (1971)]					12.5	72.5	9	4
FM [Lot 6244 A (1974)]					13	77.5	5	4.5
FM BP-6						90	10	
FM BP-6				1	18	73	8	
FM BP-6					33	63	4	
FM BP-6					7.7	74.5	5.6	
FM BP-6					24.5	79	6	
2,2',4,4',6,6' (RFR)					12	84	1	
2,2',4,4',6,6' (Aldrich)				2	24	70	4	
"Hexabromobiphenyl" (RFR)					25	67	4	
					(12–25)	(60–80)	(1–11)	(2–5)
Octanonabromobiphenyl								
Bromkal 80–9D (Kalk)	81–82.5	9	65	25	1			
Bromkal 80				72	27	1		
XN-1902 (Dow Chemical)	82	6	47	45	2			
XN-1902 (Dow Chemical)		2	34	57	7			
Lot 102–7-72 (Dow Chemical)		6	60	33	1			
"Octabromobiphenyl" (RFR)		4	54	38	2			
2,2',3,3',5,5',6,6'(RFR)		1	28	46	23	2		
FR 250 13 A (Dow Chemical)		8	49	31	1			
Decabromobiphenyl								
HFO 101 (Hexcel)	84	96	2					
Adine 0102 (Ugine Kuhlmann)	83–85	96	4					
Adine 0102 (Ugine Kuhlmann)		96.8	2.9	0.3				
"Decabromobiphenyl" (RFR)		71	11	7	4	4		
"DBB": Flammex B 10 (Berk)		96.8	2.9	0.3				

PBBs are solids with a low vapour pressure. The volatility of PBBs has a wide range and is lower than the volatility of the corresponding PCBs [5]. PBBs are almost insoluble in water, and the solubility decreases with increasing bromination [2]. Brominated compounds have a lower solubility in water than the corresponding chlorinated compounds [5]. Table 2 shows the variance in the solubility of commercial PBBs in water from different sources and qualities. Determinations of water solubility of these very hydrophobic compounds are difficult to perform. Adsorption effects on particles may influence the results. PBBs were found to be 200 times more soluble in landfill leachate than in distilled water. PBBs are soluble in various organic solvents, their solubility decreases steeply with increasing bromine content [2]. Most PBBs have a $\log K_{ow} > 7$, and are therefore regarded as superlipophilic compounds [13].

Table 2A. Chemical and physical data of commercial PBB and PBDE mixtures [2, 3, 5]

	HxBB $(C_{12}H_4Br_6)$	OcBB $(C_{12}H_2Br_8)$	NoBB $(C_{12}H_1Br_9)$	DeBB $(C_{12}Br_{10})$
Relative molecular mass[a]	627.4	785.2	864.1	943.0
Melting point (°C)	124–248	200–250	220–290 360–380 385	380–386
Decomposition Point (°C)	300–400	435	435	395 > 400
Volatility (% weigt loss)		<1% at 250°C	1–2% at 300°C	<5% at 341°C
		<10% at 330°C		<10% at 363°C
		<50% at 350°C		<25% at 388°C
Vapour Pressure (Pa)	25°C; 0.000007			<0.0000006 (temperature not given)
	90°C; 0.01			
	140°C; 1			
	222°C; 100			
Solubility H_2O	11	30–40	insoluble	<30
(µg/l; 25°C) destilled deionized pure BB 153	610 0.32 0.06 30			
Well soluble in: (g/kg solvent; at 28°C)	carbontetra- chloride; 300 chloroform; 400 benzene; 750 toluene; 970 dioxane; 1150	petroleum- ether; 18 benzene; 81	insoluble	carbontetra- chloride; 10
Log K_{ow}^a	7.20			8.58

[a] Data of BDE congeners.

2.2
PBDEs

Most preparation methods of PBDEs reported are patents describing the bromination of diphenylether in the presence of a catalyst [1]. This results in products containing mixtures of brominated diphenylethers (Table 3). PBDEs have not been reported to occur naturally in the environment, but the related polybrominated phenoxy phenols have been found in several marine organisms, e.g. in *Dysidea herbacea, Dysidea chlorea, and Phyllospongio foliascents* [3]. Vionov et al. [14] showed that the bacteria *Vibrio* sp. associated with the sponge *Dysidea* sp. is capable of producing brominated diphenylethers.

Table 2B. Chemicaland physical data of commercial PBDE mixtures [2, 3, 5]

	TeBDE ($C_{12}H_4Br_4O$)	PeBDE ($Cl_2H_2Br_6O$)	OcBDE ($C_{12}H_2Br_8O$)	DeBDE ($C_{12}Br_{10}O$)
Relative molecular mass[a]	485.82	564.75	801.47	959.22
Melting point (°C)		$-7 - -3$ (boiling 300)	200 79–89 75–125 170–220	290–306
Decomposition Point (°C)		>200		>320 >400 >425
Volatility (% weigt loss)				1% at 319 °C 5% at 353 °C 10% at 370 °C 50% at 414 °C 90% at 436 °C
Vapour Pressure (mm Hg)		22 °C; 9.3	25 °C $< 10^{-7}$	20 °C $< 10^{-7}$ 250 °C <1 278 °C; 2.03 306 °C; 5.03
Solubility H_2O (at 25 °C)		9×10^{-7} mg/l at 20 °C	<1 g/l	20–30 µg/l
Well soluble in: (g/kg solvent; at 25 °C)		methanol; 10 (and other organic solvents)	toluene, 190 (353) benzene; 200 styrene; 250	o-xylene 8.7
Log K_{ow}^a	5.87–6.16	6.64–6.97	8.35–8.90	9.97

[a] Data of BDE congeners.

Table 3. Composition of commercial PBDEs [3]

Product	Composition								
	PBDE[a]	TrBDE	TeBDE	PeBDE	HxBDE	HpBDE	OBDE	NBDE	DeBDE
DeBDE								0.3–3%	97–98%
OBDE					10–12%	43–44%	31–35%	9–11%	0–1%
PeBDE		0–1%	24–38%	50–62%	4–8%				
PBDE[b]	7.6%	–	41–41%	44.4–45%	6–7%				

[a] Unknown structure.
[b] No longer commercially produced; analysis of one single congener.

Because of the presence of an oxygen atom, there is less similarity in the molecular structure between PBDEs and PCBs than between PBBs and PCBs [3]. The commercial PBDEs are rather stable compounds with boiling points ranging between 310 and 425 °C [3], and with low vapour pressures (Table 2B) [1, 3]. The volatility of PBDEs is low and their solubility in water is very low, especially that of higher brominated diphenylethers. It was concluded that higher brominated compounds are more persistent than lower brominated compounds. PBDEs are soluble in toluene. The commercial PBDEs are lipophilic substances of which the log K_{ow} increases with increasing bromine content [3].

3
Production and Use

PBBs and PBDEs belong to a group of brominated organic compounds which are used as flame retardants. Flame retardants are valued for their ability to inhibit combustion in plastics, textiles, electric, and other materials. There are different groups of flame retardants: inorganic and organic chemicals. Usually they are divided into reactive and additive flame retardants.

Reactive flame retardants have the same functional groups as the monomer with which they react. They are covalently bound to the polymer and are therefore less likely to leach to the environment. Reactive type flame retardants offer advantages such as polymer strength permanency and solvent resistance. A disadvantage is that they are polymer specific [15].

Additive flame retardants are not chemically incorporated into the polymer molecule. The additives are only mixed with or dissolved in the material and can therefore migrate out of the product during its entire lifetime [1].

What all flame retardants have in common is that they start to decompose when heated. A critical factor in the selection of a flame retardant is therefore its thermal stability with respect to that of the polymer. The ideal situation is when the flame retardant decomposes at about 50 % below the combustion temperature of the polymer. This is the case with most organic bromine compounds and most synthetic polymers [1]. In addition, brominated flame retardants are economically feasible, and they have little effect on the flexibility of the base compounds [16]. Because of the advantages mentioned above, the commercial use of brominated compounds is attractive.

3.1
PBBs

The commercially used PBB mixtures consist mainly of HexaBB, OctaBB, NonaBB, and DecaBB (Table 1). Commercially manufactured PBBs are primarily processed as flame retardants. Further potential uses of PBBs are: in the synthesis of biphenylesters or in a modified Wurtz-Fittig synthesis; in light sensitive compositions to act as colour activators; as relative molecular mass control agents for polybutadiene; as wood preservatives; as voltage stabilizing agents in electrical insulation; as functional fluids, such as diëlectric media [17].

In the early 1970s PBBs were introduced as flame retardants. In the USA the production of HexaBB ceased as a result of the Michigan disaster (Table 4).

Table 4. Commercial production of PBBs in the USA, 1970–1976 [97]

Product	Estimated production in thousand kg							
	1970	1971	1972	1973	1974	1975	1976	1970–76
Hexabromobiphenyl	9.5	84.2	1011	1770	2221	0	0	5369
Octabromobiphenyl and decabromobiphenyl	14.1	14.1	14.6	163	48	77.3	366	702
Total PBBs	23.6	98.3	1025	1922	2269	77.3	366	6071

OctaBB and DecaBB were produced until 1979 [2]. There was no import of any PBB mixtures. Since 1975–1976 all PBBs manufactured in the USA have been exported, mainly to Europe [10]. In Japan some PBBs were imported upto 1978, but there was no production. A mixture of highly brominated PBBs (Bromkal 80–9D) was produced in Germany until mid 1985 [2]. The production of DeBB in Great Britain was discontinued in 1977 [17]. The French company Atochem is currently producing technically grade DeBB (Adine 0102) in quantities of a few hundred thousand kg/year. It is marketed in France, Great Britain, Spain and the Netherlands [18]. In the Netherlands more than 200 tonnes DeBB/year (1989) were used [19]. The use of PBB mixtures in European textile is not allowed according to EU directive 83/264 [20].

Most research of PBBs has been carried out on Fire Master BP-6 and FF-1, which were involved in the Michigan disaster. The PBB composition of Fire Master changes from batch to batch (Table 1), but also mixed bromochlorobiphenyls and polybrominated naphthalenes have been observed as minor components [2]. Approximately 20 compounds other than PBBs were tentatively identified in Fire Master [21]. An extensive study was performed on a large number of batches of Fire Master, analysed for the toxic compounds PBDDs and PBDFs. These compounds were found in only one sample of Adine 0102 [2].

3.2
PBDEs

Products based on penta-, octa-, and decaBDE are of commercial interest (Table 3). PBDEs are mainly used as a flame retardant. There are nine manufacturers who currently produce PBDEs. These are:

- Dead Sea Bromines and Eurobrome, The Netherlands
- Atochem, France
- Great Lakes Chemical Ltd, Great Britain
- Great Lakes Chemical Corporation, USA
- Albemarle, USA
- Tosoh, Japan
- Matsunaga, Japan
- Nippo, Japan

The global production of DeBDE is approximately 40,000 tons/year [1, 3, 22]. Penta, octa- en decaBDE are currently being evaluated by the European Commssion according to EU directive 793/93 [20].

4
Consumption

Due to more stringent fire regulations in many countries and the increased use of plastic materials and synthetic fibres, the use of flame retardants has increased. In 1992, 600,000 tons of flame retardants were used worldwide, 150,000 tons were brominated compounds. 50,000 Tons of these were the reactive flame retardant with TBBPA and its derivatives and 40,000 tons were PBDEs [1].

The annual global consumption of PBDEs is 40,000 tons (30,000 tons DeBDE, 6,000 tons OcBDE and 4,000 tons PeBDE) [1, 3]. Data on the usage of PBDE are [3]:

- Germany 3,000–5,000 tons/year (1991)
- Sweden 1,400–2,000 tons/year (1991) –400 tons/year (1991) [1]
- The Netherlands 3,300–3,700 tons/year (1992)
- Great Britain 2,000 tons/year (1993)

TeBDEs were not used in Japan after 1990. Inorganic flame retardants (a.o. aluminium trihydrate (ATH)) were mainly used, but there is an increase in the use of brominated organic flame retardants [1].

In the USA brominated flame retardants belong to the most widely used additive flame retardants (Table 5). It is expected that in the USA the demand for

Table 5. The flame retardant demand in 1993 and the expected flame retardant demand of 1998 in the USA, in tonnes [15]

Product	1993	1998
Additive flame retardant	367,740	476,700
Aluminatrihydrate	196,128	256,510
Phosphorus compounds	44,038	57,658
Bromine compounds	39,952	50,848
Antimony oxide	28,148	35,412
Chlorinated compounds	24,970	29,964
Boron compounds	7,264	9,080
Other additives	27,240	37,228
Reactive flame retardant	54,480	68,100
Epoxy intermediates	16,344	19,976
Polyester intermediates	12,258	14,528
Urethane intermadiates	9,080	10,896
Polycarbonate intermediates	7,264	9,988
Other intermediates	9,534	12,712
Percent Additive	87,1	87,5
Percent Reactive	12,9	12,.5
Total Demand	422,220	544,800

additive products will increase with about 5,3%/year, to 476,700 tonnes in 1998. Despite their specificity, the demand for reactive products is expected to increase with 4,6%/year to 68,100 tonnes in 1998, as new products are introduced into the market. It is expected that the brominated flame retardant consumption will increase up to over 50,000 tons in the USA [15].

Compared with data of the worldwide use in 1992, the demand for flame retardants in the USA in 1993 was extremely high. However, only a small fraction of this consisted of brominated compounds. The use of brominated compounds as flame retardants in Japan in 1993 was higher, both relatively and in absolute amounts. The 1998 USA consumption of brominated compounds (in absolute amounts) was expected to reach the same level as Japan in 1994. Unfortunately more accurate data on flame retardants demands in Europe are not available. The European consumption of brominated compounds is estimated to be at a similar level as in Japan and the USA.

5
Combustion and recycling of PBBs and PBDEs

A disadvantage of brominated flame retardants is that PBDDs and PBDFs may be formed during combustion. Little is known about the toxicity of PBDDs and PBDFs. They are estimated to be in the same order as those of PCDD and PCDF [2, 3].

Debromination reactions of higher brominated flame retardants lead to lower brominated PBDF and PBDD congeners [3]. On pyrolysis PBDEs produce larger amounts of dioxins and furans than PBBs, and in this respect PBDEs are more toxic than PBBs. Most likely is that with the oxidation of PBDE, PBDFs and PBDDs are formed in intramolecular cyclization reactions involving the attack by oxygen on the diphenylether system (Fig. 2) [23]. Most of the reports have indicated that maximum production of PBDFs and/or PBDDs were observed at temperatures of 400–800 °C, depending on the type of brominated flame retardant, and that the 2,3,7,8-substituted compounds were seen only in very low concentrations [3]. At 600 °C 2,3,7,8-TeBDD and TeBDF in concentrations of 0.01–7 ppm and 0.01–6 mg/kg, respectively are formed from plastics containing DeBDE or PBDE as flame retardant. With increasing temperature the concentration of these isomers decreases until they are no longer detectable above 800 °C (detection limit 0.01 mg/kg) [24].

In a study of 78 TV sets and 34 personal computers, 78% of the modified polystyrene housings contained PBDEs, 16% PBBs, and 3% 1,2-bis-(tribromophenoxy)ethane. Composed samples of this material with PBDEs and PBBs as flame retardants already contained traces of PBDFs and PBDDs. When PBBs were used as flame retardant, the levels of these impurities increased during the recycling process. The formation of these PBDFs and PBDDs from the primary flame retardants, should also be considered when assessing the toxic properties of PBBs and PBDEs [25].

The results of laboratory pyrolysis experiments with PBDEs and PBBs, showed that PBDFs and/or PBDDs were formed in various concentrations, depending on the type of PBDEs and PBBs, the polymer matrix, the specific

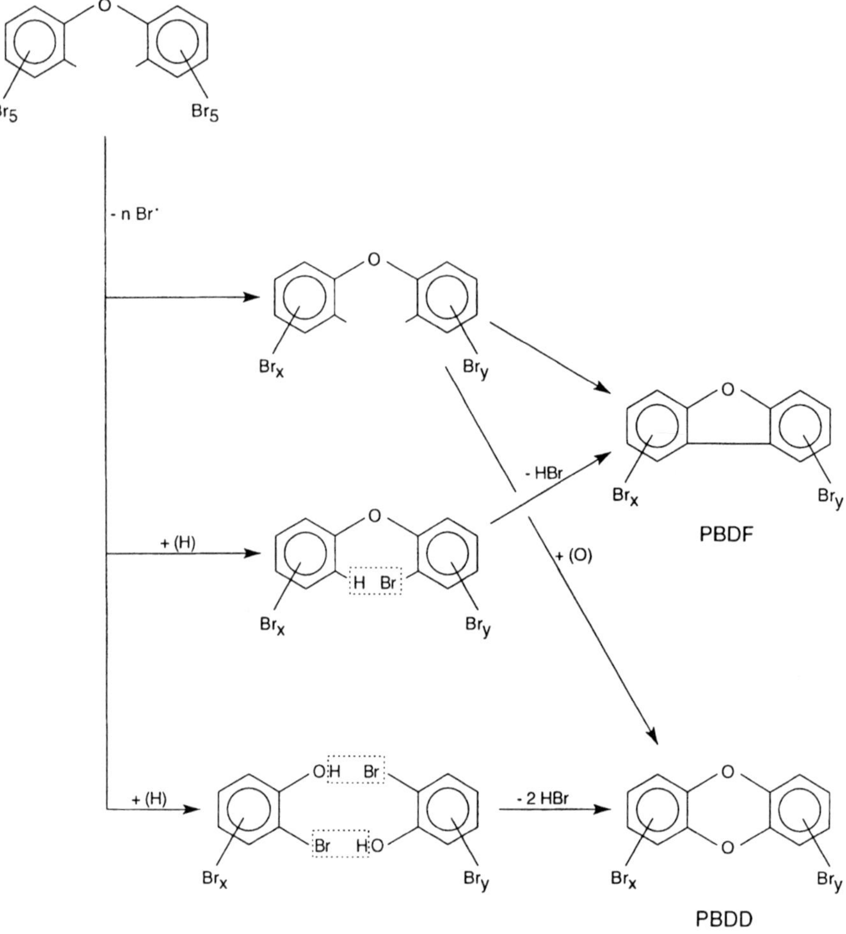

Fig. 2. Possible mechanisms for the formation of PBDFs and PBDDs from DeBDE [23]

processing conditions (temperature, presence of oxygen, etc.), and equipment used, and the presence of antimony oxide. This antimony oxide, used as a synergist in flame retardants, plays a catalytic role in the formation of PBDFs and PBDDs. Because the behaviour of PBDEs is strongly dependent upon the polymer matrix and upon the specific processing conditions, laboratory pyrolysis experiments can hardly be used as reliable models to predict behaviour in commercial moulding operations [3].

As with PCB disposal, the destruction of PBB- and PBDE-contaminated waste should be carefully controlled. For PCBs, a burning temperature above 1000 °C for 2 s is recommended [26]. Since PBDEs and PBBs are undetectable above 800 °C, the same approach might be effective for PBDEs and PBBs [2].

Recycling of plastics is difficult, because of the content of the additives PBBs and PBDEs [27]. Pyrolysis of flame retardant material of printed circuit board and electronics components (laboratory scale) produces high amounts of brominated dioxins and furans (2,3,7,8-TeBDF, 29 µg/kg: residue after quarts flask pyrolysis in N_2/H_2 atomosphere at 1100 °C) located in the condensed material. It was known that these plastics contain flame retardants to a maximum of 20 wt%. PBDEs can be extracted from plastics based on propyl-carbonate. The origin of brominated dioxins and furans detectable in propyl-carbonate extract is still to be investigated. Further recycling activities which process flame retarded plastics might produce hazardous products, an aspect that has to be investigated more closely [27].

6
Alternatives

There is a need for good alternatives for brominated flame retardants. Apart from other disadvantages, the heavy metal antimony is necessary for the production of brominated (or chlorinated) flame retardants. Antimony oxides, Sb_2O_5 (used in reactive flame retardants) and Sb_2O_3, are used as synergists to increase the efficiency of bromine and chlorine. Bromine and chlorine (together), and antimony oxides are used at a ratio of 1:3 [7].

Non-halogenated products, based mainly on phosphorus, aluminium tri-hydrate (ATH) and magnesiumhydroxide, halt flame spread without the formation of halogenated byproducts. However, at relatively high loading they can compromise the polymer's mechanical properties [7]. The additive flame retardant ATH is multifunctional, cheap, and widely used as a filler and plasticizer. It is particularly used in large quantities in carpet underlay [15]. Despite high loading requirements (20–70 wt.%) these inorganics are widely used because of their ability to suppress smoke generation and avoid the production of toxic offgasses [7].

Consumption of halogenated additives is expected to decline as the market moves towards synergistic systems that mix flame retardants such as antimony oxide, phosphorus, and zinc borate with a halogen [15].

To combine fire resistance with low smoke and gas formation, a low halogen flame retardant is produced, which contains just 25–26% bromine, which is used at loadings of 4–6%. Most of the widely used halogenated compounds contain up to 80% halogen [7].

The so-called intumescents form another class of flame retardants. These are low smoke release flame retardants which combine nitrogen and phosphorus. They are more costly than many halogenated compounds but are used in a.o. wire-and-cable and electronic housing uses where toxic smoke poses an immediate threat [7].

Because reactive types of flame retardants are polymer-specific, their application is limited. There are several reactive flame retardants, specifically produced and all different in composition. For example, there is a 25% pelletized concentrate of antimony pentoxide, bromine and polypropylene resin of various melt flow indices, which is geared to PP fibers for textiles and carpets, and also to PP thin-film applications where colour and clarity are desired. Furthermore dibromostyrene (DBS), a reactive styrene monomer, is used to make copolymers for adhesives and coatings. DBS can also be used to impart flame retardancy in polyolefins [15].

For structural applications a line of glass yarn fabrics is engineered. Woven with Advantex glass yarn it can withstand high temperatures. The silica based fabrics are inert and there is nothing to be offgassed. A new product is composed of two different forms of glass fiber, the material is fire, smoke and flame resistant, without flame retardants [15].

7
Analytical Methods

7.1
PBBs

Analytical methods for the determination of PBBs were adapted from established methods for chlorinated hydrocarbon insecticides (like DDT) and PCBs [2]. Usually hexane or iso-octane are used as solvents in the analysis of PBB mixtures and individual congeners. Gas chromatography (GC) is used for the final analysis.

Griffin and Chou [28] found that for the extraction of PBBs from soils and sediments the use of a polar organic solvent was important. The best results were obtained with hexane/acetone (9:1). Extraction was followed by further sample-clean up with Florisil.

Fehringer [29] describes the use of dichloromethane for extraction of PBBs from dry animal food. Sample-clean up is performed with Florisil columns.

Extraction of blood and serum follows pretreatment of the serum with methanol, and extraction is performed with a hexane/ether mixture. This method described by Burse et al. [30] is a standard extraction method for blood and serum, and has been used in most studies. Florisil columns are used in the sample-clean up. An analytical method was developed to quantitate PCBs and PBBs in human serum. This method includes a hexane-ethyl ether extraction of methanol-denatured serum, and an adsorption chromatography with deactivated silica gel [31].

PBBs can be co-extracted together with the fat from biological tissue samples but afterwards an intensive clean up procedure for PBBs is necessary. Various sample clean up methods such as adsorption chromatography with Florisil, gel permeation chromatography, Florisil cartridges and Unitrex have been proposed [2]. Isolation of coplanar compounds from the major compounds in the extract is achieved by using activated charcoal, since this adsorbs the planar molecules more strongly than the non-planar ones. Brominated naphthalenes, dioxins and furans, will also be separated from the non-planar compounds in these steps. To aid these separations, high performance liquid chromatography (HPLC) methods are now being adopted [2]. Soxhlet extraction with dichloromethane/n-pentane, followed by clean up over alumina columns and fractionation over silicagel columns, results in recoveries over 95% for all PBBs. Toluene may particularly be effective for extraction of decaBB [32]. Saponification may be an alternative, but decomposition of some PBB congeners may occur as in the case of PCBs [5].

Recoveries of PBBs using established methods are in the range of 80–90% [33]. The solvent system used for sample extraction can affect recovery [2]. PBBs adsorb to glass more tenaciously than other halogenated hydrocarbons, and are not easy to remove by the usual cleaning methods. Using disposable glassware prevents erroneous values in the data [34].

The 209 possible PBB congeners have a wide range of volatility, which causes complex separation problems. GC oven temperatures vary between 240 °C and 300 °C. Although most PBBs elute after PCBs, higher chlorinated PCBs may interfere with lower brominated PBBs [5]. Decachlorobiphenyl caused most interference in the analytical method for quantification of PBBs in human serum [31]. Polychlorinated terphenyls (PCTs) may also interfere with PBBs [35]. Therefore, mass spectrometry (MS) with electron impact (EI) or electron capture negative ionisation (ECNI) as ionisation methods is the preferred technique for detection of peaks after separation [5]. Quantification can be achieved by comparison with known standards [2]. In PBB detection commercial mixtures are used, since the commercial availability of pure congener standards is limited. The synthesis of pure congeners for use as standards is a prerequisite for advances in chemical analysis, as well as research into toxicological and biological effects of PBBs [2]. Although some individual PBB congeners are available as standards, the number of studies reporting individual PBB data is limited.

A recent method of detection is electron capture negative ionisation (ECNI) as ionization technique in combination with GC-MS analysis. This method is advantageous because it offers a high sensitivity for compounds with four or more bromine atoms [36]. The sensitivity of ECNI for these compounds is approximately 10 times higher than with the use of an electron capture detector (ECD) [5]. In the analytical method which was developed to quantitate PCBs and PBBs in human serum, GC/ECD was used [30]. Because the response, and therefore the sensitivity, of the ECD depends on the position of the halogen on the biphenyl nucleus as well as the number of halogen atoms, it is necessary to run a standard for each compound to be determined [2]. The use of narrow bore (0.15 mm i.d.) capillary columns is advised to obtain the required resolution [5].

Because of its low specificity and sensitivity flame ionisation detection (FID) can only be used in the analysis of standard substances [37]. The same limited application is envisaged for the method with the microwave-induced plasma emission detector, which is not sensitive enough for environmental samples [2].

7.2
PBDEs

Extraction and clean-up techniques for the analysis of PBDE residues in biological samples are similar to those developed for PBB. Table 6 shows several methods to determine PBDEs in various media. Most methods are based on extraction with organic solvents, purification of the extracts by gel permeation or

Table 6. Analytical methods for PBDE [3]

Sample	Extraction and clean-up	Separation and detection	Limit of determination	Ref.
Sewage	extract with chloroform; evaporate and dissolve residue in ethanol	GC/MS	0.06 mg/kg	[98]
Sediment	extract with acetone; clean-up on Florisil	NAA; GC/EC	<5 µg/kg <5 µg/kg	[67]
Fish	extract with acetone-hexane + hexane-ethyl ether; treatment with sulfuric acid or clean-up on alumina; chromatography on silica gel	GC/EC; GC/MS	limit of detection 0.1 mg/kg fat	[99]
Animal tissues (Multi-residue method)	homogenize; extract with *n*-hexane-acetone; treatment with sulfuric acid; gel permeation chromatography; chromatography or silica gel; chromatography or activated charcoal	GC/MS (NCl)	10 ng/kg	[37]
Fish	extract freeze-dried powdered sample with pet.ether; gel permeation chromatography; clean-up on Florisil; elute with hexane	GC/MS (NCI/SIM)	<5 µg/kg fat	[36]
Cow's milk	centrifuge; gel permeation chromatography; clean-up on Florisil; elute with hexane	GC/MS (NCI/SIM)	<2.5 µg/kg fat	[36]
Human adipose tissue	extract with methylene chloride; evaporate; clean-up on silica gel followed by clean-up alumina and on a carbon/silica gel column	HRGC/HRMS[a]	limit of detection 0.73–120 ng/kg (different congeners)	[100]
Commercial PBDE	homogenize and dissolve in tetrachloromethane for HPLC and GC/MS or *n*-hexane for TLC/UV	HPLC; GC/MS; TLC/UV	–	[101]

[a] High resolution gas chromatography/high resolution mass spectrometry.

adsorption chromatography, and determination mainly by GC, either with ECD or coupled with MS, with EI or ECNI. The recovery of the different PBDEs is generally higher than 80% [3].

Other extraction methods described for PBDEs in the literature are basically batch and Soxhlet extractions. The various clean-up methods for biological, sediment, and sewage sludge samples differ, depending on other compounds of interest which are determined simultaneously [1].

A multi-residue method has also been developed by Jansson et al. [38]. This method includes a multi-step separation enabling the determination of several polychlorinated an polybrominated pollutants in biological samples [3]. However, the recovery of 2,2′,4,4′-TeBDE with this method is only 49% [5].

The extraction of DeBDE (and also DeBB) is more difficult than other PBDEs, but a good solvent system is hexane/acetone (3:1) [1]. Toluene may also be a good alternative [31]. Yamamoto et al. [39] reported the determination of DeBDE in water and sediment samples. For the extraction of DeBDE from the sediments they used ultrasonic extraction with acetone. Rapid sample preparation techniques such as C_{18} solid-phase extraction and cartidge type Florisil extraction were used to clean up water and sediment samples. Average recoveries were 103% and detection limits were 0.12 ng/ml in water and 9.7 ng/g in sediment.

Typical GC analysis is performed using a non-polar capillary column (15–60 m) of methyltype (SE-30, OV-1, OV-101) or methyl + 5% phenyl groups (DB-5,SE-54,CP-Sil 8CB) [1].

Both GC-ECD and GC-MS with EI or ECNI may be used for the final analysis of PBDEs [5]. ECNI-MS is a very sensitive method for many halogenated compounds [1]. Using GC-MS, the type of reaction gas can influence the data. A study of PBDE residues in guillemot eggs showed an increase in levels of 2,2′,4,4′-TeBDE, an unidentified PeBDE, and 2,2′,4,4′,5-PeBDE of respectively 10–35%, 25–80%, and 0–20% after re-analysis using ammonia as reaction gas instead of methane [1].

Another variety of the GC-MS detection method is high resolution GC/high resolution MS (HRGC-HRMS). Not only human adipose tissue (Table 4) can be analysed with HRGC-HRMS. Studies are described by Loganathan et al. [40], and Takasuga et al. [41], investigating the analysis of PBDE residues in environmental samples with HRGC-HRMS. After the standard clean up a further carbon clean up stage with HPLC porous graphitic carbon was added. The mass spectrometer was operated in the selected ion monitoring (SIM) mode. Additionally, mass peak profile monitoring acquisition at high resolution and low resolution scanning were performed to identify the interferences. With this method the identification of PBDEs as interferences of heptachlorinated dibenzofurans in the analysis of routine environmental samples can be quantified [41].

In most studies the technical mixture Bromkal 70–5 DE is used as an external standard. The percentage of PBDE congeners of Bromkal 70–5 DE is 44% 2,2′,4,4′-TeBDE, 48% 2,2′,4,4′,5-PeBDE an 8% of an unknown PeBDE [1, 5, 9]. Like the PBB analysis, the analysis of PBDEs requires individual PBDE congeners as analytical standards. Synthesized pure standards of 2,2′,4,4′-TeBDE, 2,2′,4,4′,5-PeBDE and 2,2′,4,4′,5,5′-HxBDE [1, 9] are available. Comparison of the Bromkal mixture used with the 2,2′,4,4′-TeBDE, 2,2′,4,4′,5-PeBDE standards showed that the percentage of the TeBDE in this mixture was 36,1% and PeBDE 35,5% [9]. De Boer and Dao [9] made a correction of the initial estimation of –5,6% for 2,2′,4,4′-TeBDE and –8,9% for 2,2′,4,4′,5-PeBDE in their overview of BDE data in aquatic biota and sediments. More individual PBDEs have recently become available [1, 42]. Wolf and Rimkus [43] described the synthesis of 2,2′,4,4′-TeBDE for the analysis of this congener in fish. The synthesis of ^{14}C labeled 2,2′,4,4′-TeBDE and 2,2′,4,4′,5-PeBDE was described by Örn et al. [44].

8
Emission, Transformation and Distribution of PBBs and PBDEs in the Environment

Losses of PBBs and PBDEs to the environment during normal production can occur through emission into the air, waste water, losses into the soil and to land-fills. These chemicals can also enter the environment during shipping and handling. There is also a possibility of their entrance into the environment as a result of the incineration of materials containing PBBs and PBDEs as well as during accidental fires, together with the formation of other toxic chemicals, as such, or as degradation products [2, 3]. Both PBBs and PBDEs are persistent, lipophilic, and only slightly soluble in water. They tend to concentrate at non-polar surfaces of particles, and in living organisms [2, 3]. Preferential adsorption of PBB congeners was noted, depending on the characteristics of the soil (e.g. organic content) and the degree and position of bromine substitution [2]. Once introduced into the soil PBBs and PBDEs do not appear to be translocated readily. The solubility of PBBs and PBDEs in water decreases with increasing bromination, so congeners with low bromine content are more easily distri-buted in the aquatic environment [2, 3]. The principal known routes of PBBs in the aquatic environment are from industrial waste discharge and leachates from dumping sites into receiving waters and from erosion of polluted soils. Pollution of soils can originate from point sources, such as PBB plant areas and waste dumps [2, 3].

Both PBBs and PBDEs are slowly degraded in the environment. Fifteen years after the Michigan disaster (1988) cores of the Pine River sediments contained 10–12% non-Fire Master compounds indicating a partial degradation of the PBB residues in the soil. It appeared that bromines were selectively removed from the *meta-* and *para-*positions. Micro-organisms isolated from Pine River sediment were capable of debrominating Fire Master PBB compounds [5]. Organic co-contaminants like petroleum products and heavy metals inhibited *in situ* debromination in the most heavily contaminated Pine River sediment [45]. Microbial degradation of PCBs occurs by anaerobic dechlorination fol-lowed by aerobic ring fission. Since the physical and chemical properties of PBBs resembles those of PCBs (more than PBDEs) the same microbial degra-dation could be possible for PBBs and, perhaps also for PBDEs. Anaerobic micro-organisms eluted from PCB-contaminated river sediments were shown to debrominate a Fire Master mixture in a reductive process. Two bacterial strains of the genus Pseudomonas isolated from a lake sediment, using *p*-chlo-robiphenyl as a sole carbon source, were capable of degrading 2- and 4-bromo-biphenyl, but unable to degrade 4,4′ dibromobiphenyl [2]. Degradation of PBBs by purely abiotic chemical reactions, excluding photochemical reactions (pho-todegradation), is considered to be unlikely [2]. Lower brominated PBBs were found to prime microbial reductive dechlorination of Arclor 1260 in river sediment [46].

Earlier studies on photodegradation using lower brominated PBB congeners (TeBB and lower), reported a preferential loss of ortho bromines [47, 48]. On photolysis of Fire Master BP-6 (solvent cyclohexane), no preferential loss of ortho bromines was found but other congeners, known as relatively toxic (e.g. 2,3′,4,4′,5′ PeBB) were enriched [49]. The main component of Fire Master BP-6; 2,2′,4,4′,5,5′ HxBB (BB 153) was consistently found in relatively high levels, and degradation of this compound occurred more rapidly than with its hexachloro analogue. PBBs degraded readily by UVR under laboratory conditions [2]. Watanabe et al. [50] reported that DeBDE dissolved in hexane can be degraded to NoBDE, OcBDE, HpBDE and HxBDE.

Studies have been performed on the photodegradation of DeBDE in organic solvents and water. DeBDE was irradiated in hexane solution with ultra violet radiation (UVR) and sunlight [3]. A mixture of tri- to octaBDE congeners was detected. In addition, a large number of PBDFs containing 1–6 bromoatoms and small amounts of polybromobenzenes were formed. Photodegradation of PBDE in water does not lead to the formation of lower BDE or BDF [3]. Sellström et al. [51] showed a rapid degradation of DeBDE in toluene (half-life 15 minutes) after UV radiation and in sand (half-lives 12–37 h) after UV and sunlight radiation. Lower PBDEs and other compounds were found after radiation. Sellström et al. [52] reported that the photolysis of DeBDE during extraction may be prevented by the presence of coextractives from the sample. Watanabe et al. [53] showed the formation of polybrominated dibenzofurans after UV radiation of DEBDE in hexane. This possible degradation of DeBDE should be taken into account when establishing conditions for the analysis of DeBDE.

No degradation of PBBs by plants has been recorded. In contrast to plants, animals readily absorb PBBs [2]. Data on environmental fate (although limited to MBDE, DiBDE, DeBDE) suggest that biodegradation is not an important degradation pathway for PBDEs, but that photodegradation may play a significant role [3].

Environmental studies so far indicate a possible partial degradation of the original PBDEs and PBBs to lower brominated congeners. Because the carbon-bromine bond is less stable than the carbon-chlorine bond reductive debromination may be a degradative pathway of bromobiphenyls and this reaction may have toxicological consequences, not encountered with PCBs [2]. In vitro studies with hepatic microsomes of marine mammals showed a high persistency of a number of PBB and PBDE congeners [54].

9
Environmental Levels and Human Exposure

Some typical PBDE and PBB concentrations in environmental samples are given in Table 7. Most concentrations determined until now are semi-quantitative total PBB or PBDE concentrations, but recently more congener-specific data are becoming available.

9.1
PBBs

The only report [55] on PBB levels in the air concerns air samples taken in the vicinity of 3 PBB plants in the USA. Traces of HxBB (0.06–1.10 ng/m^3) were found in 2 samples. Depending on its source, the predominant PBB compounds detected in surface water were HxBB and DeBB [2].

Many studies started after the accidental contamination in 1973 in Michigan, with Fire Master FF-1 being inadvertently substituted for magnesium oxide in the production of cattle feed. Estimates on the amount of PBBs used vary between approximately 290 kg [33] to 1000 kg [56]. PBBs were mixed into feeds, distributed widely to Michigan farmers. In addition, feeds not formulated to contain magnesiumoxide became contaminated (relatively low concentrations) due to carry over of PBBs from batch to batch through mixing equipment and, on farms, through the recycling of contaminated products. The mixing error was not discovered immediately, and it was almost a year before analysis indicated that a compound of PBB was involved in the illness or death of farm animals. During this time, contaminated animals and their products entered the human food supply and the environment of the state Michigan [2]. Groundwater near local disposal sites was not contaminated by PBBs [57]. Soils from PBB industrial sites (2000 mg/kg dry weight, Fire Master plant) have in general been more heavily contaminated than Michigan soils (371 mg/kg dry weight) [33, 58]. Contamination of animal feed or foods by PBBs has been reported only in connection with the Michigan PBB incident [2].

High levels of NoBB and OcBB (besides PBDE) were present in fish from German rivers [2]. However HxBBs were predominant in fish from the North Sea and Baltic Sea [2, 5]. In all samples from the Baltic Sea 3,3',4,4',5,5' HxBB was found in relative high concentrations (maximum concentration: 36 mg/kg fat), but it was not detected in the North Sea or rivers [38]. The concentrations of other HxBBs are usually higher in fish from the Baltic Sea than in fish from the North Sea. Concentrations of BB 153 (2,4,5,2',4',5'-hexaBB) determined in marine fish ranged from 0.2–2.4 mg/kg lipid (Baltic fish) and in seals 0.4 (Northern Ice Sea) – 26 (Baltic Sea) mg/kg lipid [38, 59]. The congener pattern found in fish is quite different from that found in commercial products. Many of the major peaks could well be the result of photochemical debromination of DeBB, but this has not been confirmed [2].

Long range transport has not been proven, but the presence of these compounds in Arctic seal samples indicates a wide geographical distribution [2].

For most human populations, direct data on exposure to PBBs from various sources have not been documented. Occupational exposure was found in employees in chemical plants in the USA (skin contact and inhalation) and in farm workers (skin contact, inhalation, and contaminated food). Median serum and adipose tissue PBB levels were higher among chemical workers [2].

Recently, PBBs (and also PBDEs) have been detected in cow's milk and human milk in Germany [37]. The congener patterns in these samples differ from that found in fish. BB 153 was the most abundant component in human milk [37]. The relative concentration of BB 153 is higher in human milk (1.03 mg/kg lipid) [37] than in fish (ranged between 0.092–24 mg/kg lipid) [37,59]. Total levels found in human samples were substantially higher than levels that were detected in cow's milk (both samples, cow and human from the same region) [37]. Thus, an infant of 6 kg body weight consuming human milk will have a higher intake of PBBs than an adult consuming cow milk, respectively 0.01 mg PBB/kg body weight per day and 0.00002 mg PBB/kg body weight per day [2].

9.2
PBDEs

In Japan a large amount of PBDEs was determined in the airborne dust, DeBDE observed being dominant ($83-3060$ pg/m^3), other congeners were TeBDE, PeBDE, and HxBDE [60]. These PBDEs were also present in two ash and soil samples from a recycling plant in Taiwan, in which DeBDE was the dominant congener ($510-2500$ mg/kg ash and $260-330$ mg/kg soil) [61].

Two samples of sewage sludge from the same sewage treatment plant in Gothenburg (Sweden) were analysed. One sample was a pool of subsamples taken during a period with little rain (1) and the other was composed of sub-samples during a rainy period (2). The levels were 25 and 21 ng/g dry weight for the dry and wet period respectively, indicating that the primary PBDE sources to this matrix are household and industrial effluent and not washout from the atmosphere [1].

Surfacial sediment samples up- and downstream from a plastic industry in Sweden indicated this industry as the most likely source. The relative amounts in the analysed sewage sludge and surfacial sediments samples are quite similar to the pattern for the technical PBDE product Bromkal $70-5$ DE [1]. De Boer and Dao [9] found a PBDE pattern in sediments, which is comparable to the pattern of this technical mixture. In these sediment samples PeBDE-concentrations were higher than the TeBDE-concentrations. TeBDE, PeBDE and HxBDE were found in sediments of Osaka Bay (Japan), and in 7 of 15 riverine and estuarine samples, DeBDE was found in higher concentrations [6], indicating accumulation of higher brominated congeners in the sediment. Recently DeBDE has been detected for the first time in Sweden in some sediment samples from the river Viskan and sludge samples [1, 52]. The upper layer in a laminated sediment core from the Baltic Sea contained higher levels of TeBDE and PeBDE than lower layers, indicating an increasing burden of these compounds [62]. Other time-trend studies of Baltic sediments showed an increasing trend in the concentrations of PBDEs between 1973 and 1990 [1]. PBDEs seem to have a higher absorption to the sediment than PCBs [9]. Allchin et al. [63] found relatively high tetra-BDE concentrations in sediments of the rivers Tees and Skerne in the UK, whereas also in other UK rivers tetra and penta BDEs were detected. In the same study PBDEs were found in liver and muscle tissue of several fish species from the same rivers and estuaries (Table 7). In a survey on PBDE and PBB concentrations in sediments of 22 estuaries in Europe, high decaBDE concentrations were found in the rivers Mersey and Scheldt (Table 7) [64]. At all locations deca BDE concentrations were higher than tetraBDE concentrations.

PBB concentrations relatively low and often below the detection limits. Swedish results from freshwater fish studies (Table 7) indicate that southern Sweden may be more contaminated with PBDEs than northern Sweden [1]. Compared with levels found in terrestrial animals (rabbit, moose, reindeer respectively 0; 1.7; 0.47 ng/g lipid weight) [59] showed that the concentrations

Table 7. A selection of typical PBB and PBDE concentrations in the aquatic environment (ww: wet weight, lw: lipid weight, dw: dry weight)

Matrix	Location	PBB (µg/kg)	PBDE (µg/kg)	Ref.
Sediment	Osaka May		11–30 dw, tetra	[102]
Sediment	Japanese rivers		33–375 dw, deca	[102]
Sediment	River Tees, UK		1,348 dw, total	[62]
Sediment	Mersey, UK	0.33 dw, deca <0.01 dw, hexa	1,700 dw, deca 2.2 dw, tetra	[63]
Sediment	Western Scheldt, Netherlands	0.84 dw, deca 0.024 dw, hexa	200 dw, deca 0.42 dw, tetra	[63]
Sediment	Rhine	0.39 dw, deca 0.04 dw, hexa	15.7 dw, deca 1.4 dw, tetra	[63]
Sediment	Elbe, Germany	<0.1 dw, deca 0.83 dw, deca	<0.01 dw, hexa <0.17 dw, tetra	[63]
Mussels	Osaka Bay		15 ww, tetra	[63]
Cod liver	southern North Sea		170 lw, tetra 25 lw, pentaa	[103]
Mackerel	North Sea	0.04 ww, hexa	5.4 ww, tetra	[8]
Pike	South Sweden		27,000 lw, tetra	[104]
Herring	Baltic Sea	0.16 lw, total	528 lw, total	[58]
Herring	Skagerrak	0.27 lw, total	735 lw, total	[58]
Flounder liver	Humber, UK		217 ww, tetra 22 ww, penta	[62]
Flounder liver	Tees Bay, UK		1,294 ww, tetra 238 ww, penta	[62]
Salmon	Baltic Sea		167 lw, tetra 96 lw, penta 12.7 lw, hexa	[105]
Harbour Seal	North sea	13–61 ww, hexa <1 ww, deca	280–1200 ww, tetra 140–270 ww, penta <10 ww, deca	[8]
Grey seal	Baltic Sea		308 lw, tetra 101 lw, penta 38 lw, hexa	[105]
Whitebeaked dolphin	North Sea	13 ww, hexa 8.3 ww, penta <0.9 ww, deca	5,500 ww, tetra 2,200 ww, penta <10 ww, deca	[8]
Sperm whale	Atlantic Ocean	1.1–1.9 ww, tetra 0.4–0.9 ww, penta <0.5 ww, deca	58–95 ww, tetra 17–40 ww, penta <5 ww, deca	[8]
Pilot whale	Faroe Islands		435–1850 lw, tetra 265–940 lw, penta 140–370 lw, hexa	[106]
Cormorant liver	Rhine delta		28,000 ww, total	[107]
Sea eagle	Baltic Sea	280 lw, total	350 lw, total	[108]
Guillemot	Northern Ice Sea	50–130 lw, total		[58, 108]

a Penta: often based on two congeners.

of PBDEs are higher in aquatic organisms than in terrestrial organisms [1, 3]. Sellström [1] and Watanabe et al. [64] detected PBDEs in fresh water fish (Sweden and Japan) with TeBDEs congeners dominant in the samples. Comparing these results with Bromkal 70–5 DE the relative amount of TeBDEs is much higher [1]. In addition, de Boer and Dao [9] found relatively higher concentrations of TeBDEs in biological samples. The PBDE pattern in sediments is more comparable to the pattern of Bromkal 70–5DE than the PBDE pattern in biological samples. Relatively higher TeBDEs concentrations in fish may be caused by a more rapid uptake of lower brominated compounds. Possibly, a membrane barrier exists for higher brominated compounds due to the larger size of these molecules. Consequently, in sediments relatively more higher brominated compounds may be expected [9].

Noticeable concentrations of PBDEs were determined in carp of three age classes collected from the Buffalo river (New York, USA). TeBDEs accounted for 94–96% of total PBDE concentrations [40].

As in Europe, in Japan TeBDEs are the major component in marine and shellfish samples [60].

Contamination of PBDEs on lipid weight basis is about nine times higher in herring caught in spring than herring caught in autumn. This relationship has previously been shown for PCB, DDT and dioxins. Spring herring is caught near the breeding season and this probably affects lipid disposition and metabolism, which in turn may affect concentrations of organohalogens [1].

Several fish eating animals were studied. Extremely high PBDEs levels (upto 25,000 mg/kg TeBDE and 4,000 mg/kg PeBDE (wet weight)) were found in the liver of a cormorant from the Rhine delta. Since this is based on only a single animal, further research is necessary [9]. The PBDEs levels found in an osprey from Sweden, which also feeds on freshwater fish, were high as well (160–1900 mg/kg lipid) [1]. Baltic seals contained higher TeBDEs concentrations than the North Sea seals [9]. Time-trend studies of guillemot eggs (Stora Karslö, Baltic Sea, Sweden) indicate that the levels of PBDEs have increased since 1970 and that this increase is significant. A similar time-trend study of pike (from Lake Bolmen, Sweden) shows a similar trend. However, in guillemot eggs there are indications that the PBDEs levels may have decreased during the last years [1]. Both guillemot and grey seals show higher concentrations of the three PBDE congeners, 2,2′,4,4′ TeBDE, an unknown PeBDE, and 2,2′,4,4′,5 PeBDE than the herring they feed on. TeBDE seems to biomagnify to the relatively highest extent [1]. Biomagnification is also observed in dolphins and porpoise from the southern North Sea and the Atlantic west of Ireland. Biomagnification factors between fish and investigated marine mammals are approximately 10–30 [9].

Only long range transport through air can explain the contamination in whitefish in Lake Storvindel, Sweden. PBDEs have been found in an air sample collected near this lake. In another air sample collected at the southern part of the Baltic Sea PBDEs were detected as well [1].

In Germany cow's milk was analysed and an average concentration of 3.57 mg/kg fat (4 samples) was found and determined as Bromkal 70–5 DE, main component HxBDE. PBDEs were also detected in breast milk (Germany).

The samples contained 0.6–11.1 mg PBDE/kg fat, determined as Bromkal 70–5DE [37]. Uptake of TeBDEs and PeBDEs may occur in humans via the foodchain, e.g. by consuming fish. Exposure may also occur through skin contact (flame retardants in polymers used in textiles) and via inhalation [3].

In general environmental PBDE concentrations are considerably higher than those of PBBs. De Boer et al. [8] found PBDE concentrations 50-fold higher than PBB concentrations in sperm whales. This corresponds with the production of PBDEs which is still ongoing and increasing, whereas the PBB production is mainly restricted to DeBB and seems to be relatively small. One of the questions to be solved is the difference between the current PBDE production, which is mainly based on DeBDE with a relatively smaller amount of PeBDEs, and the patterns found in the environment which show the highest concentrations of 2,4,2′,4′-TeBDE and 2,4,5,2′,4′-PeBDE. Degradation of DeBDE could be one explanation. The former production of lower brominated PBDE mixtures may also be a possibility. The observation of PBDEs and PBBs in sperm whales indicates that PBDEs and PBBs have reached deeper waters of the Atlantic Ocean and apparently are becoming global pollutants [8]. The PBDE concentrations found in North Sea dolphins (>7 mg/kg in blubber) are not as high as PCB concentrations found in North Sea dolphins (up to 128 mg/kg), but an increasing production may soon cause an increase in the environmental PBDE levels [8, 66].

PBDEs have also been determined in human milk and tissue samples. A very interesting figure was produced by Norén and Meyronité [67] (Fig. 3). This figure shows the trend of PBDEs compared with that of the classic contaminants DDT and PCBs, showing that whereas the DDTs and PCBs have decreased in human milk from Stockholm over the last 25 years, PBDEs are exponentially increasing. The concentrations profiles of the PBDEs in these human milk samples were given by Meyronité et al. [68], showing the highest concentrations for 2,2′,4,4′-TeBDE. DeBDE was not analysed. Darnerud et al. [69] reported PBDE concentrations in human from Uppsala, Sweden. They also found 2,2′,4,4′-TeBDE in the highest concentrations (mean 2 µg/kg lipid). Mean total PBDE values were around 4 µg/kg lipid. De Boer et al. [70] found 2 µg/kg (wet weight) 2,2′,4,4′-TeBDE and 4 µg/kg 2,4,5,2′,4′ PeBDE in human adipose tissue of an Israeli who had extensively been exposed to vapours from a TV set. Lindström et al. [71] determined 2,2′,4,4′-TeBDE is human adipose tissue from 77 individuals from Sweden (1995–1997). Mean concentrations were 4–16 µg/kg lipid.

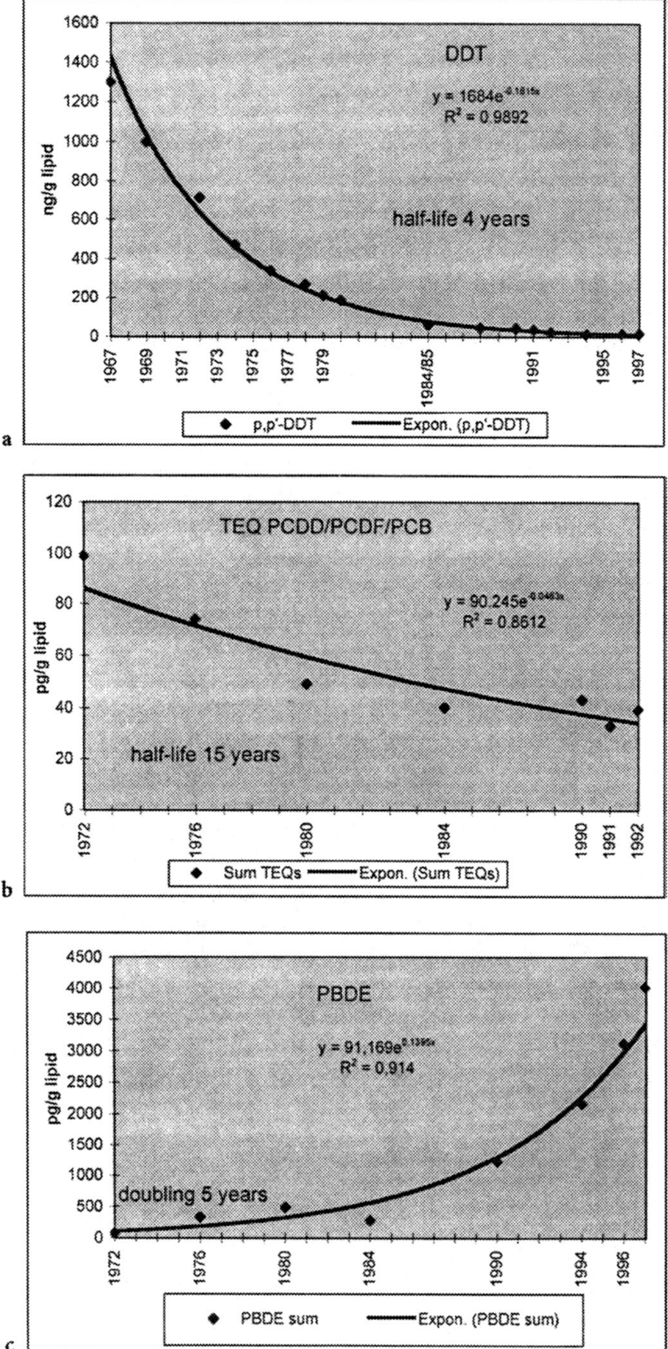

Fig. 3. Trends in DDT (**a**), PCB (**b**) and PBDE (**c**) concentrations in human milk from Stockholm, Sweden [67]

10
Toxicology

10.1
Environmental toxicity

10.1.1
PBBs

Most available data on the effects of PBBs on organisms in the environment are
data on farm animals from the Michigan disaster. The estimated average
exposure of cows at a high contaminated farm was 250 mg/kg body weight [33].
A few weeks after ingestion of contaminated cattle food, clinical signs were a
reduction of about 50% in food consumption (anorexia) and a decrease of
about 40% in milk production. Some cows showed an increased frequency of
urination and lacrimation, and developed haematomas, abscesses, abnormal
hoof growth, lameness, alopecia, hyperkeratosis, and cachexia. Several cows
died within 6 months of exposure [72]. The death rate in 6- to 18-months old
calves was much higher, 50% died within 6 weeks [72,73]. After the ingestion of
a Fire Master mixture, hyperkeratosis and hair loss were seen in cattle, and
lesions resembling chloro-acne were seen in rhesus monkeys after exposure to
50 mg/kg in the diet [12]. After 20 weeks, exposure at a dose of 2 mg/animal
twice a week, Fire Master FF-1 also caused skin papillomas in previously
initiated mice [74]. Fire Master BP-6 caused chronic and subchronic neuronal
symptoms, such as irritation, changed behaviour and a decreased muscular
control [12].

In contrast of the observed toxicity in cattle, no clear health effects on the
human population in Michigan could be correlated with PBB exposure.
However, the follow-up period might not have been long enough for the
development of cancer. In industry it appears that chloro-acne like lesions may
develop in workers producing PBB, and hypothyroidism in workers exposed to
BB-209 [2]. Further *in vitro* studies are necessary to identify toxic effects of
PBBs. Controlled long term feeding studies on cattle exposed to low doses of
Fire Master did not reveal any adverse effects as indicated by food intake,
clinical signs, clinicopathological changes, or performance. Minks, guinea-pigs,
and monkeys appeared to be more susceptible to PBB toxicity [2].

10.1.2
PBDEs

There is hardly any information available on the toxicity of PBDEs in organisms
in the environment. A case study has been described of a young man who
developed Yusho-like health effeccts after having watched to a newly bought
television set in a small non-ventilated insulated room for several hours a day
during 8 consecutive months at the age of 13. PBDEs were determined in a fat
biopsy and a blood sample of this man at the age of 21. PBDE levels above the
detection limit were only found in the fat biopsy. PBDEs, but also tetrabromo-

bisphenol-A, were also found in different parts of the TV set [70]. Since eight years had elapsed between the possible exposure to high levels and the moment of sampling, no definite answer could be given as to the role that PBDEs played in the adverse health effects of this man.

10.2
Acute toxicity

10.2.1
PBBs

Fire Master BP-6 appears to have a similar acute toxicity to rats as the PCB mixtures Aroclor 1254 and Kanechlor 500 [5].The LD_{50} values of commercial mixtures show a relatively low order of acute toxicity ($LD_{50} > 1$ g/kg body weight) in rats, rabbits and quails, following oral or dermal administration. The toxicity of PBBs was higher with multiple dose rather than single dose administration. The few studies performed with commercial octabrominated biphenyl mixtures and BB-209 did not result in mortality in rats and fish. On the basis of the limited available data, octabrominated biphenyls and BB-209 appear to be less toxic than other lower brominated PBB mixtures, probably because they are less efficiently absorbed [2].

10.2.2
PBDEs

The acute toxicity of commercial PeBDE, DeBDE and OcBDE for laboratory animals is also low ($LD_{50} > 1$ g/kg body weight). BDE-209 and OcBDEs were also not irritant to the skin. BDE-209 is neither irritant to the eyes of a rabbit nor had it effect on survival, body weight or food consumption in feeding studies on rats and mice. No gross or microscopic pathological effects have been found [3].

However in short term toxicity studies with Octa-BDEs, rats administered dietary levels of 100 mg/kg had increased liver weights and showed microscopal changes of liver tissue. These liver changes were more severe at the even higher dose levels, i.e. 1000 and 10,000 mg/kg diet. Octa-BDEs gave also minor eye irritation to rabbits [3]. In addition, Penta-BDEs increased liver/body weight ratio with 64%, Octa-BDEs with 45%, and BDE-209 with 25% in a study where a dose of 0.1 mM/kg/day was administered to male rats during 14 days [75].

10.3
Mechanisms of Toxicity

The toxicity of PBB and PBDE congeners strongly depends on their molecular structure [2, 3, 5].

Induction of the P450 1A subfamily of cytochrome P450 is the precursor of a whole spectrum of possible effects at more integrated levels of biological structure: weight loss, thymus atrophy, and changes in the liver such as proliferation of the smooth endoplasmatic reticulum (location of the P450 system), increased RNA and protein content, decreased DNA content, cell necrosis, liver enlargement, and hepatic porphyria [76–78].

10.3.1
PBBs

The more toxic PBB congeners cause a decrease in thymus and/or body weight and produce pronounced histological changes in the liver and thymus. Categorisation of halogenated biphenyls has been made on a structural basis. Category 1 comprises isomers and congeners lacking *ortho*-substituents (co-planar PBBs). Mono-*ortho*-substituted derivatives constitute the second category. Other PBBs (mainly those with two or more *ortho*-bromine substituents) have been organised into the third category. Congeners of Category 1 tend to elicit the most severe effects, while the congeners of the second and third categories show decreasing toxicological changes. Within these categories, the degree of bromination may also influence toxicity. In all combinations tested, 3,3',4,4',5,5'-hexaBB (BB-169) was found to be the most toxic PBB [2].

10.3.2
PBDEs

Penta-BDEs increased cytochrome P450 to a higher extend than Octa-BDEs, while BDE-209 did not significantly increase cytochrome P450. 2,2',4,4'-tetraBDE alone (6 and 18 mg/kg body weight) induced EROD (CYP1A1) and MROD (CYP 1A2) in a minor way compared to PCBs in rats. A combination with chlorinated paraffins showed a synergistic effect on this induction. The induction of PROD (CYP2B) by 2,2',4,4'-tetraBDE alone or in combination was as strong as that of the PCBs [79].

10.4
Genotoxicity

10.4.1
PBBs

Several reports on the carcinogenicity of PBBs and PCBs have concluded that there are strong indications that these compounds are not mutagenic in itself, but do promote the carcinogenicity of mutagenic compounds, such as nitrosamine and certain polyaromatic hydrocarbons (PAHs) [5, 12, 78, 80, 81]. This is highly relevant, since in the marine environment halogenated compounds often co-occur with PAHs. The only lifetime study with a technical nonabromobiphenyl mixture was conducted on rats and mice. The lowest (oral) dose tested that still produced carcinogenic effects on rodents was 0.5 mg/kg body weight per day, and no observed effect level in a rat was < 0.15 mg/kg body weight per day [82]. The carcinogenicity of technical octabrominated biphenyl mixture and BB-209 has not been studied, although a number of chronic effects have been observed in experimental animals at doses of around 1 mg/kg body weight/day during long term exposure [2].

10.4.2
PBDEs

In a carcinogenicity study in rats and mice, DeBDE was administered at dietary levels of up to 50 g/kg. An increased incidence of adenomas (but no carcinomas) was found in the livers of male rats receiving 25 g/kg and female rats receiving 50 g/kg. In male mice, increased incidences of hepatocellular adenomas and/or carcinomas (combined) were found at 25 g/kg and an increasing thyroid follicular cell adenomas/carcinomas (combined) at both dose levels. Female mice did not show any increase in tumour incidence. There was equivocal evidence for carcinogenicity in male and female rats and male mice only at dose levels of 25 – 50 g BDE-209/kg diet [83, 84]. Since the results of all mutagenicity tests have been negative, it was concluded that BDE-209 is not a genotoxic carcinogen [3]. In 1990 the International Agency for Research on Cancer (IARC) concluded that there was limited evidence for carcinogenicity, indicating that BDE-209, at present exposure levels, does not present a carcinogenic risk for humans [3].

The results for mutagenicity of penta-BDEs and BDE-209 were negative. Results of the mutagenicity tests of octa-BDEs including an unscheduled DNA assay, *in vitro* microbial assays, and an assay for sister chromatid exchange with Chinese hamster ovary cells were also negative [3]. Very recently, the current levels of PBDEs adipose tissue in humans from Sweden were correlated to the the occurrence of non-Hodgekin's lymphoma (NHL), malignant melanoma, and other forms of cancer or in situ changes. All 77 patients had detectable levels of PBDEs in their adipose tissue, as represented by the concentrations of BDE-47. Only NHL patients showed significantly higher levels of BDE-47 (mean of 13.1 ng/g) than people without any malignancies (mean of 5.1 ng/g) [71]. However, a mechanistic connection was not infered.

10.5
Reproductive Effects

Both classes of compounds are listed as endocrine disrupters [85].

10.5.1
PBBs

Fire Master FF-1 caused a longer sexual cycle in monkeys [5], and PBBs caused decreased egg production and nesting behaviour in Japanese quail [86]. One recent study reported that in mice PBB (di-BBs and tetra-BBs) reduced the *in vitro* fertilisation rate at higher dosages. Furthermore, an increased incidence of abnormal two-cell embryos and degenerative ococytes was observed at the 1 and 10 mg/ml concentration of PBB [87]. PBBs also affect the regulation of steroid hormones. The extent depends on the species as well as the dose and duration of exposure.

10.5.2
PBDEs

DeBDE caused no teratogenic response in fetuses of rats intubated with
10–1000 mg/kg day on gestation days 6–15. Fetal toxicity only occurred at
1000 mg/kg as subcutaneous edema and a delayed ossification of normally
developed bones of the fetal skull [88]. At high dose levels of octa-BDEs (25 and
50 mg/kg body weight) in rats, resorptions or delayed ossification of different
bones and fetal malformations were observed [3]. In rabbits there was no
evidence for teratogenic activity, but fetotoxicity was seen at a maternally toxic
dose level of 15 mg octa-BDE/kg body weight (no observed effect level 2.5 mg/
kg body weight) [89]. Test results for teratogenicity of PeBDE were negative [3].
Oral administration of a dietary dose of approximately 0.5 mg Bromkal 70–
5 DE for 3.5 months to female sticklebacks, *Gasterosteus aculeatus,* resulted in a
decreased spawning success [90].

10.6
Effects on Thyroid Hormones

The interactions of both classes of brominated flame retardant with the thyroid
hormone system seems to follow those of all major classes of polyhalogenated
aromatic hydrocarbons (PHAHs) [91]. Overall three levels of interference in the
thyroid system have been found for PHAHs, including the thyroid gland,
thyroid hormone metabolism, and thyroid hormone transport. Effects on
thyroid hormonal systems are important to follow, as these hormones play a
crucial role in the development of many organs, e.g., the brain. Several enzyme
systems involved in thyroid hormone metabolism have been found to be affected by PHAHs. A weak induction of cytochrome P450 1 A was measured after
exposure of rats to 2,2′,4,4′ tetrabromodiphenylether [79]. The reduction of
plasma levels of thyroxine (T4) seems to be caused especially by the hydroxylated metabolites of PHAHs throught the interaction with the transport
protein transthyretin (TTR). Binding to TTR not only disrupts the T4 transport,
with concomittant low plasma T4 levels, but may also results in the selective
transport of the hydroxylated PHAHs across the blood-brain barrier and the
placental barrier. TTR has been suggested to play a major role in mediating the
delivery of T4 from the mother to the fetus across the placental barrier, but also
across the blood-brain barrier, where T4 is locally converted to T3, which is
absolutely essential for e.g., brain development. Competitive inhibition of
thyroid binding to choroid plexus by several hydroxylated PCBs and a hydroxylated metabolite of was shown to occur in rats [92]. The possible impact of
the high accumulation of phenolic PHAHs on the development of fetal brain,
behaviour, and reproductive organs and functions is the focus of ongoing research. Planarity of the hydroxy-PHAH molecules is not a requirement for
binding to TTR. Many effects on the thyroid hormone system are primarily
caused by the hydroxy-metabolites of PHAHs. In this respect it is noteworthy
that of a number of PBB and PBDE congeners investigated, only 4,4′-dibromo-
biphenyl (BB-15) was metabolised in *in vitro* assays with microsomal prepara-

tions of seals, cetaceans and Laysan albatross [8, 53]. In contrast, BDE-47 and BDE-99 were rapidly metabolised in an experiment with rats and mice [93]. Thus, large species differences in sensitivity towards this mechanism of toxicity may exist.

10.6.1
PBBs

Rats and pigs showed dose related decreases in serum thyroxine and triiodo-thyronine. There was also a pronounced influence of PBBs on vitamin A storage above a no observed effect level of 0.1 mg/kg body weight/day [2]. This effect does probably also involve hydroxy-metabolites, since this is also the case with PCBs.

10.6.2
PBDEs

Hyperplasia of the thyroid has been observed [94]. Similar observations have been reported for Penta-BDEs [3] and BDE-209 [88]. Exposure of rats for 14 days to a daily dose of 18 mg 2,2',4,4'-tetraBDE/kg body weight resulted in lowered levels of free thyroxine in blood plasma. Lower doses showed no such effect [79].

10.7
Immuno-Suppression

PBBs cause immuno-suppression at levels that also cause a number of the other toxic effects described [5].

10.7.1
PBDEs

The immunotoxic potential of the PBDE mixture Bromkal 70–5 DE was compared to the technical PCB mixture Aroclor 1254 and the potential of the single congener 2,2',4,4'-tetraBDE to that of 2,3,3',4,4'-pentachlorobiphenyl (CB-105) in rats and mice. In mice, all compounds were immunotoxic, but in rats only the PCBs. This points to the existence of a considerable range of species differences [95].

10.8
Hepatic Porphyria

PBBs produced porphyria in rats and male mice at doses as low as 0.3 mg/kg body weight per day.

10.9
Conclusions

In summary, it can be concluded that the acute toxicity of brominated flame retardants is relatively low. The long term effects on the balance of endocrine systems seems to present the greatest danger of these compounds. These endocrine effects need further consideration, since the majority of animals and man are exposed to these brominated flame retardants. The exposure range for humans via the food was calculated as 0.2–0.7 mg per day [96].

References

1. Sellström U (1996) Polybrominated Diphenyl Ethers in the Swedish environment. Licentiate thesis, ITM-Rapport 1996 45, Stockholm University, Stockholm, Sweden
2. World Health Organization (1994) Polybrominated Biphenyls. IPCS, Environmental Health Criteria, Geneva, 152
3. World Health Organization (1994) Brominated Diphenyl Ethers. IPCS, Environmental Health Criteria, Geneva, 162
4. Ballschmiter K, Bacher R, Mennel A, Fischer R, Riehle U, Swerev M (1992) J High Resolut Chromatogr 15:260
5. Pijnenburg AMCM, Everts JW, Boer J de, Boon JP (1995) Rev Environ Contam Toxicol 141:1
7. Shelley S (1993) Chem Engin 100:71
8. Boer J de, Wester PG, Boon JP (1998) Nature 394:28
9. Boer J de, Dao QT (1993) Overview of bromodiphenylether data in aquatic biota and sediments. RIVO report C020/93, IJmuiden, The Netherlands
10. Brinkman UAT, Kok A de (1980) Production, properties usage. In: Kimbrough RD (ed) Halogenated biphenyls, terphenyls, naphtalenes, dibenzodioxins and related products. Elsevier, Amsterdam, Oxford, New York, pp 1–40
11. Pomerantz I, Burke J, Firestone D, McKinney, J, Roach J, Trotter W (1978) Environ Hlth Persp 24:133
12. Safe S (1984) CRC Crit Rev Toxicol 13:319
13. Prins RA, Meyer W (1996) Microbiële afbraak van Xenobiotica. Ecotoxicology. University of Groningen, Groningen, The Netherlands, pp 89–170
14. Vionov VG, El'kin YN, Kuznetsova TA, Mal'tsev II, Mikhailov VV, Sasunkevic VA (1991) J Chromatogr 586:360
15. Hairston DW(1995) Chem Engin 102:65
16. Mumma CE, Wallace DD (1975) Survey of industrial processing data. Task II: Pollution potential of polybrominated biphenyls. Washington, DC, US Environmental Protection Agency (EPA-560/3–75–004)
17. Neufeld ML, Sittenfield M, Wolk KF (1977) Market input/output studies. Task IV: Polybrominated biphenyls. Washington, DC, US Environmental Protection Agency (EPA-560/6–77–017)
18. Anon (1990) Decabromobiphenyl (Adine 0102). La Défense, Paris, France, ATOCHEM (unpublished report)
19. UBA (Federal Environmental Agency)(1989) [Stat of facts. Polybrominated dibenzodioxins (PBDD) – Polybrominated dibenzofurans (PBDF). Second supplement: Polybrominated biphenyls.] In: Polybrominated dibenzodioxins and dibenzofurans (PBDD/PBDF) from brominated flame retardants. Report of the working group on brominated flame retardants to the Environment Minister Conference. Bonn, Tederal Ministry for the Environment, Nature Conservation and Nuclear Safety, pp 95–97 (in German)
20. Bjerregaard M (1998) Response of the European Commission to questions asked by mr. D. Eisma in the European Parliament (E 3004/98 NL, 1 December)

21. Hass JR, McConnel EE, Harvan DJ (1978) J Agric Food Chem 26:94
22. Spiegelstein M (1998) Bromine Scientific Environmental Forum, Brussels, Belgium. Personal communication
23. Bieniek D, Bahadir M, Korte, F (1989) Heterocycles 28:719
24. Lahaniatis ES, Bergheim W, Bieniek D (1991) Toxicol Environ Chem 31:521
25. Riess M, Ernst T, Popp R, Schubert T, Thoma H, Vierle O, Ehrenstein GW, Eldik R van (1998) Organohalogen Compounds 35:443
26. WHO (1987) In: Rantanen JH, Silano V, Tarkowski S, and Yrjänheikki E (eds) PCB's, PCDD's and PCDF's, prevention and control of accidental and environmental exposures. World Health Organization, Copenhagen, Denmark
27. Dumler-Gradl, R, Tartler D, Thoma H, Vierle O (1995) Organohalogen Compounds 24:101
28. Griffin RA, Chou SFJ (1981) Attenuation of polybrominated biphenyls and hexachloro-benzene by earth materials. Washington, DC. US Environmental Protection Agency (EPA-600/2–81–186)
29. Fehringer NV (1975) J Assoc Off Anal Chem 58:1206
30. Burse VW, Needham, LL, Liddle JA, Bayse DD, Price HA (1980) J Anal Toxicol 4:22
31. Needham LL, Burse VW, Price HA (1981) J Assoc Off Anal Chem 64:1134
32. Ranken P (1998) Personal Communication, Albemarle, Baton Rouge, Louisiana, USA
33. Fries GF (1985) CRC Crit Rev Toxicol 16:105
34. Willet LB, Brumm CJ, Williams CL (1978) J Agric Food Chem 26:122
35. Wester PG, Boer J de (1996) Environ Sci Technol 30:473
36. Boer J de (1995) Analysis and biomonitoring of complex mixtures of persistent halo-geneted micro-contaminants. Ph D Thesis, Free University, Amsterdam, The Netherlands
37. Krüger C (1988) Polybrominated biphenyls and polybrominated biphenyl ether-detection and quantitation in selected foods. Ph D Thesis, University of Münster, Münster, Germany
38. Jansson B, Andersson R, Asplund L, Bergman, Å, Litzen K, Nylund K, Reutergardh L, Sellström, U, Uvemo U-B, Wahlberg C, Wideqvist U (1991) Fresenius J Anal Chem 340:439
39. Yamamoto H, Okumura T, Nishikawa Y, Knishi H (1998) J Assoc Off Anal Chem 80:102
40. Loganathan BG, Kannan K, Watanabe I, Kawano M, Irvine K, Kumar S, Sikka HC (1995) Environ Sci Technol 29:1832
41. Takasuga T, Inoue T, Ohi, E, Umetsu N (1995) Organohalogen Compounds 23:81
42. Örn U, Erikkson L, Jakobsson E, Bergman Å (1996) Acta Scandinavica 50:802
43. Wolff M, Rimkus GG (1985) Dtsch Tierärztl Wschr 92:174
44. Örn U, Jakobsson E, Bergman Å (1998). Organohalogen Compounds 35:451
45. Morris PJ, Quensen JF, Tiedje JM, Boud SA (1993) Environ Sci Technol 27:1580
46. Bedard DL, Dort H van, Deweerd KA (1998) Appl Environ Microbiol 64, 1786
47. Bunce NJ, Safe S, Ruzo LO (1975) J Chem Soc Perkin Trans 1:1607
48. Ruzo LO, Sundstrom G, Hutzinger O, Safe S (1976) J Agric Food Chem 24:1062
49. Robertson LW, Chynoweth DP (1975) Environment 17:25
50. Watanabe I, Kashimoto T, Tatsukawa R (1986) Bull Environ Contam Toxicol 36:839
51. Sellström U, Söderström G, Wit C de, Tysklind M (1998) Organohalogen Compounds 35:447
52. Sellström U, Kierkegaard A, Wit C de, Jansson B (1998) Environ Toxicol Chem 17:1065
53. Watanabe I, Kawano M, Tatsukawa R (1994) Organohalogen Compounds 19:235
54. Boer J de, Wester PG, Pastor I Rodriguez D, Lewis WE, Boon JP (1998) Organohalogen Compounds 35:383
55. Stratton CL, Whitlock SA (1979) A survey of polybrominated biphenyls (PBBs) neat sites of manufacture and use in Northwestern New Jersey. Washington, DC, US Environmental Protection Agency (EPA 560/13–79–002)
56. IARC (1978) Polychlorinated biphenyls and polybrominated biphenyls. Lyon, France, International Agency for Research on Cancer, 140 pp (IARC Monographs on the Evaluation of the Carcinogenic Risk of Chemicals to Humans, Vol 18)

57. Shah JD (1991) Environ Hlth Persp 23:27
58. Jacobs LW, Chou SF, Tiedje JM (1978) Environ Health Persp 23:1
59. Jansson B, Andersson R, Asplund L, Litzen K, Nylund K, Sellström U, Uvemo U-B, Wahlberg C, Wideqvist U,Odsjö T, Olsson M (1993) Environ Toxicol Chem 12:1163
60. Watanabe I, Kashimoto T, Tatsukawa R (1995) Organohalogen Compounds 24:337
61. Watanabe I, Kawano M, Tatsukawa R (1993) Consumption trend and environmental research on brominated flame retardants in Japan and the formation of polyhalogenated dibenzofurans at the metal reclamation factory. Document distributed at the OECD Workshop on the Risk Reduction of Brominated Flame Retardants, Neuchatel, Switzerland
62. Nylund K, Asplund L, Jansson B, Jonsson P, Litzen K, Sellström U (1992) Chemosphere 24:1721
63. Allchin, CR, Law, RJ and Morris, S (1999) Polybrominated diphenylethers (PBDEs) in sediments and biota downstream of potential sources in the UK. Environ Pollut, in press
64. Anon (1997) Report of the results of the one-off survey DIFFCHEM. Oslo and Paris Commissions, Working Group on concentrations, trends and effects of substances in the marine environment (SIME), London, UK
65. Watanabe I, Kashimoto T, Tatsukawa R (1987) Chemosphere 16:10
66. Boon JP, Meer J van der, Allchin, CR, Law RJ, Klungsøyr J, Leonards PEG, McKenzie C, Wells DE (1997) Arch Environ Contam Toxicol 33:298
67. Norén K, Meironyté D (1998) Organohalogen Compounds 38, 1–4
68. Meironyté D, Bergman Å, Norén K (1998) Organohalogen Compounds 35, 387
69. Darnerud PO, Atuma S, Aune M, Cnattingius S, Wernroth M-L, Wicklund-Glynn A (1998) Organohalogen Compounds 35, 411
70. Boer J de, Robertson LW, Dettmer F, Wichmann H, Bahadir, M (1998) Organohalogen Compounds 35, 407
71. Lindstrøm G, Hardell L, Bavel B van, Wingfors H, Sundelin E, Liljegren G, Linholm P (1998) Organohalogen Compounds 35:431
72. Jackson TF, Halbert FL (1974) J Am Vet Med Assoc 165:436
73. Robertson LW, Chynoweth DP (1975) Environment 17:25
74. Poland A, Palen D, Glover E (1982) Nature 300:271
75. Carlson GP (1980a) Toxicol Lett 5:19
76. Koster P, Debets FMH, Strik JJTWA (1980) Bull Environ Contam Toxicol 25:313
77. Render JA, Aust SD, Sleight SD (1982) Toxicol Appl Pharmacol 62:428
78. Jensen RK, Sleight SD, Aust SD, Goodman JI, Trosko JE (1983) Toxicol Appl Pharmacol 71:163
79. Hallgren S, Darnerud PO (1998) Organohalogen Compounds 35:391
80. Kavanagh TJ, Rubenstein C, Lui PL, Chang CC, Trosko JE, Sleight SD (1985) Toxicol Appl Pharmacol 79:91
81. Silberhorn EM, Glauert HP, Robertson LW (1990) Crit Rev Toxicol 20:440
82. Momma J (1986) Jpn Pharmacol Ther 14:11
83. Huff JE, Eustis SL, Haseman JK (1989) Cancer Metastasis 8:1
84. NTP (1986) Toxicology and carcinogenesis studies of decabromodiphenyl oxide (CAS No. 1163–19–5) in F344/N rats and B6C3F1 mice (feed studies). Research Triangle Park, North Carolina, US Department of Health and Human Services, National Toxicology Programm (NTP Technical Report Series No. 309)
85. Colborn T, Saal FS van, Soto AM (1993). Environ Health Perspect 101:378
86. Aust SD, Millis CD, Holcomb L (1987) Arch Toxicol 60:229
87. Kholkute SD, Rodriquez J, Dukelow WR (1994) Arch Environ Contam Toxicol 26:208
88. Norris JM, Kociba, RJ, Schwets BA, Rose JQ, Humiston CG, Jewett, GL, Gehring PJ, Mailhes JB (1975) Environ Hlth Persp 11:153
89. Breslin WJ, Kirk HD, Zimmer MA (1989) Fundam Appl Toxicol 12:151
90. Holm G, Norrgren L, Andersson T, Thurén A (1993) Aquat Toxicol 27:33
91. Brouwer A (1998) Organohalogen Compounds 37:225
92. Sinjari T, Darnerud PO, Hallgren S (1998) Organohalogen Compounds 37:241

93. Klasson-Wehler E, Jakobsson E, Örn U (1996) Organohalogen Compounds 28:495
94. Great Lakes Chemical Corporation (1987) Toxicity data of octabromo-diphenyloxide (DE-79). West Lafayette, Indiana, Great Lakes Chemical Corporation (Unpublished data submitted to WHO by BFRIP)
95. Darnerud PO, Thuvander A(1998) Organohalogen Compounds 35:415
96. Darnerud PO, Atuma S, Aune M, Cnattingius S, Wernroth, M-L, Wicklund-Glynn A (1998) Organohalogen Compounds 35:411
97. Di Carlo FJ, Seifter J, DeCarlo VJ (1978) Environ Health Persp 23:351
98. Kaart KS, Kokk KY (1987) Ind Lab 53:289
99. Andersson Ö, Blomkist G (1981) Chemosphere 10:1051
100. Cramer PH, Stanley JS, Thornburg KR (1990a) Mass spectral confimation of chlorinated and brominated diphenyl ether in human adipose tissues. Kansas City, Missouri, Midwest Research Institute (Prepared for the US Environmental Protection Agency, Wasington) (US NTIS report PB91-159699)
101. Kok JJ de, Kok A de, Brinkman UATh (1979) J Chromatogr 171:269
102. Watanabe I, Tatsukawa R (1990) Anthropogenic brominated aromatics in the Japanese environment. Proceedings Workshop on Brominated Flame Retardants, Skokloster, 24-26 October 1989, KEMI, National Chemicals Directorate, Sweden, pp 63-71
103. Boer J de (1989) Chemosphere 18:2131
104. Sellström U, Jansson B, Jonsson P, Nylund K, Odsjö T, Olsson M (1990) Organohalogen Compounds 2:357
105. Haglund P, Zook DR, Buser H-R, Hu J (1997) Environ Sci Technol 31, 3281
106. Lindström G, Wingfors H, Dam M, Bavel B van (1999) Identification of 19 poly-brominated diphenyl ethers, PBDEs, in long-finned pilot whale (*Globicephala melas*) from the Atlantic. Arch Environ Contam Toxicol 36:355
107. Boer J de (1990) Organohalogen compounds 2:315
108. Jansson, B, Asplund L, Olsson M (1987) Chemosphere 16, 2353-2359

CHAPTER 5

Polychlorinated Naphthalenes (PCNs)

Eva Jakobsson · Lillemor Asplund

E. Jakobsson (e-mail: eva.jakobsson@mk.su.se)
Department of Environmental Chemistry. Stockholm University, SE-106 91 Stockholm, Sweden
L. Asplund (e-mail: lillemor.asplund@itm.su.se)
Institute of Applied Environmental Research (ITM), Laboratory for Analytical Environmental Chemistry, Stockholm University, SE-106 91 Stockholm, Sweden

Polychlorinated naphthalenes (PCNs) have been commercially produced and used mainly in electrical devices, but also for impregnation of wood, paper and textiles to attain waterproofness, flame resistance and protection against insects, molds and fungi. Today, the PCNs are widespread in the environment and are to be regarded as an environmental problem. Generally, the levels are lower compared to e.g. polychlorinated biphenyls (PCBs), but high levels have been observed near point sources such as manufacturers of chlorine/soda, magnesium, copper and aluminum. PCB products and incineration processes are also sources of PCN releases. The present review summarizes data on physical properties such as melting points, solubility and log K_{ow}. Chemical reactivity and methods used for synthesis of individual congeners are briefly discussed. The analytical methods used for PCN analysis are presented as well as sources and environmental levels. Finally, the biological fate and toxicological data available are summarized.

Keywords. Polychlorinated naphthalenes, Physical properties, Analytical methods, Levels, Toxicological behavior

The Handbook of Environmental Chemistry Vol. 3 Part K
New Types of Persistent Halogenated Compounds
(ed. by J. Paasivirta)
© Springer-Verlag Berlin Heidelberg 2000

List of Abbreviations

Ah	aryl hydrocarbon
AHH	aryl hydrocarbon hydroxylase
BCF	bioconcentration factor
BMF	biomagnification factor
CZP	3-(N-carbazolyl)propylsilylated silica
ECNI-MS	electron-capture negative ionization-mass spectrometry
EI	electron ionization
EROD	7-ethoxyresorufin-O-deethylase
GC	gas chromatography
GPC	gel permeation chromatography
HPLC	high-performance liquid chromatography
LR-MS	low-resolution mass spectrometry
MFO	mixed function oxygenases
MS	mass spectrometry

OHS	organohalogen substances
PCBs	polychlorinated biphenyls
PCDDs	polychlorinated dibenzo-p-dioxins
PCDFs	polychlorinated dibenzofurans
PCNs	polychlorinated naphthalenes
PCTs	polychlorinated terphenyls
PYE	2-(1-pyrenyl)ethyldimethylsilylated silica
RP	reversed-phase
SIM	selected ion monitoring
TCDD	2,3,7,8-tetrachlorodibenzo-p-dioxin
TEFs	toxic equivalency factors
TEQ	toxic equivalents

1
Introduction

1.1
General Remarks

Polychlorinated naphthalenes (PCNs) have been commercially produced and used in a variety of applications due to their dielectric, water-repellent, flame-retardant, and fungus-resistant properties in combination with high stability and compatibility with other materials. The industrial production of PCNs dates back to the beginning of the 20th century. During the period 1930–1950, PCNs were most extensively used, especially in cable and capacitor production [1, 2]. Today, most manufacturers of PCNs have stopped their production. Official regulations relating to PCNs differ however considerably between countries. In Japan, PCNs are prohibited entirely [2] while most countries have no regulations for the use of PCNs. The reasons for the declined production of PCNs are most probably their toxic effects and bioaccumulation in the environment. Also, new materials such as polyesters and polycarbonates have been introduced as substitutes for chlorinated naphthalenes in the capacitor and cable industries [2].

Besides the industrial production of PCNs there is also a release of PCNs to the environment via polychlorinated biphenyl (PCB) commercial products, in which PCNs are present as minor contaminants [3]. PCNs are also formed in various incineration processes [4, 5] and industrial processes such as the production of magnesium [6], copper [7] and in chloroalkali production [8, 9]. PCNs are currently widespread in the environment and are to be regarded as an environmental problem [4]. In general, PCNs are present in biota at ng g^{-1} levels (on lipid weight basis). High levels (2.4 µg g^{-1}) have been reported in e.g., white-tailed sea eagle from Poland [5].

PCNs consist of congeners with widely differing toxicological behavior. Some congeners have been identified as highly persistent and bioaccumulating, whereas most congeners seem to be quite readily metabolized. Studies of the binding of PCNs to the Ah (aryl hydrocarbon) receptor and induction of the enzymes AHH/EROD (aryl hydrocarbon hydroxylase/7-ethoxyresorufin-O-deethylase) have for some PCNs resulted in TCDD (2,3,7,8-tetrachlorodibenzo-p-dioxin) toxic equivalency factors (TEFs) comparable with those of the mono-*ortho*-PCBs, i.e. these PCNs are three orders of magnitudes less toxic compared to TCDD [10–12].

Selective retention of two hexachloronaphthalenes (1,2,3,4,6,7- and 1,2,3,5,6,7-hexaCN) in the liver has been reported in rats [13, 14] and in harbor porpoises *(Phocoena phocoena)* [15]. In humans, the PCN levels are slightly higher in the liver compared to the adipose tissue [16]. There are relatively few studies on the metabolism of PCNs. Generally, congeners of higher chlorination degree seem to be more slowly metabolized than lower chlorinated congeners [17–19]. However, congeners with chlorine substituents in the peri-positions, such as for example octaCN, are probably more rapidly metabolized, cf. Sect. 2.2. Metabolism occurs mainly by hydroxylation [18, 20–24]. Covalent binding to macromolecules (including DNA) in the liver, kidney and lung has been reported in rats dosed with ^{14}C-PCNs [25, 26].

Chloracne and liver damage have been recognized as the most common health effects observed after exposure to PCNs. The penta- and hexaCNs seem to be the most toxic congeners in this respect. In the 1930s and 1940s, there were some large outbreaks of chloracne among workers producing PCNs and there were also some fatalities [1]. Currently, however, there is little knowledge of the effects the PCNs have on wildlife and humans at the levels generally found in biota.

Brinkman and Reymer (1976) have written an extensive review on PCNs, covering production and uses, synthesis of individual congeners, physical properties, UV and IR spectral data, analytical methods for PCNs as well as toxicity and metabolism data published before 1976 [1]. Kover [27] and later Crookes and Howe [4] published comprehensive environmental hazard assessment reports on PCNs [4, 27]. Toxicity data have also been summarized in a report to the US EPA Federal Register [28].

1.2
Commercial Production and Uses

PCNs are commercially produced via chlorination of molten naphthalene with chlorine gas at a temperature slightly above the melting point of the desired product (the melting point increases with increasing chlorine content) in the presence of catalytic amounts of iron(III) or antimony(V) chloride [1]. The crude product is then treated with sodium carbonate or sodium hydroxide, fractionated under reduced pressure and purified with activated clay [4, 29]. The final products are generally complex mixtures of chlorinated naphthalenes, ranging from liquids to waxes with high melting points. The major commercial PCN products are Halowaxes, Nibren Waxes, Seekay Waxes, Clonacire Waxes and Cerifal Materials, cf. Table 1. A gas chromatogram of Halowax 1014 is shown in Fig. 1.

The synthesis of PCNs was first published in 1833 [30]. It was, however, not until the beginning of the 20th century that the production and use of PCNs

Table 1. Commercial PCN products, approximate composition and manufacturers[a]

Trade name		Approximate composition	Manufacturer
Halowax	1031	mono-diCN (22 % Cl)	Koppers Co Pittsburg, PA, USA
	1000	mono-diCN (26 % Cl)	
	1001	di-penta (50 % Cl)	
	1099	di-pentaCN (52 % Cl)	
	1013	tri-pentaCN (56 % Cl)	
	1014	tetra-hexaCN (62 % Cl)	
	1051	hepta-octaCN (70 % Cl)	
Basileum	SP-70	mono-diCN (80 % PCN)	Desowag-Bayer, Germany
Nibren wax	D88	(50 % Cl, estimated from melting point)	Bayer Leverkusen, Germany, formerly I. G. Farbenindustrie
	D116 N	(50 % Cl, estimated from melting point)	
	D130	(60 % Cl, estimated from melting point)	
Seekay wax	R68	(46.5 % Cl)	ICI Runcorn, Great Britain
	R93	(50 % Cl)	
	R123	(56.6 % Cl)	
	R700	(43 % Cl)	
	RC93	(50 % Cl)	
	RC123	(56.5 % Cl)	
Clonacire wax	95	(50 % Cl, estimated from melting point)	Prodelec, Paris, France
	115		
	130		

[a] The data are from references 1 and 4.

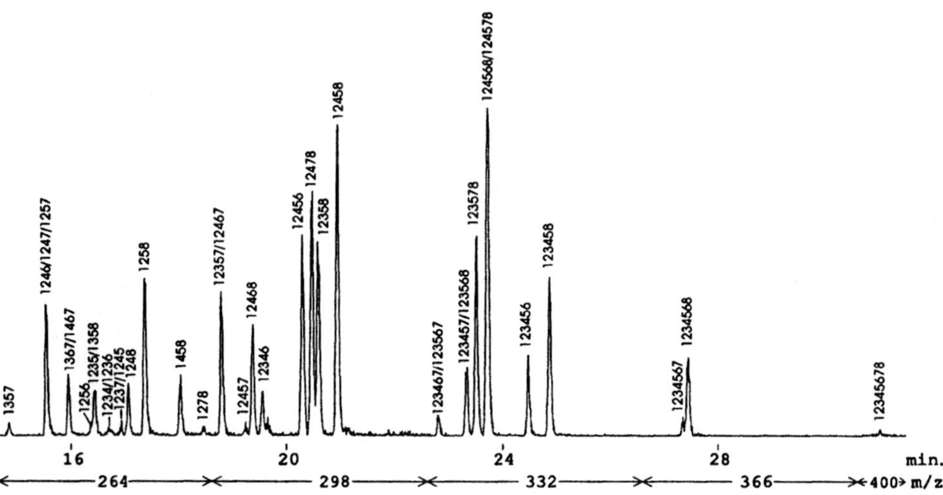

Fig. 1. GC-MS (EI) mass fragmentograms of the molecular ions for tetraCN to octaCN in Halowax 1014 on a phenyl-methylpolysiloxane (5%) column

became important. PCNs have been used mainly in the electrical industry as separators in storage batteries, capacitor impregnates, as binders for electrical grade ceramics and sintered metals, and in cable covering compositions. PCNs have also been used for impregnation of wood, paper and textiles to attain waterproofness, flame resistance and protection against insects, molds and fungi. Furthermore, PCNs have been used as additives in oils for lubrication in gear and cutting oils, in lacquers and underwater paints and as raw material for dyes [1].

The information of the worldwide production of PCNs is scarce and seems highly uncertain. It has been estimated that the annual total world production was about 9000 tons in the 1920s. In the United States, 3200 tons were produced in 1956 which had decreased to 320 tons/year in 1978 and was stopped in 1980 [4, 28]. Today, the production is most likely minimal, since most of the well-known manufacturers have discontinued production of PCNs. However, there are indications that PCNs have still been used in the 1990s. For example, capacitors from 1991 have been shown to contain PCNs [31] and as late as in 1989, workers were exposed to PCNs in a German factory [32].

Most probably, it is mainly lower chlorinated naphthalenes that are still used today. These have been used, for example, as glues for plywood in Finnish plywood plants [33]. Lower chlorinated naphthalenes were used in a product for wood preservation against bugs in Sweden until 1987 when it was stopped, as it was shown to contain higher chlorinated naphthalenes than declared [Asplund, unpublished data]. From our experience, chlorinated naphthalenes mixtures of low chlorination degree also contain small amounts of more highly chlorinated congeners [Asplund and Jakobsson, unpublished].

2
General Physical and Chemical Properties

2.1
Nomenclature and Structure

Naphthalene is an aromatic compound that may be substituted with up to eight substituents. The numbering of the naphthalene nucleus is shown in Fig. 2. In older literature the 1- and 2-positions were named α- and β-position, respectively, and this nomenclature is still used to some extent.

There are theoretically 75 chlorinated naphthalenes (CNs), i.e. two monoCNs, 10 diCNs, 14 triCNs, 22 tetraCNs, 14 pentaCNs, 10 hexaCNs, two heptaCNs and one octaCN. In Table 2 all structures are listed. Numbering of the congeners is according to Wiedmann and Ballschmiter (1993) and Chemical Abstract Systematic (CAS) numbers are also given in the table [34].

Fig. 2. Numbering scheme of the naphthalene nucleus

Table 2. Physical properties and references to methods used for synthesis of individual PCN congeners

PCN No.	Substitution	CAS registry No.	Range of melting points (°C)[a]	Water solubility (ug/l)[b]	Log K_{ow}[b]	Reference to synthesis and melting points
MonoCNs		25586–43–0				
1	1-	90–13–1	−17 to −2.3	2870	3.90	[35[d], 36–41]
2	2-	91–58–7	56–61	924	3.98, 4.07[c]	[41, 42[d], 43–58]
DiCNs		28699–88–9				
3	1,2-	2050–69–3	34–37	137	4.42	[59[d]–67]
4	1,3-	2198–75–6	56–62			[59[d], 61 68–70]
5	1,4-	1825–31–6	56–72	314, 309[c]	4.66, 4.88[c]	[65, 66, 71[d]–80]
6	1,5-	1825–30–5	104–107	396	4.67	[61, 65, 69, 73, 80, 81[d]–85]
7	1,6-	2050–72–8	47.5–49			[47, 69, 73, 86[d]–90]
8	1,7-	2050–73–9	59–64			[69, 82, 91[d]–96]
9	1,8-	2050–74–0	83–89.5	590, 309[c]	4.19, 4.41[c]	[47, 61, 65, 68, 82, 91, 97[d]–99]
10	2,3-	2050–75–1	119.5–121.5	862, 85[c]	4.51, 4.71[c]	[47, 100[d]-102]
11	2,6-	2065–70–5	134.5–141			[65, 71[d], 103–110]
12	2,7-	2198–77–8	90–116	240[c]	4.81[c]	[61, 82, 95[d], 110–112]
TriCNs		1321–65–9				
13	1,2,3-	50402–52–3	75–84			[65, 68, 74[d], 113–115]
14	1,2,4-	50402–51–2	88–92			[68, 74, 100[d], 115, 116]
15	1,2,5-	55720–33–7	74–79			[74[d], 95, 97, 117]
16	1,2,6-	51570–44–6	90–92.5			[74, 91, 95[d], 97, 118]
17	1,2,7-	55720–34–8	88			[91, 95[d]]
18	1,2,8-	55720–35–9	83			[91, 97[d]]
19	1,3,5-	51570–43–5	94–103			[74, 100[d], 119–122]
20	1,3,6-	55720–36–0	80.5–81			[74[d], 88]
21	1,3,7-	55720–37–1	111–113	64.4, 65[c]	5.35, 5.59[c]	[74, 96, 123[d]–126]
22	1,3,8-	55720–38–2	89.5			[91, 123[d]]
23	1,4,5-	2437–55–0	130–133			[74[d], 76, 77, 98, 121, 127]
24	1,4,6-	2437–54–9	64–68			[74, 96, 100[d], 122, 125]
25	1,6,7-	55720–39–3	109–109.5			[74, 100[d]]
26	2,3,6-	55720–40–6	90.5–91, 145	16.7	5.12	[74[d], 100, 108]

Table 2. (continued)

PCN No.	Substitution	CAS registry No.	Range of melting points (°C)[a]	Water solubility (ug/l)[b]	Log K_{ow}[b]	Reference to synthesis and melting points
TetraCNs		1335–88–2				
27	1,2,3,4-	20020–02–4	54–55, 186–201	4.2	5.75, 5.50[c]	[74[d], 102, 128[d]–144]
28	1,2,3,5-	53555–63–8	138–142	3.7	5.77	[74[d], 85, 129, 145, 146[e]]
29	1,2,3,6-	149864–78–8				
30	1,2,3,7-	55720–41–7	115			[74[d]]
31	1,2,3,8-	149864–81–3	128			[e]
32	1,2,4,5-	6733–54–6	139			[147]
33	1,2,4,6-	51570–45–7	111			[74[d], 148[d]]
34	1,2,4,7-	67922–21–8	143–144			[74[d], 149[d,e]]
35	1,2,4,8-	6529–87–9	98			[147]
36	1,2,5,6-	67922–22–9	159.5–164			[74[d], 150[d]–151[e]]
37	1,2,5,7-	67922–23–0	114			[74[d], 148[d]]
38	1,2,5,8-	149864–80–2	180–181			[e]
39	1,2,6,7-	149864–79–9				
40	1,2,6,8-	67922–24–1	125–127			[152[d]]
41	1,2,7,8-	149864–82–4				
42	1,3,5,7-	53555–64–9	178–181	4.0, 4.3	6.19, 6.38[c]	[74[d], 152–153]
43	1,3,5,8-	31604–28–1	130–131	8.2, 8.3	5.76, 5.96[c]	[74[d], 113]
44	1,3,6,7-	55720–42–8	120			74[d]]
45	1,3,6,8-	150224–15–0				
46	1,4,5,8-	3432–57–3	180–190			[145[d], 153, 154[d], 155[d], 156–158]
47	1,4,6,7-	55720–43–9	139	8.1	5.81	[74[d]]
48	2,3,6,7-	34588–40–4	200–205			[153[d], 159[d]–162]
PentaCNs		1321–64–8				
49	1,2,3,4,5-	67922–25–2	168.5, 214–215			[128[d], 163[d], 164[d]]
50	1,2,3,4,6-	67922–26–3	144.5–151			[74[d], 128[d], 165]
51	1,2,3,5,6-	150224–18–3				
52	1,2,3,5,7-	53555–65–0	171–172			[74[d,e]]

53	1,2,3,5,8-	150224–24–1	174–176			[e]
54	1,2,3,6,7-	150224–16–1				[159d]
55	1,2,3,6,8-	150224–23–0	112–114			[e]
56	1,2,3,7,8-	150205–21–3	115–117			[166d,e]
57	1,2,4,5,6-	150224–20–7	136–139			[e]
58	1,2,4,5,7-	150224–19–4				
59	1,2,4,5,8-	150224–25–2	150–151			[e]
60	1,2,4,6,7-	150224–17–2				
61	1,2,4,6,8-	150224–22–9				
62	1,2,4,7,8-	150224–21–8				
HexaCNs		1335–87–1				
63	1,2,3,4,5,6-	58877–88–6	132–134			[18d,e]
64	1,2,3,4,5,7-	67922–27–4	164–166			[128d]
65	1,2,3,4,5,8-	103426–93–3	163.5–164.5			[128d]
66	1,2,3,4,6,7-	103426–96–6	205–206			[128d, 167d]
67	1,2,3,5,6,7-	103426–97–7	234–235			[159d, 167d]
68	1,2,3,5,6,8-	103426–95–5	153–154			[159d, e]
69	1,2,3,5,7,8-	103426–94–4	148–149			[e]
70	1,2,3,6,7,8-	17062–87–2	158–160			[166d, 167d, 168d, 169d, e]
71	1,2,4,5,6,8-	90948–28–0	175–177			[153d, 166d, 170d, e]
72	1,2,4,5,7,8-	103426–92–2	136–139.5			[170d, e]
HeptaCNs		32241–08–0				
73	1,2,3,4,5,6,7-	58863–14–2	160–184.5			[68, 166d, 171d, 172d, 173d, 175, e]
74	1,2,3,4,5,6,8-	58863–15–3	110–110.5, 176–194			[166d, 171d, 172d, 174, 176d, e]
OctaCN						
75	1,2,3,4,5,6,7,8-	2234–13–1	185–203	0.08	6.42	[156, 167d, 171, 177–183e]

[a] The range of melting points reported in the literature is given.
[b] The data are from reference 187 except where indicated.
[c] The data are from reference 188.
[d] Reference that describes the synthesis.
[e] Nikiforov V, unpublished data.

2.2
Geometry

In contrast to benzene, in which all bonds have the same lengths (1.40 Å), the (C_1)-(C_2) bond in naphthalene is somewhat shorter (1.364 Å) than the (C_2)-(C_3) bond (1.415 Å) [184]. This is due to an unequal distribution of the π-electrons over the molecule with a higher electron density over (C_1)-(C_2).

The naphthalene molecule is planar. The distance between substituents in the 1- and the 8-positions (and in the equivalent 4- and 5-positions) is shorter compared to substituents in e.g., the 1- and 2-positions. The 1-/8- and 4-/5-positions are often called the *peri*-positions after the Greek word "*peri*" which means "near". Bulky substituents in these positions will sterically interact and cause a distortion of the molecule. For example, in the crystal structure of 1,4,5,8-tetraCN (CN-46) the chlorine atoms are positioned slightly above and under the plane of the naphthalene ring [157]. In octaCN (CN-75) the aromatic ring system is also remarkably distorted [185, 186]. These distortions, as well as the unequal distribution of the π-electrons, most probably influence the physical properties and enhance the chemical as well as the metabolic reactivity of the compound.

2.3
Solubility, Melting Points, and UV Absorption

Chlorinated naphthalenes are lipophilic compounds with log K_{ow} values ranging from 3.90 for 1-monoCN up to 6.42 for octaCN, cf. Table 2 [187, 188]. Most PCNs are readily soluble in most organic solvents, e.g., diethyl ether, dichloromethane, hexane, toluene and isooctane. The solubility in more polar solvents such as methanol is low (ng μL^{-1}), but high enough for e.g., liquid chromatography on analytical columns [189]. Mono- and diCNs are slightly soluble in water (0.1–3 mg L^{-1}) while the higher chlorinated naphthalenes have very low solubility in water (0.08 $\mu g\ L^{-1}$) [187, 188].

Except for 1-monoCN, which is a liquid at room temperature, the chlorinated naphthalenes are crystalline compounds. The melting point increases with increasing chlorine substitution with considerable variation within each homologue group, cf. Table 2. The commercial PCN products, which occur as complex mixtures of isomers and homologues, are generally waxes with high compatibility with other materials. The solid products melt to liquids of extremely low viscosity [1].

PCNs have strong absorbance maxima between 220 and 275 nm and weaker maxima between 275 and 345 nm. The absorption maxima are shifted towards longer wavelengths as the chlorine degree increases. Brinkman and Reymer [1976] have reported detailed information on UV spectra of PCNs [1].

2.4
Chemical Reactivity and Synthesis

The resonance energy for naphthalene (61 kcal mol^{-1}), i. e. the energy difference between the hypothetical compound with localized π-bonds and the authentic aromatic compound, is not twice the resonance energy of benzene (36 kcal mol^{-1}), which might be expected, but 11 kcal mol^{-1} lower. Naphthalene loses its aromaticity in one ring in the rate determining step in e. g. electrophilic aromatic substitutions but will still have one intact benzenoid ring. The activation energy for the rate-determining step is 25 kcal mol^{-1} for naphthalene compared to 36 kcal mol^{-1} for benzene. Consequently, naphthalene is more reactive than benzene toward e.g., electrophilic substitution, oxidation and reduction [190–192]. However, generally the chlorinated naphthalenes still possess a high degree of chemical stability. The stability increases with the number of chlorine substituents [1]. They are resistant to concentrated acids, except concentrated nitric acid, and generally also to bases [Jakobsson, unpublished].

Electrophilic and nucleophilic substitution as well as radical attack occur predominantly in the α-positions [193]. Consequently, since PCN is commercially produced by chlorination of naphthalene the CNs formed are predominately chlorinated in the α-positions. For example, among the hexaCNs in Halowax 1014 the major compounds are 1,2,4,5,6,8-hexaCN (CN-71) and 1,2,4,5,7,8-hexaCN (CN-72) while the minor compounds are 1,2,3,4,6,7-hexaCN (CN-66) and 1,2,3,5,6,7-hexaCN (CN-67) [194, 195], cf. Fig. 1.

References to the synthesis of the individual PCNs are given in Table 2. The preparation of individual chlorinated naphthalenes dates back to the nineteenth century when several of the diCNs and triCNs were prepared [123]. A large number of the tri- and tetraCNs were synthesized in the 1930s and 1940s [74]. The structures of several of the chlorinated naphthalenes synthesized at this early stage, were not confirmed but were predicted based on directing properties of substituents in the starting material. An impressive number of these early determined structures have later been confirmed. In 1976, Brinkman and Reymer [1] summarized the synthesis of different CNs and found that 55 of the 75 theoretical PCN congeners had been described. Since then, the synthesis of several CNs have been reported [128, 159, 166, 167, 170, 172]. The starting materials have mainly been chloronitronaphthalenes [128, 159, 166, 170], chloronaphthalene sulfonyl chlorides [128], methyl naphthalenes [172] or aminochloronaphthalenes [153, 166]. A *peri*-bridged 1,8-diselenole or 1,8-dithiole derivative was used for the synthesis of 2,3,6,7-tetraCN (CN-48) [159] and 1,2,3,6,7,8-hexaCN (CN-70) [166–169]. Dechlorination of octaCN was used to prepare 1,2,3,4,5,6,7-heptaCN (CN-73) [173], 1,2,3,4,6,7-hexaCN (CN-66) and 1,2,3,5,6,7-hexaCN (CN-67) [166, 167]. The synthesis of eight ^{13}C10-labelled PCNs, namely 1,2,3,4-tetraCN (CN-27), 1,3,5,7-tetraCN (CN-42), 1,2,3,5,7-pentaCN (CN-52), 1,2,3,4,5,7-hexaCN (CN-64), 1,2,3,4,5,8-hexaCN (CN-65), 1,2,3,5,6,7-hexaCN (CN-67), 1,2,3,4,5,6,7-heptaCN (CN-73) and octaCN (CN-75) have been reported [196]. OctaCN (CN-75) and a mixture of 1,2,3,4,6,7- and 1,2,3,5,6,7-hexaCN (CN-66 and CN-67) have also been prepared with ^{13}C8 labelling [197].

3
Analytical Methods

3.1
General Considerations

The analysis of organohalogen substances (OHS) in biological samples includes extraction of the OHS and the lipids from the matrix, removal of the lipids, separation of various classes of OHS and finally, detection and quantification. As pointed out above, PCNs have physical and chemical properties similar to the PCBs and the methods used to analyze PCBs and PCNs are therefore rather similar.

After extraction, sulfuric acid may be used to degrade the lipids in the sample. The PCNs should thereafter be separated from other OHS such as PCBs, which are generally present at much higher levels, otherwise these will interfere with the PCNs in most chromatographic systems. One early approach to solve this problem was to use different dechlorination [198, 199] or perchlorination methods [200, 201] to convert all PCNs into naphthalene or octaCN, respectively, which could then more easily be quantified. A large drawback with these methods was that information on concentrations of individual PCN congeners was lost. More recently, liquid chromatographic methods have been developed to isolate PCNs from other OHS.

3.2
Extraction

Traditional extraction methods which extract OHS from biological matrixes with high reproducibility can also be used for PCN analysis [202, 203]. Since PCNs are planar compounds and may adsorb rather strongly to carbon particles, methods that are especially effective at extracting planar OHS from the matrix are recommended, e.g., soxhlet extraction with toluene, for samples with a high carbon content [204].

3.3
Liquid Chromatography

3.3.1
Gel Permeation Chromatography

Gel permeation chromatography (GPC) on styrene-divinylbenzene copolymer (PLgel, Polymer Laboratories) columns has been found to separate PCN from e.g., PCB. The gel has a defined pore size of 50 Å and may be used also at high pressures. The method was first applied to the isolation of PCNs in commercial PCB products [3].

3.3.2
Charcoal

Charcoal has frequently been used to separate planar and non-planar substances and also to separate PCNs and PCBs [205–207]. Planar compounds form electron donor-acceptor complexes with the aromatic graphite structure and are in this way retained on the charcoal. Generally, the sample is dissolved in a non-aromatic solvent and transferred to an open column filled with charcoal. Using hexane as the mobile phase, non-planar compounds such as most of the PCBs will elute from the column while PCNs and other planar compounds are retained on the column. The mobile phase is then changed to an aromatic solvent such as toluene, which liberates the planar compounds from the charcoal. Different types of charcoal mixed with Celite, Chromosorb or dispersed on polyurethane foam or glass wool have been used in open column chromatography [208–210]. Activated carbon (Amoco PX-21) dispersed on a LiChrosher RP-18 phase has been used in high-performance liquid chromatography (HPLC) for PCN analysis [211, 212].

The use of an aromatic solvent makes it difficult to utilize UV detection. This is of course unfortunate since a continuous detection of the compounds enables a closer observation of the analysis and helps in the determination of cutting points between fractions.

3.3.3
PYE Column

During the 1990s, the use of HPLC and 2-(1-pyrenyl)ethyldimethylsilylated silica (PYE) to separate planar compounds has become more and more common, cf. Fig. 3. For example, PCNs and non-*ortho*-PCBs have been separated from the bulk of PCBs using the PYE column with n-hexane as the mobile phase [8, 213]. The PYE phase and similar electron-donor-bonded phases such as 3-(*N*-carbazolyl)propyl-silyl (CZP) are hitherto the only ones reported to separate 1,2,3,4,6,7-hexaCN (CN-66) and 1,2,3,5,6,7-hexaCN (CN-67) [14, 166, 167, 214, 215]. It is notable that the CZP phase separates 1,2,3,4,6,7-hexaCN (CN-66) and 1,2,3,5,6,7-hexaCN (CN-67) with the opposite elution order compared to the PYE phase [215].

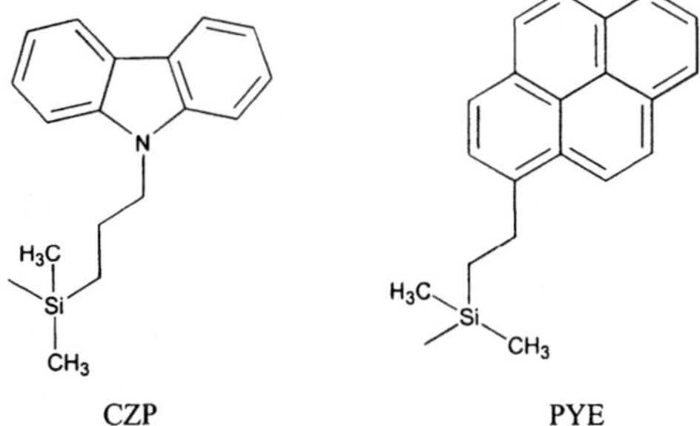

Fig. 3. Structures of the two LC phases 2-(1-pyrenyl)ethyldimethylsilyl (PYE) and 3-(N-carbazolyl)propylsilyl (CZP) [215]

3.3.4
Straight-Phase and Reversed-Phase Liquid Chromatography

Brinkman and Reymer (1976) have summarized the behavior of PCNs in straight-phase (silica gel) and reversed-phase chromatography [1]. Several CNs, e.g., the two heptaCNs and several of the hexaCNs, have been separated on reversed-phase RP-8 and RP-18 columns [1, 166]

3.4
Detection/Quantification

3.4.1
Gas chromatography

Järnberg et al. (1994) tested several GC stationary phases for separation of the individual CN congeners in the Halowax mixtures [194]. On phenyl-methyl-polysiloxane (5%) columns, frequently used in PCN analysis, the CN congeners elute in homologue groups. In general, more polar stationary phases give wider retention time intervals for each homologue group. On the most polar stationary phases the different chlorination degrees overlap each other and the elution order is sometimes altered [189, 194]. To completely resolve the CN congeners, different types of gas chromatographic stationary phases need to be used. Using three types of cyclodextrine columns and a liquid crystalline column (SB-Smectic) eight of the ten hexaCN congeners have been separated [216]. However, 1,2,3,4,6,7-hexaCN (CN-66) and 1,2,3,5,6,7-hexaCN (CN-67) remained unresolved also on these phases.

The PCN profiles of Halowax and fly ash seen in gas chromatograms are very different, cf. Sect. 4.1. When the PCNs in an unknown sample are to be identified, it is advisable to compare the PCN profile in the unknown sample with chromatograms of both Halowax 1014 and fly ash.

3.4.2
Mass Spectrometry

In order to avoid interferences from coeluting OHS, PCNs are preferably analyzed by gas chromatography-mass spectrometry (GC-MS). The response in electron-capture negative ionization (ECNI-MS) is high for penta- up to octaCN, but varies considerably between congeners. Electron ionization (EI) is therefore the advantageous method when several PCNs are included in the analysis. For quantitative analysis, selected ion monitoring (SIM) detecting molecular ions should be used to obtain highest sensitivity. For samples with high levels of PCNs, low-resolution mass spectrometry (LR-MS) could be sufficient [8] but for low levels, or when high selectivity is necessary, high-resolution mass spectrometry (HR-MS) is the preferred method [217, 218].

There are currently no commercially available ^{13}C-labeled PCN congeners and therefore ^{13}C-labeled PCBs are frequently used as internal standards also for PCN analysis. It should however be noted that the major hexaCNs in technical products i.e. 1,2,4,5,6,8-hexaCN (CN-71) and 1,2,4,5,7,8-hexaCN (CN-72) (minor contaminants in biota) co-elute in GC with 3,3′,4,4′,5-pentaCB (CB-126). Due to the presence of ^{35}Cl and ^{37}Cl, the ions the hexaCN forms (m/z 332; 334; 336; 338) will interfere also in MS with the ^{13}C12-labelled CB-126 (m/z 336). This problem may be overcome by the use of a GC column with a high number of theoretical plates or by HR-MS with a resolution of at least 4000 [217]. There are also other PCBs as well as other substances e.g., polychlorinated dibenzofurans (PCDFs) and polychlorinated terphenyls (PCTs) that may interfere with PCN congeners in MS [217]. However, by carefully checking that the GC-retention times and the isotope ratios of the measured ions are in agreement with the theoretical values, this problem can be solved.

3.4.3
Quantification

A troublesome circumstance in PCN analysis has been the limited commercial availability of authentic standards. Quantification has to be performed with few individual congeners and should therefore preferably be performed with GC-MS and EI, which gives similar responses for all CN congeners of the same chlorination degree [8]. Wiedmann and Ballschmiter (1993) have suggested a method for quantification of PCNs based on only one or two CN congeners [34]. They found that the molar response for CNs in EI increased with increasing degree of chlorination. However, by normalizing the ion yield for the ionization cross section (Q) of the molecule, constant values for the molar mass to signal intensity were obtained [219]. The various methods influence the result of the quantification and this has to be considered when levels and profiles from different publications are compared.

4
Sources and Environmental Releases

4.1
CN Patterns from Various Sources

An important source of PCNs in the environment is probably leakage from landfills containing electrical equipment [4]. In addition to the leakage of commercially produced PCNs, there is also a release of PCNs to the environment through the release of PCBs [3, 220]. Several industrial processes such as waste incineration and other incineration processes [4, 5], chloroalkali processes [8, 9, 204, 223], production of magnesium [6], copper [7] and aluminum [222] have also been identified as sources of PCNs. Interestingly, the CN patterns found from industrial activities are considerably different from the patterns of CNs in commercial PCN products, cf. Fig. 4. This might be explained by either (i) degradation of less stable CN congeners or (ii) the formation of CNs via other mechanisms in these processes. Some of the major CN congeners in the out-puts from the industrial processes as well as in PCB products, are the most toxic PCNs with regard to dioxin-like toxicity. The PCN sources that have been identified are summarized below.

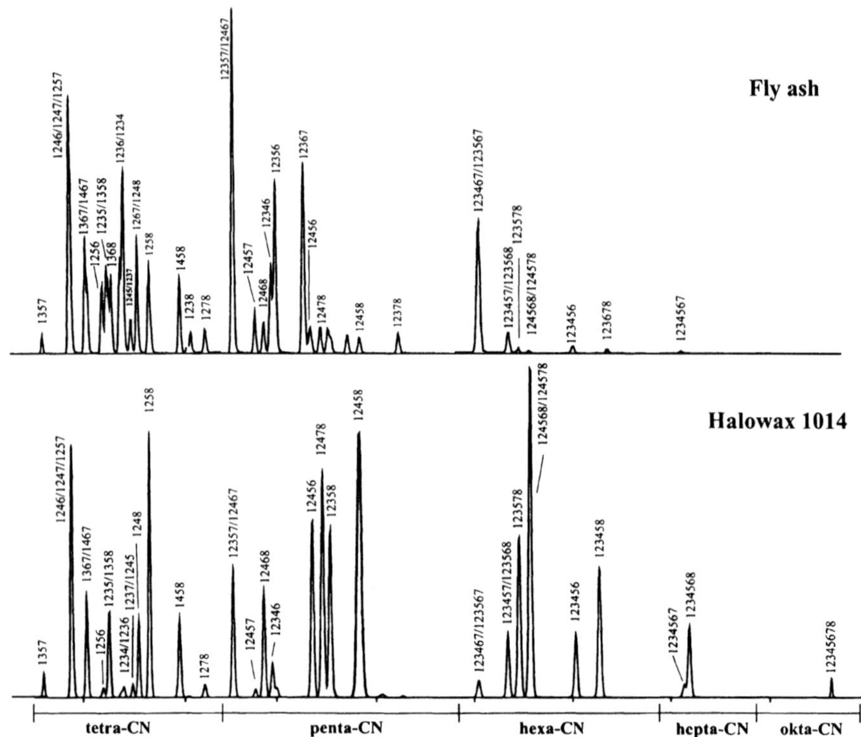

Fig. 4. GC-MS(EI) mass fragmentograms of the molecular ions for tetraCN to octaCN in a sample of fly ash (particle phase) collected in a municipal waste combustion incinerator and Halowax 1014 on a phenyl-methylpolysiloxane (5%) column [Dr. Rasha Ishaq, Stockholm University, Sweden]

4.2
Technical PCB Products

Haglund et al. (1993) quantified CN congeners in nine different commercial PCB products (Aroclor and Clophen products) [3]. The PCN levels in the products were found to vary between 1.8 and 870 µg g^{-1}. Congeners substituted with two to six chlorine atoms were the most abundant. The origin of the PCNs may be chlorination of naphthalene reported to be present in the biphenyl used for PCB production [221]. However, the CN congener pattern in the PCB products differs from the pattern found in e.g., the commercial PCN products Halowax 1014 [3]. The CN profile in the PCB products has similarities with the profile in fly ash samples, cf. Fig. 4.

High levels of PCBs and PCNs have been found in Sweden in Lake Järnsjön outside a paper mill handling recycled paper [8]. The source was suggested to be carbon-less copying paper, which previously has been shown to contain lower chlorinated PCBs.

4.3
Industrial Processes

Several industrial processes that release CNs have been identified. Baumann Ofstad et al. have reported that PCNs are released in the effluents from the production of magnesium [6]. In Germany, copper roasting processes have been shown to release PCNs [7] and emission of octaCN has been detected from an aluminum refinery [222].

Several reports have shown that high levels of CNs are present in graphite sludge from chloroalkali plants [8, 9, 223]. Järnberg et al. (1997) reported that mainly hexa- and heptaCNs are found in the sludge [204]. The dominating congener was 1,2,3,4,5,6,7-heptaCN (CN-73), followed by 1,2,3,4,5,6,8-heptaCN (CN-74), 1,2,3,4,6,7- and/or 1,2,3,5,6,7-hexaCN (CN-66/CN-67). Also this CN profile has similarities with the profile of fly ash samples, cf. Fig. 4. The highly chlorinated CNs in the graphite sludge might originate from the use of highly chlorinated PCB or PCN products as binders for the graphite electrodes [27]. Kannan et al. (1998) reported a slightly different pattern for PCNs in sediment collected near a chloroalkali plant [223]. Hepta- and octaCN dominated the CN congener pattern in this study with 1,2,3,4,5,6,8-heptaCN (CN-74) as the main congener. The authors suggested that the CNs are formed in the chloroalkali process.

4.4
Incineration

The presence of CNs in samples of fly ash from municipal waste incinerators has been reported in several investigations [34, 204, 224–229] as reviewed by Crookes and Howe (1993) [4] and Falandysz (1998) [5].

Mono- to octaCN are present in the fly ash with di- to pentaCNs being most abundant. The lower chlorinated CNs (mono- and triCNs) dominate in the flue gas. The CN patterns in both fly ash and flue gas are very different compared to the pattern in commercial PCN products e.g., Halowax 1014, cf. Fig. 4. The formation of CNs has also been shown in several laboratory combustion and incineration experiments [230–233].

5
Levels and Congener Patterns in the Abiotic Environment

5.1
General Remarks

PCNs are generally present at low levels (pg g^{-1}) in sediment samples from background areas [204, 212] as summarized by Falandysz (1998) [5]. The CN profile seems to be a combination of technical PCNs and PCNs originating from PCB products and/or incineration processes. The pattern in Swedish background sediment samples has been proposed to be a result of atmospheric weathering of mainly PCB products [204].

High PCN levels have been reported in abiotic samples from contaminated areas [4, 234, 235]. Remarkably high PCN concentrations (61000 ng g^{-1} dry weight) have been reported in sediment samples from Trenton located on the Detroit River (Michigan, USA) where several chemical manufacturing plants are located [236].

5.2
Ambient Air

PCNs have recently been reported in ambient air from Europe, North America and the Arctic [204, 210, 237, 238]. The concentrations are generally higher in urbanized areas compared to remote areas. The average PCN concentration in Chicago air (winter 1995) was reported to be 68 pg m^{-3} compared to 12 pg m^{-3} in air from the East Arctic Ocean (summer 1996) [210, 238]. The PCN concentrations in e.g., Chicago air are approximately one order of magnitude lower than typical PCB values for urban centers [210]. The main part of the CNs in ambient air from background locations is found in the gas phase (98%) and only small amounts on the particles (2%) [237]. Mainly tri- to pentaCNs are found in the gas phase and predominantly hexa- to octaCN in the particle phase [210].

6
Biological Fate and Transformation *in Vivo*

6.1
Congener Pattern and Levels in Biota

PCNs are generally found in all types of biological samples, also from remote areas as reviewed by Crookes and Howe [4] and later described in several studies [8, 204, 212, 218, 239–244], cf. Table 3.

In general, sediment samples and biological samples from the lower trophic level (e.g., plankton, mussel and plankton feeding fish) collected from the same location have similar CN patterns. Biological processes seem to alter the com-

Table 3. Concentrations of the sum of tetra- up to heptaCNs in some biological samples

Sample type (site)	Conc. (ng g^{-1} lipid weight) of the sum of tetra- to heptaCNs	References
Plankton (Gdansk Basin)	7.5–20	[218]
Mussel (Gdansk Basin)	80–110	[241]
Herring (Baltic Sea)	0.98–29	[5]
Fish, whole (Baltic Sea, Gulf of Gdansk)	6.3–260	[212]
Black cormorants, muscle (Gulf of Gdansk)	75–160	[242]
White-tailed sea eagle, muscle (Poland)	25–1400	[239]
Guillemot, egg (Baltic Sea)	84–220	[8]
White-tailed sea eagle, egg (Baltic Sea)	120–13	[204]

position and the CN pattern in species higher in the food web (e.g., in the guillemot) so they no longer reflect the CN pattern in the sediment. In biological background samples at lower trophic levels tetraCNs are the most abundant, while a few congeners i.e. 1,3,5,7-tetraCN (CN-42), 1,2,3,5,7-/1,2,4,6,7-pentaCN (CN-52/CN-60) and 1,2,3,4,6,7-/1,2,3,5,6,7-hexaCN (CN-66/CN-67) dominate in samples from higher trophic levels [8]. According to Falandysz et al. (1998), the pentaCNs are most abundant in fish predators such as perch, cod as well as adult white-tailed sea eagle [5]. In general, the PCNs are found at ng g^{-1} levels (lipid weight) in uncontaminated areas, but high levels are occasionally reported. For example, 1400 ng g^{-1} (lipid weight) PCN was reported in a muscle sample from white-tailed sea eagle, collected in Poland [239].

6.2
Congener Pattern and Levels in Human Samples

The CN profile in humans is similar to the pattern generally found in biological samples from higher trophic levels. Weistrand and Norén (1998) reported the pentaCNs (1,2,3,5,7-1,2,4,6,7-pentaCNs, CN-52/CN-60) and the hexaCNs (1,2,3,4,6,7-/1,2,3,5,6,7-hexaCNs, CN-66/CN-67) to be the most abundant CN congeners in human adipose and liver tissues followed by several tetraCN congeners i.e. 1,3,5,7-tetraCN (CN-42), 1,2,4,6-tetraCN (CN-33), 1,2,4,7-tetraCN (CN-34), and 1,2,5,7-tetraCN (CN-37) [16]. This is in agreement with other Swedish studies of PCNs in human milk [246] and human plasma [217], cf. Fig. 5.

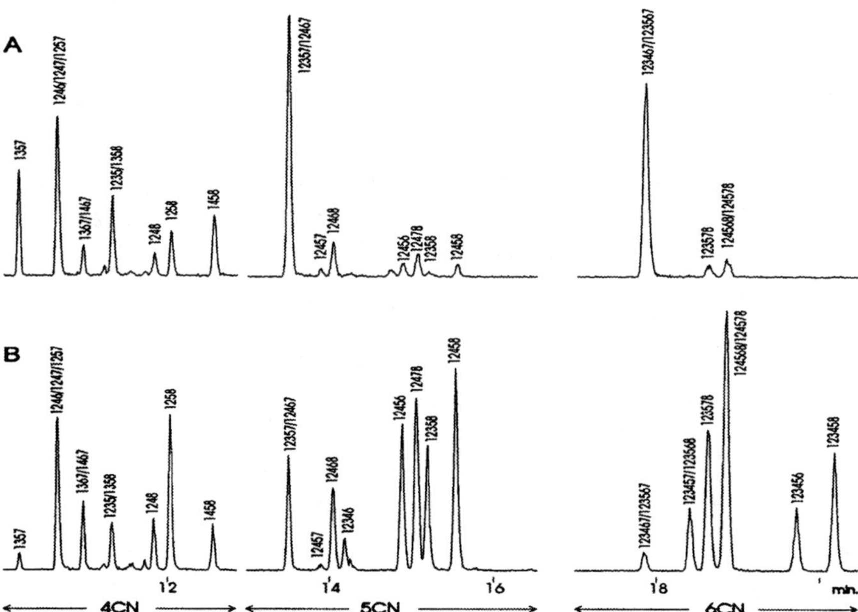

Fig. 5. GC-MS(EI) mass fragmentograms of the molecular ions for tetraCN to octaCN in a sample of human plasma on a phenyl-methylpolysiloxane (5%) column [217]

A similar CN pattern has been reported in adipose tissue from a Yusho patient [247].

In Canada, the concentrations of PCNs in human adipose tissue have been reported to be 0.1–25 ng g^{-1} and 0.1–2.4 ng g^{-1}, for the pentaCNs and hexaCNs, respectively [214]. Weistrand and Norén (1998) compared the PCN levels in human adipose and liver tissues from Swedish subjects and found that concentrations in the adipose tissue ranged from 0.19 to 1.2 ng g^{-1} and for liver, from 0.39 to 1.09 ng g^{-1} (lipid weight) for the pentaCNs and the hexaCNs, respectively [16]. The CN concentrations were in general, slightly higher in the liver compared to the adipose tissue. However, one subject had 20 times higher concentrations of the two hexaCNs in the liver compared to the adipose tissue. A specific and strong retention of these congeners in rat liver has previously been reported [217], cf. Sect. 7.3.1.

PCN concentrations have been determined in human plasma samples from Swedish people with varying intakes of fish [217]. 1,2,3,5,7-/1,2,4,6,7-pentaCN (CN-52/CN-60) and 1,2,3,4,6,7-/1,2,3,5,6,7-hexaCN (CN-66/CN-67) were found at concentrations ranging from 0.3 to 32 ng g^{-1} and 0.4 to 4.1 ng g^{-1} (lipid weight), respectively [217]. For comparison, the concentration of one of the most abundant PCBs (2,2′,4,4′,5,5′-pentaCB, CB-153) was reported to be 220–2400 ng g^{-1} lipid weight [245].

6.3
Bioaccumulation/Biomagnification

The uptake of CNs from the diet has been studied in rainbow trout (*Oncorhynchus mykiss*) fed with pike (*Esox lucius*) injected with CNs. The uptake was reported to be 63% for 1,2,3,4,6,7-/1,2,3,5,6,7-hexaCN (CN-66/67), 78% for 1,2,4,5,6,8-hexaCN (CN-71), 68% for 1,2,3,4,5,6,7-heptaCN (CN-73) and 35% for octaCN (CN-75) [248].

Bioconcentration factors (BCFs) have been experimentally determined for PCNs in fish by Opperhuizen et al. (1985) [249]. Falandysz (1998) has calculated biomagnification factors (BMFs) from environmental levels in white-tailed sea eagle related to their food [5]. The highest BMF (50) were reported for 1,3,5,7-tetraCN (CN-42) and 1,2,3,5,7-/1,2,4,6,7-pentaCN (CN-52/60). A BMF of about 30 was found for 1,2,3,4,6,7-/1,2,3,5,6,7-hexaCN (CN-66/67).

7
Toxic Effects of PCNs

7.1
Enzyme Induction

PCN exposure to rats causes enzyme induction in the liver, kidney and lung [250–252]. Mixed-function oxygenases (MFO) are involved in the hydroxylation of xenobiotica. The induction potency of a compound depends on the chlorination degree, substitution pattern and planarity of the molecule. The compounds 3-methylcholanthrene and phenobarbital induce two different types of MFO: P4501A and P4502B, respectively. Generally, P4501A enzymes are induced by planar and P4502B by non-planar compounds. Halowaxes containing PCN of lower chlorination degree are P4502B inducers in rat liver, whereas Halowaxes with PCNs of higher chlorination degree are mixed P4501A/P4502B-type of inducers [253].

7.2
Toxic Equivalency Factors (TEFs)

Many of the toxic and biological effects induced by polychlorinated dibenzo-*p*-dioxins and furans (PCDD/Fs) and PCBs such as carcinogenesis, reproductive disturbances and immunotoxic effects are believed to be mediated via the hepatic cytosolic aryl hydrocarbon receptor (Ah receptor) [254, 255]. Based on *in vitro* and *in vivo* studies, the toxicity of individual organochlorines have been determined relative to 2,3,7,8-tetrachlorodibenzo-*p*-dioxin (TCDD) and expressed as toxic equivalency factors (TEFs) [254, 256]. In addition to PCDD/F, structurally related PCBs and PCNs bind to the Ah receptor. After binding to the Ah-receptor, the receptor-ligand complex is transferred into the nucleus where it binds to specific DNA sequences and causes transcription of structural genes, which in turn causes synthesis of various cytochrome P4501A1-dependent enzymes such as ethoxyresorufin *O*-deethylase (EROD) and aryl hydrocarbon hydroxylase (AHH). TEFs for PCNs have been estimated from enzyme-induction assays of EROD and AHH [10, 257] and Luciferase assays in rat cells [12] cf. Table 4.

Table 4. Toxic equivalence factors (TEFs) for PCNs

Compound	Luciferase assay[a]	EROD induction[b,c]
2,3,7,8-TCDD	1	1
2-MonoCN (CN-2)	–	ND
1,4-DiCN (CN-5)	$2.04 \cdot 10^{-7}$	ND
1,5-DiCN (CN-6)	–	ND
2,3-DiCN (CN-10)	–	ND
1,2,7-TriCN (CN-17)	–	ND
1,2,3,4-TetraCN (CN-27)	–	ND
1,2,4,6-TetraCN (CN-33)	–	ND
1,2,5,6-TetraCN (CN-36)	–	ND
1,2,6,8-TetraCN (CN-40)	$1.60 \cdot 10^{-5}$	ND
2,3,6,7-TetraCN (CN-48)	–	ND
1,2,3,5,8-PentaCN (CN-53)	–	ND
1,2,3,6,7-PentaCN (CN-54)	$1.69 \cdot 10^{-4}$	ND
1,2,3,4,5,6-HexaCN (CN-63)	ND	$2 \cdot 10^{-3}$
1,2,3,4,6,7-HexaCN (CN-66)	$3.85 \cdot 10^{-3}$	ND

Table 4 (continued)

Compound	Luciferase assay[a]	EROD induction[b,c]
1,2,3,5,6,7-HexaCN (CN-67)	$1.00\ 10^{-3}$	ND
50:50 mix 1,2,3,4,6,7- and 1,2,3,5,6,7-hexaCN (CN-66/67)	$1.25\ 10^{-3}$	$2\ 10^{-3}$
1,2,3,4,5,7- and/or 1,2,3,5,6,8-hexaCN (CN-64/68)	ND	$2\ 10^{-5}$
1,2,3,5,6,8-HexaCN (CN-68)	$1.53\ 10^{-4}$	ND
1,2,3,5,7,8-HexaCN (CN-69)	ND	$2\ 10^{-3}$
1,2,3,6,7,8-HexaCN (CN-70)	$5.87\ 10^{-4}$	ND
1,2,4,5,6,8-HexaCN (CN-71)	–	$7\ 10^{-6}$
1,2,3,4,5,6,7-HeptaCN (CN-73)	$1.01\ 10^{-3}$	$3\ 10^{-3}$
1,2,3,4,5,6,8-HeptaCN (CN-74)	–	ND
1,2,3,4,5,6,7,8-OctaCN (CN-75)	–	ND

[a] Blankenship et al., 1999 [12].
[b] Hanberg et al. (1990) [10]. The structures of several hexaCNs were not determined in 1990, but have later been solved, see e.g., reference [194] for structures.
[c] ND indicates that no value was determined in the study.

TEFs are used in the risk assessment of substances with the same toxic mechanism as TCDD. By multiplying the concentration of each substance in a tissue with its TEF, the concentration of TCDD toxic equivalents (TEQ) can be estimated. It should be emphasized that the TEFs are based only on the Ah receptor mediated toxicity and the assumption that the toxic effects of TCDD-like substances are additive. Some of the toxic effects of PCNs are likely to be due to binding to the Ah receptor, but there are most probably also other independent mechanisms that influence the toxicity as well.

7.3
Tissue Distribution and Metabolism

7.3.1
Specific Retention in the Liver

An uptake of all PCNs was found in rats orally dosed with Halowax 1014. Thereafter, there was a highly selective and strong retention of especially 1,2,3,4,6,7-hexaCN (CN-66) and 1,2,3,5,6,7-hexaCN (CN-67) in the liver of the animals [13–14]. The half-lives of the two hexaCNs were remarkably long and comparable with those reported for several 2,3,7,8-chlorinated dibenzofurans [258]. Ahlborg and Hanberg (1992) proposed that the strong retention of PCDFs in the liver is due to binding to microsomal proteins [256]. No correlation between the strong retention in the liver and induction of AHH and EROD has however been found. Specific retention of 1,2,3,4,6,7- and/or 1,2,3,5,6,7-hexaCN has also been reported in the liver of harbor porpoise (*Phocoena phocoena*) [15]. In humans, the PCN levels seem to generally be only slightly higher in the liver compared to adipose tissue (on lipid weight bases) [16]. One of the seven subjects included in the study had though 20 times higher concentrations of 1,2,3,4,6,7-hexaCN (CN-66) and 1,2,3,5,6,7-hexaCN (CN-67) in the liver compared to the adipose tissue.

7.3.2
Metabolism

Naphthalene is metabolized via an arene oxide intermediate in the 1,2-position which results in the formation of hydroxylated metabolites [259, 260]. There are relatively few studies on the metabolism of PCNs. Congeners of higher chlorination degree seem to be more slowly metabolized than lower chlorinated congeners [17–19]. Metabolism occurs mainly by hydroxylation [18, 20–24], but also via the mercapturic acid pathway [17, 18, 26]. The CN congeners with adjacent hydrogen atoms seem to be most easily metabolized. Generally tetraCNs with vicinal hydrogen atoms in $\beta\beta$-positions are more easily metabolized than tetraCNs with vicinal hydrogen atoms in $\alpha\beta$-positions [5].

Metabolism of xenobiotics may occasionally result in the formation of reactive metabolites that become covalently bound to macromolecules and might cause damage to the tissue. In a study of the distribution of a mixture of ^{14}C-labelled tetra-, penta- and hexaCNs in the rat it was indicated that PCNs are metabolized in the tissues via the formation of reactive intermediates that bind to macromolecules (including DNA) in the tissues [25, 26]. The distribution of the total concentration of ^{14}C (based on the fresh weight) in the rat tissues 5 days after the oral dose was: liver (10 pmol mg^{-1}), lung (1.5 pmol mg^{-1}), kidney (3.0 pmol mg^{-1}) and abdominal fat (9.2 pmol mg^{-1}). A remarkably large part of the radioactivity in the tissues was found to be non-extractable. In the liver, as much as 75% of the radioactivity was non-extractable. In the kidneys, lungs and abdominal fat the amount non-extractable radioactivity were 69%, 56% and 1%, respectively [25–26]. The irreversible binding of PCNs and binding to macromolecules as well as the specific retention in the liver may be of importance for the toxic effects PCNs causes in the liver, cf. below.

7.4
Health Effects in Animals and humans

The most common toxic effects observed after acute, subchronic and chronic exposure of animals to mixtures of PCNs are chloracne, liver damage and effects on epithelial tissues. Weight loss, loss of appetite and jaundice are also reported as well as effects on the kidney, vitamin A levels, reproductive system, gastrointestinal tract, salivary glands, pancreas, lung, spleen and thymus. Penta-, hexa- and heptaCN appear to be more toxic than PCNs of lower chlorination degree. Some toxic effects persist for a long period after exposure has ceased (for reviews see [1, 27]). Oral LD$_{50}$ values in the rat, mouse and guinea pig indicate also that the penta- and hexaCNs are the most toxic PCNs followed by the tri-, tetra- and heptaCNs. No information was however given about the experimental conditions [261].

In cattle, PCNs have been associated with X-disease (bovine hyperkeratosis). Other symptoms observed in cattle are vitamin A deficiency, excessive salivation, and thickening of skin followed by loss of hair and occasionally death [1, 262]. Chickens exposed to PCNs showoedema, enlarged fibrous livers and lack of feather pigmentation [263].

In occupationally exposed humans, chloracne, dermal effects and liver damage are the most common health effects. Some cases with fatal outcome have been reported. Symptoms of PCN poisoning include jaundice, constipation, abdominal pain, abdominal distension, anorexia, nausea, vomiting, anemia, skin problems, eye irritation, headache, fatigue and vertigo. Some of the toxic effects persisted for a long period after the exposure had ceased. In severe cases the patients died [1, 27, 264–266]. Studies of experimentally induced chloracne show that tetra-, penta-, and hexaCNs have acnegenic properties when applied on the human skin [263, 267].

References

1. Brinkman UATh, Reymer HGM (1976) J Chromatogr 127:203
2. Ullmann's Encyclopedia of Industrial Chemistry Vol A6 (1986) Campbell FT, Pfefferkorn R, Rounsaville JF (eds) VCH, Weinheim, Germany
3. Haglund P, Jakobsson E, Asplund L, Athanasiadou M, Bergman Å (1993) J Chromatogr 634:79
4. Crookes MJ, Howe PD (1993) Environmental hazard assessment: halogenated naphthalenes TSD/13. Department of the Environment, London, UK
5. Falandysz J (1998) Environmental Pollution 101:77
6. Baumann Ofstad E, Lunde G, Martinsen K, Rygg B (1978) Sci Total Environ 10:219
7. Theisen J, Maulshagen A, Fuchs J (1993) Chemosphere 26:881
8. Järnberg U, Asplund L, de Wit C, Grafström A-K, Haglund P, Jansson B, Lexén K, Strandell M, Olsson M, Jonsson B (1993) Environ Sci Technol 27:1364
9. Lutz G, Otto W, Schönberger H (1991) Chlorchemie und Dioxine/Furane – Hand in Hand von Anfang an Chloralkalielektrolyse – Ursache für hochgradig PDCF–belastete Rückstände und damit Altlasten. Report, Regierungspräsidium Freiburg, Abt Wasserwirtschaft; Freiburg, Germany
10. Hanberg A, Waern F, Asplund L, Haglund E, Safe S (1990) Chemosphere 20:1161
11. Engwall M, Brunström B, Jakobssson E (1994) Arch Toxicol 68:37
12. Blankenship A, Kannan K, Villalobos S, Villeneuve D, Falandysz J, Imagawa T, Jakobsson E, Giesy J (1999) Relative potencies of Halowax mixtures and individual polychlorinated naphthalenes (PCNs) to induce Ah reseptor-mediated responses in rat hepatoma H4IIE-Luc cell bioassay Environ Sci Technol (in press)
13. Asplund L, Jansson B, Sundström G, Brandt I, Brinkman UATh (1986) Chemosphere 15:619
14. Asplund L, Jakobsson E, Haglund P, Bergman Å (1994) Chemosphere 28:2075
15. Ishaq R, Karlsson K, Näf C (1999) Tissue distribution of polychlorinated naphthalenes (PCNs) and non-*ortho* chlorinated biphenyls (non-*ortho* PCBs) in harbor porpoises (*Phocoena phocoena*) from Swedish waters (manuscript
16. Weistrand C, Norén K (1998) J Toxicol Environ Health Part A 53:293
17. Oishi H, Oishi S (1983) Toxicol Lett 15:119
18. Ruzo L, Jones D, Safe S, Hutzinger O (1976) J Agric Food Chem 24:581
19. Cornish H, Block W (1958) J Biol Chem 231:583

20. Chu I, Secours V, Villeneuve DC, Viau A (1977) Bull Environm Contam Toxicol 18:177
21. Ruzo LO, Safe S, Jones D, Platonow N (1976) Bull Environ Contam Toxicol 16:233
22. Ruzo LO, Safe S, Hutzinger O, Platonow N, Jones D (1975) Chemosphere 4:121
23. Chu I, Secours V, Viau A (1976) Chemosphere 5:439
24. Chu I, Villeneuve DC, Secours V, Viau A (1977) J Agric Food Chem 25:881
25. Jakobsson E (1994) Synthesis and Analysis of Chlorinated Naphthalenes Biological and Environmental Implications (Thesis) Dept of Environmental Chemistry, Stockholm University, Sweden
26. Klasson-Wehler E, Jakobsson E, Örn U (1996) Organohalogen Compounds 28:495
27. Kover FD (1975) Environmental hazard assessment report: Chlorinated naphthalenes EPA 560/8-75-001 Environmental Protection Agency Washington, DC, USA
28. Federal Register (1983) USEPA 48:20668
29. Kirk-Othmer (1980) In: Encyclopedia of Chemical Technology, 3rd edn. Wiley, New York, USA
30. Brinkman UA Th and de Kok A (1980) In: Kimbrough (ed.) Halogenated biphenyls, terphenyls, naphthalenes, dibenzodioxins and related products. Elsevier/North-Holland Biomedical Press
31. Weistrand C, Lundén Å, Norén K (1992) Chemosphere 24:1197
33. Kauppinen T (1986) Ann Occup Hyg 30:19
34. Wiedmann T, Ballschmiter K (1993) Fresenius J Anal Chem 346:800
35. Varma PS, Parekh NB, Subramanium VK (1939) J Indian Chem Soc 16:460
36. Menschutkin (1912) Chem Zentralbl 83:1436
37. Silberman, Raschewsskaja, Martynzewa (1937) Chem Zentralbl 108:4786
38. Wibaut, Bloem (1950) Recl Trav Chim Pays-Bas 69:586
39. Dow Chem Co (1930) US 1917822
40. Parts (1930) Z Phys Chem (Leipzig) 10:264
41. Khanna MS, Khetarpal SC Lal K, Bharnagar HL (1981) Indian J Chem Sect A 20:544
42. Wahl H, Bassilios H (1974) Bull Soc Chim Fr 482
43. Rymarenko (1876) Chem Ber 9:666
44. Jefremow (1923) Chem Zentralbl 94:380
45. Roux (1887) Ann Chim (Paris) 12:349
46. Gockel (1935) Z Phys Chem (Leipzig) 29:79
47. Hampson, Weissberger (1936) J Chem Soc 393
48. Heumann, Koechlin (1883) Chem Ber 16:1627
49. Liebermann, Palm (1876) Justus Liebigs Ann Chem 183:270
50. Nefedow et al. (1960) Chem Abstr 24567
51. Soloveichik et al. (1976) J Org Chem USSR (Engl. Transl) 12:2136
52. Blum (1966) Tetrahedron Lett 3041
53. van Miltenburg JC, Verdonk ML (1991) J Chem Thermodynamics 23:273
54. Schein, Ignatow (1966) J Org Chem USSR (Engl. Transl) 2:706
55. Masterton et al. (1971) J Chem Thermodynamics 3:243
56. Nguyen-Ba-Chanh (1970) J Chim Phys Phys Chim Biol 67:1198
57. Makosza M, Gajos I (1974) Rocz Chem 48:1883
58. Kozlov et al. (1963) J Gen Chem USSR (Engl. Transl.) 33:658
59. Armstrong HE, Wynne WP (1896) Chem News J Ind Sci 73:55
60. Cleve (1887) Chem Ber 20:1991
61. Weissberger, Saengewald, Hampson (1934) Trans Faraday Soc 30:884
62. Clemo, Cockburn, Spence (1931) J Chem Soc 1265
63. Trotter (1964) Acta Crystallogr 17:63
64. Baeckstroem (1887) Chem Ber 20:1991
65. Wilson, Zehr (1978) J Org Chem 43:1768
66. de la Mare et al. (1966) J Chem Soc B 834
67. Cleve (1878) Bull Soc Chim Fr 29:415
68. Sundström G (1976) Chemosphere 5:191
69. Koptyug VA et al. (1966) J Org Chem USSR (Engl. Transl) 2:467

70. Freidlina et al. (1977) Dokl Chem (Engl. Transl) 236:563
71. Beattie, Whitmore (1933) J Amer Chem Soc 55:1546
72. Meyer H (1915) Monatsh Chem 36:728
73. Erdmann (1888) Justus Liebigs Ann Chem 247:351
74. Turner EG, Wynne WP (1941) J Chem Soc 243
75. von Auwers, Freuling (1921) Justus Liebigs Ann Chem 422:195
76. Farbenind IG (1929) Patent US1822982
77. Farbenind IG (1928) Fortschr Teerfarbenfabr Verw Industriezweige 17:670
78. Doyle MP et al. (1977) J Org Chem 42:2426
79. Suzuki et al. (1965) Chem Abstr 63:4422f
80. de la Mare PBD, Suzuki H (1967) J Chem Soc C 1586
81. Ferrero P, Bolliger G (1928) Helv Chim Acta 11:1144
82. Krollpfeiffer (1923) Justus Liebig Ann Chem 430:198
83. Woroshzow, Lisizyn (1960) J Gen Chem USSR 30:2794
84. Frater, Havinga (1970) Recl Trav Chim Pays-Bas 89:273
85. de la Mare PBD, Suzuki H (1968) J Chem Soc C 1159
86. Armstrong HE (1888) Chem News 58:295
87. Forsling (1887) Chem Ber 20:2105
88. Armstrong HE, Wynne WP (1890) Chem News J Ind Sci 62:164
89. Schroeter (1930) Chem ber 63:1308
90. Friedlaender, Kielbasinski (1896) Chem Ber 29:1981
91. Armstrong HE, Wynne WP (1895) Chem News J Ind Sci 71:253
92. Claus, Volz (1885) Chem Ber 18:3157
93. Arnell (1886) Bull Soc Chim Fr 45:184
94. Erdmann (1893) Justus Liebigs Ann Chem 275:257
95. Armstrong HE, Wynne WP (1889) Chem News J Ind Sci 59:188
96. Komagorov AM et al. (1967) J Org Chem 3:147
97. Cleve (1893) Chem-Ztg Chem Appar 17:398
98. Atterberg A (1876) Chem Ber 9:1732
99. Rees CW, Storr RC (1969) J Chem Soc C 760
100. Armstrong HE, Wynne WP (1890) Chem News J Ind Sci 61:273
101. Leeds, Everhart (1880) J Amer Chem Soc 2:211
102. Wege D, Wilkinson SP (1973) Aust J Chem 26:1751
103. Claus, Zimmermann (1881) Chem Ber 14:1483
104. Vesely, Jakes (1923) Bull Soc Chim Fr 33:948
105. Weissberger, Saengewald (1933) Z Phys Chem (Leipzig) 20:145
106. Fosling (1887) Chem Ber 20:81
107. Jacobs et al. (1946) J Org Chem 11:27
108. Franzen, Staeuble (1921) J Prakt Chem 103:354
109. Chozjanowa, Strutschkow (1964) J Struct Chem (Engl. Transl.) 5:375
110. Rosowsky A et al. (1969) J Chem Soc 1376
111. Bayer, Duisberg (1887) Chem Ber 20:1432
112. Dziewonski, Majawicz, Schimmer (1936) Bull Acad Pol Sci Ser Sci Chim A:43
113. Faust, Saame (1871) Justus Liebig Ann Chem 160:71
114. Cum G et al. (1967) J Chem Soc B 244
115. Burton, de la Mare (1970) J Chem Soc B 897
116. Cleve (1888) Chem Ber 21:893
117. Hellström (1889) Översikt Kgl Svenska Vetenskaps Akad Förhandl 116
118. Forsling (1888) Chem Ber 21:3498
119. Buffle, Corbaz (1932) Arch Sci Phys Nat 14:149
120. Atterberg A (1876) Chem Ber 9:317
121. Friedlaender, Karamessinis, Schenk (1922) Chem Ber 55:47
122. Kozlov VV, Korneeva NA (1976) J Org Chem USSR 12:1514
123. Armstrong HE, Wynne WP (1897) Chem News J Ind Sci 76:69
124. Alen (1884) Chem Ber 17:437

125. Armstrong HE, Wynne WP (1890) Chem News J Ind Sci 61:93
126. Kozlov VV et al. (1960) J Gen Chem USSR (Engl. Transl.) 30:2696
127. Woroshtzow, Koslov (1936) Chem Ber 69:412
128. Nikiforov V, Auger P, Wightman R, Malaiyandi M, Williams D (1992) Organohalogen Compounds 8:123
129. Wynne (1946) J Chem Soc 61
130. von Braun (1923) Chem Ber 56:2337
131. Danish et al. (1954) J Amer Chem Soc 76:6144
132. Hales NJ, Heaney H, Hollinshead, JH, Ley SV (1995) Tetrahedron 51:7755
133. Gribble GW et al. (1976) Tetrahedron Lett 3673
134. Duerr H, Scheppers G (1970) Justus Liebigs Ann Chem 734:141
135. Klemm LH, Stevens MP, Tran LK, Sheley J (1988) J Hetrocycl Chem 25:1111
136. Heaney H, Jablonski JM (1968) J Chem Soc 1895
137. Howard, Gilbert (1962) J Org Chem 27:2685
138. Roedig, A et al. (1979) Chem Ber 112:2730
139. Martmeau, Dejongh (1977) Can J Chem 55:34
140. Nakayama J, Kuroda M, Hoshino M (1986) Heterocycles 24:1233
141. Moser GA et al. (1970/1971) Organomet Chem Synth 1:99
142. Vernon JM et al. (1978) J Soc Perkin Trans I 837
143. Wilt, Vasilianskes (1970) J Org Chem 35:2410
144. Ranken, Battiste (1971) J Org Chem 36:1996
145. Atterberg A, Widman O (1877) Chem Ber 10:1841
146. Burton et al. (1974) J Chem Soc Perkin Trans II 1914
147. Ward ER, Hardy A (1966) J Chem Soc C 1038
148. Cencelj L (1960) Chem Ber GR 93:988
149. Hardy A, Ward ER, Day LA (1956) J Chem Soc 1979
150. Piggott HA, Slinger FH (1952) J Chem Soc 259
151. Alen (1881) Bull Soc Chim Fr 36:435
152. Chatt J, Wynne WP (1943) J Chem Soc 33
153. Reimlinger H, King G (1962) Chem Ber 95:1043
154. Clar E, Marschalk C (1950) Bull Soc Chim Fr 433
155. Atterberg A, Widman O (1877) Bull Soc Chim Fr 28:513
156. Gafner G, Herbstein FH (1960) Acta Crystallogr 13:702
157. Gafner G, Herbstein FH (1962) Acta Crystallogr 15:1081
158. Bassilios HF et al. (1962) Recl Trav Chim Pays-Bas 81:209
159. Nikiforov VA, Karavan VS, Miltsov SA, Tribulovich VG (1993) Organohalogen Compounds 14:229
160. Amer Cyanamid (1970) US3636048
161. Amer Cyanamid Comp (1972) US 3636048 Chem Abstr 76:142391
162. Levy LA (1983) Synth Commun 13:639
163. Graebe (1869) Justus Liebigs Ann Chem 149:1
164. Claus A, van de Lippe (1883) Chem Ber 16:1016
165. Blangey et al. (1951) Helv Chim Acta 34:501
166. Auger P, Malaiyandi M, Wightman RH, Bensimon C, Williams DT (1993) Environ Sci Technol 27:1673
167. Jakobsson E, Eriksson L, Bergman Å (1992) Acta Chem Scand 46:527
168. Klingsberg E (1972) Tetrahedron 28:963
169. Klingsberg E (1973) US Patent 3769279
170. Jakobsson E, Lönnberg C, Eriksson L (1994) Acta Chem Scand 48:891
171. Clark J, Maynard R, Wakefield BJ (1976) J Chem Soc Perkin trans II 73
172. Garcia R, Riera J, Carilla J, Julia L, Molins E, Miravitlles C (1992) 57:5712
173. Brady JH, Tahir N, Wakefield BJ (1984) J Chem Soc Perkin Trans I 2425
174. Claus, Wenzlik (1886) Chem Ber 19:1165
175. Consortium für Elektrochemische Industrie GmbH (1970) Chem Abstr 72:100490q
176. Tahir NA (1975) M Sc Thesis Univ Salford

177. Schwemberger, Gordon (1932) Zh Obshch Khim 2:924
178. Ruoff (1876) Chem Ber 9:1486
179. Deacon GB, Farquharson GJ (1977) Aust J Chem 30:1701
180. Epshtein et al. (1977) SU 2338962
181. Farbenfabr Bayer AG (1953) US 2734927 Chem Abstr 1956:10782
182. Farrar, Storms (1968) J Chem Eng Data 13:284
183. Ginsberger et al. (1962) Tetrahedron Lett 779
184. Cruickshank DWJ, Sparks RA (1960) Proc Roy Soc A 258:270
185. Gafner G, Herbstein FH (1963) Nature 200:130
186. Herbstein FH (1979) Acta Cryst B35:1661
187. Opperhuizen A (1987) Toxicol Environ Chem 15:249
188. Isnard P and Lambert S (1988) Chemosphere 17:21
189. Järnberg U (1997) Analytical methods for studying polychlorinated naphtalene congener profiles and levels in the environment. Thesis Stockholm University, Sweden
190. Marsh J (1985) Advanced organic chemistry John Wiley & Sons, New York
191. Clar E, Zander M (1958) J Chem Soc 1861
192. Wehland G (1942) J Am Chem Soc 64:900
193. Campbell N (1978) In: Coffey S (ed) Rodd's Chemistry of Carbon Compounds, vol III part G. Elsevier, Amsterdam, p 99
194. Järnberg U, Asplund L, Jakobsson E (1994) J Chromatogr A 683:385
195. Jakobsson E, Asplund L, Haglund P, Bergman Å (1990) Organohalogen compounds 4:187
196. Nikiforov VA, Karavan VS, Miltsov SA, Tribulovich VG (1998) Organohalogen Compounds 35:159
197. Auger P, Malaiyandi M, Wightman RH (1993) J Labelled Comp Radiopharmaceut 33:263
198. de Kok A, Geerdink RB, Brinkman UATh (1983) Anal Chem Symp Ser (Chromatogr Biochem Med Environ Res) 13:203
199. Stojkovski S, Markovec LM, Magee RJ (1991) J Chem Technol Biotechnol 51:407
200. Hutzinger O, Safe S, Zitko V (1972) J Environ Anal Chem 2:95
201. Brinkman UATh, de Vries G, de Kok A, de Jonge AL (1978) J Chromatogr 152:97
202. Jensen S, Reutergårdh L, Jansson B (1983) FAO Fish Tech Pap 212:21
203. Bligh EG, Dyer WJ (1959) Can J Biochem Physiol 37:911
204. Järnberg U, Asplund L, de Wit C, Egebäck A-L, Wideqvist U, Jakobsson E (1997) Arch Environ Contam Toxicol 32:232
205. Stalling DL, Smith LM, Petty JD (1979) Approaches to comprehensive analyses of persistent halogenated environmental contaminants. In: van Hall CE (ed) Measurement of organic pollutants in water and wastewater ASTM STP 686. American Society for Testing and Materials
206. Jensen S, Sundström G (1974) Ambio 3:69
207. Kuehl DW, Dougherty RC, Tondeur Y, Stalling DL, Smith LM, Rappe C (1980) Negative chemical ionization studies of polychlorinated dibenzo-p-dioxins, dibenzofurans and naphthalenes in environmental samples. In: McKinney JD (ed) Environmental Health Chemistry – The Chemistry of Environmental Agents as Potential Human Hazards. Ann Arbor Science Publishers, Ann Arbor, MI
208. Jansson B, Andersson R, Asplund L, Bergman Å, Litzén K, Nylund K, Reutergårdh L, Sellström U, Uvemo U-B, Wahlberg C, Widequist U (1991) Fresenius J Anal Chem 340:439
209. Falconer R, Bidleman T, Cotham W (1995) Envirin Sci Technol 29:1666
210. Harner T, Bidleman T (1997) Atmospheric Environmenent 31:4009
211. Lundgren K, Bergqvist PA, Haglund P, Rappe C (1992) Organohalogen Compounds 10:103
212. Falandysz J, Strandberg L, Bergqvist P-A, Kulp SE, Strandberg B, Rappe C (1996) Environ Sci Technol 30:3266
213. Haglund P, Asplund L, Järnberg U, Jansson B (1990) J Chromatogr 507:389
214. Williams DT, Kennedy B, LeBel GL (1993) Chemosphere 27:795

215. Kimata K, Hosoya K, Kuroki H, Tanaka N, Barr JR, McClure PC, Patterson DG, Jakobsson E, Bergman Å (1997) J Chromatogr A 786:237
216. Imagawa T, Yamashita N (1997) Chemosphere 35:1195
217. Asplund L (1994) Development and application of methods for determination of polychlorinated organic pollutants in biota. (Thesis) Dept of Environmental Chemistry, Stockholm University, Stockholm, Sweden
218. Falandysz J, Rappe C (1996) Environ Sci Technol 30:3362
219. Fitch WL, Sauter AD (1983) Anal Chem 55:832
220. Vos JG, Koeman JH, van der Maas HL, ten Noever de Brauw MC, de Vos RH (1970) Fd Cosmet Toxicol 8:625
221. Albro PW and Parker CE (1979) J Chromatogr 169:161
222. Vogelgesang J (1986) Z Wasser-Abwasser-Forsch 19:140
223. Kannan K, Imagawa T, Blankenship A, Giesy J (1998) Environ Sci Technol 32:2507
224. Tong HY, Shoree DL, Karasek FW, Helland P, Jellum E (1984) J Chromatogr 285:423
225. Oehme M, Manø S, Mikalsen A (1987) Chemosphere 16:143
226. Alarie Y, Iwasaki M, Stock MF, Pearson RC, Shane BS, Lisk DJ (1989) J Toxicol Environ Health 28:13
227. Imagawa T, Yamashita N, Miyazki A (1993) J Environ Chem 3:221
228. Imagawa T, Yamashita N (1994) Organohalogen Compounds 19:215
229. Schneider M, Stieglitz R, Will R, Zwick G (1998) Chemosphere 37:2055
230. VanDell RD, Shadoff LA (1984) Chemospere 13:1177
231. Ballschmiter K, Kirschmer P, Zoller W (1986) Chemosphere 15:1369
232. VanDell RD, Boggs GU (1987) Chemosphere 16:973
233. Yasuhara A, Morita M (1988) Environ Sci Technol 22:646
234. Crump-Wiesner HJ, Feltz HR, Yates ML (1973) J Research US Geol Survey 1:603
235. Erickson MD, Michael LC, Zweidinger RA, Pellizzari ED (1978) J Assoc Off Anal Chem 61:1335
236. Furlong ET, Carter DS, Hites RA (1988) J Great Lakes Res 14:489
237. Dörr G, Hippelein M, Hutzinger O (1996) Chemosphere 33:1563
238. Harner T, Kylin H, Bidleman T, Halsall C, Strachan W, Barrie L, Fellin P (1997) Environ Sci Technol 32:3257
239. Falandysz J, Strandberg L, Kulp SE, Strandberg B, Bergqvist P-A, Rappe C (1996) Chemoshere 33:51
240. Falandysz J, Rappe C (1997) Chemosphere 35:1737
241. Falandysz J, Strandberg L, Bergqvist P-A, Strandberg B, Rappe C (1997) The Science of the total Environment. 203:93
242. Falandysz J, Strandberg B, Strandberg L, Bergqvist P-A, Rappe C (1997) The Science of the Total Environment 204:77
243. Falandysz J, Strandberg L, Strandberg B, Rappe C (1998) Chemosphere 37:2473
244. Paasivirta J, Rantio T (1991) Chemosphere 22:47
245. Asplund L, Svensson G, Nilsson A, Eriksson U, Jansson B, Jensen S, Wideqvist U, Skerfving S (1994) Arch Environ Health 49:477
246. Lundén Å (1996) Licentiate Thesis Karolinska Institutet Sweden
247. Haglund P, Jakobsson E, Masuda Y (1995) Organohalogen Compounds 26:405
248. Burreau S, Axelman J, Broman D, Jakobsson E (1997) Environm Toxicol Chem 16:2508
249. Oppenhuizen A, van der Velde EW, Gobas RAPC, Liem DAK, van der Steen JM, Hutzinger O (1985) Chemosphere 14:1871
250. Ahotupa M, Aitio A (1980) Biochemical Biophysical Research Communications 93:250
251. Ahotupa M, Hietanen E, Mäntylä E (1982) J Appl Toxicol 247:53
252. Mäntylä E, Aitio A, Ahutopa M (1983) Effect of polychlorinated naphtalenes and biphenyls on polysubstrate monooxygenase and UDP-glucuronosyltransferase activities in the rat lung. In: Rydström J et al. (eds) Elsevier, Amsterdam, p 223
253. Cockerline R, Shilling M, Safe S (1981) Gen Pharmac 12:83
254. Safe S (1990) CRC Crit Rev Toxicol 21:51
255. Hankinson O (1995) Annu Rev Pharmacol Toxicol 35:307

256. Ahlborg UG, Hanberg A (1992) Toxic Substances Journal 12:197
257. Hanberg A, Ståhlberg M, Gergellis A, de Wit C, Ahlborg UG Pharmacol Toxicol 69:442
258. Ahlborg UG, Håkansson H, Lindström G, Rappe C (1990) Chemosphere 20:1235
259. Bakke JE, Struble C, Gustafsson J-Å, Gustafsson B (1985) Proc Natl Acad Sci USA 82:668
260. Horning MG, Stillwell WG, Griffin GW, Tsang W-S (1980) Drug Metab and Dispos 8:404
261. Reggiani G, (1982) In: Hutzinger et al. (eds) Chlorinated dioxins and related compounds. Impact on the environment, Pergamon Press, Oxford 463
262. Olson C (1969) Bovine hyperkeratosis (X-disease, highly chlorinated naphthalene poisoning). Historical Review. In: Brandly CA, Cornelius OE (eds) Advances in Veterinary Sciences and Comparative Medicine. Academic Press, New York 13:101
263. Pudelkiewicz WJ, Boucher RV, Callenbach EW, Miller RC (1959) Poultry Sci 38:424
264. Drinker CK, Field Warren M, Bennett GA (1937) J Ind Hyg Toxicol 19:283
265. Kleinfeld M, Messite J, Swencicki R (1972) J Occupational Med 14:377
266. Greenburg L, Mayers MR, Smith AR (1939) J Ind Hyg Toxicol 21:29
267. Shelley W, Kligman A (1957) AMA Archives of dermatology 75:689

CHAPTER 6

Non- and Mono-*ortho* Chlorinated Biphenyls

N. Kannan

N. Kannan (e-mail: nkannan@ifm.uni-kiel.de)
Department of Chemistry, Institute for Marine Research, Düsternbrooker Weg 20,
D 24105 Kiel, Germany

Polychlorinated biphenyls (PCBs) are a complex mixture, identified in every component of the global ecosystem. If potential toxicity, environmental prevalence, and potential risk to humans and wildlife are considered, non- and mono-*ortho* substituted chlorinated biphenyls become extremely important. They resemble 2,3,7,8-tetrachloro dibenzo-*p*-dioxin (TCDD) in their biological action, causing hepato-, immuno-, reproductive, and dermal toxicities, and inducing mixed function oxidase (MFO) enzymes. This is the basis behind TCDD toxic equivalency (TEQs) as an approach towards risk assessment. Improvements in chromatographic techniques have yielded an array of analytical methods for these congeners that resulted in the determination of coplanar non-*ortho* and mono-*ortho* CBs in air, water, soil, and in organisms. Long range trans-hemispheric transportation of these substances in air and ocean currents have been demonstrated. Their contents in commercial PCBs have been studied and their formation during incineration is shown. Long term studies in river sediments demonstrate reductive dechlorination. PCB clearance from animals occurs both by non-metabolic and metabolic routes. Their flux in air, water, and sediments is estimated. OH reactions in the atmosphere is proposed as a principal source of loss along with permanent burial in marine and fresh water sediments.

Keywords. Non-*ortho*, Mono-*ortho*, Coplanar, PCBs, Review

The Handbook of Environmental Chemistry Vol. 3 Part K
New Types of Persistent Halogenated Compounds
(ed. by J. Paasivirta)
© Springer-Verlag Berlin Heidelberg 2000

1
Introduction

Polychlorinated biphenyls (PCBs) by virtue of their structural complexity vary in their physico-chemical and biological characteristics [1]. Biphenyl molecules with few chlorine substitutions (mono-, di, and tri-chloro isomers) differ widely in their environmental persistence, trans-hemispheric transportation, biotransformation, pharmaco kinetics, and toxic expressions from moderately chlorinated congeners (penta-, hexa, and hepta chlorobiphenyls) and highly chlorinated congeners (octa- to deca chlorobiphenyls). In general, the former PCBs are more water soluble, vaporizable, easily biodegradable, and are less potent inducers of mixed function oxidases (MFOs). The last group of congeners is less available for bioaccumulation and transportation because they congeners bind strongly to soils and sediments. The concern for environmental contamination of PCBs arises mostly from moderately chlorinated congeners which are persistent, bioaccumulative, less degradable, but potent inducers of Cytochrome P450 isozymes. They are 112 congeners among the theoretically possible 209 congeners. Still fewer congeners are demonstrably or potentially toxic [2].

Although PCBs, unlike dioxins, are inherently non-planar with opposing aromatic rings twisted to each other, the *meta*- and *para*-substituted compounds have sufficiently low rotational barriers to enable a small fraction to assume planar conformation [3]. These planar congeners are known inducers of Cytochrome P450 IA isozymes and exhibit toxic properties similar to polychlorinated dibenzo-*p*-dioxins (PCDDs) and polychlorinated dibenzofuranes (PCDFs), limited only by their potency [4]. Addition of *ortho* substitution in the molecule makes them a mixed type of inducer (Cytochrome P450IA and IIB enzymes) and substitution in both *para* and at least two *ortho* positions leave them as pure Cytochrome P450IIB inducers [5]. The group of congeners having enzyme inducing potencies and potential toxicity of high concern are four non-*ortho* coplanar congeners (IUPAC Nos. 77, 81, 126, and 169) and their mono-*ortho* structural analogs (Table 1) [6, 7].

When PCBs were found as environmental contaminants in the late 1960s [8], the then available analytical techniques were inadequate for the determination of non-*ortho* CBs and several important mono-*ortho* CBs [9]. Later studies on Quantitative Structure Activity Relationship (QSAR) and subsequent improvements in analytical chemistry of these compounds made it possible to deter-

Table. 1. Induction equivalency factors (IEFs), early life stage mortality factors (ELS) and toxic equivalency factors (TEFs) for non-, mono- and di-*ortho* PCBs in mammals, birds, and fish

	IUPAC #	Mammals H411E[b,h] bioassay			TEF[i]		Birds CEH[g] bioassay	Fish	
					Int./Safe[c]	WHO/IPCS[d]	IEF[a]	1[e]	2[f]
		IEF[a] Safe	IEF[a] Tillitt	IEF[a] Clemons				IEF[a]	ELS[j]
2,3,7,8-T4CDD		1.0	1.0	1.0	1.0	1.0	1.0	1.0	1.0
1,2,3,7,8-P5CDD		0.01	NV	1.1	0.5	0.5	1.1	NV	0.73
2,3,7,8- T4CDF		0.09	0.0064	0.03	0.1	0.1	1.1	NV	0.028
Non-*ortho* PCBs									
3-CB	2	NV	NV	NV	NV	NV	0.00001	NV	NV
4-CB	3	NV	NV	NV	NV		0.0	NV	NV
3,3'-DCB	11	NV	NV	NV	NV		0.0	NV	NV
3,4-DCB	12	NV	NV	NV	NV	NV	0.0008	NV	NV
3,3',4-T3CB	35	NV	NV	NV	NV	NV	0.0007	NV	NV
3,4,4'-T3CB	37	NV	NV	NV	NV	NV	0.0004	NV	NV
3,3',4,4'-T4CB	77	0.0009	0.000018	0.0008	0.01	0.0005	0.03	0.0006	0.00016
3,3',4,5-T4CB	78	NV	NV	NV	NV	NV	0.0001	NV	NV
3,3',4,5'-T4CB	79	NV	NV	NV	NV	NV	0.004	NV	NV

[a] IEF (Induction Equivalency Factors) values based on rat hepatoma.
[b] H411E bioassay are from [34, 49, 184–186].
[c] The International Safe TEFs are from [4, 49].
[d] The WHO/IPCS TEFs are from [39].
[e] Fish 1 TEFs are from [186].
[f] Fish 2 TEFs are from [41].
[g] CEH = Chicken Embryo Hepatocyte bioassay.
[h] H411E = rat hepatoma bioassay.
[i] TEF = Toxic Equivalency Factor.
[j] ELS = Early Life Stage mortality study bioassay.
NI = No Induction.
NV = No Value.

Table. 1 (continued)

	IUPAC #	Mammals H411E[b,h] bioassay			TEF[i]		Birds CEH[g] bioassay	Fish	
		IEF[a] Safe	IEF[a] Tillitt	IEF[a] Clemons	Int./Safe[c]	WHO/IPCS[d]	IEF[a]	1[e] IEF[a]	2[f] ELS[j]
3,3',5,5'-CB	80	NV	NV	NV	NV	NV	0.00008	NV	NV
3,4,4',5-T4CB	81	0.00004	0.0019	0.007	NV	NV	0.2	NV	NV
3,3',4,4',5-P5CB	126	0.3	0.022	0.1	0.1	0.1	0.3	0.0014	0.005
3,3',4,5,5'-P5CB	127	NV	NV	NV	NV	NV	0.005	NV	NV
3,3',4,4',5,5'-H6CB	169	0.003	0.00047	0.001	0.05	0.01	0.02	0.0003	NV
Mono-*ortho* PCBs									
2,3',4,4'-T4CB	66	NV	NV	NV	NV	NV	0.002	NV	NV
2,3',4',5-T4CB	70	NV	NV	NV	NV	NV	0.0004	NV	NV
2,3,3',4,4'-P5CB	105	0.0007	0.000008	0.00003	0.001	0.0001	0.005	NI	< 0.00007
2,3,4,4',5-P5CB	114	–	< 0.000001	NV	–	–	NV	NV	NV
2,3',4,4',5-P5CB	118	0.000009	0.00000035	0.00001	0.0001	0.0001	0.001	NI	< 0.00007
2',3,3',4,5-P5CB	122	NV	NV	NV	NV	NV	0.00002	NV	NV
2',3,4,4',5-P5CB	123	–	0.000012	NV	–	–	NV	NV	NV
2,3,3',4,4',5-H6CB	156	0.00009	0.000055	0.00005	0.0004 *	0.0005	0.001	NI	NV
2,3,3',4,4',5'-H6CB	157	0.00006	0.000015	NV	0.0003 *	0.0005	0.002	NV	NV
2,3',4,4',5,5'-H6CB	167	NV	0.000009	NV	0.001	0.00001	0.002	NV	NV
2,3,3',4,4',5,5'-H7CB	189	NV	0.000010	NV	–	–	NV	NI	NV
Di-*ortho* PCBs									
2,2'-.DCB	4	NV	NV	NV	NV	–	0.0	NV	NV
2,2',4,4'-T4CB	47	NI	NV	NV	NV	–	0.0	NV	NV
2,2',5,5'-T4CB	52	NV	NV	NV	NV	–	0.0	NV	NV
2,2',4,5,5'-P5CB	101	NV	NV	NV	NV	–	0.0	NV	NV
2,3,3',4',6-P5CB	110	NV	NV	NV	NV	NV	0.00005	NV	NV
2,2',3,3',4,4'-H6CB	128	NV	NV	NV	0.00002	NV	0.001	NV	NV
2,2',3,4,4',5'-H6CB	138	NV	NV	NV	0.00002	NV	0.001	NV	NV
2,2',4,4',5,5'-.H6CB	153	NI	NV	NV	NV	–	0.0	NV	NV

2,2',3,3',4,4',5-H7CB	170	NV	NV	NV	NV	NV	0.0002	NV	NV
2,2',3,4,4',5,5'-.H7CB	180	NV	NV	NV	0.00002	0.00001	0.0002	NV	NV
2,2',3,3',4,4',5,5'-OCB	194	NV	NV	NV	0.00002	NV	0.00005	NV	NV
Tri-*ortho* PCBs									
2,2',3,5',6-P5CB	95	NV	NV	NV	NV	NV	0.0	NV	NV
2,2',3,4,4',6-H6CB	139	NV	NV	NV	NV	NV	0.0006	NV	NV
2,2',3,4',5,5',6-H7CB	187	NV	NV	NV	NV	NV	0.0	NV	NV
Tetra-*ortho* PCBs									
2,2',6,6'-T4CB	54	NV	NV	NV	NV	NV	0.0	NV	NV
2,2',3,3',6,6'-.H6CB	136	NV	NV	NV	NV	NV	0.0	NV	NV
2,2',3,3',4,4',5,5',6,6'D10CB	209	NV	NV	NV	NV	NV	0.0	NV	NV

[a] IEF (Induction Equivalency Factors) values based on rat hepatoma.
[b] H411E bioassay are from [34, 49, 184–186].
[c] The International Safe TEFs are from [4, 49].
[d] The WHO/IPCS TEFs are from [39].
[e] Fish 1 TEFs are from [186].
[f] Fish 2 TEFs are from [41].
[g] CEH = Chicken Embryo Hepatocyte bioassay.
[h] H411E = rat hepatoma bioassay.
[i] TEF = Toxic Equivalency Factor.
[j] ELS = Early Life Stage mortality study bioassay.
NI = No Induction.
NV = No Value.

mine these important substances in the environment and to assess the associated risks [10–14].

The non- and mono-*ortho* CBs have been quantitated accurately in the principal source, namely, commercial PCB mixtures [15, 16]; additional environmental sources such as incineration have been identified [17, 18]; their presence in every ecosystem including the pristine polar regions has been shown [19, 20]; estimates of their flux in air, water, soil, and the removal mechanisms such as OH reactions in atmosphere and sediment burial in rivers and oceans have been proposed [21]; their microbial degradation and biotransformation in organisms have been studied [22, 23]; a battery of in-vitro and in-vivo bioassays using mammalian, avian, and piscian models for the benefit of risk-assessment of these CBs have been developed [24]. Studies like these in the last decade have resulted in a new awareness of these important class of industrial contaminants.

Reviews from yesteryears on the biochemistry, toxicology, and environmental occurrence of these compounds are available [2, 7, 10, 25–27].

2
Biological Effects and the Concept of 2,3,7,8-TCDD Toxic Equivalency

Investigations on the biological effects of PCBs in experimental animals during the period 1971–1979 (as summarized in [28]) revealed the following syndromes: (i) decreased reproductive efficiency; (ii) changes in liver morphology; (iii) changes in plasma lipid concentrations; (iv) hepatic porphyria; (v) decreased immuno competence; (vi) dermatological effects; (vii) production of tumors in the liver.

These effects are preceded by the induction of numerous enzymes including the hepatic and extra hepatic drug metabolizing enzymes. Poland and Glover [5] were among the first to note that the most potent inducers of Aryl-hydrocarbon hydroxylase (AHH) were also the most toxic. They showed the induction potential of several polychlorinated compounds and the importance of halogen substitution at the lateral positions (3,4,5,3′,4′,5′ positions of PCBs) in inducing Cytochrome P-450IA1 and Cytochrome P-450IA2 hemoproteins, and found that 2,3,7,8-tetrachloro dibenzo-*p*-dioxin (TCDD) was a potent enzyme inducer as well the most toxic compound. They showed that Aroclor 1254 induced not only these enzymes but also other forms of Cytochrome P-450 (such as Cytochrome P-450IIB1, P-450IIB2, P-450IIA1, and P-450IIIA1).

Researchers from the US Food and Drug Administration demonstrated in the following years that a simple bioassay, based upon measurement of the induction of Cytochrome P 450-associated monooxygenase activity in the H4IIE rat hepatoma cell line, could be a useful tool in determining the presence of TCDD and similar compounds in environmental samples [29]. In subsequent studies, Safe and co-workers [30] have demonstrated that induction of either AHH or 7-ethoxyresorufin-O-dethylase (EROD) activity in the H4IIE cell line by PCBs, PCDFs, and PCDD congeners, either singly or in combination, correlates well with the in vivo toxicity of these compounds to rats.

These developments in biochemistry had given the needed impetus to develop analytical methods for measuring non- and mono-*ortho* PCBs in commercial PCB mixtures and in environmental samples [15, 31, 32]. These authors have first used a TEF approach for assessing the toxic equivalence (TEQs) of PCBs in the commercial mixtures and in environmental extracts. The TEF values adopted for the non- and mono-*ortho* coplanar PCBs were derived from the relative potencies of these compounds as inducers of AHH or EROD activities in rat hepatoma H4IIE cells compared to 2,3,7,8-TCDD as the toxic reference standard [6, 11, 12]. They applied this to determine the contribution of PCB congeners to the total TEQs in a sample containing PCBs, PCDDs, and PCDFs [12, 33]. It was shown that in most environmental extracts coplanar PCBs contributed significantly higher percentage of the total TEQs than the PCDDs and PCDFs combined.

However, in this TEF evaluation, there is no assurance that an additive model is the best predictor of toxicity of mixtures of planar and non-planar PCBs. Studies with different combinations of PCBs, PCDFs, and PCDDs suggest that interactions among congeners may range from antagonism to additivity to, perhaps, synergism [34]. From a hazard assessment standpoint, a simple measure of the biological potency of complex mixtures of PCBs and/or complex mixtures of PCBs, PCDFs, and PCDDs would be extremely useful in obviating problems associated with the assumptions involved in TCDD "traditional TEF" evaluations. An integrated biochemical measure of potency would be a reasonable proximate predictor of the biological effects of these types of compounds. Subsequently, Zacharewski et al. [35] and Tillitt et al. [34] demonstrated the utility of the H4IIE bioassay for assessing the potency of various combinations of complex mixtures of PCBs, PCDFs, and PCDDs in samples of fishes and birds from Great Lakes. These studies, however, confirmed that PCBs contributed significantly more to the TEQs (mainly from 3,3',4,4',5-pentaCB (CB-126) and 2,3,3',4,4'-pentaCB (CB-105)) than the PCDDs plus PCDFs. Moreover, there was a correlation between the reproductive failure in the wildlife populations and the PCB-TEQs.

The development, validation, and application of TEFs have been reviewed regularly since 1984 by Safe [25, 36, 37]. The limitations of this approach have been identified [38]. To assess the relative potencies and to derive consensus on TEFs for PCDDs, PCDFs, and dioxin-like PCBs, the WHO-European Centre for Environmental and Health (WHO-ECEH) and the International Program on Chemical Safety (IPCS) have initiated a project and published their interim report [39]. The most important limitations, namely possible synergism or antagonism among dioxins and its stereo isomers and the lack of pharmacokinetic consideration, have been addressed in this report. USEPA researchers [40] showed that the relative potency of PCBs, PCDFs, and 2,3,7,8-TCDD is tissue specific and thus estimates of TEFs based on hepatic EROD activity is limited. Additionally, limitations of TEFs based on mammalian models to aquatic species have been demonstrated [41]. The international initiative on TEFs has recognized these pitfalls and recommended development of TEFs for fish and other wildlife.

The induction of Cytochrome P-450 forms in fish has been well characterized and considered as a biomarker for exposure to PAHs, PCBs, and other toxic compounds [42–44]. Such induction in fish from contamination has been detected by use of catalytic assay, antibodies to fish P-450, and cDNA probes that hybridize with P-450 messenger RNA. However, the potential of this approach to direct-acting carcinogens and tumor promoters that are non-inducers of Cytochrome P-450 isozymes is limited. A study conducted over the course of a year to determine the induction of hepatic Cytochrome P450IA (CYP1 A) in three species of benthic fish collected from a contaminated site compared to fish sampled from a less-contaminated site clearly showed site-specific induction with higher values for fish from contaminated site [44].

A recent investigation on the leopard frog *Rana Pipens* [45] suggests that ethoxy- (EROD), methoxy- (MROD), benzyloxy- (BROD), and pentoxy-resorufin-*O*-dethylase (PROD) activities are not sensitive biomarkers to amphibians.

Using the photo affinity ligand 2-azido-3(^{125}I) iodo-7,8-dibromo dibenzo-*p*-dioxin (N$_3$ (125I) Br$_2$DD), Hahn and co-workers [46] have studied the presence and properties of the Ah receptor in 20 species of aquatic vertebrates and invertebrates. They confirmed the presence of this receptor in teleosts and elasmobranch fish but not a vertebrate equivalent in invertebrate species. However, this does not rule out the possibility of drug induction in invertebrates. That is what a Japanese team [47] has shown in crabs by measuring glutathione *S*-transferase activity in hepatopancreas.

Avians, among other wildlife fauna, first drew the attention of PCB researchers, and induction of microsomal enzymes in pigeon had been measured as early as 1968 [48]. EROD inducing capacity of coplanar PCBs in chicken embryos has been standardized [49]. AHH and EROD activities were measured in black-headed and black-tailed gulls [50], Antarctic penguins and south polar skua. TEFs for birds have been proposed [49].

Species specific differences in Cyt. P-450 mediated monooxygenase activities have been demonstrated in seal species, dolphins, and whales [51, 52]. EROD, PROD, and BROD activities have been measured recently in three species of sea lions from the South-West Atlantic [53].

The available screening methods for detecting potential (anti-) estrogenic/androgenic chemicals in wildlife have been reviewed recently [24]. The available TEF values for fish, bird and mammal are given in Table 1. It is obvious from this table that new TEFs should be developed for different aquatic organisms.

While TEF analysis is based mainly on dioxin-like compounds, non-dioxin-like PCBs also have their own independent toxicities which, in certain cases, may be as important as those associated with dioxin-like compounds (e.g., cancer, neurotoxicity). PCB-induced alteration of Ca^{2+} homeostasis in neuronal transmission has been demonstrated by USEPA researchers [54]. The SAR among 3 PCB mixtures, 24 individual congeners, and in vitro perturbed cellular Ca^{2+} homeostasis and protein kinase C (PKC) translocation reveals that: (i) congeners with *ortho*-chlorine substitution such as 2,2'-DCB or *ortho* lateral (*meta, para*) chlorine substitution such as 2,2',5,5'-TeCB are most potent; (ii) congeners with only *para*-substitution such as 4,4'-DCB or high lateral content in the absence of *ortho* substitution such as 3,3',4,4',5,5'-HCB are not effective; (iii) increased chlorination was not clearly related to the effectiveness of these congeners, although hexa- and hepta chlorination was less effective than di-

and tetra chlorination. Low lateral substitution, especially without *para*-substitution, or lateral content in the presence of *ortho*-substitution, may be the most important structural requirement for the in vitro activity of these PCB congeners in neuronal preparations [54].

3
Biotransformation

As a general rule, PCB congeners with unsubstituted adjacent *meta* and *para* positions, and congeners with unsubstituted adjacent *ortho* and *meta* positions and no *ortho*-chlorine are most susceptible to oxidative metabolism, catalyzed predominantly by Cytochrome P450-dependent monooxygenases [52]. Some congeners (for example, CB-153 (2,2′,4,4′,5,5′-hexa CB) are hardly metabolizable and are retained in the bodies of Yusho patients for more than 20 years [55]. In general, the CYP1A subfamily shows a high affinity for planar molecules, but the CYP2B family shows a high affinity for globular molecules. The rate of biotransformation varies among species. For example, dog CYP2B2 can metabolize CB-153 at a much faster rate than rat and humans. The polar bear shows an ability to metabolize CBs without any vicinal H atoms which has not been reported for any other marine mammal species [56]. For an excellent overview on the biotransformation of PCBs in marine mammals see [52].

The metabolism of CBs in an organism has a direct bearing on their toxicity. For example, Yoshimura et al. [57] demonstrated that the toxic or biological effects such as thymus, spleen, and liver hypertrophy and increase of liver lipids or CYP1 A enzymes by planar CBs are mainly due to unmetabolized parent compounds whereas the toxicity/biological effects arising from globular CBs (*ortho* substituted) are due to their metabolites. However, the hydroxy metabolites of planar CBs have been associated with alterations in vitamin A and thyroid hormone metabolism while methyl sulphone metabolites have been implicated as possible toxic agents in the lung and adrenals and foetuses [58].

Studies on wildlife species suggest that metabolite-related toxicity is only observed in species that have the capacity to produce hydroxy metabolites of relatively planar congeners (harbour seals, cormorants, marmoset monkey, mice, and rats). In some species CYP1A induction is probably not accompanied by the capacity to metabolize planar CBs (eider ducks), while in another species the lack of CYP1A induction is accompanied by metabolite production of planar CBs (DBA/2 mouse) [58].

There are consistent attempts in the literature to interpret specific biotransformation of PCBs that occurred in an organism from the relative ratio of X-CB to that of a reference CB-Y. The later could be CB-153, CB-138 (2,2′,3,4,4′,5-hexa CB), and CB-180 (2,2′,3,4,4′,5,5′-hepta CB) [52]. Such structure-dependent metabolism studies predicted induction of Cytochrome P450 isozymes in the marine food chain [59], small cetaceans [52], beluga and narwhal [60], river dolphins [61], pinnipeds [52], aquatic fauna including invertebrates [62], and in humans [63]. Comprehensive characterization of xenobiotic metabolizing enzymes in whales and pinnipeds has been carried out to show the utility of this approach [51, 64].

Using molar X/153 ratios in organisms of a coastal ecosystem Kannan et al. [59] constructed "metabolic slopes" for various PCB congeners and predicted ecotoxic stress. According to these authors the "metabolic slopes" are useful in assessing CYP1A and CYP2B isozymes using precise chemical data, even in the absence of non-*ortho* CB measurement that needs expensive instrumentation. The metabolic slope risk assessment is proposed as alternative or complimentary to the widely used 2,3,7,8-TCDD TEFs risk assessment which needs precise determination of non-*ortho* CBs.

4
Ecoepidemiology

Several recent studies correlating the species extinction with coplanar PCB contamination come from the Great Lakes region of Northern America. These ecotoxicological approaches takes into account the exploitation of fish stocks, sea lamprey predation, and physical habitat destruction in the case and effect scenario as well. For example, the extinction of Lake Ontario deep water sculpin (*Myoxocephalus quadricornis*) between 1953 and 1964 occurred at the time that loading of 2,3,7,8-TCDD and other organochlorine compounds were exponentially increasing from US chemical manufacturing along the banks of Niagara River [65]. Recent studies on egg and sac fry mortality indicate that this fish species is the most sensitive among all the vertebrates so far tested. Isomer-specific chemical analysis revealed that maternally transferred PCBs reduce egg hatchability of Lake Michigan lake trout and that fry survival in lake trout is controlled by a severe mortality syndrome affecting fry just before swim-up and is reminiscent of the syndrome and associated behavioral signs observed 35 years ago by New York State researchers [66]. Signs of PCB-126 toxicity in lake trout early life stages were yolk-sac edema, multifocal hemorrhages, craniofacial malformations, and mortality identical to toxicity caused by 2,3,7,8-TCDD. Non-*ortho* coplanar CB-77 and CB-126 are synergistic or additive in causing these syndromes with TCDD [67].

Intraperitonial injection of white perch (*Morone americana*) with CB-77 reduced gonadal somatic indices (GSIs) and caused complete failure of gonadal maturation in some individuals [68]. Similar study in *Fundulus heteroclitus* of CB-77, CB-126, and mono-*ortho* CBs (CB-105, CB-118, CB-156, CB-157, CB-167, and CB-189) mimicking the mixture found in fish collected from New Bedford Harbour, Massachusetts, USA revealed 58% female mortality, reduced egg production by 77% at the highest dose compared to control. Significant residue effects linkages were found with TCDD toxic equivalents emerging as a potential indicator of adverse effects [69]. The same authors have studied accumulation of coplanar PCBs, survival, and reproduction of the same fish along a gradient of increasing sediment contamination and found out that significantly greater mortality (30% and 23%) was observed in a contaminated site compared to a less contaminated site (0%). Progeny of fish from the most contaminated station exhibited significantly reduced survival (49%) and greater incidence of spinal abnormalities (26%) compared to the other site (70% survival, 7% spinal abnormalities) [70].

Common tern (*Sterna hirundo*) breeding in the sedimentation area of Rhine and Meuse rivers, were affected by non- and mono-*ortho* CB contamination, compared to the clean inland colony near Zeawolde [71]. Studies on double crested cormorants (*Phalacrocorax auritus*) using CB-126, 2,3,7,8-TCDD, and extract derived from field collected eggs revealed that chicken (*Gallus domesticus*) is at least 70 times more sensitive to developmental abnormalities such as pronounced edema [72].

In order to investigate properly the causes of injury to wildlife, an ecoepidemiological approach involving population ecologists, veterinary pathologists, experimental toxicologists, analytical chemists, biochemists, and sedimentologists is needed [73]. This needs new initiative as people from these disciplines have never worked collaboratively.

5
Congener-Specific Analysis

The historic breakthrough in 1966 [8] in finding PCBs as analytical interferences of pesticide analyses has reversed the importance from DDTs to PCBs and methodologies were then devised to separate PCBs from pesticides such as DDTs (DDT, DDE and DDD), toxaphene, lindane, and dieldrin efficiently.

The findings on the structure-biological activity of 2,3,7,8-TCDD and its stereoisomers have made it necessary for analytical chemists to devise methods to determine non-*ortho* PCBs and its *ortho* analogs. However, the general mood in the early 1980s was of scepticism as exemplified in a statement in the Outlook column of Environmental Science and Technology [74]: "The pure P-448 inducers, 3,3′,4,4′-tetra and 3,3′,4,4′,5,5′-hexa chlorobiphenyls are present in commercial mixtures in the low ppb quantities – if at all – too low to have biological significance."

An early attempt to isolate PCBs into structural classes began in 1974 when Jensen and Sundström [9] showed that activated charcoal can separate PCBs based on their structure. The need for high resolution congener-specific analysis of PCBs was mentioned in a paper in 1980 [75]. The inadequacy of quantifying PCBs in environmental samples using Aroclor mixtures as standards was shown. Ballschmiter and Zell [76] in the same year presented a detailed analyses of seven technical mixtures and proposed a scheme for systematic numbering of PCBs. The Columbia National Fisheries Researchers [77] have devised an analytical scheme including non-*ortho* PCBs, PCDD/Fs, and pesticides as part of their pesticide monitoring program. However, no actual measurements of non-*ortho* PCBs were measured in environmental samples. Using a narrow bore SE-54 capillary column, Mullin et al. [78] characterized 187 of 209 PCB congeners. The significance of congener-specific analysis for assessing the potential environmental and human health impact of PCBs was shown by measuring several important mono-*ortho* PCBs in human milk [79]. In the absence of an easy method to measure non-*ortho* CBs and a risk assessment analysis, Safe et al. [79] observed that non-*ortho* CBs such as CB-77, CB-126, and CB-169 were unlikely to be the sole contributors to the toxicity of PCB mixtures.

However, this situation changed when Tanabe and his co-workers devised an analytical method with rigorous clean-up, enrichment of non-*ortho* CBs using charcoal chromatography, and congener-specific determination using HRGC-ECD and HRGC-MS [31, 32]. They quantified these toxic PCBs in technical mixtures [15], in wildlife [11], and in humans [80].

In the same period, researchers at the University of Kiel devised a method using multidimensional gas chromatography-ECD for direct measurement of non-*ortho* CBs in commercial mixtures [81]. This method utilized the high resolution separation power of two capillary columns in series and permitted a selected portion of a mixture to be moved from the first column to the second without loss using a pneumatically controlled live-T piece (heart-cut). Using this method, they [13] resolved the 11 pairs of unresolved congeners from the study of Mullin et al. [78] and corrected minor errors in the Ballschmiter and Zell [76] scheme of PCB numbering.

The analytical chemistry of non- and mono-*ortho* PCBs was reviewed in the early 1990s [7, 14] and 40 methods had been reported up to then, stressing the importance of these compounds.

The subsequent development in analytical chemistry went in two directions, one in enriching these compounds using adsorbent materials and the other in utilizing the separation power of capillary gas chromatography for a direct measurement using sensitive detectors.

The often used adsorbents for the enrichment of non-*ortho* CBs are: charcoal/carbon dispersed on glass fiber or silica [82], porous graphitic carbon (HPLC grade) [83], and florisil or alumina [84, 85]. A liquid chromatographic method using 2-(1-pyrenyl) ethyldimethylsilylated silica (PYE) in environmental analysis has been shown [86]. Most of these materials have been successfully incorporated into automated analytical procedures in recent times. Porous graphitic carbon HPLC has been hyphenated with GC-ECD and GC-MS [87]. The widely used Amoco PX-21 has been newly dispersed on C18 (octadecylsilane) sorbent to suit automation [88]. This material has been used in a supercritical fluid extraction instrument as a solid phase trap and hyphenated with GC-MS [89]. A coupled column HPLC system consisting of a nitrophenylpropylsilica (nitro) column (Nucleosil) and PYE (Cosmosil) column has been described [90] for subsequent GC-MS analysis of PCBs, PCDD/Fs, PAHs, and related compounds in complex environmental matrices. This separating principal based on aromaticity followed by planarity has been utilized in a recent MDGC-ECD analysis [91]. An off-line PYE HPLC-HRGC-MS (NCI) combination has been described for non- and mono-*ortho* CBs in herring and polecat [92]. An amino column (aromaticity) and hypercarb (planarity) column in combination with GC-MS has also been reported [93].

The second approach is the direct measurement of non-*ortho* CBs using high resolution GCs (multidimensional) and sensitive detectors. In line with Mullin et al. [78] and Duinker et al. [81], de Boer et al. [94] have described retention times for 51 CBs in 7 narrow bore capillary columns. Larsen et al. [95] analyzed 140 CB congeners in technical mixtures using five narrow bore columns. Their choice of the stationary phase for CB-77, CB-126, and CB-169 analysis is 5% diphenyl dimethyl siloxane (SIL-8) and 1,2-dicarba-*closo*-dodecaborane dimethyl siloxane (HT-5) [96]. Frame et al. [97] have reported relative retention times of all 209 congeners on 27 HRGC-ECD or HRGC-MS instruments using 20 different stationary phases. A method for AHH-inducing PCBs using HRGC-MS (NCI) points out that even mass specific analysis is no guarantee for error free quantitation if coelution occurs at gas chromatographic level [98]. The need for a correction factor in mass spectrometric analysis is mentioned as well [82]. The resolution power of MDGC in combination with tandem MS (mass selective and NCI) has been the thrust of a Finnish team [99].

A recent development in this field is the introduction of comprehensive two dimensional gas chromatography [100] where the entire chromatogram eluting from the primary column is submitted to a secondary column. An on-column thermal modulator collects the sample portions from one to another. The resulting two-dimensional chromatogram has peaks scattered about a plane rather than a line (orthogonal projection). This method has been tested for PCBs in an unpublished study [101].

A recent review examines the background contamination by coplanar PCBs in trace level HRGC-HRMS analytical procedures [102]. De Voogt et al. [103] have reported the results of an inter laboratory exercise on non- and mono-*ortho* PCB analysis and pointed out that the identification and quantitation of these compounds was a challenging task. A collaborative study between Japan and Germany [104] in evaluating non- and mono-*ortho* PCB analysis has pointed out that choice of the adsorbent for enrichment is the first critical step, as many charcoal preparations are from petroleum or of plant origin with inherent background signals that might interfere in the analysis. The choice of the solvent and volume used in the separation procedure is important as well. The widely used solvents such as dichloromethane, benzene, ethyl acetate, and toluene are capable of absorbing PCBs from the laboratory air. While the automation techniques are time and labor saving, they use large volumes of these solvents with background concentrations, thereby reducing the sensitivity of the measurement [105]. In this respect, PYE column HPLC utilizes only hexane as a solvent with very low solvent volumes, suitable for non- and mono-*ortho* PCB analysis [91]. Recovery of toxic PCBs from different environmental matrices is still a daunting problem while recovery studies based on artificial spiking do not truly represent environmental samples. A matrix integrated quality control procedure has been proposed [91].

Larsen [106] has reviewed the field of PCB analysis using high resolution chromatography. Frame [16], while reviewing the need for congener-specific analysis of PCBs, criticizes the continued usage of Aroclor and Kanechlor mixtures as laboratory standards for radically altered mixtures of PCBs typically found in the environment. This review further points out that it is not enough in environmental or biological monitoring studies of PCBs to focus on few target congeners; instead, knowing the levels of other congeners or even all of them is important and may hold the key to answering new questions.

6
Source and Formation

It has been estimated that 1,054,800 tonnes of PCBs had been produced by the western industrialized nations to the end of 1980 [1]; the USA contributed 61.4%. The other major producers of PCBs and their percentage contributions are Germany 12.4, France 9.6, UK 6.3, Japan 5.6, Spain 2.4, and Italy 2.2. The estimated cumulative production and consumption of Sovol in the former USSR during the period from the 1940s to the 1990s was about 100,000 tonnes [107].

The total load that has escaped into the environment is estimated to be about 370,000 tonnes. The rest, roughly 780,000 tonnes, is still in use, mainly in electrical equipment [108]. Since commercially produced PCBs are the ultimate source of CBs in the environment, the contents of 3,3',4,4'-TCB have been reported in a few mixtures [109–111]. The first successful attempt in determining all the three non-*orthos* (CB-77, CB-126, and CB-169) in four Japanese and American mixtures came from Kannan et al. in 1987 [15]. A study in Kiel quantitated CB-77 and CB-169 when they occurred at > 0.05 wt% in German and American mixtures [13]. The most recent attempt at characterizing 17 Aroclor mixtures using a standard containing all the 209 CB congeners is from Frame et al. [97].

Table 2 gives the reported content of four major non-*ortho* CBs, namely CB-77, CB-81, CB-126, and CB-169 in several Aroclor mixtures. Their content in Kanechlor (Japanese) and Chlorofen (Polish) mixtures have been reported [15, 112]. Several mono- and di-*ortho* CBs in Sovol have been reported [107, 113]. The content of all the mono- and di-*ortho* congeners in Clophen and Aroclors have been reported in [13, 95, 97].

Recent studies indicate that non-*ortho* congeners are produced de novo during incineration processes [17, 18]. Production of toxic CBs through successive dechlorination by sunlight irradiation has been shown under laboratory conditions [114].

7
Global Occurrence, Transportation, and Fate

7.1
Air, Water, and Sediment

Congener-specific analysis of PCBs in air was not feasible until recently [20] due to levels being often in the sub-picogram per cubic meter range. The reported levels of CB congeners in urban and remote regional air are given in Table 3.

The PCB flux and airborne concentration is much greater in urban areas than in non-urban areas [115]. Iwata et al. [116] observed PCBs over Lake Baikal in the range 8.7–23 pg m^{-3} and concluded that the magnitude of PCB volatilization was smaller than in Great Lakes. A study conducted a year later [117] refuted this finding and showed that air concentrations are comparable to that in Great Lakes. The implication is that larger lakes like Baikal and Great Lakes

Table. 2. Non-*ortho* coplanar congeners in Aroclor mixtures (wt% contribution) – a comparison[a]

No.		1996[b]	1993[c]	1992[d]	1990[e]	1989[f]	1987[g]	1985[h]	1981[i]	1980[j]	1979[k]
	CB-77										
1	1016	NM	NM	NM	NM	< 0.05	NM	NM	NM	ND	NM
2	1221	0.01	NM	NM	NM	NM	NM	NM	NM	NM	NM
3	1242	0.27 – 0.33	0.22	0.25	0.22	0.45	0.51	0.35		0.24	0.23
4	1248	0.41 – 0.52	0.35	0.4	Nm	NM	0.62	0.65	0.47	0.34	0.24 – 0.33
5	1254	0.03 – 0.2	0.022	0.12	0.25	< 0.05	0.062	0.11	0.12	0.02	ND
6	1260	ND	0.006	0.0038	0.017	< 0.05	0.026	0.11	0.04	ND	ND
	CB-81										
1	1016	NM	NM	NM	NM	< 0.05	NM	NM	NM	NM	NM
2	1221	ND	NM	NM	NM	NM	NM	NM	NM	NM	NM
3	1242	0.01	0.014	0.016	NM	< 0.05	NM	NM	NM	NM	NM
4	1248	0.01 – 0.02	0.026	0.027	NM	NM	NM	NM	NM	NM	NM
5	1254	ND	0.001	0.0062	NM	< 0.05	NM	NM	NM	NM	NM
6	1260	ND	< 0.0005	< 0.0001	NM	< 0.05	NM	NM	NM	NM	NM
	CB-126										
1	1016	ND	NM	NM	NM	< 0.05	NM	ND	NM	ND	NM
2	1221	ND	NM	NM	NM	NM	NM	ND	NM	NM	NM
3	1242	ND	0.002	0.0037	0.003	< 0.05	0.0019	ND	NM	< 0.025	NM
4	1248	ND	0.0074	0.011	NM	NM	0.0052	ND	ND	< 0.025	NM
5	1254	0.02	0.0033	0.027	0.016	< 0.05	0.0038	ND	0.16	< 0.025	NM
6	1260	ND	< 0.0003	0.0004	< 0.007	< 0.05	0.00032	ND	1.59	< 0.025	NM
	CB-169										
1	1016	ND	NM	NM	NM	< 0.05	NM	NM	NM	NM	NM
2	1221	ND	NM	NM	NM	NM	NM	NM	NM	NM	NM
3	1242	ND	< 0.0001	< 0.0001	< 0.007	< 0.05	< 0.000008	NM	NM	NM	NM
4	1248	ND	< 0.0002	< 0.0001	< 0.007	NM	< 0.000008	NM	NM	NM	NM
5	1254	ND	< 0.0001	< 0.0001	< 0.007	< 0.05	0.000051	NM	NM	NM	NM
6	1260	ND	< 0.0001	< 0.0001	< 0.007	< 0.05	< 0.000008	NM	NM	NM	NM

[a] These values are accurate to the lot tested. Lot-to-lot variation in the content of CB congeners are reported. ND = not detected; NM = not measured; [b] [97]; [c] [187]; [d] [188]; [e] [189]; [f] [13]; [g] [15]; [h] [190]; [i] [191]; [j] [110]; [k] [109].

Table 3. Concentration (pg m^{-3}) of PCBs in ambient air from urban and pristine areas

Reference	1[a]	2[b]	3[c]	4[d]	5[e]	6[f]	7[g]
Country	USA[h]	USA[i]	Nether-lands	Germany	Russia	Norway	Canada[j]
Location	Urban[l]	Urban[l]	Urban	Urban	Lake Baikal[l]	Arctic air[k]	S.Ontario[l]
Non-*ortho* CBs							
77			1.85				
126			0.08				
169			0.013				
Mono-*ortho* CBs							
66	19.11	75.05					6.5
70	12.24	47.36			6.9		4.6
105			2.21			0.79	0.16
114	1.42	1.81	0.18				1.2
118	2.25	6.97	6.03	34.2	8.3	2.5	2.3
123			0.10				
156			1.04			0.15	0.07
157	0.02		0.17				
167			0.53				0.13
189			0.22				0.01
Di-*ortho* CBs							
4	2.16	7.01					
47	5.18	37.85					8.3
52	9.92	69.47		114.6		24	16
101	4.67	20.16		83.5		10	6.4
110	11.33	18.65			11.3		4.0
128	0.74	1.01					0.30
138	2.92	7.72		28.6	5.6	2.10	2.8
153	4.05	14.80		24.6	9.2	3.0	3.2
170	1.16	1.49					0.48
180	1.36	3.89		2.0	0.5	0.35	1.10
194	0.73	0.30					0.11

[a] [192].
[b] [118].
[c] [193].
[d] [123].
[e] [117].
[f] [20].
[g] [194].
[h] Western Green Bay.
[i] Southern Green Bay.
[j] Annual mean.
[k] Jul 6–8, 1993.
[l] Not strictly congener-specific.
Coelution occurred with CB-4 (+10), CB-47 (+48), CB-52 (+43), CB-66 (+95), CB-70 (+76), CB-114 (+134), CB-118 (+149), CB-153 (+105 and +132), CB-138 (+163), CB-170 (+190). Cells without values are not measured.

act as secondary (at times major) source of PCBs to the local atmosphere [118]. On the global transport of these contaminants, the concentrations are still higher in the northern hemisphere compared to the southern hemisphere, although the distribution pattern suggests a shift or expansion of their major sources from mid to low latitudes over the last decade [19, 119, 120].

Although it is generally believed that the world ocean is the ultimate sink of these chemicals there are only a few reliable reports on the concentration of PCBs in ocean water. This is largely due to the fact that reliable sampling has proven to be difficult as their concentrations in ocean water are extremely low [121].

The improvement in water sampling techniques such as the development of the Kiel *in-situ* Pump (KISP) [122] and high resolution gas chromatography (such as MDGC-ECD) has resulted in a series of papers on the concentration of di-, mono-, and occasionally non-*ortho* CB levels in world waters. This is detailed in Table 4.

It is obvious from this table that the concentrations of non-*ortho* CBs and, to a certain extent, mono-*ortho* CBs are extremely low ($<$ ppq). On the other hand, di-*ortho* CBs are measurable. In general, PCB concentrations are higher in rain, well, tap, and bottled waters which are close to a contamination source (urban environment) [123–126]. Congener-specific reports of PCBs in inland lakes such as Great Lakes and Baikal have been published [117, 127].

Regardless of solubility or chlorination, the majority of PCBs are associated with the particulate phase. Congener-specific PCB data on suspended particulate matter are available in [128–130]. Toxic potencies of settling particulate matter in a PCB-contaminated lake in Sweden has been reported recently [131].

Suspended and bottom sediments are widely regarded as a sink for PCBs released into aquatic systems [108]. Sorption of non-*ortho* CBs is stronger than other CBs with the same degree of chlorination [132]. Sediments fairly reflect water pollution in a region from the past. Trends in assessment of PCBs in lake sediments of the Großer Arbersee (Germany) over the past 130 years (1860–1990) showed that levels of non- and mono-*ortho* CBs were found to be highest in around 1968–1972. The di-*ortho* CBs reached their maxima earlier in 1968–1972 [133].

PCB concentrations in air and water bodies have declined since efforts to reduce their escape into the environment were enacted in the 1970s. With falling PCB fugacities in the air, soils, and sediments may release previously deposited PCBs to water bodies and to the atmosphere [134]. This aspect needs further consideration in a congener-specific manner.

7.2
Aquatic Organisms

7.2.1
Invertebrates

A transplantation experiment with green lipped mussels in Hong Kong waters raised concern over the bioaccumulation potential of coplanar PCBs and their

Table. 4. Concentration of non-, mono-, and di-*ortho* CBs in salt and sweet waters around the world

Concentration	Salt water							Sweet water			
	(fg dm^{-3}	(pg dm^{-3})					(ng dm^{-3})	(pg dm^{-3})			
Reference	1[a]	2[b]	3[c]	4[c]	5[d]	6[e]	7[f]	8[g]	9[h]	10[i]	11[j]
Location	Japan Sea	N. Atlantic	North Sea	Skagerrak	Baltic	Mediterranean	N.B. Harbour	Germany	Germany	Japan	France
Sample	500 m	250 m	Surface	Surface	Surface	Surface	Surface	Well Water	Rain Water	Tap water	Bottled Water
Non-*ortho* CBs											
77	<0.05	< 0.05	0.3	0.1	< 0.1	< 0.01 – 0.05	900	1.6	NM	1.35	NM
126	<0.05	< 0.05	< 0.1	< 0.1	< 0.1	< 0.01 – 0.05	30	0.1	NM	0.07	NM
169	<0.05	< 0.05	< 0.1	< 0.1	< 0.1	< 0.01 – 0.05	< 30	<0.1	NM	0.03	NM
Mono-*ortho* CBs											
70	69	0.61	24.4	9.1	1.3	1.9	NM	19.5	NM	NM	NM
105	<2.0	< 0.05	4	0.1	< 0.1	< 0.01 – 0.05	2000	8.3	NM	NM	NM
118	58	0.6	11.6	3	3.1	0.5	4400	26.1	241.4	NM	NM
156	6.4	< 0.05	1.3	0.7	1.0	< 0.01 – 0.05	NM	1.7	NM	NM	NM
157	<2.0	< 0.05	0.5	< 0.1	< 0.1	< 0.01 – 0.05	NM	<0.1	NM	NM	NM
Di-*ortho* CBs											
52	76	0.9	28	43.3	3.0	3.1	25000	46.2	639	NM	740
101	110	1.7	29.8	18.7	4.7	1.9	7100	<0.1	535.1	NM	1200
110	121	1.13	30.2	11	2.1	1.0	NM	24.1	NM	NM	NM
128	2.6	0.6	3.9	0.7	< 0.1	1.0	200	6.2	NM	NM	NM
138	58	4.5	29.1	11.5	11.5	2.9	1300	52.6	936.7	NM	1200
153	56	2.7	17.1	12.6	1.3	1.2	2000	38.3	736.6	NM	870
170	7.2	0.8	22.6	0.3	1.8	< 0.01 – 0.05	200	7.9	NM	NM	NM
180	10	1.7	22.6	1.4	2.9	< 0.01 – 0.05	100	17.2	261.3	NM	920
194	<2.0	< 0.05	2	< 0.1	< 0.1	< 0.01 – 0.05	NM	<0.1	NM	NM	NM

[a] [195]; [b] [121]; [c] [128]; [d] [129]; [e] [130]; [f] [138]; [g] [124]; [h] [123]; [i] [125]; [j] [126]; NM = Not measured.

health impact to humans in 1989 [135]. Follow up studies in other species of bivalves and in other parts of the world not only confirmed this concern but showed that:

1. The depuration half-lives of non-*ortho* CBs are much longer than observed before, leading to enrichment. For example, CB-126 from 0.26% to 0.62% and CB-169 from 0.3% to 1.4% of the total PCB load in Galveston and Tampa Bays, USA [136].
2. Zebra mussels were shown to accumulate these toxic congeners from food and sediments at greater levels in Great Lakes waters [137].
3. Blue mussels showed that steady-state concentration of CB-77 and CB-126 was settled faster than of non-planar CBs in uptake studies [138].

To understand the toxicological significance of sediments in Saginaw River water shed (Michigan, USA) that contain PCBs, a sediment-based food chain accumulation study using emergent aquatic insects, eggs, and nestlings of tree swallows was conducted [139]. TEQs derived from several non-mono- and di-*ortho* CBs in these organisms showed that toxic potentials of PCB mixtures change as a function of trophic level in the following order: invertebrates $(0.3) \gg$ tree swallow egg $(0.8) \gg$ tree swallow nestlings (1.0). This study also showed that biota-sediment accumulation factors (BSAFs) values measured in the field were close to the predicted values using fugacity theory.

7.2.2
Fish

Elimination rate constants study for non-, mono-, and di-*ortho* CBs in rainbow trout showed that substitution pattern influenced the elimination kinetics; for example, CB-126 was more slowly eliminated than di- and mono-*ortho* congeners of similar k_{ow}, supporting the conclusion that toxic CBs declined more slowly in biota than other congeners [140]. Non- and mono-*ortho* CBs have been determined in 16 species of fishes from Netherland's coast and inland waters [141]. In all these species, potential toxicity (as represented TEQs) arise from CBs-126, CB-156, and CB-118 and the TEQs are at least four times higher than those derived from PCDD/Fs. Reports carrying congener-specific PCB data on various fish species from all around the world are available.

7.2.3
Aquatic Mammals

A series of mass mortalities in aquatic mammals occurred in the last decade. The North Sea harbor seal epizootic of 1988, the US bottle nose dolphin die-off of 1987–1988, mass mortality of Baikal seals in 1987–1988, and mortality in Mediterranean striped dolphin in this decade (1990–1991) are well known. Though the primary cause of death is known to have been infection with a virus, it is highly possible that immuno suppression induced by toxic CBs may have contributed additionally [123]. Congener-specific data on marine mammals to elucidate this fact appeared only in the mid-1980s.

High levels of non-, mono-, and di-*ortho* CBs in mammalian tissues had been reported by Ehime University researchers consistently from the late 1980s [11, 12, 15]. Very high contents of non-*ortho* CBs have been reported from stranded whales, dolphins, porpoises, and seals in Scottish waters (ranging from 120 ng/g to 1300 ng/g wet weight) [142]. Dolphins affected by Mediterranean epizootic contained 4–42 ng/g wet weight of non-*ortho* CBs [143, 144]. Biomagnification (in lipid weight) of toxic CBs in a fresh water ecosystem having otter as the top predator has been shown recently [145]. Selective enrichment of CB-126 in this food web and CB-169 in Mediterranean dolphins has been reported.

Among marine mammals, seals are well studied. Congener-specific data are available for common seals in North Sea [52], ringed, and grey seals from Finish waters [146, 147], harbor seals from the Danish coast [148], ringed seals in the Savalbard area [149], harbour seals from US inland waters [150], and ringed seals from Lake Baikal, Russia [151].

Small cetaceans from the North Sea and Atlantic Ocean, North Atlantic coast of US, Baltic Sea, Scandinavian waters, and Atlantic coast of Canada were analyzed for specific CB congeners and a few large whale species have been analyzed thoroughly for PCBs, such as the sperm whale, beaked whale, killer whale, minke, fin, blue, and humpback whales [52].

Among the marine mammals polar bears have the highest capacity to metabolize CBs leaving non-detectable amounts of non-*ortho* CBs [152].

In general, among the non-*ortho* CBs in marine mammals, CB-126 contribute the most to TEQs with additional contributions from mono-*ortho* CBs such as CB-105, CB-118, and CB-156. TEQs in aquatic mammals (Fig. 1) reported in this decade show that animals inhabiting the North Sea ecosystem have one of the highest toxic CB burdens (50–93 ng/g wet weight Σ TEQs) followed by otters in Netherlands inland waters (Σ TEQs = 13 ng/g wet weight) which are much higher than the no-observed effect level of 1.0 ng/g of TCDD for humans and the body burden of Yusho and Yu-cheng victims [153].

7.3
Birds

The causal relationship between the occurrence of non-*ortho* CBs and other dioxin-like compounds and reproductive impairment in Forster's tern was firmly established in 1989 by US Great Lakes researchers, [154, 155]. An improvement in reproductive performance of the same population between 1983 and 1988 was recorded [156]. This was due to low river flow and consequent reduction in contamination input to the Lake ecosystem. TEQs derived from congener-specific PCB analysis showed 42% reduction in chemical toxicity. Subsequent studies in Green Bay, Wisconsin, USA by USEPA established uptake rates for specific CB congeners in Forster's tern chicks. For example, 70 ng/day, 200 ng/day, 6.5 ng/day, and 0.14 ng/day for CB-77, CB-105, CB-126, and CB-169 respectively and 270 pg/day for TEQs [157]. Determination of PCDD/Fs and PCBs in California peregrine falcons and their eggs indicated that CB-126 was largely responsible for the TEQs derived. The other CBs of importance were CB-126 < CB-77 < CB-156 < CB-105 < CB-118 < CB-169 [158]. Occurrence of non-,

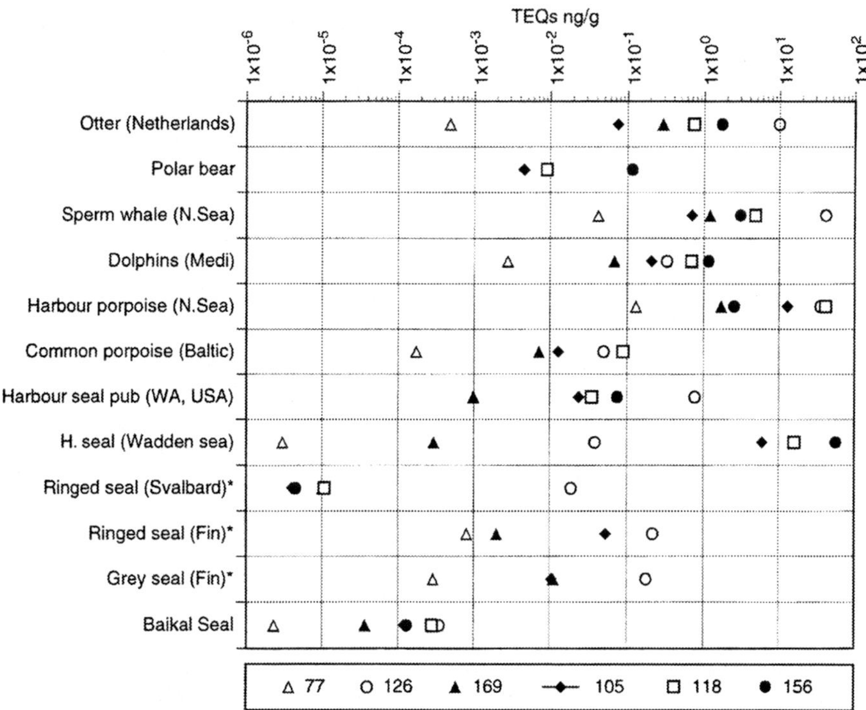

Fig. 1. TEQs (ng/g wet weight, except in cases marked by * (lipid weight) in aquatic mammals. The TEF values are after EPA. The CB contents in animal tissues (fat) are obtained from the following references: otter [145], polar bear [152], sperm whale, harbor porpoise [142], dolphins [144, 143], common porpoise [200], harbor seal pup (USA) [150], harbor seal (Wadden Sea) [148], ringed seal (Svalbard) [149], ringed and gray seals (Finland) [201], Baikal seal [151]

mono-, di-*ortho* CBs, and PCDD/Fs have been detected in albatrosses living in pristine locations such as Midway Atoll in the Pacific Ocean [159]. Their study confirmed an earlier prediction by Kannan et al. [12] that impact of planar PCBs and PCDFs will be greater than PCDDs in organisms living far away from land. Common cormorants from Biwa Lake, Japan showed unusual biomagnification patterns. The highly biomagnified congeners were CB-126, CB-153, CB-169, CB-180, and CB-194 [160]. Similarly, a peculiar pattern of TEQs in tree swallows was noticed, where CB-169 gave greater values than CB-126, CB-77, and CB-105 [161]. A high degree of biomagnification of CB congeners with *meta, para* vicinal H atoms in cormorants suggests activity of CYPIIB enzymes. In fact a very high correlation (r = 0.94) between PROD activity and total TEQs was shown in the Japanese study. In addition, a significant increase in EROD activity was measured as well [160]. The appropriateness of the TEF concept in cormorants using the following end points has been evaluated recently [162], namely, mortality at hatching, the lowest dose at which EROD activity was significantly greater than control activity, and the lowest dose to have an effect on spleen weights. The TEF value of 0.02 proposed for CB-126 in cormorants is close to TEFs reported for the chicken, namely 0.05 and 0.07. The authors concluded from this study that TEF concept could be applied across species even when sensitivity differences occur between species to a given compound.

7.4
Humans

A number of reports appeared after Kannan et al. in 1988 [80] showed that the most toxic non-*ortho* PCBs bioaccumulated in the Japanese population.

The highest CB-77, CB-126, and CB-169 human body burden has been shown in a French study [163]. These levels are higher than what was reported in Yusho victims [33]. A Dutch study involving 198 mothers [164] and a Swedish study involving 1072 mothers (between 1972 and 1989) [165] are, so far, the largest sampled works. A study in Finland involving 237 mothers [166] indicates that the body burden in Nordic countries is more or less the same and is one of the lowest reported (Table 5). Potentially very toxic CB-126 has been reported in all the studies on human beings. Persistency of CB-169 has been confirmed and the metabolizability of CB-77 has been shown.

A significant decline in residue levels of several PCB congeners, including non-*ortho* CBs, in breast milk between the first and third breastfed child has been shown, indicating lactational transfer of PCBs in human beings [167]. A clear difference in CB contents among rural and urban Dutch population is recorded [164]. Higher content of toxic CBs in occupationally exposed workers than that of the common public is reported [63, 168]. A German study [169] shows that coplanar PCB content in human milk is 10 to 50 times lower than their values in bovine milk.

Using congener ratios in adipose tissue, liver, and whole blood it was shown that Cytochrome P450IA1 and IIB enzymes are inducible in the common public, Yusho victims and capacitor workers [33, 63, 80]. Lucier [170] showed

Table. 5. Content of non-*ortho* coplanar CBs in humans

Country	Tissue	Unit	(n)	CB-77	CB-126	CB-169	Reference
France	Milk	pg/g fat	20	39,000	45,000	29,000	[163]
Japan	Fat (Yusho)	pg/g wet	1	700	720	380	[33]
Japan	Adipose	pg/g wet	12	345	333	90	[80]
Japan	Milk (normal)	pg/g fat	9	12	183	66	[196]
Japan	Milk (Yusho)	pg/g fat	1	12	160	780	[196]
Japan	Liver (Yusho)	pg/g wet	1	130	54	50	[33]
Italy	Fat	pg/g wet	23	323	393	236	[197]
Sweden	Milk	pg/g fat	140	27	98	47	[165]
Finland	Milk	pg/g fat	47	18	171	74	[166]
Germany	Milk	pg/g fat	4	93	240	140	[169]
Germany	Fat	pg/g fat	1	< 1	100	40	[63]
Canada	Fat	pg/g fat	16	8	80	33	[198]
Canada	Fat	pg/g fat	7	< 20	375	200	[199]
USA	Whole blood	pg/g fat	50	79	104	46	[171]

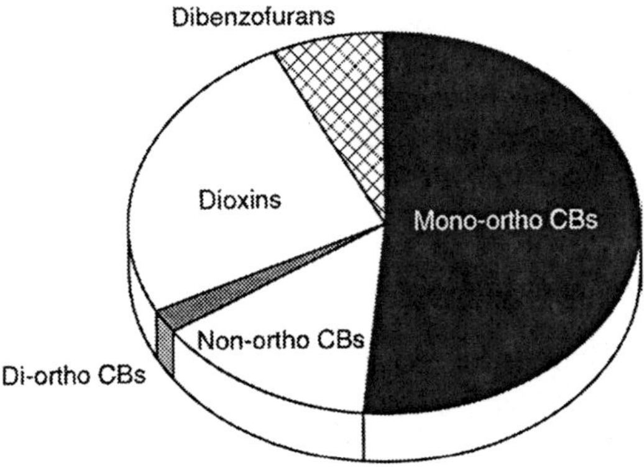

Fig. 2. Average percent contribution to TEQs in blood of Michigan veteran (after [171])

that humans are more sensitive than rats to the P450 inductive actions of the toxic halogenated aromatics. Toxic evaluation of these residues in comparison to PCDD and PCDF residues using TEFs indicate that PCBs are potentially more toxic to humans than dioxins and furans [80, 171] (Fig. 2). However, it is dibenzofurans that played a major role in Yusho and Yu-Cheng victims [33, 153]. A Dutch study on the effects of perinatal exposure to PCBs on early human development [164] concludes that elevated levels of PCBs can alter human thyroid hormone status, bring forth hypotonia, and affect developing human brain. Evidence from the Michigan maternal/infant cohort related to PCB exposure and alterations in neonatal health status suggests a causal relationship [172].

8
Time Trend

Several interesting publications appeared in past years on temporal trend of PCBs. Analysis of herbage samples (1965–1989) in rural England has shown a decreasing trend of lower chlorinated CBs up to a factor of 50 during 1965–1969 and 1985–1989. High molecular weight CBs have also decreased in concentration but not to such a great extent [173]. Studies on lake sediments in middle Europe showed that PCB concentration increased between 1946 and 1972 with their peak concentrations in 1968–1972 which was 19 times the background in 1884–1890. PCB concentrations have decreased since but even in 1991–1993 they were still 15 times higher than background [133]. Temporal trend in cod-liver oil from southern Baltic proper between 1971 and 1989 showed a decrease at a very slow rate [174]. A similar study on human adipose tissue from Japan (1928–1985) revealed a steady state even after two decades of a ban on production [175]. Levels of non- and mono-*ortho* congeners in Swedish sediment and biota have shown little decline since 1970 [131].

A process of global fractionation of PCBs along the latitude "condensing" at different ambient temperatures has been proposed recently [176]. This has been verified by three independent studies in Norway [177], Canada [178], and Germany [179]. These studies showed that Arctic regions receive a high proportion of lower chlorinated congeners and temporal trends in deposition is delayed relative to temperate regions, as predicted by the model. Future studies may reveal the partition of non-*ortho* CBs in this process.

9
PCB Transformation and Removal

PCBs deposited in fresh water or marine sediment undergo biotransformation. This is well documented in Hudson River sediments where reductive dechlorination by anaerobic microorganisms caused *o,m,p*-dechlorination (primarily determined by reductive potential) and *m,p*-dechlorination (determined by molecular shape) [22]. These authors noted that these processes removed PCBs that are of toxicological concern or bioaccumulative in higher animals. Further, the PCBs formed are known to be removed oxidatively by aerobic microbes. A quantitative mother-daughter analysis of key hexa- and hepta-CBs that were decreased relative to Aroclor 1260 indicated that the most extensively dechlorinated samples in Woods Pond, MA had lost 11–19% of *meta* chlorine and 2–7% of *para* chlorine [180]. A similar process was shown to occur in coastal sediments as well [181]. A reductive dechlorination process, designated as Process H, selectively removed non-*ortho* chlorines to yield a half-life of 8 ± 2 years for CB-77 and CB-126. Incubation of CB-77 with Hudson River microorganisms removed them completely and yielded 3,3′-DCB [23]. Using the Ah-receptor mediated bioassay the toxic potencies of Aroclor 1242 and 1254, before and after dechlorination by microorganisms from PCB-contaminated river and lake sediments, revealed reduction in potencies after dechlorination [182]. No unexpected synergistic interactions occurred among components of the dechlorinated mixtures.

These studies are encouraging in the sense that dioxin-like PCBs are reduced in the environment. At the same time, it is disturbing to note that the *ortho*-substituted CBs that are formed due to microbial dechlorination are exactly the ones that are neurotoxins [54]. Recent studies in Baltic sediments suggest that microbial dechlorination does not occur universally and PCBs in these sediments have been unchanged for the last five decades [183].

Though microbial biotransformation is an important removal process, it is not complete in the sense that the biphenyl backbone is not broken. Moreover, recent studies [116–118] indicate PCBs that flow to terrestrial, fresh water, and ocean surfaces are returned to the atmosphere. So it is important to identify "permanent" PCB sinks; that is, sinks that result in the complete removal of the PCB molecule from atmospheric cycling. Burial of PCBs in fresh water or marine sediments below the resuspension layer and transformation processes initiated by OH attack in the troposphere are identified as major permanent sinks.

A recent mass-balance estimation shows that PCB loss due to fresh water sediment burial is about 2 tonnes per year and to marine sediments about 240 tonnes per year whereas total chemical loss rate for PCBs from the atmosphere is 8300 tonnes per year [21]. The atmospheric loss is substantially larger than the sediment burial. This pathway needs to be further investigated due to its extreme importance.

References

1. Rantanen JH, Silano V, Tarkowski S, Yrjänheikki E (eds) (1987) PCBs, PCDDs, and PCDFs: prevention and control of accidental and environmental exposures. World Health Organisation, Copenhagen
2. McFarland VA, Clarke JU (1989) Environ Health Perspect 81:225
3. McKinney JD (1996) Environ Health Perspect 104:810
4. Safe S (1990) Crit Rev Toxicol 21:51
5. Poland A, Glover E (1977) Mol Pharmacol 13:924
6. Kannan N, Tanabe S, Tatsukawa R (1988) Bull Environ Contam Toxicol 41:267
7. De Voogt PM, Wells DE, Reutergårdh L, Brinkman UDT (1990) Intern J Environ Anal Chem 40:1
8. Jensen S (1966) New Sci 32:612
9. Jensen S, Sundström G (1974) Ambio 3:70
10. Safe S, Safe L, Mullin M (1987) In: Safe S, Hutzinger O (eds) Environmental toxin series. Springer, Berlin Heidelberg New York, p 1
11. Tanabe S, Kannan N, Subramanian A, Watanabe S, Tatsukawa R (1987) Environ Pollut 47:147
12. Kannan N, Tanabe S, Ono M, Tatsukawa R (1989) Arch Environ Contam Toxicol 18:850
13. Schulz DE, Petrick G, Duinker JC (1989) Environ Sci Technol 23:852
14. Creaser CS, Krokos F, Startin JR (1992) Chemosphere 25:1981
15. Kannan N, Tanabe S, Wakimoto T, Tatsukawa R (1987) J Assoc Off Anal Chem 70:451
16. Frame G (1997) Anal Chem 69:468A
17. Sakai S, Hiraoka M, Takeda N, Shiozaki K (1993) Chemosphere 27:233
18. Miyata H, Aozasa O, Mase Y, Ohta S, Khono S, Asada S (1994) Chemosphere 3:458
19. Iwata H, Tanabr S, Sakai N, Tatsukawa R (1993) Environ Sci Technol 27:1080
20. Oehme M, Haugen JE, Schlabach M (1996) Environ Sci Technol 30:2294
21. Anderson PN, Hites RA (1996) Environ Sci Technol 30:1756
22. Brown JFJ, Bedard DL, Brennan MJ, Carnahan JC, Feng H, Wagner RE (1987) Science 236:709
23. Rhee YG, Sokol RC, Bethoney CM (1993) Environ Toxicol Chem 12:1829
24. Ankley G, Mihaich E, Stahl R, Tillitt D, Colborn T, McMaster S, Miller R, Bantle J, Campbell P, Denslow N, Dickerson R, Folmar L, Fry M, Giesy J, Gray LE, Guiney P, Hutchinson T, Kennedy S, Kramer V, LeBlanc G, Mayes M, Nimrod A, Patino R, Peterson R, Purdy R, Ringer R, Thomas P, Touart L, Van Der Kraak G, Zacharewski T (1998) Environ Toxicol Chem 17:68
25. Safe S (1984) CRC Critic Rev Toxicol 13:319
26. Safe S, Bandiera S, Sawyer T, Robertson L, Safe L, Parkinson A, Thomas PE, Ryan DE, Reik LM, Levin W, Denomme MA, Fujita T (1985) Environ Health Perspect 60:47
27. Parkinson A, Safe S (1987) In: Safe S, Hutzinger O (eds) Environmental toxin series. Springer, Berlin Heidelberg New York, pp 49
28. Neal RA (1985) Environ Health Perspect 60:41
29. Bradlaw JA, Casterline JL (1979) J Assoc Off Anal Chem 62:904
30. Leece B, Denomme MA, Towner R, Li SMA, Safe S (1985) J Toxicol Environ Hlth 16:379
31. Kannan N, Tanabe S, Wakimoto T, Tatsukawa R (1987) Chemosphere 16:1631
32. Tanabe S, Kannan N, Wakimoto T, Tatsukawa R (1987) Intern J Environ Anal Chem 29:199
33. Tanabe S, Kannan N, Wakimoto T, Tatsukawa R, Okamoto T, Masuda Y (1989) Toxicol Environ Chem 24:215

34. Tillitt DE, Giesy JP, Ankley GT (1991) Environ Sci Technol 25:87
35. Zacharewski T, Safe L, Safe S, Chittim B, DeVault D, Wilberg K, Berqvist P, Rappe C (1989) Environ Sci Technol 23:730
36. Safe SH (1986) Annual Rev Pharmacol Toxicol 26:371
37. Safe S (1992) Chemosphere 25:61
38. Neubert D, Golar G, Neubert R (1992) Chemosphere 25:65
39. Ahlborg UG, Becking GC, Birnbaum LS, Brouwer A, Derks HJGM, Feeley M, Golar G, Hanberg AL, Larsen JC, Liem AKD, Safe SH, Schlatter C, Wœrn F, Younes M, Yrjänheikki E (1994) Chemosphere 28:1049
40. De Vito MJ, Maier WE, Diliberto JJ, Birnbaum LS (1993) Fund Appl Toxicol 20:125
41. Walker MK, Peterson RE (1991) Aquat Toxicol 21:219
42. Stegeman JJ, Lech JJ (1991) Environ Health Perspect 90:101
43. Goksøyr A, Förlin L (1992) Aquat Toxicol 22:287
44. Collier TK, Anulacion BF, Stein JE, Goksøyr A, Varanasi U (1995) Environ Toxicol Chem 14:143
45. Huang Y-W, Melancon MJ, Jung RE, Karasov WH (1998) Environ Toxicol Chem 17:1564
46. Hahn ME, Poland A, Glover E, Stegeman JJ (1992) Mar Environ Res 34:87
47. Ishizuka M, Sakiyama T, Iwata H, Fukushima M, Kazusaka A, Fujita S (1998) Environ Toxicol Chem 17:1490
48. Risebrough RW, Riche P, Peakall DB, Herman SG, Kirven MN (1968) Nature 220:1098
49. Kennedy SW, Lorenzen A, Norstrom RJ (1996) Environ Sci Technol 30:706
50. Yamashita N, Shimada T, Tanabe S, Yamazaki H, Tatsukawa R (1992) Mar Pollut Bull 24:316
51. Watanabe S, Shimada T, Nakamura S, Nishiyama N, Yamashita N, Tanabe S, Tatsukawa R (1989) Mar Environ Res 27:51
52. Boon JP, Arnhem EV, Jansen S, Kannan N, Petrick G, Schulz D, Duinker JC, Reijnders PJH, Goksøyr A (1992) In: Walker CH, Livingstone DR (eds) Persistent pollutants in marine ecosystems. Pergamon, London, p 119
53. Fossi MC, Marsili L, Junin M, Castello H, Lorenzani JA, Casini S, Savelli C, Leonzio C (1997) Mar Pollut Bull 34:157
54. Kodavanti PRS, Ward TR, McKinney JD, Tilson HA (1995) Toxicol Appl Pharmacol 130:140
55. Koga N, Yoshimura H (1996) In: Kuratsune M, Yoshimura H, Hori Y, Okumura M, Masuda Y (eds) Yusho – a human disaster caused by PCBs and related compounds. Kyushu University Press, Fukuoka, p 105
56. Muir DCG, Norstrom RJ, Simon M (1988) Environ Sci Technol 22:1071
57. Yoshimura H, Ishida C, Hanioka N, Koga N (1991) In: Hodgson E, Roe RM, Motoyama N (eds) Pesticides and the future: toxicological studies of risks and benefits: reviews in pesticide toxicology. North Carolina University, Raleigh, p 73
58. Brouwer A (1991) In: Walker, CH (ed) The role of enzymes in regulating the toxicity of xenobiotics. University of Reading, UK, p 731
59. Kannan N, Reusch TBH, Schulz-Bull DE, Petrick G, Duinker JC (1995) Environ Sci Technol 29:1851
60. Norstrom RJ, Muir DCG, Ford CA, Simon M, Macdonald CR, Beland P (1992) Mar Environ Res 34:267
61. Kannan K, Tanabe S, Tatsukawa R, Sinha RK (1994) Toxicol Environ Chem 42:249
62. Brown JFJ (1992) Mar Environ Res 34:261
63. Kannan N, Schulz-Bull DE, Petrick G, Macht-Hausmann M, Wasserman O (1994) Arch Environ Health 49:375
64. Goksøyr A, Beyer J, Larsen HE, Andersson T, Förlin L (1992) Mar Environ Res 34:113
65. Gilbertson M (1992) Can J Fisheries Aquat Sci 49:1078
66. Mac MJ, Edsall CC (1991) J Toxicol Environ Health 33:375
67. Janz DM, Metcalf CD (1991) Chemosphere 23:467
68. Monosson E, Fleming WJ, Sullivan CV (1994) Aquat Toxicol 29:1

69. Black DE, Gutjahr-Gobell R, Pruell RJ, Bergen B, McElroy AE (1998) Environ Toxicol Chem 17:1396
70. Black DE, Gutjahr-Gobell R, Pruell RJ, Bergen B, McElroy AE (1998) Environ Toxicol Chem 17:1405
71. Bosveld ATC, Gradener J, Murk AJ, Brouwer A, van Kampen M, Evers EHG, van den Berg M (1994) Environ Toxicol Chem 14:99
72. Powell DC, Aulerich RJ, Meadows JC, Tillitt DE, Powell JF, Restum JC, Stromborg KL, Giesy JP, Bursian SJ (1997) Environ Toxicol Chem 16:1450
73. Gilbertson M (1997) Environ Toxicol Chem 16:1771
74. Miller S (1983) Environ Sci Technol 17:11A
75. Duinker JC, Hillebrand MTJ, Palmork KH, Wilhelmsen S (1980) Bull Environ Contam Toxicol 25:956
76. Ballschmiter K, Zell M (1980) Fresenius Z Anal Chem 302:20
77. Ribick MA, Smith LM, Dubay GR, Stalling DL (1981) Applications and results of analytical methods used in monitoring environmental contaminants. American Society for Testing and Materials, Philadelphia, PA, USA, Special Technical Publication. 737, pp 249–269
78. Mullin MD, Pochni CM, McCrindle S, Romkes M, Safe SH, Safe LM (1984) Environ Sci Technol 18:468
79. Safe S, Safe L, Mullin M (1985) J Agr Food Chem 33:24
80. Kannan N, Tanabe S, Tatsukawa R (1988) Arch Environ Health 43:11
81. Duinker JC, Schulz DE, Petrick G (1988) Anal Chem 60:478
82. Kuehl DW, Butterwoth BC, Libal J, Marquis P (1991) Chemosphere 22:849
83. de Boer J, Stronck CJN, van der Valk F, Wester PG, Daudt MJM (1992) Chemosphere 25:1277
84. Storr-Hansen E, Cleemann M, Cederberg T, Jansson B (1992) Chemosphere 24:323
85. Harrad SJ, Sewart AS, Boumphrey R, Duarte-Davidson R, Jones KC (1992) Chemosphere 24:1147
86. Leonard PEG, Van Hattum B, Cofino WP, Brinkman UAT (1994) Environ Toxicol Chem 13:129
87. Echols K, Gale R, Tillitt D, Schwartz T, O'Laughlin J (1994) Environ Toxicol Chem 16:1590
88. Feltz KP, Tillitt DE, Gale RW, Peterman PH (1995) Environ Sci Technol. 29:709
89. van Bavel B, Järemo M, Karlsson L, Lindström G (1996) Anal Chem 68:1279
90. Bandh C, Ishaq R, Broman D, Näf C, Rönquist-Nii Y, Zebühr Y (1996) Environ Sci Technol 30:214
91. Kannan N, Petrick G, Bruhn R, Schulz-Bull DE (1998) Chemosphere 37:2385
92. Bagheri H, Leonards PEG, Ghijsen RT, Brinkman UAT (1993) Int J Environ Anal Chem 50:257
93. Zebühr Y, Näf C, Bandh C, Broman D, Ishaq R, Pettersen H (1993) Chemosphere 27:1211
94. de Boer J, Dao QT, van Dortmond R (1992) J High Resol Chromatogr 15:249
95. Larsen B, Bowadt S, Tilio R (1992) Int J Environ Anal Chem 47:47
96. Larsen B, Bøwadt S, Tilio R, Facchetti S (1992) Chemosphere 25:1343
97. Frame GM, Wagner RE, Carnahan JC, Brown JF Jr, May RJ, Smullen LA, Bedard DL (1996) Chemosphere 33:603
98. Schmidt LJ, Hesselberg RJ (1992) Arch Environ Contam Toxicol 23:37
99. Himberg KK, Sippola E (1993) Chemosphere 27:17
100. Phillips JB, Xu J (1995) J Chromatogr 703:327
101. Xu J (1997) PhD thesis, Southern Illinois University
102. Ferrario J, Byrne C, Dupuy AE Jr (1997) Chemosphere 34:2451
103. De Voogt P, Haglund P, Reutergärdh LB, De Wit C, Waern F (1994) Anal Chem 66:305A
104. Kannan N, Petrick G, Schulz D, Duinker J, Boon J, Arnhem EV, Jansen S (1991) Chemosphere 23:1055
105. Kannan N, Petrick G, Schulz-Bull DE, Duinker JC (1993) J Chromatogr 642:425
106. Larsen R (1995) J High Resol Chromatogr 18:141
107. Ivanov V, Sandell E (1992) Environ Sci Technol 26:2012

108. Tatsukawa R, Tanabe S (1990) In: Baumgartner DJ, Duedall IW (eds) Oceanic processes in marine pollution. Krieger, Malabar, Florida, p 39
109. Kamops LR, Trotter WJ, Young SJ, Smith AC, Roach JAG, Page SW (1979) Bull Environ Contam Toxicol 23:51
110. Huckins JN, Stalling DL, Petty JD (1980) JAOAC 63:750
111. Stalling DL, Huckins JN, Petty JD (1980) In: Mackay D, Abqhan BM (eds) Hydrocarbons and halogenated hydrocarbons. Plenum, New York, p 131
112. Flandysz J, Yamashita N, Tanabe S, Tatsukawa R (1992) Int J Environ Anal Chem 47:129
113. Kannan N, Schulz-Bull DE, Petrick G, Duinker JC (1992) Intern J Environ Anal Chem 47:201
114. Lepine FL, Milot SM, Vincent NM, Gravel D (1991) J Agr Food Chem 39:2053
115. Holsen TM, Noll KE, Liu S, Lee W (1991) Environ Sci Technol 25:1075
116. Iwata H, Tanabe S, Ueda K, Tatsukawa R (1995) Environ Sci Technol 29:792
117. Mcconnell LL, Kucklick JR, Bidleman TF, Ivanov GP, Chernyak SM (1996) Environ Sci Technol 30:2975
118. Hornbuckle KC, Achman DR, Eisenreich SJ (1993) Environ Sci Technol 27:87
119. Oehme M (1991) Ambio 20:293
120. Bidleman TF, Cotham WE, Addison RF, Zinck ME (1992) Chemosphere 24:1389
121. Duinker JC, Schulz-Bull DE, Petrick G (1993) IOC Manual and Guides, UNESCO 27:1
122. Petrick G, Schulz-Bull DE, Martens V, Scholz K, Duinker JC (1996) Mar Chem 54:97
123. Duinker JC, Bouchertal F (1989) Environ Sci Technol 23:57
124. Petrick G, Schulz-Bull DE, Duinker JC (1992) Z Wasser Abwasser Forsch 25:115
125. Miyata H, Aozasa O, Ohta S, Chang T, Yasuda Y (1993) Chemosphere 26:1527
126. Chevreuil M, Granier L (1991) Chemosphere 23:1637
127. Eadie BJ, Morehead NR, Klump JV, Landrum PF (1992) J Great Lakes Res 18:91
128. Schulz-Bull DE, Petrick G, Duinker JC (1991) Mar Chem 36:365
129. Schulz-Bull DE, Patrick G, Kannan N, Duinker JC (1995) Mar Chem 48:245
130. Schulz-Bull DE, Petrick G, Johannsen H, Duinker JC (1996) CCACAA 70:309
131. Engwall M, Broman D, Ishaq R, Näf C, Zebühr Y, Brunström B (1996) Environ Toxicol Chem 15:213
132. Cortes A, Riego J, Payaperez AB, Larsen B (1991) Toxicol Environ Chem 31–2:79
133. Bruckmeier BFA, Jüttner I, Schramm K, Winkler R, Steinberg CEW, Kettrup A (1997) Environ Pollut 95:19
134. Chiarenzelli JR, Scrudato RJ, Wunderlich ML (1997) Environ Sci Technol 37:597
135. Kannan N, Tanabe S, Tatsukawa R, Phillips DJH (1989) Environ Pollut 56:65
136. Sericano JL, Wade TL, El-Husseini AM, Brooks JM (1992) Mar Pollut Bull 24:537
137. Brieger G, Hunter RD (1993) Ecotoxicol Environ Safety 26:153
138. Bergen BJ, Nelson WG, Pruell RJ (1996) Environ Toxicol Chem 15:1517
139. Froese KL, Verbrugge DA, Ankley GT, Niemi GJ, Larsen CP, Giesy JP (1998) Environ Toxicol Chem 17:484
140. Coristine S, Haffner GD, Ciborowski JJH, Lazar R, Nanni ME, Metcalfe CD (1996) Environ Toxicol Chem 15:1382
141. de Boer J, Stronck CJN, Traag WA, van der Meer J (1993) Chemosphere 26:1823
142. Wells DE, Echarri I (1992) Int J Environ Anal Chem 47:75
143. Kannan K, Tanabe S, Borrell A, Aguilar A, Focardi S, Tatsukawa R (1993) Arch Environ Contam Toxicol 25:227
144. Corsolini S, Focardi S, Kannan K, Tanabe S, Borrell A, Tatsukawa R (1995) Mar Environ Res 40:33
145. Leonards PEG, Zierikzee Y, Brinkman UAT, Cofino WP, Van Straalen NM, Van Hattum B (1997) Environ Toxicol Chem 16:1807
146. Koistinen J, Stenman O, Haahti H, Suonperä M, Paasivirta J (1997) Chemosphere 35:1249
147. Koistinen J (1990) Chemosphere 20:7
148. Storr-Hansen E, Spliid H (1993) Arch Environ Contam Toxicol 25:328
149. Daelemans FF, Mehlum F, Lydersen C, Schepens PJC (1993) Chemosphere 27:429
150. Hong C, Calambokidis J, Bush B, Steiger GH, Shaw S (1996) Environ Sci Technol 30:837

151. Nakata H, Tanabe S, Tatsukawa R, Koyama Y, Miyazaki N, Belikov S, Boltunov A (1998) Environ Toxicol Chem 17:1745
152. Bernhoft A, Wiig Ø, Skaare JU (1997) Environ Pollut 96:159
153. Masuda Y (1996) In: Kuratsune M, Yoshimura H, Hori Y, Okumura M, Masuda Y (eds) Yusho – a human disaster caused by PCBs and related compounds. Kyushu University Press, Fukuoka, p 49
154. Kubiak TJ, Harris HJ, Smith LM, Schwartz TR, Stalling DL, Trick JA, Sileo L, Docherty DE, Erdman TC (1989) Arch Environ Contam Toxicol 18:706
155. Yamashita N, Tanabe S, Ludwig JP, Kurita H, Ludwig ME, Tatsukawa R (1993) Environ Pollut 79:163
156. Harris HJ, Erdman TC, Ankley GT, Lodge KB (1993) Arch Environ Contam Toxicol 25:304
157. Ankley GT, Niemi GJ, Lodge KB, Harris HJ, Beaver DL, Tillitt DE, Schwartz TR, Giesy JP, Jones PD, Hagley C (1993) Arch Environ Contam Toxicol 24:332
158. Jarman WM, Burne SA, Chang RR, Stephens RD, Norstrom RJ, Simon M, Linthicum J (1993) Environ Toxicol Chem 12:105
159. Jones PD, Hannah DJ, Buckland SJ, Day PJ, Leathem SV, Porter LJ, Auman HJ, Sanderson JT, Summer C, Ludwig JP, Colborn TL, Giesey JP (1996) Environ Toxicol Chem 15:1793
160. Guruge KS, Tanabe S (1997) Environ Pollut 96:425
161. Jones PD, Ankley GT, Best DA, Crawford R, DeGalan N, Giesy JP, Kubiak TJ, Ludwig JP, Newsted JL, Tillitt DE, Verbrugge DA (1993) Chemosphere 26:1203
162. Powell DC, Aulerich RJ, Meadows JC, Tillitt DE, Kelly ME, Stromborg KL, Melancon MJ, Fitzgerald SD, Bursian SJ (1998) Environ Toxicol Chem 17:2035
163. Bordet F, Mallet J, Maurice L, Borrel S, Venent A (1993) Bull Environ Contam Toxicol 50:425
164. Koopman-Esseboom C (1994) Dissertation, Erasmus University, Rotterdam
165. Noren K, Lunden Å (1991) Chemosphere 23:1895
166. Vartiainen T, Saarikoski S, Jaakkola JJ, Tuomisto J (1997) Chemosphere 34:2571
167. Mes J, Davies DJ, Doucet J, Weber D, McMullen E (1993) Environ Technol 14:555
168. Luotamo M, Patterson DG Jr, Needham LL, Aitio A (1993) Chemosphere 27:171
169. Böhm V, Schulte E, Thier H (1993) Z Lebensm Unters Forsch 196:435
170. Lucier GW (1991) Environ Toxicol Chem 10:727
171. Schecter A, McGee H, Stanley J, Boggess K (1993) Chemosphere 27:241
172. Swain WR (1991) J Toxicol Environ Health 33:587
173. Jones KC, Sanders G, Wild SR, Burnett V, Johnston AE (1992) Nature 356:137
174. Kannan K, Falandysz J, Yamashita N, Tanabe S, Tatsukawa R (1992) Mar Pollut Bull 24:358
175. Loganathan BG, Tanabe S, Hidaka Y, Kawano M, Hidaka H, Tatsukawa R (1993) Environ Pollut 81:31
176. Wania F, Mackay D (1993) Ambio 22:10
177. Lead WA, Steinnes E, Jones KC (1996) Environ Sci Technol 30:524
178. Muir DCG, Omelchenko A, Grift NP, Savoie DA, Lockhart WL, Wilkinson P, Brunskill GJ (1996) Environ Sci Technol 30:3609
179. Bruhn R, Kannan N, Petrick G, Schulz-Bull DE, Duinker JC (1999) Sci Total Environ (in press)
180. Bedard DL, May RJ (1996) Environ Sci Technol 30:237
181. Brown JF Jr, Wagner RE (1990) Environ Toxicol Chem 9:1215
182. Quensen JF III, Mousa MA, Boyd SA, Sanderson JH, Froese KL, Giesey JP (1998) Environ Toxicol Chem 17:806
183. Blanz T (1996) PhD thesis, Christian-Albrechts-Universität, Kiel
184. Bandiera S, Farrell K, Mason G, Kelley M, Romkes M, Bannister R, Safe S (1984) Chemosphere 13:507
185. Mason G, Farrell K, Keys B, Piskorska-Pliszczynska J, Safe L, Safe S (1986) Toxicology 41:21
186. Newsted JL, Giesy JP, Ankley GT, Tillitt DE, Crawford RA, Gooch JW, Jones PD, Denison MS (1995) Environ Toxicol Chem 14:861

187. Schwartz TR, Tillitt DE, Feltz KP, Peterman PH (1993) Chemosphere 26:1443
188. Larsen B, Bøwadt S, Facchetti S (1992) Intern J Environ Anal Chem 47:147
189. Himberg K, Sippola E (1990) In: Hutzinger O, Fiedler H (eds) Dioxin 90. Eco-Informa, Bayreuth, Germany, p 183
190. Mullin MD (1985) Congener-specific PCB analysis. EPA Large Lakes Research Station, Gross IL, p 38
191. Albro PW, Corbett JT, Schroeder JL (1981) Ambio 205:103
192. Sweet CW, Murphy TJ, Bannasch JH, Kelsey CY, Hong J (1993) J Great Lakes Res 19:109
193. Garcia AL, Den Boer AC, De Jong APJM (1996) Environ Sci Technol 30:1032
194. Hoff RM, Muir DCG, Grift NP (1992) Environ Sci Technol 26:266
195. Kannan N, Yamashita N, Petrick G, Duinker JC (1998) Environ Sci Technol 32:1747
196. Matsueda T, Iida T, Hirakawa H, Fukamachi K, Tokiwa H, Nagayama J (1993) Chemosphere 27:187
197. Corsolini S, Focardi S, Kannan K, Tanabe S, Tatsukawa R (1995) Arch Environ Contam Toxicol 29:61
198. Dewailly E, Weber J, Gingras S, Laliberte C (1991) Bull Environ Contam Toxicol 47:491
199. Williams DT, LeBel GL (1991) Chemosphere 22:1019
200. Falandysz J, Tanabe S, Tatsukawa R (1994) Sci Total Environ 145:207
201. Koistinen J (1990) Chemosphere 20:7

CHAPTER 7

Polychlorinated Diphenyl Ethers (PCDE)

Jaana Koistinen

J. Koistinen (e-mail: Jaana.Koistinen@ktl.fi)
National Public Health Institute, Division of Environmental Health, P.O. Box 95,
FIN-70701 Kuopio, Finland

Polychlorinated diphenyl ethers (PCDE) are common impurities in chlorophenol formulations, which were earlier used as fungicides, slimicides, and as wood preservatives. PCDEs are structurally and by physical properties similar to polychlorinated biphenyls (PCB). They have low water solubility and are lipophilic. PCDEs are quite resistant to degradation and are persistent in the environment. In the aquatic environment, PCDEs bioaccumulate. These compounds are found in sediment, mussel, fish, bird, and seal. PCDEs show biomagnification potential, since levels of PCDEs increase in species at higher trophic levels. PCDEs are also detected in human tissue. Despite the persistence and bioaccumulation, the significance of PCDEs as environmental contaminants is uncertain. The acute toxicity and Ah-receptor-mediated (aryl hydrocarbon) activity of PCDEs is low compared to those of polychlorinated dibenzo-p-dioxins (PCDD) and dibenzofurans (PCDF). Due to structural similarity to thyroid hormone, PCDEs could bind to thyroid hormone receptor and alter thyroid function. Furthermore, PCDEs might be metabolized to toxic metabolites. In the environment, it is possible that photolysis converts PCDEs to toxic PCDDs and PCDFs.

Keywords. PCDE, Sources, Toxicology, Levels, Fate

The Handbook of Environmental Chemistry Vol. 3 Part K
New Types of Persistent Halogenated Compounds
(ed. by J. Paasivirta)
© Springer-Verlag Berlin Heidelberg 2000

List of Symbols and Abbreviation

Ah	Aryl hydrocarbon
AHH	Aryl hydrocarbon hydroxylase
APDM	Aminopyridine N-demethylase
BCF	Bioconcentration factor
DCM	Dichloromethane
dw	Dry weight
EROD	Ethoxy resorufin *o*-deethylase
fw	Fresh weight
GC	Gas chromatography
GPC	Gel permeation chromatography
HAH	Halogenated aromatic hydrocarbons
HPLC	High performance liquid chromatography
HRMS	High resolution mass spectrometry
HRGC	High resolution gas chromatography
K_{ow}	*n*-Octanol/water partition coefficient
LC	Liquid chromatography
MC	Methyl chloranthrene
MS	Mass spectrometry
NMR	Nuclear magnetic resonance
OH-PCDE	Hydroxyl-derivates of polychlorinated diphenyl ethers
PAHE	Petroleum ether: acetone: hexane: diethyl ether
PB	Phenobarbital
PBDE	Polybrominated diphenyl ethers
PCB	Polychlorinated biphenyls
PCDD	Polychlorinated dibenzo-*p*-dioxins
PCDE	Polychlorinated diphenyl ethers
PCDF	Polychlorinated dibenzofurans
PCPA	Polychlorinated phenoxyanisoles
PCPP	Polychlorinated phenoxyphenols
RI	Retention index
RRT	Relative retention time
SFE	Supercritical fluid extraction
TEF	Toxic equivalency factor

1
Structures and Nomenclature

Polychlorinated diphenyl ethers (PCDE), also called bis(chlorophenyl) ethers and chlorodiphenyl oxides, are neutral aromatic compounds which resemble structurally polychlorinated biphenyls (PCB), except that one oxygen atom connects the phenyl rings. The empirical formula of the PCDEs is $C_{12}H_{10-n}Cl_nO$, where n = 1 – 10. The structural formula of the PCDEs is given in Fig. 1.

Similarly to PCBs, there are 209 possible PCDE congeners, in which the number of chlorines varies from one to ten. The number of possible isomers at each chlorination degree (congener groups) from mono to decachlorinated PCDEs

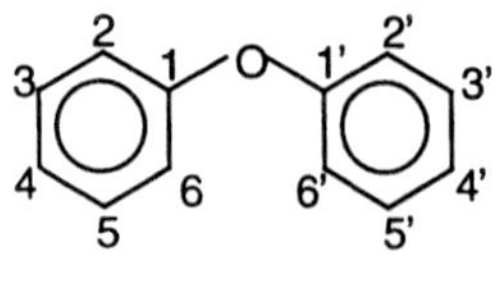

Fig. 1. The structure of PCDEs

Table 1. The number of PCDE congeners according to the degree of chlorination

Congener group	Abbreviation	Formula	Number of isomers
Monochlorodiphenyl ether	MonoCDE	$C_{12}H_9Cl_1O$	3
Dichlorodiphenyl ether	DiCDE	$C_{12}H_8Cl_2O$	12
Trichlorodiphenyl ether	TriCDE	$C_{12}H_7Cl_3O$	24
Tetrachlorodiphenyl ether	TetraCDE	$C_{12}H_6Cl_4O$	42
Pentachlorodiphenyl ether	PentaCDE	$C_{12}H_5Cl_5O$	46
Hexachlorodiphenyl ether	HexaCDE	$C_{12}H_4Cl_6O$	42
Heptachlorodiphenyl ether	HeptaCDE	$C_{12}H_3Cl_7O$	24
Octachlorodiphenyl ether	OctaCDE	$C_{12}H_2Cl_8O$	12
Nonachlorodiphenyl ether	NonaCDE	$C_{12}H_1Cl_9O$	3
Decachlorodiphenyl ether	DecaCDE	$C_{12}H_0Cl_{10}O$	1
Total number of congeners			209

are presented in Table 1. The PCDEs congeners are numbered analogously to PCBs [1] following the systematic numbering presented by Ballschmiter and Zell [2], except that according to the numbering rules the order of a few congeners should be different [3, 4]. The chlorine substitution and numbering of PCDEs is presented in Table 2.

Crystal structures of PCDEs have shown that, depending on the number of *ortho*-substituents, PCDEs adopt a skew and twist conformation [5–8]. For example, PCDE 138 (2,2′,3,4,4′,5′-hexaCDE) has twist conformation with the twist angle 85.46(8)° between phenyl rings and the angle (C-O-C) 117.2(3)° between the symmetry-related phenyl rings [7]. Conformational energy maps calculated using semi-empirical AM1 method have shown that PCDEs are flexible molecules and not completely rigid, and thus planar conformations are disfavored [8].

2
Production and Formation

PCDEs can be prepared by direct chlorination of diphenyl ether and they were intended to be used for similar purposes as commercial PCB mixtures, chlorination mixtures of biphenyl. Direct chlorination of diphenyl ether at room temperature in acetic acid solution without catalyst yields mainly a monochlorinated diphenyl ether (4-monoCDE) [9], whereas direct halogenation of diphenyl ether with Lewis acid catalysts such as AlCl₃ results a complex mixture of PCDEs [10]. The synthesis of most individual PCDE congeners can be performed via coupling of biaryliodonium salts and phenols [10].

Table 2. Suggested numbering of PCDEs [1]

Monochloro		Tetrachloro		Pentachloro	
1.	2	40.	2,2',3,3'	82.	2,2',3,3',4
2.	3	41.	2,2',3,4	83.	2,2',3,3',5
3.	4	42.	2,2',3,3'	84.	2,2',3,3',6
		43.	2,2',3,5	85.	2,2',3,4,4'
Dichloro		44.	2,2',3,5'	86.	2,2',3,4,5
4.	2,2'	45.	2,2',3,6	87.	2,2',3,4,5'
5.	2,3	46.	2,2',3,6'	88.	2,2',3,4,6
6.	2,3'	47.	2,2',4,4'	89.	2,2',3,4,6'
7.	2,4	48.	2,2',4,5	90.	2,2',3,4',5
8.	2,4'	49.	2,2',4,5'	91.	2,2',3,4',6
9.	2,5	50.	2,2',4,6	92	2,2',3,5,5'
10.	2,6	51.	2,2',4,6'	93.	2,2',3,5,6
11.	3,3'	52.	2,2',5,5'	94.	2,2',3,5,6'
12.	3,4	53.	2,2',5,6'	95.	2,2',3,5',6
13.	3,4'	54.	2,2',6,6'	96.	2,2',3,6,6'
14.	3,5	55.	2,3,3',4	97.	2,2',3',4,5
15.	4,4'	56.	2,3,3',4'	98.	2,2',3',4,6
		57.	2,3,3',5	99.	2,2',4,4',5
Trichloro		58.	2,3,3',5'	100.	2,2',4,4',6
16	2,2',3	59.	2,3,3',6	101.	2,2',4,5,5'
17.	2,2',4	60.	2,3,4,4'	102.	2,2',4,5,6'
18.	2,2',5	61.	2,3,4,5	103.	2,2',4,5',6
19.	2,2',6	62.	2,3,4,6	104.	2,2',4,6,6'
20.	2,3,3'	63.	2,3,4',5	105.	2,3,3',4,4'
21.	2,3,4	64.	2,3,4',6	106.	2,3,3',4,5
22.	2,3,4'	65.	2,3,5,6	107.	2,3,3',4',5
23.	2,3,5	66.	2,3',4,4'	108.	2,3,3',4,5'
24.	2,3,6	67.	2,3',4,5	109.	2,3,3',4,6
25.	2,3',4	68.	2,3',4,5'	110.	2,3,3',4',6
26.	2,3',5	69.	2,3',4,6	111.	2,3,3',5,5'
27.	2,3',6	70.	2,3',4',5	112.	2,3,3',5,6
28.	2,4,4'	71.	2,3',4',6	113.	2,3,3',5',6
29.	2,4,5	72.	2,3',5,5'	114.	2,3,4,4',5
30.	2,4,6	73.	2,3',5',6	115.	2,3,4,4',6
31.	2,4',5	74.	2,4,4',5	116.	2,3,4,5,6
32.	2,4',6	75.	2,4,4',6	117.	2,3,4',5,6
33.	2',3,4	76.	2',3,4,5	118.	2,3',4,4',5
34.	2',3,5	77.	3,3',4,4'	119.	2,3',4,4',6
35.	3,3',4	78.	3,3',4,5	120.	2,3',4,5,5'
36.	3,3',5	79.	3,3',4,5'	121.	2,3',4,5',6
37.	3,4,4'	80.	3,3',5,5'	122.	2',3,3',4,5
38.	3,4,5	81.	3,4,4',5	123.	2',3,4,4',5
39.	3,4',5			124.	2',3,4,5,5'
				125.	2',3,4,5,6'
				126.	3,3',4,4',5
				127.	3,3',4,5,5'

Table 2 (continued)

Hexachloro		Heptachloro		Octachloro	
128.	2,2,3,3′,4,4′	170.	2,2′,3,3′,4,4′,5	194.	2,2′,3,3′,4,4′,5,5′
129.	2,2′,3,3′,4,5	171.	2,2′,3,3′,4,4′,6	195.	2,2′,3,3′,4,4′,5,6
130.	2,2′,3,3′,4,5′	172.	2,2′,3,3′,4,5,5′	196.	2,2′,3,3′,4,4′,5′,6′
131.	2,2′,3,3′,4,6	173.	2,2′,3,3′,4,5,6	197.	2,2′,3,3′,4,4′,6,6′
132.	2,2′,3,3′,4,6′	174.	2,2′,3,3′,4,5,6′	198.	2,2′,3,3′,4,5,5′,6
133.	2,2′,3,3′,5,5′	175.	2,2′,3,3′,4,5′,6	199.	2,2′,3,3′,4,5,6,6′
134.	2,2′,3,3′,5,6	176.	2,2′,3,3′,4,6,6′	200.	2,2′,3,3′,4,5′,6,6′ [a]
135.	2,2′,3,3′,5,6′	177.	2,2′,3,3′,4′,5,6	201.	2,2′,3,3′,4,5,5′,6′ [a]
136.	2,2′,3,3′,6,6′	178.	2,2′,3,3′,5,5′,6	202.	2,2′,3,3′,5,5′,6,6′
137.	2,2′,3,4,4′,5	179.	2,2′,3,3′,5,6,6′	203.	2,2′,3,4,4′,5,5′,6
138.	2,2′,3,4,4′,5′	180.	2,2′,3,4,4′,5,5′	204.	2,2′,3,4,4′,5,6,6′
139.	2,2′,3,4,4′,6	181.	2,2′,3,4,4′,5,6	205.	2,3,3′,4,4′,5,5′,6
140.	2,2′,3,4,4′,6′	182.	2,2′,3,4,4′,5,6′		
141.	2,2′,3,4,5,5′	183.	2,2′,3,4,4′,5′,6	Nonachloro	
142.	2,2′,3,4,5,6	184.	2,2′,3,4,4′,6,6′	206.	2,2′,3,3′,4,4′,5,5′,6
143.	2,2′,3,4,5,6′	185.	2,2′,3,4,5,5′,6	207.	2,2′,3,3′,4,4′,5,6,6′
144.	2,2′,3,4,5′,6	186.	2,2′,3,4,5,6,6′	208.	2,2′,3,3′,4,5,5′,6,6′
145.	2,2′,3,4,6,6′	187.	2,2′,3,4′,5,5′,6		
146.	2,2′,3,4′,5,5′	188.	2,2′,3,4′,5,6,6′	Decachloro	
147.	2,2′,3,4′,5,6	189.	2,3,3′,4,4′,5,5′	209.	2,2′,3,3′,4,4′,5,5′,6,6′
148.	2,2′,3,4′,5,6′	190.	2,3,3′,4,4′,5,6		
149.	2,2′,3,4′,5′,6	191.	2,3,3′,4,4′,5′,6		
150.	2,2′,3,4′,6,6′	192.	2,3,3′,4,5,5′,6		
151.	2,2′,3,5,5′,6	193.	2,3,3′,4′,5,5′,6		
152.	2,2′,3,5,6,6′				
153.	2,2′,4,4′,5,5′				
154.	2,2′,4,4′,5,6′				
155.	2,2′,4,4′,6,6′				
156.	2,3,3′,4,4′,5				
157.	2,3,3′,4,4′,5′				
158.	2,3,3′,4,4′,6				
159.	2,3,3′,4,5,5′				
160.	2,3,3′,4,5,6				
161.	2,3,3′,4,5′,6				
162.	2,3,3′,4′,5,5′				
163.	2,3,3′,4′,5,6				
164.	2,3,3′,4′,5′,6				
165.	2,3,3′,5,5′,6				
166.	2,3,4,4′,5,6				
167.	2,3′,4,4′,5,5′				
168.	2,3′,4,4′,5′,6				
169.	3,3′,4,4′,5,5′				

[a] PCDEs 200 and 201 should be reversed according to the systematic numbering system.

According to the literature survey of Sundström and Hutzinger [11], there have been many patents for PCDEs and desired applications have included use as hydraulic fluids, electric insulators, flame retardants, lubricants, and plasticizers. Ethers have had wide use in industry and they appear in heat transfer agents like a mixture of diphenyl ether (73.5%) and biphenyl (26.5%), Dowtherm A [12]. Lower chlorinated PCDEs, mono- and dichlorinated, have aroused interest as chemical intermediates [13] and 4-monoCDE has been desirable as a synthetic intermediate in agricultural and topical medicinal applications [14]. Higher chlorinated PCDEs have been used in the electrical industry, but their toxicity hindered their use as plasticizers and in high-pressure greases [13]. Tetra- through hexaCDEs were suspected of causing handling hazards, since they showed high toxicity when fed to guinea pigs.

According to Rappe et al. [15], PCDEs have been used as thermostable heat exchange and hydraulic liquids, but their production has been much smaller than that of PCBs. The production of PCBs had been about 1.5 million tonnes since 1929 until their production and sales decreased markedly in the 1970s [16]. In contrast to PCDEs, brominated diphenyl ethers (PBDE) have been widely used as flame retardants in a variety of materials and products including electronic devices [17]. Commercial products, penta-, octa-, and deca-BDE mixtures, have been produced globally at about 40,000 tonnes a year.

PCDEs have also been suggested for other applications such as for pesticides and synergistics in pesticides [11, 18]. A dichlorodiphenyl ether was among 34 most effective materials out of 5963 tested for clothing treatments against chigger mites, but it was suspected of being unsafe for use on clothing worn for an extended period [19]. One di-CDE showed acaricidal activity against the citrus red mite when 100 organic compounds related to DDT were tested, but no insecticidal activity against the greenhouse thrips [20]. The toxicity of a mono-CDE was tested among 106 compounds as an insecticide to body lice, but it proved to be not very toxic [21]. Insecticidal activity of mono- through trichlorodiphenyl ethers has not been high to house flies, but they have acted as synergists when mixed with pyrethrins [22]. OCDE has been synergistic with insecticides such as Sumithion, DDT, malation, and bromophos [23].

Derivatives of PCDEs have obviously had more applications than PCDEs. A butylated monochloro diphenyl ether (Dow XFS-4169L) was developed for capacitor dielectric fluid at the beginning of the 1970s [24]. Hydroxy-derivatives (OH-PCDEs) like 2,4,4′-trichloro-2′-hydroxy-diphenyl ether (5-chloro-2-(2,4-dichlorophenoxy)phenol), known by the tradenames Triclosan and Irgasan, have been used as antimicrobial agents [18]. Triclosan has been widely used in personal care products, since it is neither acute oral toxicant nor carcinogen, teratogen, or mutagen [25]. Triclosan with the trade name CH 3565 was presented as a broad-spectrum antimicrobial agent, which could be applied in soaps, deodorants, and cosmetic products as well as fabric softeners [26]. The antiseptic activity of Triclosan (Irgasan DP 300) was tested as an alternative skin disinfectant to hexachlorophene [27]. OH-PCDEs and their derivatives have also had applications as household chemicals (polishes) and in industrial chemicals (food preservatives, paper, textiles) [18].

Many herbicides are nitro-PCDEs such as nitrofen and DNP [28, 29]. Different diphenyl ether herbicides are presented in Table 3. The use of nitrofen, 2,4-dichloro-4′-nitro-diphenyl ether, was restricted to crops without detectable residues at harvest in the 1980s, since it has caused neonatal deaths in rats [30]. A aminoderivative of PCDEs, Aminofen, is an intermediate in the chemical industry and has been used in the manufacture of herbicides Illoxan and Diclofop-methyl [4]. Amino-PCDEs are also impurities of Eulan WA New that contains

Table 3. Examples of diphenyl ether herbicides [28, 29]

Common name	Chemical name	Structure
Nitrofen	2,4-dichlorophenyl-4′-nitrodiphenyl ether	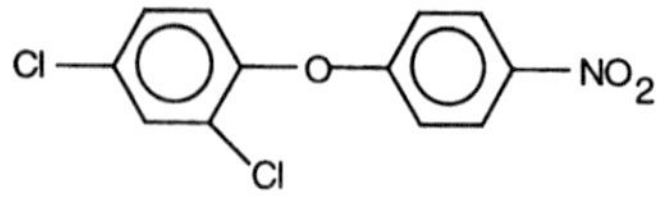 Cl, Cl, O, NO$_2$
CNP	2,4,6-trichloro-4′-nitrodiphenyl ether	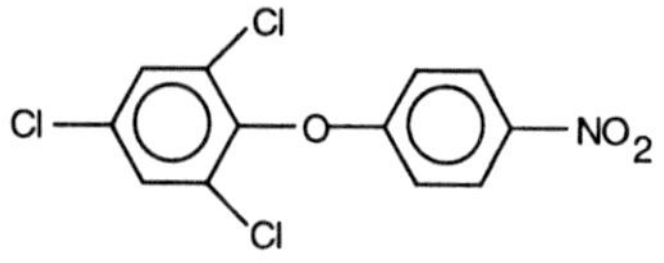 Cl, Cl, Cl, O, NO$_2$
CFNP	4,6-dichloro-2-fluoro-4′-nitrodiphenyl ether	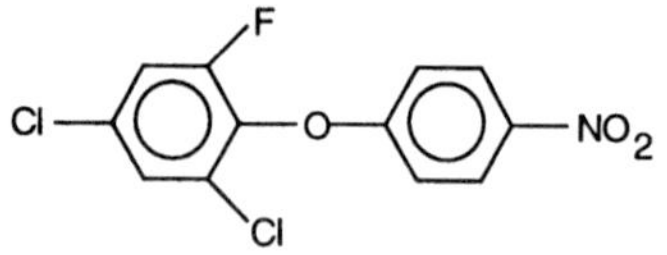 F, Cl, Cl, O, NO$_2$
Bifenox	2,4-dichloro-3′-methoxycarbonyl-4′-nitrodiphenyl ether	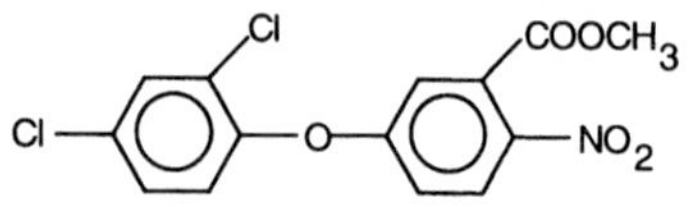 Cl, Cl, O, COOCH$_3$, NO$_2$

polychloro-2-(chloromethylsulfonamid)diphenyl ethers (PSCD) as active ingredients [31]. PSCDs have been used in the textile industry as mothproofing agents.

Nitro-trifluoromethyl-PCDEs and hydroxy-trifluoromethyl-PCDEs have been detected as impurities in 3-trifluoromethyl-4-nitrophenol (TFM), which is a field formulation of a lampricide [32]. This lampricide has been applied to protect salmonids from sea lampreys in streams since the 1950s.

Like other chlorinated compounds, PCDEs can be formed during combustion and as unwanted byproducts during manufacturing processes of chlorinated compounds such as chlorophenols or in processes where chlorine is used. Based on low contents of PCDEs in sediments near a Finnish pulp mill, chlorine bleaching of pulp is probably not a significant source of PCDEs [33]. PCDEs can also be formed by perchlorination of industrial or sewage effluents containing diphenyl ether [34]. Combustion is a well known source of various groups of organic compounds [35] and the occurrence of PCDEs in combustion wastes was reported by Paasivirta et al. in 1986 [36]. The major sources of PCDEs in the environment, however, are technical chlorophenol formulations in which PCDEs have been detected as abundant impurities [37–43]. Chlorophenols or their sodium or potassium salts were earlier used as wood preservatives, slimicides, fungicides, and herbicides as well as key intermediates in the production of chlorinated phenoxyacetic acid herbicides [15, 37, 40]. Chlorophenols can be prepared by the chlorination of phenol or by alkaline hydrolysis of the corresponding halobenzene [15].

Based on the similarity of PCDEs with PCBs, Sundström and Hutzinger [11] suggested in the 1970s that leakage of PCDEs into the environment could cause problems similar to those of PCBs. After the discovery of relatively high amounts of PCDEs in chlorophenols and the possible contamination of the environment with these substances, the synthesis of model substances of PCDEs was started for environmental and toxicological studies. PCDEs, however, have not been commercially available until recently, which has hindered environmental and toxicological studies of these compounds. The most common route to synthesize individual PCDE congeners is the coupling of biaryliodonium salts with phenols, although PCDEs can be synthesized via several different routes [10, 11]. Nowadays over 100 PCDE congeners have been synthesized [4, 44].

3
Occurrence in Products and Emissions

3.1
Chlorophenol Formulations

Technical chlorophenol formulations are the major sources of PCDEs in the environment [4, 40, 45, 46]. Hydroxy chlorodiphenyl ethers (OH-PCDE), also called polychlorinated phenoxyphenols (PCPP), are the main impurities of chlorophenols [39, 47, 48]. 2-OH-PCDEs are called predioxins, since they are phenolic precursors to polychlorinated dibenzo-p-dioxins (PCDD) [47]. The levels of 2-hydroxy-nonaCDEs have varied between 0.6 mg kg^{-1} and 1100 mg kg^{-1} in commercial pentachlorophenol preparations (sodium salts) [47]. 3,4,5,6-tetrachloro-2-(2,3,4,5,6-pentachlorophenoxy)phenol) is the main impurity of commercial pentachlorophenols [48].

Depending on the chlorophenol formulations, the total amount of PCDEs in chlorophenols and chlorophenol-based wood preservatives have varied from 4.4 mg kg^{-1} to 1000 mg kg^{-1} [37–43]. PCDEs have not been observed in dichlorophenols [37] and the level reported for a trichlorophenol was low [4]. Examples of the content of PCDEs in different chlorophenol formulations have been presented in Table 4. The concentrations of the most toxic by-products in chlorophenols, PCDDs, are also at mg kg^{-1} levels or higher [33, 37, 39–42, 47]. The content of by-products in chlorophenols including PCDEs depend on the production method [42].

Table 4. The amount of PCDEs in technical chlorophenols and chlorophenol-based wood preservative formulations

Chlorophenol or product[a]	Origin	Total PCDEs (mg kg^{-1})	Ref.
Na-2,4,5-TCP	Germany	4.4	[43]
2,3,4,6-TeCP	Germany	213	[43]
	Finland	853	[42]
Na-2,3,4,6-TeCP/wood preservative	Finland	15–90	[33, 42]
PeCP/wood preservative	Germany	21–33	[43]

[a] TCP = trichlorophenol; TeCP = tetrachlorophenol; PeCP = pentachlorophenol.

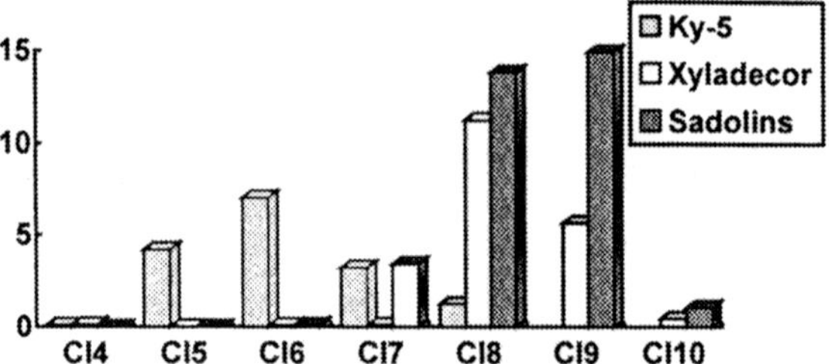

Fig. 2. The profile of tetra- through decaCDEs in wood preservatives

As the total amount of PCDEs vary, the congener profiles of PCDEs also vary between different chlorophenols depending on the manufacturing process of chlorophenol formulations. Hexachlorinated PCDE congeners have dominated in a Finnish tetrachlorophenol wood preserving formulation Ky-5 [33, 41] and octa- and nonachlorinated PCDE congeners in German pentachlorophenol formulations Sadolins and Xyladecor [43] (Fig. 2). This is most likely due to the different production route of these formulations. Pentachlorophenols are prepared from hexachlorobenzene and tetrachlorophenols from the chlorination of phenol [15].

The concentrations of PCDE congeners in a Finnish chlorophenol-based wood preservative, Ky-5, which consisted mainly of a sodium salt of 2,3,4,6-tetrachlorophenol, have ranged from < 0.03 mg kg^{-1} to 2.1 mg kg^{-1} dry weight (dw) [33]. The PCDEs identified in Ky-5 and their concentrations have been presented in Table 5. The concentrations of PCDE congeners in a German tetrachlorophenol have varied from 0.002 mg kg^{-1} to 77 mg kg^{-1} dw [43].

The use of chlorophenols was restricted after these formulations were shown to contain toxic PCDDs and PCDFs as impurities and nowadays their use has totally discontinued in many countries [49]. The estimated total amount of chlorophenols used yearly in many countries was thousands of tonnes before they were banned. At the end of the 1970s the annual world production of chlorophenols was 150,000 tonnes [15]. de Boer and Denneman [46] have estimated, based on production figures of chlorophenols and the amount of PCDEs in chlorophenols, that 250–2500 tonnes of PCDEs have been released into the environment via this route.

3.2
PCBs

The production of PCBs and use of PCB formulations are possible sources of PCDEs. PCDEs have been measured in a PCB transformer fluid, which contained 51% of Aroclor 1260 [50]. The production and use of PCBs has been quite extensive but, since there are no figures on the concentrations of PCDEs in PCB formulations, one cannot estimate their role as a source of PCDEs.

Table 5. PCDEs (mg kg^{-1}) detected in a Finnish chlorophenol-based wood preservative Ky-5 [33]

Congener		Concentration (mg kg^{-1})
PCDE 47	2,2′,4,4′-tetraCDE	0.06
PCDE 49/PCDE 51	2,2′,4,5′-/2,2′,4,6′-tetraCDE	0.04
PCDE 85	2,2′,3,4,4′-pentaCDE	0.54
PCDE 99	2,2′,4,4′,5-pentaCDE	2.4
PCDE 100	2,2′,4,4′,6-pentaCDE	1.3
PCDE 128	2,2′,3,3′,4,4′-hexaCDE	0.24
PCDE 137	2,2′,3,4,4′,5-hexaCDE	0.67
PCDE 138	2,2′,3,4,4′,5′-hexaCDE	0.95
PCDE 140/PCDE 167	2,2′,3,4,4′,6′-/2,3′,4,4′,5,5′-hexaCDE	1.1
PCDE 147/PCDE 153	2,2′,3,4′,5,6-/2,2′,4,4′,5,5′-hexaCDE	1.9
PCDE 154	2,2′,4,4′,5,6′-hexaCDE	2.1
PCDE 163	2,3,3′,4′,5,6-hexaCDE	0.05
PCDE 170	2,2′,3,3′,4,4′,5-heptaCDE	0.18
PCDE 180/PCDE 181	2,2′,3,4,4′,5,5′-/2,2′,3,4,4′,5,6-heptaCDE	0.67
PCDE 182	2,2′,3,4,4′,5,6′-heptaCDE	1.5
PCDE 184	2,2′,3,4,4′,6,6′-heptaCDE	0.83
PCDE 195	2,2′,3,3′,4,4′,5,6-octaCDE	0.07
PCDE 196	2,2′,3,3′,4,4′,5,6′-octaCDE	0.55
PCDE 197	2,2′,3,3′,4,4′,6,6′-octaCDE	0.39
PCDE 203	2,2′,3,4,4′,5,5′,6-nonaCDE	0.10
PCDE 204	2,2′,3,4,4′,5,6,6′-nonaCDE	0.13
Total PCDEs		16

3.3
PCDE Derivatives

Hydroxylated derivatives of PCDEs might also contain traces of PCDEs. There are no reports on the concentration of PCDEs in OH-PCDEs, but the manufacture of Triclosan was suggested as a source of PCDEs in Narrangesett Bay, Rhode Island [51]. PCDEs were observed in the waste water of a chemical manufacturing plant. OH-PCDEs can be prepared via five different routes [52] and one route, the coupling of chloroguaiacol and chlorodiphenyliodonium chloride, can be used in the preparation of PCDEs.

In addition to hydroxy-PCDEs, nitro- and aminoderivatives of PCDEs could be possible sources of PCDEs in the environment. Nitrofen is prepared by base-catalyzed condensation of 2,4-dichlorophenol (2,4-DCP) with 1-chloro-4-nitrobenzene [30]. PCDEs have not been detected in 2,4-DCP and 2,4,5-TCP [37] but, in theory, just as the occurrence of PCDE by-products in chlorophenols is most likely due to the condensation of chlorophenols, the condensation of 2,4-dichlorophenol during the manufacture of nitrofen might lead to the formation of PCDEs. Aminofen is produced from nitrofen by catalytic reduction [4].

Amino-PCDEs are impurities of Eulan Wa New and they have been detected as metabolites of Eulan Wa New in fish [53].

Since there are no data on levels of PCDEs in the formulations of PCDE-derivatives, it is difficult to estimate their role as PCDE sources. The amount of PCDEs released via PCDE derivatives is thought to be low compared to chlorophenol formulations [46].

3.4
Combustion

After PCDEs were detected in a fly ash from a municipal waste incinerator in Finland [36], the occurrence of PCDEs in combustion wastes has not been studied much. PCDEs could be formed during incomplete combustion by condensation from chlorophenols as has been indicated for PCDDs [54], but de novo synthesis is also possible [55]. The formation of chlorinated compounds is always possible during combustion in the presence of organic material and chloride. The formation of PCDEs de novo in combustion has been described in the literature review of Kurz's thesis [4]. Briefly, diphenyl can be formed from the phenoxy radical and benzene which in turn can be formed from alkene radicals. If the formed molecule does not already contain chlorine, chlorination of diphenyl ether can occur, e.g., in the presence of HCl. It has been suggested, however, that PCDEs, in contrast to PCDDs and PCDFs, are not formed to a great extent de novo on solid surfaces or in the gas phase in thermal processes during metal reclamation processes [56]. When PCDEs were analyzed in emission samples of a metal reclamation plant in Finland, all PCDEs were below 4 ng m^{-3}.

Different PCDE congeners have been found as combustion products compared to the chlorination of phenol [57]. PCDEs 85, 99, 100, 137, 138, 140/167, 147/153, 154, 180/181, 182, 184, 196, and 197 are typical congeners in Ky-5 [33, 57], but most of these were not detected in one Finnish waste combustion fly ash [57]. The content of PCDEs 47, 67, 77, 167, 180, and 196 ranged from 0.1 µg kg^{-1} to 3.8 µg kg^{-1} in this fly ash. The PCDEs in a German municipal waste incinerator fly ash were at the same level [43].

4
Properties and Reactions

Properties of PCDEs, including physicochemical ones, are not well known as the literature reviews of PCDEs have shown [4, 11, 40, 46]. PCDEs resemble PCBs structurally and in their chemical and physical properties, which, like PCDDs, PCDFs, and related compounds, are known to be stable and resistant to breakdown by heat, hydrolysis, bases, and acids. PCBs are also quite stable to oxidation under moderate conditions [3], but there is not much data about PCDEs concerning their stability. There is some evidence that PCDEs are resistant to bases and acids and the occurrence of PCDEs in the environment indicates that PCDEs are persistent and bioaccumulating compounds. The study of Firestone et al. [37] already showed that PCDEs are quite stable, since PCDEs could be measured in chlorophenol extracts after sulfuric acid treatment. Tetra- and octachlorinated PCDE congeners were later proven resistant in treatment with concentrated sulfuric acid [58]. Chemical treatment of PCDE-containing wastes with NaClO and NaOH has not affect the content of PCDEs [42]. Reactions of PCDEs that are significant from the environmental point of view include photochemical reactions, thermal reactions, hydrolysis in water, and oxidation by air, but they have not been studied intensively.

4.1
Vapor Pressures

Physico-chemical properties of PCDEs have been collected in Table 6. Most PCDEs synthesized are solids and their vapor pressures are low [4, 59]. The vapor pressures of sub-cooled liquids ($-\log P_L$) of PCDEs, determined from gas chromatographic (GC) data for 106 PCDE congeners at 30 °C, have ranged beween 0.27 and 5.80 Pa, being similar to those of PCBs [4, 59].

Henry's law constants for PCDEs, which describe the distribution between air and water and can be calculated from vapor pressure and water solubility, increase with increasing chlorination degree [59]. The values of Henry's law constant for PCDEs calculated based on chromatographic data at 30 °C have ranged from 2.95 Pa m^3 mol^{-1} to 14125.37 Pa m^3 mol^{-1} [59]. The Henry's law constants determined experimentally at 25 °C by Dunnivant et al. [60] for 17 PCBs from di- through hexachlorinated PCBs ranged from 3.04 Pa m^3 mol^{-1} to 90.89 Pa m^3 mol^{-1}. The Henry's law constants of PCBs increase with increasing degree of *ortho*-chlorination, but are not directly related to congener molecular weight [60].

Table 6. Physico-chemical properties of polychlorinated diphenyl ethers

	$-\log P$[a]	H[b]	Bp[c]	$-\log S$[d]	log Kow[e]
Ref.	[59]	[59]	[60]	[59]	[59]
Temp. EEC	30	30		30	30
MonoCDEs	0.27–0.36	8.13–32.36		4.21–4.78	4.45–4.70
DiCDEs	0.91–1.13	4.17–30.90		4.63–5.52	4.64–5.25
TriCDEs	1.26–1.86	2.24–112.20		4.95–6.77	4.96–5.88
TetraCDEs	1.99–2.57	17.38–151.36	oil or 69–71	6.64–7.52	5.64–6.36
PentaCDEs	2.66–3.30	12.02–208.93	oil or 64–65	7.12–8.25	5.98–6.83
HexaCDEs	3.19–4.06	12.02–199.53	68–139	7.88–8.94	6.47–7.11
HeptaCDEs	3.92–4.42	50.12–295.12	88–119	8.89–9.64	6.98–7.55
OctaCDes	4.50–4.80	234.42–562.34	125–205	10.10–10.55	7.63–7.84
NonaCDEs	5.16	1949.84		11.45	8.07
DecaCDE	5.80	14,125.37	220–2	12.95	8.16

[a] P = vapor pressure.
[b] H = Henry's law constant.
[c] Bp = boiling point.
[d] S = solubility.
[e] K_{ow} = *n*-octanol-water partition coefficient.

4.2
Melting and Boiling Points

Nevalainen et al. [44, 61] have measured melting points for 54 PCDE congeners. The melting points of congeners from tetra- to decachlorinated ranged from 69°C to 222°C. These melting points are lower than those of PCBs, which have ranged between 297°C and 579°C [62].

There are few data on boiling points of PCDEs. The boiling points of three earlier more common PCDEs, mono-, di-, and hexaCDE, were 153°C, 168.2°C, and 230–260°C, respectively [13]. At room temperature, mono- through hexaCDEs vary from clear liquids to white to yellowish waxy semisolids [13].

4.3
Water Solubility

Water solubility is one of the major parameters which affect the fate and distribution in the environment. Hydrophobic compounds with high octanol-water partition coefficients tend to bioaccumulate. Opperhuizen and Voors [63] have shown that hydrophobicity of PCDEs determines the bioconcentration factor of PCDEs and that bioconcentration kinetics of PCDEs resemble those of PCBs.

The aqueous solubilities ($-\log S$) of mono- through decaCDEs (106 PCDE congeners) range between 4.21 mol l^{-1} and 12.95 mol l^{-1} [59]. These values were calculated from high performance liquid chromatography (HPLC) data at 30°C. PCDEs are more soluble in water than PCBs, which have extremely low water solubilities. The water solubilities of PCBs range from 7×10^3 µg l^{-1} to 0.02 µg l^{-1} [3].

Water solubilities of environmentally important chemicals can also be calculated from the mobile order of thermodynamics [64]. The predicted aqueous solubility of 35 PCDE congeners at 25°C have been in good agreement with the values from HPLC data [64].

n-Octanol/water partition coefficient (log K_{ow}), the distribution between organic phase and aqueous phase, is dependent on the water solubility and describes the bioaccumulation potential of a compound. The values of log K_{ow} of mono- through decaCDEs as determined from HPLC data have ranged between 4.45 and 8.16 [59]. These values of log K_{ow} show that PCDEs have the potential for bioaccumulation. The values are similar to those of PCBs (4.3–8.4) [3]. Values for log K_{ow} depend on solute-solvent interactions and can be predicted accurately [65].

4.4
Chromatographic, MS, and NMR Properties

Gas chromatographic retention of PCDEs has been studied on different stationary phases [4, 41, 44, 66]. Kurz and Ballschmiter [4, 66] have studied GC-retention of 106 PCDE congeners on 5 different stationary phases and Nevalainen et al. [44] GC-retention of 54 PCDE congeners on 2 stationary phases. The relative retention of PCDEs depends on the substitution, but is different to that of PCBs. The retention times of PCDEs increase with the increasing number of chlorines and within a series of isomers those which have more vicinal chlorines have higher retention times [44]. 3,5-Substitution reduces the retention and 2,3-substitution enhances the retention of PCDEs [66]. Most PCDEs can be chromatographically separated from each other.

A method has been developed to calculate relative retention times (RRT) for those PCDEs not analyzed by GC [44]. Calculated RRTs have been in good agreement with measured values on SE-54 and OV-1701 columns [44].

In addition to GC properties, Kurz [4] has investigated HPLC retention of 106 PCDE congeners. Retention indexes (RI) were measured on three reversed phase columns having phenyl, cyanopropyl, and octadecyl phases. Like HRGC retention, HPLC retention of PCDEs depends on the structure. LC data of PCDEs have been used to determine water solubility and octanol-water partition coefficients [66].

Mass spectral studies of PCDEs have shown that PCDE isomers have quite similar mass spectra [4, 41, 44, 67]. The two most abundant peaks in mass spectra of PCDEs are the chlorine clusters of molecular ion ($M^{\cdot+}$) and fragment ion ($M^{\cdot+}$-2U:$M^{\cdot+}$-70). The (M-2Cl) is the most abundant peak, except in the case of PCDEs with no *ortho*-chlorines, and typical fragments for all PCDEs are m/z 74 and m/z 109 [44]. The fragment ion (M-2Cl) of PCDEs is the same as the molecular ion of PCDFs. It was suggested in the early 1970s that PCDEs may interfere in MS analysis of PCDFs, since *o,o'*-substituted PCDEs can rearrange to PCDFs via fragmentation [38, 68]. Predioxins, 2-OH-PCDEs, can form PCDDs in a gas chromatograph [47, 48].

NMR analyses of PCDEs, which can be used to verify structures, have been performed, e.g., by Nevalainen et al. [61, 69, 70]. They have recorded ^1H and ^{13}C NMR spectra and studied relationship between conformation and NMR parameters. The steric interactions of two adjacent chlorine atoms and the intramolecular ring effect are the most important factors that affect chemical shifts of PCDEs [69].

4.5
Photochemical Reactions

PCDEs can be dechlorinated or they can form chlorinated dibenzofurans by photodegradation [71]. The conversion of PCDEs into PCDFs can happen under environmental conditions via natural photolytic reactions. The photochemical ring closure yielding PCDFs happens to PCDEs which have at least one *ortho*-chlorine. In this reaction *ortho*-chlorine is lost. Photochemical reactions of some PCDEs have been presented in Fig. 3.

Photochemical reactions of PCDEs have been studied in different solvents, but no difference in the amount of PCDFs has been observed [72, 73]. In the case of PCBs, the rate of dechlorination is faster in polar solvents than in hydrocarbon solvents [3]. Dehalogenated PCDEs have dominated in photolysis in *n*-hexane [72], but photolysis of PCDEs in acetone solution inhibits dechlorination and favors photocyclization [73]. Therefore, it has been suggested that reactions in the environment in conditions near to acetone solution could have significance in the formation of PCDFs from PCDEs [73]. Photolysis in acetone solution can be used to synthesize PCDF congeners [74]. Carbonium ion is a possible intermediate in ring closure of PCDEs to PCDFs [75]. PCDFs can be synthesized from PCDEs

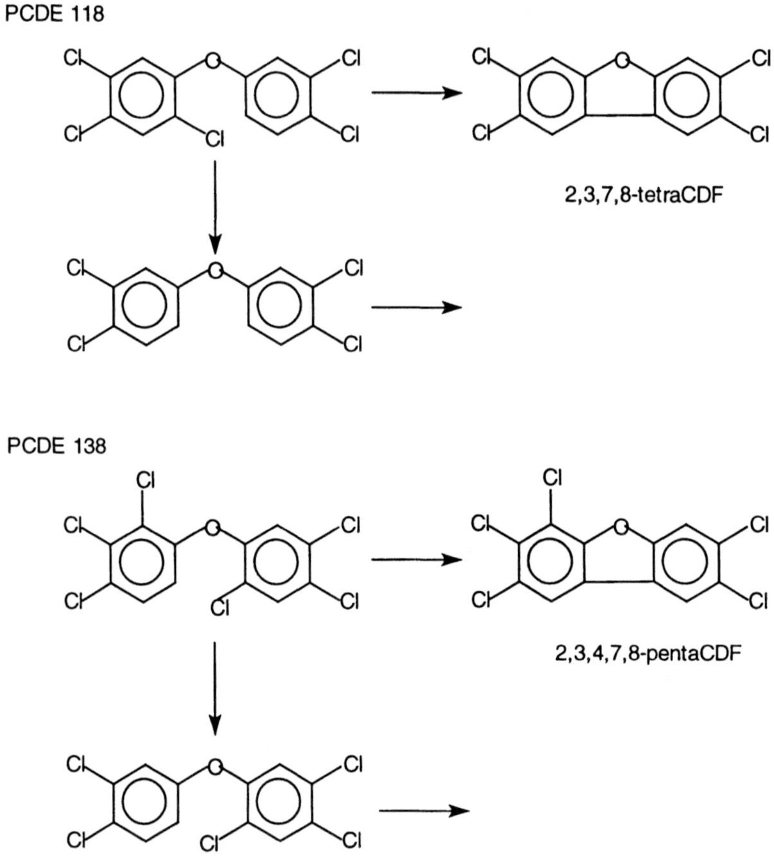

PCDE 118

PCDE 138

2,3,7,8-tetraCDF

2,3,4,7,8-pentaCDF

Fig. 3. Photochemical reactions of two PCDEs

using a catalyst (Pd(II) acetate) for ring closure [76]. *Ortho*-chlorine is lost after nucleophilic attack by palladium on both *ortho*-carbons.

OH-PCDEs form PCDDs by irradiation [77]. The main reaction in photolysis of higher chlorinated OH-PCDEs in methanol is dechlorination, whereas lower chlorinated compounds like Triclosan form methoxychlorophenoxyphenol as a main product.

4.6
Thermal Reactions

Thermal breakdown of PCDEs is possible at high temperatures, but PCDEs can also be converted into PCDFs during combustion. The pyrolytic behavior of PCDEs was investigated at different temperatures in the 1970s, but no PCDFs were detected then [72]. Later studies have verified that PCDEs can form PCDFs at elevated temperatures. Thermal conversion of PCDEs into PCDFs is at a similar rate to that of PCBs into PCDFs [15].

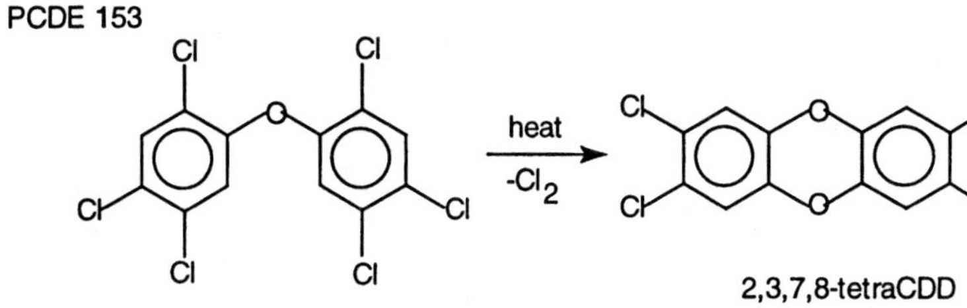

Fig. 4. Conversion of PCDE 153 to 2,3,7,8-TCDD [78]. Obvious involvement of oxygen is not drawn, while its mechanism is unknown

The pyrolysis experiments of PCDEs in sealed ampoules have shown that, in addition to PCDFs, PCDDs are also formed from PCDEs [78]. The loss of *ortho*-H_2 and *ortho*-HCl are the major routes in the formation of PCDFs, but the loss of Cl_2 is also possible: 1,2,3,4,6,7,8-heptaCDF was formed from 2,2',3,3',4,4',5,6-octaCDE (PCDE 195) via loss of HCl, whereas hexaCDFs, as well as 1,2,3,4,6,7-hexaCDD, were evidently formed via the loss of *ortho*-Cl_2. The most toxic dioxin, 2,3,7,8-tetraCDD, can be formed from 2,2',4,4',5,5'-hexaCDE (PCDE 153) via loss of two *ortho*-Cls and (unexplained in [78]) contribution of oxygen (Fig. 4).

Open fire burning of PCDE-containing material might also form PCDFs. Paasivirta et al. [42] reported that the concentrations of PCDDs and PCDFs were increased during open fire burning of chlorophenols while PCDEs were removed.

Derivatives of PCDEs can also form PCDFs and PCDDs. When hydroxy-PCDEs are heated, they can be converted into PCDDs [47, 48, 77]. Pyrolysis of the potassium salt of 4,5,6-trichloro-2-(2,4-dichlorophenoxy)phenol has given higher amounts of PCDDs than pyrolysis of the parent compound [77].

5
Toxicology and Metabolism

Halogenated aromatic hydrocarbons (HAH) cause common toxic and biochemical responses and their mechanism of action is also common [79], but the toxicology of PCDEs is not well known [4, 40, 46]. Weight loss, hepatic porphyria, chloroacne, and impairment of liver function are typical symptoms caused by HAHs [79]. Toxic effects of higher chlorinated PCDEs may appear as chloroacne and liver damage [12].

The binding of halogenated aromatic hydrocarbons to cytosolic Ah-receptor protein leads to toxic effects and biochemical responses including induction of drug-metabolizing enzymes [79]. A correlation has been noticed between toxic responses of HAHs in animals and the induction of metabolizing enzymes such as cytochrome P-4501A1 proteins and hepatic monooxygenases in mammalian cell cultures. HAHs can induce enzymes that metabolize steroid hormones and alter thyroid hormone levels. McKinney [80] has studied the structural relationship of toxic HAHs and the thyroid hormones and presented a multifunctional receptor model for toxic action of HAHs.

5.1
Acute Toxicity

Acute toxicity of PCDEs has been studied in guinea pig, rabbit, rat, and trout [13, 81–83]. Guinea pig is quite sensitive to PCDEs. Lethal doses of tetra- through hexaCDEs were between 0.05 g kg^{-1} and 0.1 g kg^{-1} after 30 days in a single dose oral feeding study [12, 13]. Repeated oral feeding (20 doses) of rabbits with hexaCDEs have caused liver injuries at 0.001 g kg^{-1} (1 mg kg^{-1}) dose [12, 13].

PCDEs 100, 132, and 139 are moderately toxic in rat [81]. Chu et al. [81] performed a short-term study with these tri-*ortho*-chlorinated compounds by administering them to rats at 0.04, 0.4, 4.0, and 40 mg kg^{-1} per day for a period of 28 days. Suppression of growth or food consumption was not detected, but the highest dose of 40 mg kg^{-1} increased liver weights. Thyroid gland was also affected by these PCDEs.

PCDEs 99, 153, and 184 have also been considered as moderately toxic in rats [82]. PCDEs 99 and 153 are di-*ortho*-chlorinated and PCDE 184 is tetra-*ortho*-chlorinated. The toxicity of PCDEs 99, 153, and 184 was studied in a 28 day study by feeding rats with 4 doses: 0.5, 5.0, 50, and 500 mg kg^{-1}. The highest dose caused decreased food consumption and increased liver weights. The no-observable-effects-levels (NOEL) of these PCDE congeners were 5–50 mg kg^{-1} in diet.

Acute toxicity and toxicokinetics of PCDEs have been reported to be dependent on the number and position of chlorine [83]. Lower chlorinated PCDEs, PCDE 3 (4-monoCDE), PCDE 7 (2,4-diCDE), PCDE 28 (2,4,4'-triCDE), and PCDE 74 (2,4,4',5-tetraCDE), were studied in brook trout (*Salvelinus fontinalis*). The LC$_{50}$ (24 h) of PCDEs 3 and 7 were 1.4 mg l^{-1} and 1.24 mg l^{-1}, respectively, but the toxicities of PCDEs 28 and 74 were less.

Toxicity of non-*ortho*- and mono-*ortho*-PCDEs to fish has recently been studied with early life stages of Japanese medaka [84]. PCDEs 77, 105, and 118 were shown to be embryotoxic to medaka, but the potencies relative to 2,3,7,8-TCDD were low: the toxic equivalency factors (TEF) were 0.00001–0.00056. The PCDE fraction isolated from a Lake Ontario lake trout was also embryotoxic and PCDEs in trout were suggested to have the potential to induce toxic effects in early life stages of fish, although not blue-sac disease.

Fish-specific TEFs of halogenated diphenyl ethers have been low also in a rainbow trout early life stage mortality bioassay [85]. Halogenated diphenyl ethers were inactive in this bioassay, whereas polyhalogenated dibenzo-*p*-dioxins, dibenzofurans, and biphenyls reduced growth and produced yolk sac edema and craniofacial malformations. PCDEs 71, 77, 102, 118, and 126 and PBDEs 47, 85, and 99 were analyzed in this bioassay.

5.2
Genotoxicity, Carcinogenicity, Mutagenicity, and Teratogenicity

There is not much data on genotoxicity, carcinogenicity, mutagenicity, and teratogenicity of PCDEs. Fahrig et al. [86] did not find any genetic activity for two tetra-CDE isomers at a concentration up to 1000 mg l^{-1}. Butylated monoCDE has not been mutagenic or teratogenic [12, 24], but nitrofen has shown mutagenic, carcinogenic and teratogenic properties [30]. It has been suggested that nitrofen exerts its teratogenic effect via alterations in thyroid hormone status [87].

5.3
Immunotoxicity and Developmental Toxicity

Immunotoxicity of PCDE has been studied in rat and mice [82, 88, 89]. Chu et al. [82] reported that PCDE 184 is immunosuppressive in rats. Immunotoxicity of PCDEs has also been reported in mice [88, 89]. Immunotoxicity of PCDEs correlated with enzyme induction activities in the study of Howie et al. [88], who measured the dose-response suppression of the splenic plaque-forming cell response to sheep red blood cells in Ah-responsive C57BL/6 mice. Immunotoxicity of PCDEs was concluded to be mediated by the Ah-receptor. Of eight PCDE congeners tested, only PCDE 167, did not cause dose-response immunosuppressive effects. PCDEs were >200 times less immunotoxic than 2,3,7,8-TCDD. In contrast to PCBs, non-*ortho* congeners PCDEs 77 and 126 were less potent than mono-*ortho*-congeners PCDEs 118 and 156. The potencies followed the order PCDE 156 > PCDE 126 > PCDE 118 > PCDE 77 > PCDE 101 > PCDE 153 > PCDE 154. Of PCBs, only non- and mono-*ortho*-coplanar congeners are immunotoxic [79].

Unexpectedly high immunotoxicity of higher chlorinated PCDEs compared to lower chlorinated PCDEs has been reported in mice [89]. Immunotoxicity of PCDEs 206, 207, 208, and 209 were measured in C57BL/6 mice and in less Ah-responsive DBA/2 mice. Harper et al. [89] suggested that some immunotoxic effects could be independent from Ah-receptor. Another study also refers that immunotoxicity of PCDEs may not be mediated by Ah-receptor. Kverkliet et al. [90] studied humoral immunotoxicity of a PCDE fraction isolated from technical grade pentachlorophenol in C57BL/6 mice and observed that this fraction was not immunosuppressive at any dose level tested. The PCDE fraction contained hexa-, hepta-, octa-, and nonaCDEs.

Nevalainen and Kolehmainen [91], who developed QSARs for PCDEs, have reported that immunotoxicity of PCDEs correlates with their electronic properties. 3,4-Substituted isomers have the highest immunotoxicity and lower frontier orbital energy gap.

PCDEs have shown developmental toxicity [92] like nitro-derivatives of chlorodiphenyl ethers, nitrofen analogs [30, 93]. Fetotoxicity of PCDEs has been studied in Swiss-Webster mice and Sprague-Dawley rats [92]. PCDEs 71 (2,3',4',6-tetraCDE) and 154 (2,2',4,4',5,6'-hexaCDE), two of nine PCDEs tested for maternal and perinatal toxicity in outbred mice, decreased the number of pups born and the number of pups surviving. PCDEs 102 (2,2',4,5,6'-pentaCDE) and 153 (2,2',4,4',5,5'-hexaCDE) decreased the number of litters without decreasing postnatal survival. In outbred rats (Sprague-Dawley), PCDE 102 decreased the number of litters born and the survival of litters, but PCDE 71 only suppressed postnatal weight gain. The developmental toxicity did not correlate with induction of cytochrome P450 or with residues in tissues.

5.4
Ah-Receptor-Mediated Enzyme Induction

The toxic responses and induction of enzyme activities such as AHH (aryl hydrocarbon hydroxylase) have been suggested to be Ah-receptor-mediated [79]. Enzyme induction of PCDEs has been studied more than acute toxicity of PCDEs. Poland et al. [94] reported that in vitro binding affinity of PCDE 77 (3,3',4,4'-tetraCDE) for hepatic cytosol from C57BL/6 J mice and biological potency in the chick embryo were low. The highest dose tested in chicken embryo was 4.7×10^{-7} mol kg^{-1}. PCDE 77 was inactive in both systems and did not induce AHH activity, suggesting that the structure-activity of PCDEs is different from PCBs. PCBs that can adopt a planar configuration, non-*ortho*-substituted PCBs (coplanar PCBs) having substituents only in *para* and *meta* positions, are the most toxic and the most potent inducers of enzyme activities among PCBs [79].

Low induction potency of PCDE 77 has also been reported by Carlson et al. [95]. They investigated induction of AHH by PCDEs in Sprague-Dawley rats. PCDEs were administered to rats 2.7 1 mg kg^{-1} day^{-1} (10 µmols kg^{-1} day^{-1}) for three days after which the liver was measured for AHH. PCDE 66 (2',3,4,4'-tetraCDE) and PCDE 99 (2,2',4,4',5-pentaCDE) induced AHH, but PCDE 77 (3,3',4,4'-tetraCDE) and PCDE 209 (decaCDE) did not. Di- and triCDEs were also inactive. The position and degree of chlorination were suggested to be important to the extent and type of induction. Compared to mixtures of brominated diphenyl ethers, PCDEs were much poorer inducers.

The PCDE fraction isolated from technical grade pentachlorophenol and studied for the humoral immune response in mice has not induced enzyme activity characteristic to toxic Ah-interactive compounds [90]. This PCDE fraction contained hexa-, hepta-, octa-, and nonachlorodiphenyl ethers.

Chui et al. [96] have reported that PCDE 74 (2,4,4',5-tetraCDE) induce 7-ethoxycoumarin O-deethylase activities in trout and PCDE 28 (2,4,4-triCDE) and PCDE 74 (2,4,4',5-tetraCDE) in rats. The effects of PCDEs were studied by administering PCDEs 100 mg kg^{-1} day^{-1} to rats and trout for three days. Chui et al. [96] classified PCDE 28 as a phenobarbital (PB)-type inducer and PCDE 74 a mixed-type inducer. Due to the fact that PCDEs cannot adopt planar configuration, it was suggested that PCDEs cannot act as 3-methyl chloranthrene(MC)-type inducers unlike non-*ortho*-PCBs. PCBs that are not acutely toxic can still induce toxic and biochemical responses and are PB-type inducers of hepatic drug-metabolizing enzymes [79].

Induction of other enzyme activities such as aminopyrine N-demethylase (APDM) and ethoxyresorufin O-deethylase (EROD) have been detected in toxicity studies of PCDEs in rats [81, 82, 97]. Iverson et al. [97] tested twelve PCDEs including tetrachlorinated and higher congeners for monooxygenase activities in Sprague-Dawley rats and reported that all PCDEs tested increased monooxygenase activities and P450 levels, resembling PCBs. Chu et al. [81] reported that PCDE 139 increased hepatic APDM in male rats and aniline hydroxylase activity in female rats fed with 40 mg kg^{-1} of PCDE 139. PCDE 99 increased APMD activity, PCDE 153 APDM, aniline hydroxylase, and EROD activities, and PCDE 184 EROD activities in Sprague-Dawley rats.

Induction of AHH and EROD by PCDEs has also been studied in mice [88, 89]. EROD activity in C57BL/6 N mice has been measured for PCDEs 77, 101, 118, 126, 153, 154, 156, 167, 206, 207, 208, and 209 [88, 89]. Howie et al. [88] reported that the structure-activity of PCDEs was different to that of PCBs. Coplanar PCDE congeners were less potent than their mono-*ortho*-analogs and increasing *ortho*-substitution was less effective in reducing the activity of PCDEs compared to *ortho*-effects of PCBs. Based on the results of immunotoxicity and induction of enzyme activity of the above PCDEs, Safe [79] proposed a similar TEF value (0.001) to that of mono-*ortho*-PCBs for non- and mono-*ortho*-PCDEs.

Koistinen et al. [98], who studied the induction of EROD activity of 29 PCDE congeners in vitro using H4IIE rat hepatoma cells, have reported that PCDEs are inactive as inducers of EROD. A similar type of activity as that reported by Howie et al. [88] was only seen with PCDE congeners before cleanup by Florisil column chromatography. After the removal of PCDD and PCDF impurities using Florisil the activity was lost, except for PCDE 156 which was a weak inducer (TEF 1.2 × 10^{-5}). It is possible that effects such as chloroacne reported for higher chlorinated PCDEs [12], could be due to toxic PCDF impurities. 2,3,7,8-Substituted PCDFs have been measured at ng ml^{-1} levels in 10 µg ml^{-1} PCDE solutions [98].

Based on low induction of EROD activity of PCDEs in H4IIE rat hepatoma cells [98], it has been suggested that the mechanism of action of PCDEs is not Ah-receptor-mediated. Sulfur analogs of PCDEs, polychlorinated diphenyl sulfides (PCDS), were also suggested be weak inducers of EROD [98]. 3,3',4,4'-TetraCDS has been reported not to induce EROD or AHH in Hepa-1 mouse hepatoma cell line bioassay [99]. The results of Hornung et al. [85] on the rainbow trout early life stage confirm that the toxicity and induction of enzyme activity of PCDEs might be different from that of PCDDs and related compounds. No TCDD-like toxicity was reported at concentrations near 50 µg g^{-1} in rainbow trout eggs with PCDEs 71, 77, 102, 118, and 126.

Even though the studies of PCDEs as inducers of enzyme activities have given somewhat controversial results, all results refer to relatively low activity of PCDEs as enzyme inducers. This is not surprising, since NMR and X-ray studies of PCDEs have shown that PCDEs are flexible molecules [8, 61, 69, 70]. In contrast to PCBs, non-*ortho*-substituted PCDEs cannot adopt a planar conformation due to steric hindrance [70]. Therefore they should not bind strongly to the Ah-receptor.

The results of a quantitative structure-activity relationships (QSAR) model, which can be used to predict Ah-receptor binding of non-tested congeners, have verified that the mechanism of toxicity of PCDEs is not similar to that of PCBs [91]. No relationship between structure and biological activity of PCDEs, except for immunotoxicity, were detected when a semi-empirical molecular orbital method (AM1 Hamiltonian) was used to calculate electronic parameters and investigate biological responses (Ah-receptor binding data and immunotoxicity data) as a function of electronic parameters. Some studies like the study of immunotoxicity in mice [88] and that of neurotoxicity of PCDEs in rats [100] have demonstrated that the structure-activity of PCDEs and PCBs is different. Like non-coplanar PCBs, PCDEs increased [^{3}H]-phorbol ester binding in rat cerebellar granule cells and inhibited $^{45}Ca^{2+}$ sequestration in rat cerebellum.

5.5
Thyroid Hormone Receptor-Mediated Effect

PCDDs, PCDFs, and PCBs have been detected altering thyroid hormone function and inducing enzymes which metabolize steroid hormones [79]. 2,3,7,8-TCDD is a potent thyroxine agonist [101]. Although nitrofen and analogs of nitrofen, which like PCDEs structurally resemble thyroid hormones [79], were reported to cause hypothyroidism in adult female mice in the early 1980s [102], thyrotoxicity of PCDEs has not been studied much. PCDEs fulfill the structural requirements for thyroid binding having ligands with an angle between the two ring systems [103].

Chu et al. [81, 82] observed changes in the thyroid when rats were administered PCDEs. Rosiak et al. [104] have lately reported that some PCDEs can cause hypothyroidism in rats. PCDEs 71 and 102, which have closest resemblance to thyroxine (T_4), were shown to cause hypothyroidism in pregnant rats and PCDEs 71, 102, and 153 in juvenile rats exposed in utero. PCDEs 71 and 102 have been fetotoxic in mice [92]. Rosiak et al. [104] suggested that, e.g., the inhibition of the release of T_4 from thyroid follicular cells by PCDEs or binding of PCDEs to thyroid hormone binding protein, transthyretin, could decrease T_4 levels. Triiodothyronine (T3) levels were not altered by PCDEs 71, 102, and 153.

5.6
Estrogen Receptor-Mediated Effects

There is not much data available on the endocrine-disrupting properties of PCDEs, but it is possible that PCDEs and their metabolites bind to estrogen receptor. PCBs belong to estrogen-mimetic xenobiotics, which have been thought to have toxicological and environmental effects by interfering with the action of hormones which are important for developmental and reproductive health of wildlife and humans, but it is not clear whether this really happens [79, 105, 106]. Furthermore, not much is known about mechanisms of xeno-estrogens, including receptor interactions, nor about indirect hormonal mechanisms. Coplanar PCBs like TCDD are possible antiestrogens, which can act as antagonist for estrogen receptor and inhibit estrogen action, whereas *ortho*-PCBs and *para*-hydroxylated PCBs may act as estrogens. The receptor affinities of OH-PCBs measured in vitro using human breast cancer cells, however, have been weak compared to 17β-estradiol, the natural estrogen hormone. Xenobiotics might also affect the metabolism of steroid hormones by inducing cytochrome P450 isoenzymes.

5.7
Metabolism and Excretion

Just as diphenyl ether is metabolized to hydroxydiphenyl ether derivatives [107], PCDEs can be metabolized to hydroxy-PCDEs [18, 34]. Tulp et al. [18] studied the metabolism of PCDEs in rats and observed that the major metabolic route is hydroxylation, but chlorophenols can also be formed by scission of the ether bond. They detected mono-, di-, and trihydroxy metabolites in rats and hydroxy-PCDEs were suggested to be less bioaccumulative than PCDEs. In addition to hydroxy derivatives, biotransformation of diphenyl ether structures can lead to methoxylated species [107].

PCDEs are converted to toxic PCDFs in photochemical reactions [71] and metabolism of PCDEs could in theory also lead to the formation of PCDDs when both 2- and 2'-positions are hydroxylated and then dehydrated. No PCDDs and PCDFs, however, were detected in the urinary and fecal extracts of rats fed with PCDEs [18]. Metabolic formation of PCDDs or PCDFs was neither observed in the case of a predioxin, Triclosan.

Metabolism studies of PCDEs in rats and fish have shown that PCDEs are metabolized more and excreted faster in rats compared to fish. When metabolism, tissue distribution and excretion of PCDEs was studied in male Sprague-Dawley rats with PCDE 99 (2,2',4,4',5-pentaCDE), 55% of PCDE 99 orally fed was excreted in feces in seven days and 64% of PCDE 99 was excreted unchanged [34]. Half-lives of hepta- through nonaCDEs in various tissues of rat have ranged between 5.7 days and 13.4 days [108].

The metabolism of PCDEs in fish has not been studied much, but apparently the metabolism in fish is low and similar to that of PCBs. Hydroxy-PCDEs were not detected in guppy (*Poecilia reticulata*) exposed to tri- and tetraCDEs [63]. Low metabolism and slow excretion leads to persistence and bioaccumulation. According to a study of Zitko and Carson [109], tri- through pentaCDEs are somewhat more persistent in fish than the corresponding PCBs. The excretion half-lives of one trichloro (PCDE 28; 2,4,4'-), one tetrachloro (PCDE 66; 2,3',4,4'-) and one pentachloro (PCDE 99; 2,2',4,4',5-) were 15, 55, and 55 days in juvenile Atlantic salmon (*Salmo salar*), respectively. Half-lives of PCDEs were near to those of PCBs. The depuration half-lives of mono- through tetraCDEs have varied from 4 to 63 days in brook trout (*Salvelinus fontinalis*) [83] and those of tri- through decachlorinated PCDEs between 46 and 100 days in rainbow trout (*Salmo gairdneri*) [110].

6
Analytical Methodology

There are no validated methods for analysis of PCDEs. The same general principles that apply to analysis of PCDDs and PCDFs, including recommendations for sampling procedures, sample storage, sample pretreatment, extraction, and analytical instruments, should also be advisable for PCDEs. Analysis of organochlorines including PCDEs [1], PCDDs, and PCDFs [111, 112] has previously been reviewed. PCDEs as lipid soluble compounds are extracted together with PCDDs and PCDFs and analytical methods have aimed to allow group separation of these and other organochlorines. Analytical methods of PCDEs have been collected in Table 7.

Table 7. Analytical methods of PCDEs[a]

Matrix	Drying method	Extraction method	Extraction solvent	Cleanup (lipid removal)	Isolation	Analysis technique	Ref.
Sediment	Freeze-drying	Ultrasonic	hex:acet (1:1)	L-L	Florisil	HRGC-ECD	[115]
	Freeze-drying	Soxhlet (6 h)	toluene	H_2SO_4	1) Florisil 2) carbon (SK-4)	HRGC-LRMS (EI, SIM)	[33, 113, 114]
Mussel		Homogenizer	1) acetone 2) Freon 113	silica	silica	GC-MS (EI, Scan)	[51]
	Freeze-drying	Soxhlet (6 h)	PAHE	H_2SO_4	1) Florisil 2) carbon (SK-4) 3) silica	HRGC-LRMS (EI, SIM)	[125]
Fish							
Muscle	Na_2SO_4	Soxhlet (8 h)	hex:DCM (1:1)	Celite + H_2SO_4	Carbon (Amoco PX-21) + glass	HRGC-MS (EI, Scan, SIM)	[116]
Muscle	Na_2SO_4	Soxhlet (24 h)	hex:acet (1:1)	GPC		HRGC-LRMS (NCI)	[117, 118]
Muscle	Na_2SO_4	Soxhlet	DCM	GPC	Silica	HRGC-HRMS (EI, SIM)	[119]
Muscle	Na_2SO_4	Soxhlet (16 h)	hex:DCM (1:1)	H_2SO_4	1) silica +H_2SO_4 2) Florisil 3) alumina (basic)	HRGC-HRMS (EI, SIM)	[120]
Muscle liver, spawn	Na_2SO_4	Soxhlet (6 h)	PAHE	H_2SO_4	Carbon (SK-4)	HRGC-LRMS (EI, SIM)	[122]
Muscle	Na_2SO_4	Soxhlet (6 h)	PAHE	H_2SO_4	1) Florisil 2) carbon	HRGC-LRMS (EI, SIM)	[123]
Muscle, liver	Freeze-drying	Soxhlet (6 h)	PAHE	H_2SO_4	1) Florisil 2) carbon (SK-4)	HRGC-LRMS (EI, SIM)	[33, 113, 114]
Liver		separatory funnel	hexane	H_2SO_4	1) Florisil 2) carbon (SK-4)	HRGC-LRMS (EI, SIM)	[123]
Muscle	Na_2SO_4		DCM	GPC	Silica	HRGC-ECD	[129]

Table 7 (continued)

Birds						
Muscle, egg	Na_2SO_4	Soxhlet (8 h)		Florisil	SilicAR	HRGC-HRMS (EI, Scan) [121]
Muscle, liver, fat		homogenizer	hex:acet (1:2)	L-L	1) Florisil, 2) HPLC	HRGC-ECD, HRGC-LRMS (EI, Scan) [126]
Muscle, egg	Na_2SO_4	Soxhlet (6 h)	PAHE	H_2SO_4	1) Florisil, 2) carbon (SK-4)	HRGC-LRMS, (EI, SIM) [124]
Seals						
Blubber	Na_2SO_4	ultrasonic	hex:acet (1:1)	H_2SO_4	1) Florisil, 2) carbon (SK-4)	HRGC-LRMS, (EI, SIM) [33, 113]
Humans						
Adipose tissue			hex:acet (1:5)	GPC	1) Florisil, 2) carbon	HRGC-LRMS, (EI, SIM) [131]
			DCM	silica+H_2SO_4	1) silica, 2) silica + carbon (Amoco AX-21)	HRGC-HRMS, (EI, SIM) [132]
	Na_2SO_4	Soxhlet	PAHE	H_2SO_4	1) Florisil, 2) carbon (SK-4)	HRGC-LRMS, (EI, SIM) [123]

[a] acet = acetone; hex = hexane; DCM = dichloromethane; PAHE = petroleum ether:acetone:hexane:diethyl ether mixture; L-L = liquid-liquid chromatography; GPC = gel permeation chromatography; HR = high resolution; GC = gas chromatography; ECD = electron capture detector; LR = low resolution; MS = mass spectrometry; EI = electron impact ionization; SIM = selected ion monitoring.

6.1
Extraction

6.1.1
Abiota

6.1.1.1
Sample Pretreatment

Pretreatment steps before extraction include drying of the sample, which can be performed as air-drying, freeze-drying, or drying by mixing the sample with sodium sulfate. Sediment samples have been air-dried or freeze-dried before extraction [33, 113–115].

6.1.1.2
Extraction

Soxhlet extraction is a preferred method for extraction of PCDDs and PCDFs in particulates [111] and has been applied to extract PCDEs from sediment [33]. Koistinen et al. [33] extracted sediments for analyses of PCDEs, PCDDs, and PCDFs with toluene for 48 h. Coburn and Comba [115] used ultrasonic extraction with a mixture of hexane:acetone (1:1). Suspended particulates were extracted with dichloromethane for 2 h [51].

As with sediment, Soxhlet extraction has been preferred for extraction of PCDEs from fly ash. Before extraction of fly ash, acid treatment with HCl increases the recoveries of PCDDs and PCDFs, and was applied by Kurz and Ballschmiter [43]. Koistinen et al. [57] did not use pretreatment with HCl when they extracted a fly ash for analysis of PCDEs.

For analyses of PCDEs in chlorophenol formulations, the sample has been dissolved in 1 N NaOH before extraction with petroleum ether or hexane [37–43]. Firestone et al. [37] warmed the sample with water and 1 N NaOH before extraction with petroleum ether. The extract was washed first with water until neutral and then with 0.1 N NaOH. Paasivirta et al. [42] mixed the sample with methanol and NaOH aqueous solution before extraction with petroleum ether. Nilsson and Renberg [39] dissolved the samples in 1 mol l^{-1} KOH and extracted with hexane:diethyl ether (1:1) mixture. The extract was washed with 1 mol l^{-1} KOH and then with water. Koistinen et al. [33] dissolved the sample in acetone, added hexane, and 0.1 mol l^{-1} K$_2$CO$_3$. After shaking the organic layer was separated for analyses.

6.1.2
Biota

6.1.2.1
Sample Pretreatment

Fish and bird tissues are homogenized before extraction and usually dried with sodium sulfate for PCDE analyses [116–121]. Homogenization can also be performed by mixing with sodium sulfate at the same time [122–124]. Fish tissues analyzed for PCDEs have included muscle, liver, and spawn and, of birds, muscle and eggs have been studied. Biota samples (fish, mussels) have also been freeze-dried before solvent extraction [33, 57, 58, 114, 125]. Chicken tissue has been homogenized with a homogenizer in an extraction solvent [126]. Acid or base digestion can be used to break down the sample matrix before solvent extraction, but it has not been used for PCDE analyses.

6.1.2.2
Extraction

Extraction of biota samples can be performed in a Soxhlet apparatus or by column extraction. Column extraction, which is carried out in glass columns packed with the dried biota sample (free-floating mixture of anhydrous sodium sulfate and the sample), has not been used in analysis of PCDEs.

Soxhlet extraction of PCDEs has been performed using one solvent or mixtures of different solvents. Fish and bird samples (muscle, egg) have been extracted for 6 h using a mixture of petroleum ether:acetone:hexane:diethyl ether (PAHE) (18:11:5:2 by volume) [33, 57, 58, 114, 122–124] in Finnish studies. This solvent mixture has been used since it was reported to give the best recoveries of PCBs in Soxhlet extraction when different solvent combinations were tested [127]. Soxhlet extraction gave better results for PCBs than those obtained by column extraction.

Kuehl et al. [116] have extracted fish (dried with sodium sulfate) in a Soxhlet apparatus for 8 h with hexane:dichloromethane (1:1) mixture. Birkholz et al. [120] extracted fish liver with the same solvent mixture, but used a longer extraction time (16 h). Acetone:hexane (2:1) mixture [128] and dichloromethane [117–119, 129] have also been used to extract PCDEs in fish. Jaffe et al. [117] and Jaffe and Hites [118] extracted fish with acetone:hexane (1:1) in a Soxhlet for 24 h. Fish liver oil has been analyzed for PCDEs after being dissolved in hexane [43].

Bird eggs and carcasses have been extracted after homogenization and drying with sodium sulfate in a Soxhlet apparatus for 8 h [121]. Koistinen et al. [124] have extracted bird eggs and muscle tissues after sodium sulfate drying in a Soxhlet with PAHE mixture for 6 h.

In addition to Soxhlet extraction, PCDEs have been extracted from biota samples by mixing with a solvent in a homogenizer [126]. This method was developed for determination of PCDEs in chicken tissue. Chicken samples (muscle, liver, and fat) were homogenized in a Silverson homogenizer for 1 min in acetone:hexane (2:1). The filtered extract was partitioned with water in a separation funnel and then hexane treated further. The same kind of procedure has been applied for edible marine organisms [51]. Lake et al. [51] extracted mussel, clam, and lobster by polytron homogenization with acetone and Freon 113, and partitioned the acetone extract with water and Freon. Extracts of Freon were cleaned further for analysis of PCDEs. Koistinen et al. [125] have extracted mussel tissue after freeze-drying in a Soxhlet apparatus using PAHE mixture.

Ultrasonic extraction is an alternative extraction method for abiota and biota samples. Blubber samples of seal have been extracted with an acetone:hexane mixture (1:1) in a water bath by ultrasonication [33, 113]. The mixture of blubber and sodium sulfate was extracted twice with the solvent mixture.

Human adipose tissue has been extracted with 15 % acetone/hexane mixture [130, 131], dichloromethane [132, 133] and by using PAHE mixture [123].

6.2
Cleanup

Removal of interferences is a critical step before analyses of trace level contaminants such as PCDDs and related compounds [111]. Coextractives can occur at much high levels than studied compounds and can complicate identification and quantitation. Extensive sample cleanup techniques which include bulk matrix removal and liquid-solid chromatography are needed to remove coextractives that can interfere with analysis by causing false positives and raising detection limits.

The cleanup methods used for analyses of PCDEs have been developed simultaneously with the cleanup methods of PCDDs and PCDFs. Since PCDEs may interfere in the GC-MS determination of PCDFs [38, 68], the sample preparation procedures for analyses of PCDDs and PCDFs have been developed aiming to isolate PCDEs from these compounds.

Chromatography on an alumina microcolumn [67] and a Florisil column [134] can be used to separate PCDEs from PCDFs. PCBs are also separated from PCDDs and PCDFs on Florisil and alumina [111, 112]. Group separation on these materials can be achieved by using sequential elution with solvents of increasing polarity. Florisil column chromatography technique has been used in most PCDE studies [33, 57, 58, 113, 114, 120, 123–126, 128, 130, 131] and alumina chromatography less [120]. Fractionation has also been performed on silica gel [51, 119] and carbon [116, 131, 132].

During extraction and cleanup, concentration of sample extracts can be performed using similar techniques as in the analysis of organochlorines, PCDDs, and PCDFs [1, 111]. Vacuum evaporation and evaporation under nitrogen flow are also typical concentration techniques in PCDE analyses.

6.2.1
Bulk Matrix Removal

Bulk matrix removal aims to remove material such as lipids which can disturb final analysis. This can be performed by acid treatment of the extract or by liquid-solid chromatography. Alumina fractions of chlorophenol extracts have been purified with concentrated sulfuric acid [37, 38] and it has been used to remove lipid and organic coextractives in sediment, biota, and human extracts [33, 43, 57, 58, 113, 114, 120, 122–125]. The sulfuric acid treatment of PCDEs has been reported not to affect their recoveries [58].

Bulk matrix removal by liquid-solid chromatography has been performed on Florisil, silica, modified silica gel, and modified celite [51, 108, 115, 116, 121, 126, 128, 132]. Silica gel (activated) [51] and the Florisil column chromatography technique [108, 121, 126, 128] have been used to remove lipids from fish muscle, bird eggs and tissues (muscle, liver, fat), and rat tissues (muscle, liver, blood, skin) and for bulk matrix removal of sediment extracts. Extracts have been eluted with hexane from 2% deactivated Florisil [108, 126]. Stafford [121] removed lipids, eluting the sample extract with 6% diethyl ether in hexane through a partially deactivated Florisil column (60/100 mesh).

Kuehl et al. [116] have used Celite-545 coated with concentrated sulfuric acid for lipid removal. They eluted fish extract using hexane through this column to a cesium silicate column. Acid treated silica gel is a common material for lipid removal of biota and abiota samples for PCDD/PCDF analyses [111, 112] and has been applied to human adipose tissue extracts for analysis of PCDE [132]. Silica impregnated with concentrated sulfuric acid packed on silica gel in a column has been applied as a second cleanup step [43, 120, 132]. A multi-silica gel column consisting of layers of silica (silica gel 60), silica impregnated with sulfuric acid (44%), silica, silica impregnated with sodium hydroxide (1 mol l^{-1}), silica, silica with silver nitrate (10%), and silica has been used for fish extracts by Birkholz et al. [120]. The extract was eluted with 2% dichloromethane in hexane and was further purified on Florisil.

Gel permeation chromatography (GPC) has been applied to remove lipids for PCDE analyses in fish muscle and human adipose tissue [110, 117–119, 129–131]. A mixture of dichloromethane:cyclohexane is a typical eluent for automated GPC systems. GPC was used by Jaffe et al. [117] for fish extracts based on a method of Stalling et al. [135]. The column contained SX-2 (200–600 mesh) Biobeads swollen in the mixture which was used as an eluent (cyclohexane: dichloromethane mixture (3:2)).

6.2.2
Adsorption Chromatography

6.2.2.1
Silica Gel

Silica gel has mostly been used for bulk matrix removal, but it has also been used to fractionate PCDE-containing extracts [51, 119, 121, 129]. When silica gel chromatography is used for fractionation of PCDE-containing extracts, PCBs are often eluted with PCDEs. Lake et al. [51] eluted PCDEs from a silica column (deactivated with 5% water) together with PCBs using pentane (50 ml) as an eluent. A 2-(1-pyrenyl)ethyldimethylsilylated silica column has been used to separate PCDEs and PCBs from PCDDs and PCDFs [43].

Koistinen et al. [125] have used an activated silica gel column (1 g) as an additional cleanup after Florisil and carbon column chromatography. They separated impurities that disturbed MS analysis of PCDEs on silica.

6.2.2.2
Florisil

In addition to lipid removal, Florisil column chromatography can be used for chromatographic cleanup and to separate PCDEs from other compounds such as PCBs, PCDDs, and PCDFs. Florisil has been used activated and after deactivation with water. An activated Florisil column has been reported to separate PCDEs from PCBs [112], but in most cases Florisil has been used deactivated for separation of PCDEs from PCDDs and PCDFs [33, 57, 58, 113, 114, 123, 124, 130, 134].

Coburn and Comba [115] used activated Florisil (30 g column) and eluted PCDEs with 6% ethyl ether in hexane (200 ml) after they had collected PCBs with hexane (200 ml). Birkholz et al. [120] used a 0.8% deactivated Florisil column (12 g) as an additional cleanup after multisilica column chromatography. PCDEs and PCBs were eluted with 100 ml of hexane from Florisil. PCDEs were then separated from PCBs using a basic alumina column. Newsome and Shields [126] used Florisil (12 g column) deactivated with 2% water and eluted PCDEs with hexane (85 ml) for further cleanup by HPLC using acetonitrile as an eluent. Stafford [121] has also used deactivated Florisil.

Ryan et al. [134] have used a 1.5 g column of Florisil (60–100 mesh; activated at 130°C for 24 h) for separation of PCDEs from PCDDs and PCDFs. They eluted first PCDEs and PCBs with 2% hexane in dichloromethane (20 ml) and then PCDD/PCDFs with dichloromethane (35 ml). They further fractionated the extract on a silica column.

Koistinen et al. [58] have applied a Florisil cleanup which uses a partially deactivated (1.25%) Florisil microcolumn (1 g) to separate PCDEs from PCDDs and PCDFs in abiota and biota extracts [33, 57, 113, 114, 123–125]. After sulfuric acid cleanup, PCDEs are eluted through a Florisil column together with PCBs using hexane (15 ml). PCDDs and PCDFs are retained in the column and can be eluted with dichloromethane (12 ml).

6.2.2.3
Alumina

Alumina column chromatography has been a common cleanup method in PCDD/PCDF analyses [111, 112].Activated basic aluminum oxide has been used to isolate PCDEs from PCDDs and PCDFs [37, 43, 67], as well as from PCBs [120].

Activated alumina (260°C) column has been used for isolation of PCDEs, PCDDs, and PCDFs from chlorophenol extracts [37, 38]. Four fractions were collected from an alumina column (50 g): petroleum ether (400 ml), 5% ethyl ether in petroleum ether (200 ml), 25% ethyl ether in petroleum ether (400 ml), and ethyl ether (400 ml). PCDEs, PCDDs, and PCDFs were determined in the third and fourth fractions after sulfuric acid cleanup.

Buser [67] developed a microcolumn method in which a Pasteur pipette is packed with 1.0 g of dry basic alumina. The column is first eluted with 2% dichloromethane in hexane (10 ml) and then with 50% dichloromethane in hexane (10 ml). PCDEs and PCBs are eluted in the first fraction and PCDDs and PCDFs in the second fraction. This method has been applied in some earlier studies [122, 136].

The quality of alumina seems to vary, since there are some reports on unsuccessful separation of PCDEs from PCDDs and PCDFs on alumina. Huestis and Sergeant [129] were not able to separate PCDEs from PCDDs and PCDFs using alumina activated at 240°C. Koistinen et al. [58] reported similar results with basic alumina microcolumns. The activation of alumina is crucial for effective separation of PCDEs from PCDDs and PCDFs [111].

A basic alumina column (5 g) has been reported to separate PCDEs from PCBs [120]. PCBs can be first eluted with hexane (100 ml) and then PCDEs with dichloromethane (100 ml). Kurz and Ballschmiter [43] have also used a similar kind of basic alumina and collected PCDEs with hexane:dichloromethane (1:2; v/v; 50 ml) after eluting the column with hexane (50 ml).

6.2.2.4
Carbon

A carbon column was initially presented by Jensen and Sundström [137] for separation of non-planar and planar PCBs. It is often applied in the cleanup of PCDDs and PCDFs as presented by Smith et al. [138]. Kuehl et al. [116] used a carbon-glass column to separate non-planar and planar compounds. They prepared the column by blending AMOCO PX-21 carbon (50 mg) with a shredded glass filter pad (600 mg) in dichloromethane and packed this slurry into a glass column. The sample was fractionated by eluting with dichloromethane (50 ml), dichloromethane:benzene (1:1; 50 ml), and finally in reverse with toluene (50 ml). Non-planar compounds, loosely bound planar compounds, and tightly bound planar compounds were isolated in these three fractions. PCDEs have been reported to elute in the same fraction as PCDDs and PCDFs on activated carbon [36].

Carbon fiber column cleanup has been successfully used for separation of PCDEs from PCDDs and PCDFs in a HPLC system [129]. The separation can be achieved using three solvent system, in which dichloromethane:cyclohexane (1:1; v/v) is the first eluent, ethyl acetate:benzene (1:1; v/v) the second, and toluene the third. PCDEs are eluted in the first fraction and PCDDs and PCDFs in the third, which is collected using reversed elution.

Carbon column chromatography has also been reported to separate PCDEs from PCBs in some studies [36, 131]. Paasivirta et al. [36] eluted PCBs with a mixture of dichloromethane and hexane (1:1; v/v) from an activated carbon column and then PCDEs using reversed elution with toluene. Williams et al. [131] used a carbon/celite column for human adipose tissue extract. An eluate from a Florisil column was allowed to flow through carbon/celite and PCBs were first eluted with 3% dichloromethane in cyclohexane (35 ml) and the PCDEs with 70% dichloromethane in cyclohexane (12 ml).

Carbon column chromatography has been utilized as an additional cleanup after Florisil [33, 57, 58, 113, 114, 123–125]. Impurities are first washed with hexane (10 ml) and PCDEs are then eluted with toluene (10 ml) using reversed elution.

6.3
Analysis

Gas chromatography combined with mass spectrometry (GC-MS) is the preferred technique in the analysis of PCDDs and PCDFs and the only technique that can also be used to determine PCDE congeners reliably [1]. High resolution gas chromatography (HRGC) with capillary columns is the most commonly used GC technique nowadays and allows the separation of most of the PCDE congeners [4, 44, 66]. Before capillary columns were introduced, PCDE analyses were also performed on packed columns [115].

HRGC-ECD has been used in some studies for PCDE analysis [115, 117, 126, 129], but GC-MS has been the most often used method in analysis of PCDEs [33, 43, 57, 58, 113, 114, 116, 118–126, 128, 130–133, 139]. Analysis of PCDEs has mostly been performed in selected ion monitoring mode (SIM) [33, 57, 58, 113, 114, 119, 120, 122–125, 130, 132, 139], but full scan mode has also been used in some studies [121, 126, 131]. MS instruments are usually operated in electron impact mode (EI). EI mass spectra of most PCDEs have an abundant molecular ion $M^{\cdot+}$ and one abundant fragmention ($M^{\cdot+}$-2Cl). The fragmention ($M^{\cdot+}$-70), which is the same as the molecular ion of PCDFs, is more abundant than the molecular ion, but the latter has mostly been used in SIM analyses of PCDEs. Table 8 presents masses of molecular ions of PCDEs.

PCDE analyses can be performed by high resolution gas chromatography (HRGC) combined with low resolution mass spectrometry (LRMS) or using HRGC combined with high resolution mass spectrometry (HRMS). For example, a Hewlett Packard 5970 mass selective detector system, which is a quadrupole mass analyzer, has been used as an LRMS instrument in Finnish studies [33, 57, 113, 114, 122–125, 139]. Compared to LRMS, HRMS is more sensitive, allows less sample cleanup, and eliminates interference more effec-

Table 8. Theoretical ion masses and abundances of molecular ion clusters of PCDEs

Congener group	Molecular cluster ions	Ion masses (m/z values)	Abundance (%)
Mono	$M^{\cdot+}, (M+2)^{\cdot+}$	204.0342, 206.0315	100, 33
Di	$M^{\cdot+}, (M+2)^{\cdot+}$	237.9952, 239.9924	100, 65
Tri	$M^{\cdot+}, (M+2)^{\cdot+}, (M+4)^{\cdot+}$	271.9562, 237.9534, 275.9506	100, 97, 32
Tetra	$M^{\cdot+}, (M+2)^{\cdot+}, (M+4)^{\cdot+}$	305.9173, 307.9144, 309.9116	78, 100, 49
Penta	$M^{\cdot+}, (M+2)^{\cdot+}, (M+4)^{\cdot+}$	339.8783, 341.8754, 343.8726	62, 100, 65
Hexa	$M^{\cdot+}, (M+2)^{\cdot+}, (M+4)^{\cdot+}$	373.8393, 375.8364, 377.8335	52, 100, 81
Hepta	$(M+2)^{\cdot+}, (M+4)^{\cdot+}, (M+6)^{\cdot+}$	409.7975, 411.7946, 413.7917	100, 97, 52
Octa	$(M+2)^{\cdot+}, (M+4)^{\cdot+}, (M+6)^{\cdot+}$	443.7585, 445.7556, 447.7527	89, 100, 64
Nona	$(M+2)^{\cdot+}, (M+4)^{\cdot+}, (M+6)^{\cdot+}$	477.7195, 479.7166, 481.7137	78, 100, 75
Deca	$(M+2)^{\cdot+}, (M+4)^{\cdot+}, (M+6)^{\cdot+}$	511.6805, 513.6776, 515.6747	69, 100, 86

tively. Therefore, HRMS instruments, which are double-focusing mass spectrometers with both an electric and magnetic sector, are preferred in the analysis of PCDDs and PCDFs [111, 112], although they are more expensive than LRMS instruments. The same resolution of 10,000 as is typically used for PCDD/PCDF analyses has been utilized for HRMS analysis of PCDEs [119, 120, 132].

The use of isotope-labeled internal standards, which is recommended in the mass spectrometric analysis of PCDDs and PCDFs, to monitor the efficiency of extraction and cleanup has not been common in PCDE analyses due to the limited number of commercially available labeled standards. A D_5-labeled pentaCDE has been used as an internal standard in Finnish studies [33, 113, 114, 123–125, 139] and $^{13}C_{12}$-labeled PCDEs have been used only in few studies [120, 131]. Labeled compounds behave like the corresponding endogenous compounds during cleanup steps and would be the best internal standards for reliable results in PCDE analysis.

There are only a few reports about recoveries of PCDEs during analytical procedures. Recovery studies with fortified human adipose fat extracts have given recoveries for PCDEs that range between 46% and 104% depending on the congener [130, 131]. The overall recoveries of PCDEs in fortified chicken tissue (fat, liver, muscle), which had been extracted with acetone:hexane (2:1; v/v) for 1 min in a homogenizer, ranged between 64% and 104% [126]. Stafford [121] reported recoveries from 96% to 112% for PCDEs from fortified chicken eggs.

It is difficult to conclude what the best analytical method is for PCDEs but, based on the literature, microcolumns such as Florisil [57] seem to be effective in separation of PCDEs from PCDDs and PCDFs. Furthermore they are fast and inexpensive. Before PCDEs can be analyzed by MS, however, an additional cleanup step is needed. This has been performed using column chromatography on carbon, since silica gel and neutral alumina microcolumns have not worked well with fish extracts for this purpose [57]. Activated silica and alumina microcolumns, however, could possibly be alternatives for a carbon column. An activated silica column has been used an additional cleanup step after Florisil and carbon column chromatography in the case of mussel extracts [123]. It is not necessary to separate PCBs from PCDEs, since PCBs have been reported not to interfere in the MS analysis of PCDEs [57, 130].

7
Environmental Levels

Environmental levels of PCDEs have not been intensively studied like those of PCBs or PCDDs and PCDFs due to fact that model substances of PCDEs have not been commercially available until recently. Therefore, PCDEs have not been routinely monitored in any aquatic environment. Most data on PCDEs obtained using 17 or more PCDE standards are from Great Lakes fish [119] and from fish and sediment from a Finnish river, the Kymijoki River, [33, 58, 114]. Baltic Sea fish, birds, and seals have also been studied for PCDEs [113, 124, 139].

7.1
Sediment

Traces of a triCDE and a tetraCDE (0.03 ng l^{-1} and 0.06 ng l^{-1}) have been reported in suspended particulate material from Narrangesett Bay, Rhode Island, a disposal site for municipal and industrial wastes [51]. Significantly elevated levels of PCDEs have been measured in Whitby Harbor sediments in the Great Lakes [115] and in sediments of a Finnish river contaminated by a chlorophenol formulation [33, 114]. The total contents of PCDEs varied from 0.05 mg kg^{-1} to 4.9 mg kg^{-1} in Whithby harbor sediments [115].

The levels of total PCDEs have ranged between 0.001 mg g^{-1} and 1.0 mg g^{-1} dw in surface sediments from the Kymijoki River depending on the sampling site [33, 114]. Kymijoki River is a river in Finland, which has been contaminated by the earlier production of Ky-5, a chlorophenol-based wood preservative. The highest contents in surface sediments have been measured in sediments near the earlier manufacturing site of Ky-5 and the contents have descended downstream in the river. Hexa- and heptaCDEs have constituted the majority of PCDEs in the sediments near the manufacturing site. Sediment core analyses have shown that the PCDE contents in deeper sediments can be much higher than at the surface. Like the sediments, the concentrations of PCDEs have been elevated in mosses (*Fontinalis dalecarlica*) from the Kymijoki River downstream from the manufacturing site of Ky-5 [140].

The study of the Kymijoki River has indicated that this river has been contaminated tens of kilometers downstream from the manufacturing site with PCDEs as well as with PCDDs and PCDFs originating as impurities from Ky-5 [114]. The concentrations of PCDEs in reference site sediments upstream from the manufacturing site have been low, as well as those in Baltic sediments [33, 113]. PCDE congeners were below 0.4 ng g^{-1} dw in sediments from the Gulf of Finland, near Gotland and the Bothnian Bay.

7.2
Aquatic Organisms

7.2.1
Mussels

Lake et al. [51] first reported the occurrence of PCDE residues in biota, when they measured PCDEs in mussels (*Mytilus edulis*), clams (*Mercenaria mercenaria*), and lobster (*Homarus americanus*) from Narrangesett Bay, Rhode Island. A triCDE was detected between 3.6 ng g^{-1} and 125 ng g^{-1} and a tetraCDE between 24 ng g^{-1} and 416 ng g^{-1} dw in mussels. The sampling sites were from parts of the bay which was a disposal site for municipal and industrial wastes.

PCDEs have been detected in mussels (*Anodonta piscinalis*) incubated in the Kymijoki River contaminated by chlorophenol formulation Ky-5 [125]. The average content of PCDEs was 5.5 ng g^{-1} lipid weight (lw) in mussels incubated near the earlier manufacturing site of Ky-5 in 1995. The following PCDEs, which are typical to Kymijoki River sediments as well [33, 114], were detected in the mussels: PCDEs 99, 100, 147/153, 154, 180/181, 182, 184, and 197.

7.2.2
Fish

Fish are good indicators of PCDE contamination. Elevated contents of PCDEs have been measured in fish from the Great Lakes, watersheds near the Great Lakes, and in the Kymijoki River in Finland [33, 58, 114, 116–119, 129]. Kuehl et al. [116] found a tetraCDE in channel catfish caught from Lake St. Clair in Michigan in 1980. Jaffe et al. [117] measured PCDEs in carp from tributaries and embayments of Lake Huron including Chippewa River, Flint River, Saginaw River, Saginaw Bay, and Tittabawassee River. The contents of PCDEs ranged from 19 ng g^{-1} to 270 ng g^{-1} lw, being highest in samples from Saginaw River.

PCDEs have been detected in different fish species caught from Niagara River and tributaries of Lake Ontario in 1984 [118]. The concentration of decaCDE in common carp, catfish, goldfish, and sucker ranged from 15 ng g^{-1} to 900 ng g^{-1} lw. High concentrations of PCDEs in Great Lakes fish have also been reported in later studies [119, 129]. The levels of PCDEs have ranged between 768 ng g^{-1} and 14,005 ng g^{-1} in northern pike and carp from Whitby Harbor, Lake Ontario [129]. Niimi et al. [119] measured tetra- through decaCDEs in lake trout caught from Lakes Ontario, Huron, and Superior and in walley caught from Lake Erie in 1992. The total PCDEs ranged between 54.2 ng g^{-1} and 303.4 ng g^{-1} fw (mean 824 ng g^{-1} lw) in lake trout from Lake Ontario. The concentrations of individual PCDE congeners ranged between 0.4 ng g^{-1} and 36 ng g^{-1} lw. Fish from Lakes Huron and Superior had lower contents of PCDEs. Fish size showed no correlation with the content of PCDEs. PCDEs 99, 138, 139, 153, 154, and 183 have been among the dominating PCDEs in Lake Ontario trout. HexaCDEs accounted for 45–60% of total PCDEs in fish from Ontario, Erie, and Huron and sum of hexa- through heptaCDEs 80–90% of total PCDEs. Most PCDEs observed in Great Lakes fish were substituted at both *para*-positions and two or more *ortho*-positions.

PCDE contents have been high in Finnish fish from the Kymijoki River [33, 58, 114]. The total PCDEs in fish from this river have ranged from 2 ng g^{-1} to 33,704 ng g^{-1} lw depending on the species, tissue, and sampling site. The highest contents have been measured in liver of burbot caught near the earlier manufacturing site of Ky-5. The major identified PCDE congeners in the Kymijoki River fish have been PCDEs 99, 147/153, 154, 167, 182, 184, and 197. As with Great Lakes fish, hexaCDEs have accounted for most of the PCDEs measured and PCDE congeners detected have chlorine at both *para*-positions and at least two chlorines at *ortho*-positions. The same PCDEs detected in Kymijoki River fish have been measured in Ky-5 and in Kymijoki River sediments.

PCDEs have also been measured in fish from other sites in Finland and in Baltic fish [33, 36, 57, 122, 136, 139]. The Arctic environment has been suggested to be less polluted with PCDEs, since the concentrations of PCDEs have been lower in Atlantic salmon from the Tenojoki River compared to those in Baltic salmon (*Salmo salar* L.) [57]. The occurrence of PCDEs in Atlantic salmon from the Tenojoki River was suggested to indicate that PCDEs could be transported via air, since there are no known point sources. The occurrence of PCDEs in Arctic cod livers from Vestertana Fjord, Norway, has supported this [123]. The total amount of PCDEs in Arctic cod liver was 14 ng g^{-1} lw. Higher PCDE contents have been measured in oil from North Atlantic cod caught in 1993 (49 ng g^{-1} lw) and in cod liver oil of National Institute of Standards and Technology that had been prepared in 1983 (659 ng g^{-1} lw) [43].

The average content of PCDEs in lipid of salmon muscle from Lake Saimaa in Finland and from the Baltic Sea has been lower than in contaminated Kymijoki River pike. The average contents of PCDEs in salmon collected in 1990 from Lake Saimaa in Finland and from the Simojoki River were 0.67 ng g^{-1} fw (23 ng g^{-1} lw) and 2.5 ng g^{-1} fw (14 ng g^{-1} lw) [57], respectively. The average content of PCDEs in salmon collected from the Simojoki River in 1991 was 89 ng g^{-1} lw [139]. Elevated concentrations of PCDE congeners in Lake Saimaa salmon have been suggested to indicate pollution by chlorophenols in Lake Saimaa. The dominating congeners in Baltic salmon, Lake Saimaa salmon, and salmon from the Tenojoki River have been different from each other and differ from those in fish from the Kymijoki River, but PCDEs typical to Ky-5 have also been detected in these salmon [33, 57]. According to the congener profile of PCDEs in Lake Saimaa salmon, chlorophenols have been suggested as a source of PCDEs in Lake Saimaa [57]. In the case of Baltic salmon, the most dominant PCDE congener has been PCDE 47 [57, 139] and there are most likely many sources of PCDEs to the Baltic Sea. PBDE 47 is the most abundant PBDE congener detected in Baltic biota [17].

The content of PCDEs in pike from the Bothnian Bay has been low and PCDE 77 has been among the few PCDEs detected in Bothnian Bay pike [33]. Most PCDE congeners have been non-detectable in Bothnian Bay pike (<3–<34 ng g^{-1} lw depending on the congener).

7.2.3
Seals

The only data on PCDEs in seals has been reported for Baltic seals [113] and seals from Lake Saimaa in Finland [33]. Some PCDE congeners have also been measured in a seal (*Phoca sibirica*) from Lake Baikal [33]. The total amount of PCDEs measured in Lake Saimaa seals (*Phoca hispida saimensis*) have ranged between 0.22 µg g^{-1} and 0.46 µg g^{-1} lw [33] and those in Baltic gray seals (*Halichoerus grypus*) and ringed seals (*Phoca hispida botnica*) from 0.03 µg g^{-1} to 0.38 µg g^{-1} lw [113]. The levels of PCDEs were greater than those of toxic PCDDs and PCDFs in Baltic seals.

PCDE profiles in juvenile and adult Baltic seals have been different and higher chlorinated PCDEs have dominated in old seals [113]. There have also been differences between gray seals and ringed seals most likely being due to differences in metabolism and/or diet. Because the isomer patterns were similar between juvenile gray seals and salmon, fish has been suggested as one of the most likely sources of PCDEs in seals [113].

7.3
Birds

PCDEs have been reported in birds from the US, Canada, and the Baltic Sea [36, 121, 124, 136]. Some PCDE congeners were measured between 11 ng g^{-1} and 900 ng g^{-1} fw in carcasses and eggs of fish-eating birds from the US and Canada [121]. The chemical industry was suggested as a source of PCDEs in common tern eggs and carcasses from Rhode Island.

The total PCDEs in breast muscle of three Baltic white-tailed sea eagles (*Haliaeetus albicilla* L.) collected between 1988 and 1991 have varied between 962 ng g^{-1} and 50,360 ng g^{-1} lw (80 ng g^{-1} and 770 ng g^{-1} fw) [124]. The concentration of PCDE congeners ranged from 7 ng g^{-1} to 13,000 ng g^{-1} lw in these eagles. PCDE residues have also been measured in Baltic black guillemots (*Cepphus grylle* L.) from the Quarken area [124]. The content of total PCDEs varied between 189 ng g^{-1} and 298 ng g^{-1} lw (on average 29 ng g^{-1} fw) in three black guillemot eggs studied. The concentrations of PCDE congeners ranged from 4 ng g^{-1} to 79 ng g^{-1} lw in these eggs. PCDEs 47, 99, 147/153, 180/181, and 196 have been the dominating PCDEs in Baltic bird samples. The differences in congener profiles between different species of fish, seals, and birds, have been thought to indicate different contamination sources and/or metabolism between species [124].

7.4
Humans

Human adipose tissue has been reported to contain PCDE residues [123, 131–133]. Hexa- through decaCDEs have been measured to be 20–2000 pg g^{-1} lw in human adipose tissue from the U.S. [132]. NonaCDEs and octaCDEs have been measured at a mean level of 1.53 ng g^{-1} and 0.38 ng g^{-1} lw, respectively, in human adipose tissue from Canadian municipalities [131]. Males were reported to have higher concentrations of nonaCDEs than females. PCDEs have also been detected in Finnish human adipose tissue, but in liver and testis from the same individuals PCDEs were non-detectable [123].

8
Environmental Fate

8.1
Bioaccumulation

The occurrence of PCDE residues in biota shows that PCDEs have potential for bioaccumulation. The same PCDE congeners have been detected in biota compared to sediment from the same study area and elevated levels of PCDEs are detected in aquatic environment contaminated with PCDEs [33] (Fig. 5). Higher chlorinated PCDEs like octaCDEs seem to bioaccumulate, as well, since they have been measured in fish, seals, and birds [33, 57, 113, 124] (Figs. 5 and 6).

Ky-5

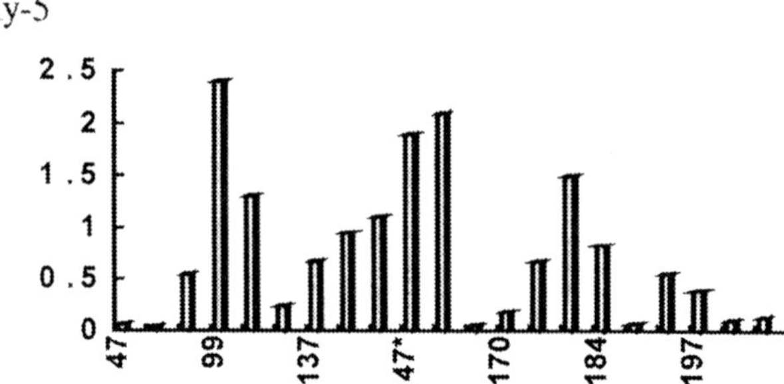

Sediment

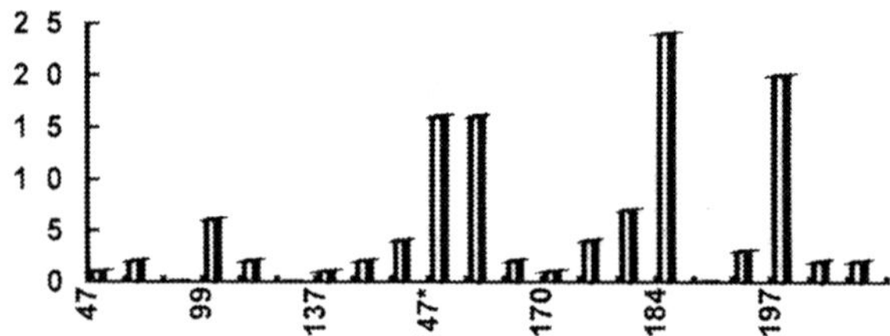

Pike

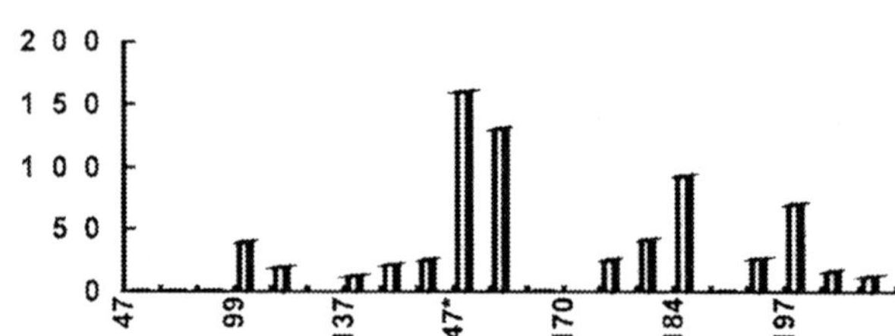

Mussel

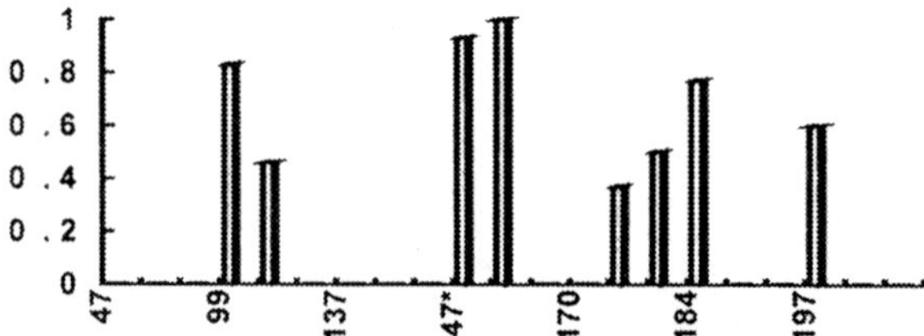

Fig. 5. PCDE congener profiles in Ky-5 compared to sediment (KRSE2 [33]) and pike (KRP2 [33]) collected from the Kymijoki River in 1993 and mussels [125] incubated in the Kymijoki River in 1995

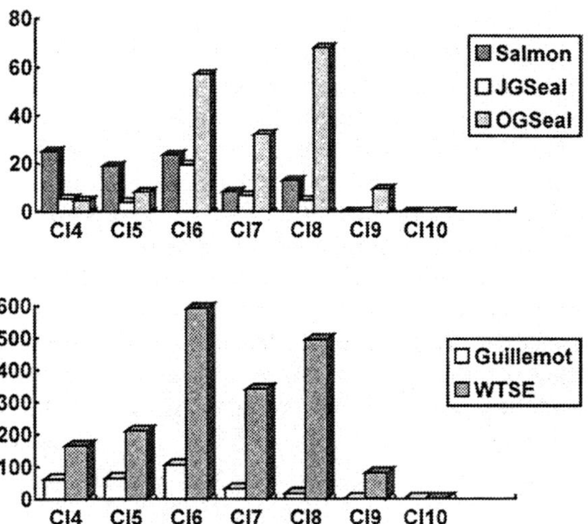

Fig. 6. The profile of tetra- through decaCDEs in salmon muscle (1991 [139]), blubber of juvenile gray seals (GS1–4 [113]), an old gray seal (GS6 [113]), black guillemots eggs (GU1–3, [124]), and WTSE muscle (EA1 [124])

It is not surprising that PCDEs are detected in aquatic biota, because their reported half-lives are sufficient for bioaccumulation and near to those of PCBs. Tri- through pentaCDEs have been reported to be somewhat more persistent in fish than the corresponding PCBs [110]. Half-lives of PCDEs in fish have been higher than those in rats.

Based on uptake and metabolism studies, the accumulation and tissue distribution patterns of PCDEs have been suggested to be similar to those of PCBs [63, 109]. According to uptake studies of PCDEs in trout and Atlantic salmon, the uptake of PCDEs in fish is similar to that of PCBs [83, 109]. When the uptake of mono through tetraCDEs was studied in brook trout, it was observed that the uptake of PCDEs was rapid ranging from 2.4 µg day^{-1} to 48.9 µg day^{-1} and did not reach a steady state during a seven day exposure [83]. The accumulation coefficients of PCDEs 28, 66, and 99 were 0.31, 0.33, and 0.36 in Atlantic salmon, respectively [109].

It has been suggested that PCDEs have non-congener specific uptake and elimination [63], but there could be differences in accumulation of PCDEs between species. The profiles of PCDE congeners in pike and sediment from a Finnish river, Kymijoki River, contaminated by a chlorophenol formulation (Fig. 5) have suggested that PCDEs are not as selectively accumulated and/or metabolized in fish as PCDDs and PCDFs [33]. All PCDEs detected in sediment from this river have also been observed in fish [33, 114]. Of PCDDs and PCDFs, mainly toxic 2,3,7,8-chlorosubstituted congeners are detected in fish, and hepta- and octaCDFs which are the dominating PCDDs and PCDFs in Kymijoki river sediments have not been detected in fish [33, 114]. In the case of PCDEs, higher chlorinated congeners have also been accumulated in Kymijoki River fish, although less than hexaCDEs (Fig. 5). When Chu et al. [82] measured PCDE tissue residues (fat, liver) in rats after a short-term toxicity study, they observed that PCDE 153 was accumulated more than PCDEs 184 or 99.

There could be some differences between PCDE congeners in their persistence. Based on the differences between the profiles of tetra- through decaCDEs in Ky-5, sediment, and pike [33] (Fig. 5), lower chlorinated PCDEs could be more easily degraded than higher chlorinated ones. There seems to be differences between higher chlorinated PCDEs in their persistence. For example, PCDEs 196 and 197 were at a similar level in a sample of Ky-5, whereas PCDE 197 dominated in sediments [33]. The congener profile of PCDEs in Ky-5, however, could differ between different production batches. In Baltic seals and white-tailed sea eagles, PCDE 196 has been more abundant than PCDE 197 [124]. This congener might originate from other sources in addition to chlorophenols, like combustion, since it has been detected in a fly ash [57].

8.2
Biomagnification

Biomagnification of PCDEs in food chains has not been studied much. Their food-chain bioaccumulation potential, however, is high because their n-octanol-water coefficients are high [59]. Log K_{ow} values of PCDEs 29 and 77 determined by Opperhuizen and Voors [63] were 5.44 and 5.78, respectively. The bioconcentration factors (ml g^{-1}) were 1.5×10^4 and 3.2×10^4, respectively, being similar to those of corresponding PCBs.

PCDE 47 has higher log K_{ow} (7.62) than PCDE 77 and its measured bioconcentration factor (log BCF), the ratio of the equilibrium concentration of a substance in animal compared to that in water, was 4.09 [141]. This PCDE has been the most abundant PCDE congener in Baltic salmon [57]. The study of Neely et al. [141] verified that bioconcentration can be predicted from partition coefficient, which is the ratio of the equilibrium concentration of a substance between an organic phase and water (n-octanol and water).

PCDEs are stable and lipophilic compounds and biomagnification of PCDEs in Baltic species at higher trophic levels has been observed. The white-tailed sea eagle (WTSE) at the top of the food chain has the highest content of PCDEs [124] (Fig. 6). The black guillemot, at a lower level of the food chain, has similar profiles of tetra- through decaCDEs in its eggs as has been detected in the blubber of juvenile gray seals, whereas the profiles of tetra- through decaCDEs have been similar in the blubber of Baltic gray seals as in the muscle of WTSE. OctaCDEs have clearly biomagnified in old gray seals compared to juvenile gray seals and salmon in which hexaCDEs dominate.

9
Future Research

Low acute toxicity and induction of enzyme activity of PCDEs refer to non-TCDD-like mechanisms of action of PCDEs, but more data is needed on other possible adverse effects of PCDEs in wildlife and humans including the role of PCDEs as hormone disrupters. More data is also needed on PCDE levels in the environment.

Analytical methodology of PCDEs should be developed utilizing new environmental friendly methods. Supercritical fluid extraction (SFE) is a new promising extraction technique which is environmentally friendly and can allow simultaneous sample cleanup [142]. It allows minimization of the use of organic solvents and reagents as well as analysis costs and time. Automatic SFE instruments, however, are quite expensive and it is possible that the PCDE fraction cannot be analyzed straight away and further cleanup is still needed. If conventional extraction techniques are utilized, instead of liquid-solid chromatography and GPC, which demand relatively large amount of solvents and adsorbents to work effectively, bulk matrix removal could be performed by shaking with concentrated sulfuric acid. This and the use of microcolumns for further cleanup will minimize use and costs of solvent and reagents. HPLC works well for separation of compound groups, but they also consume relatively large amount of solvents.

The possible adverse effects of PCDEs as well as their metabolites should be studied more. Since antiestrogenic effects may be Ah-receptor-mediated, PCDEs having low Ah-receptor binding affinity are unlike antioestrogenics, but hydroxymetabolites of PCBs have been reported to bind to estrogen receptor [106]. The same could be true of hydroxymetabolites of PCDEs. Hydroxy-PCDEs are also the major impurities in chlorophenols and their methoxyderivatives, methoxylated PCDEs, have been observed as abundant biomethylation products in surface soil [145]. Methoxylated PCDEs might have toxic properties similar to HAHs like PCDDs, since they follow PCDDs and PCDFs in analytical cleanup steps [125].

More information is needed on teratogenicity of PCDEs as well as the role of thyrotoxicity in the developmental toxicity of PCDEs [104]. Teratogenicity caused by nitro-PCDEs has been suggested to involve alterations in thyroid hormone status [87] and transient alterations in maternal thyroid status might affect development of the fetus [143]. Effects of PCDEs after prenatal exposure have been similar to those of PCBs [104].

Since the toxicology data on PCDEs is still limited, PCDEs should be considered as compounds having possible adverse effects on wildlife and humans until more data is provided. PCDEs have shown low toxic potency in fish, but this is also true for PCBs, some of which are quite toxic in mammals. Mono-*ortho*-PCBs have not caused rainbow trout early life stage mortality and non-*ortho*-PCBs have shown unexpectedly low early life stage mortality when comparing Ah-receptor binding affinities in mammals [144].

It is possible that due to interactions of PCDEs with other xenobiotic compounds, the adverse effects of xenobiotics on wildlife and humans might be enhanced. It has been recommended that when risk estimation of halogenated pollutants is performed, possible interactions between PCDEs and abundant environmental contaminants such as PCBs should be considered [104].

References

1. Paasivirta J, Koistinen J (1994) Chlorinated ethers. In: Kiceniuk JW, Ray S (eds) Analysis of contaminants in edible aquatic resources, book II. VCH Publishers, New York, chap 6
2. Ballschmiter K, Zell M (1980) Fresenius Z Anal Chem 302:20
3. Ballschmiter K, Rappe C, Buser HR (1989) Chemical properties, analytical methods and environmental levels of PCBs, PCTs, PCNs and PBBs. In: Kimbrough RD, Jensen AA (eds) Halogenated biphenyls, terphenyls, naphthalenes, dibenzodioxins and related products. Elsevier Science Publishers BV (Biomedical Division), Chap 2
4. Kurz J (1994) PhD thesis. University of Ulm
5. Singh P, McKinney JD (1980) Acta Cryst B 36:210
6. Rissanen K, Valkonen J, Virkki L (1988) Acta Cryst C44:1644
7. Rissanen K, Virkki L (1989) Acta Cryst C45:1408
8. Nevalainen T, Rissanen K (1994) J Chem Soc Perkin Trans 2:271
9. Brewster RQ, Stevenson G (1940) J Am Chem Soc 62:3144
10. Gará A, Andersson K, Nilsson C-A, Norström Å (1981) Chemosphere 10:365
11. Sundström G, Hutzinger O (1976) Chemosphere 5:305
12. Kirwin CJ Jr, Sandmeyer EE (1981) Ethers. In: Clayton GD, Clayton FE (eds) Patty's industrial hygiene and toxicology, 3rd rev edn, vol 2 A. Wiley, New York, chap 35
13. Hake CL, Rowe VK (1963) Chlorinated phenyl ethers. In: Patty FA (ed) Industrial hygiene and toxicology, 2nd rev edn, vol II. Interscience, New York, p 1706
14. Sloane HJ, Bradley KB (1962) Chem Abstr 58:7174e (US Pat 3,022,353, Feb. 20)
15. Rappe C, Buser HR, Bosshardt H-P (1979) Ann N Y Acad Sci 320:1
16. De Voogt P, Brinkman UAT (1989) Production, properties and usage of polychlorinated biphenyls. In: Kimbrough, Jensen (eds) Halogenated biphenyls, terphenyls, naphthalenes, dibenzodioxins and related products. Elsevier Science Publishers BV (Biomedical Division), Amsterdam, chap 1
17. Nordic Council of Ministers (1998) Polybrominated diphenyl ethers: food contamination and potential risks. TemaNord 1998:503
18. Tulp MThM, Sundström G, Martron LBJM, Hutzinger O (1979) Xenobiotica 9:65
19. Cross HF, Snyder FM (1949) Soap and Sanitary Chemicals 25:135
20. Metcalf RL (1948) Journal Economic Entomology 41:875
21. Blanton FS (1953) Journal of the New York Entomological Society 61:217
22. Gersdorff WA, Schecter MS (1948) Soap and Sanitary Chemicals 24:155
23. Yana A (1969) Chem Abstr 70:27933p (PhD thesis, Inst Nat Rech Agron Tunisie 1966)
24. Branson DR (1977) A new capacitor fluid – a case study in product stewardship. In: Mayer FL, Hamelink (eds) Aquatic toxicology and hazard assessment, STP 634. American Chemical Society for Testing and Materials, Philadelphia, PA, p 44 (Proceedings of the first annual symposium on aquatic toxicology. ASTM 634, p 44)
25. Bhargava HN, Leonard PA (1996) AJIC 24:209
26. Lyman FL, Furia T (1969) Ind Med 38:64
27. Lilly HA, Lowbury EJL (1974) Br Med J 4:372
28. Matsunaka S (1975) Diphenyl ethers. In: Kearney PC, Kaufman DD (eds) Herbicides, 2nd rev edn, vol 2. Marcel Dekker, New York and Basel, chap 14
29. Jäger G, Bayer AG (1977) Unkrautbekämpfungsmittel (Herbizide). In: Büchel KH (ed) Pflanzenschutz und Schändlingsbekämpfung. Thieme, Stuttgart, p 166
30. Burke SS, Smith JM, Hayes AW (1983) Toxicology 29:1
31. Wells DE (1979) Anal Chim Acta 104:253
32. Hewitt LM, Munkittrick KR, Scott IM, Carey JH, Solomon KR, Servos MR (1996) Environ Toxicol Chem 15:894
33. Koistinen J, Paasivirta J, Suonperä M, Hyvärinen H (1995) Environ Sci Technol 29:2541
34. Komsta E, Chu I, Villeneuve DC, Benoit FM, Murdoch D (1988) Arch Toxicol 62:258
35. Junk GA, Ford CS (1980) Chemosphere 9:187
36. Paasivirta J, Tarhanen J, Soikkeli J (1986) Chemosphere 15:1429
37. Firestone D, Ress J, Brown NL, Barron RP, Damico JN (1972) J AOAC 55:85

38. Villanueva EC, Burse VW, Jennings RW (1973) J Agr Food Chem 21:739
39. Nilsson C-A, Renberg L (1974) J Chromat 89:325
40. Becker M, Phillips T, Safe S (1991) Toxicol Environ Chem 33:189
41. Humppi T, Heinola K (1985) J Chromat 331:410
42. Paasivirta J, Lahtiperä M, Leskijärvi T (1982) In: Hutzinger O, Frei RW, Merian E, Pocchiari F (eds) Chlorinated dioxins and related compounds: Impact on the environment. Pergamon, Oxford and New York, p 191
43. Kurz J, Ballschmiter K (1994) Fresenius J Anal Chem 351:98
44. Nevalainen T, Koistinen J, Nurmela P (1994) Environ Sci Technol 28:1341
45. Koistinen J (1998) Toxicol Environ Chem 66:27
46. Boer J de, Denneman M (1998) Rev Environ Contam Toxicol 157:131
47. Jensen S, Renberg L (1972) Ambio 1:1
48. Rappe C, Nilsson C-A (1972) J Chromatgr 67:247
49. Tuomisto J (1998) Pentachlorophenol. In: Guidelines for drinking-water quality, 2nd edn, addendum to vol 2. World Health Organization, Geneva, p 229
50. Albro PW, Parker CE (1980) J Chromatgr 197:155
51. Lake JL, Rogerson PF, Norwood CB (1981) Environ Sci Technol 15:549
52. Nilsson C-A, Andersson K (1977) Chemosphere 5:249
53. Westöö G, Noren K (1977) Ambio 6:232
54. Dickson LC, Karasek FW (1987) J Chromatgr 389:127
55. Choudhy GG, Olie K, Hutzinger O (1982) In: Hutzinger O, Frei RW, Merian E, Pocchiari F (eds) Chlorinated dioxins and related compounds. Impact on the environment. Pergamon, Oxford and New York, p 275
56. Aittola J, Paasivirta J, Vattulainen A, Sinkkonen S, Koistinen J, Tarhanen J (1996) Chemosphere 32:99
57. Koistinen J, Vuorinen PJ, Paasivirta J (1993) Chemosphere 27:2365
58. Koistinen J, Paasivirta J, Lahtiperä M (1993) Chemosphere 27:149
59. Kurz J, Ballschmiter K (1999) Chemosphere 38:573
60. Dunnivant FM, Coates JT, Elzerman AW (1988) Environ Sci Technol 22:448
61. Nevalainen T, Kolehmainen E, Säämänen M-L, Kauppinen R (1993) Magnetic Resonance in Chemistry 31:100
62. Shiu WY, MacKay D (1986) J Phys Chem Ref Data 15:911
63. Opperhuizen A, Voors PI (1987) Chemosphere 16:2379
64. Ruelle P, Kesselring UW (1997) Chemosphere 34:275
65. Kamlet MJ, Doherty RM, Carr PE, Mackay D, Abraham MH, Taft RW (1988) Environ Sci Technol 22:503
66. Kurz J, Ballschmiter K (1994) Fresenius J Anal Chem 349:533
67. Buser H-R (1975) J Chromat 107:295
68. Crummett WB, Stehl RH (1973) Environ Health Perspect 5:15
69. Nevalainen T, Kolehmainen E (1994) Magnetic Resonance in Chemistry 32:480
70. Nevalainen T, Kolehmainen E, Vilen E (1995) Magnetic Resonance in Chemistry 33:355
71. Norström Å, Andersson K, Rappe C (1976) Chemosphere 5:21
72. Norström Å, Andersson K, Rappe C (1976) Chemosphere 6:241
73. Choudhry GG, Sundström G, Ruzo LO, Hutzinger O (1977) J Agric Food Chem 25:1371
74. Choudhry GG, Sundström G, van der Wielen FWM, Hutzinger O (1977) Chemosphere 6:327
75. Mamantov A (1985) Chemosphere 14:905
76. Norström Å, Andersson K, Rappe C (1976) Chemosphere 6:419
77. Nilsson C-A, Anderson K, Rappe C, Westermark S-E (1974) J Chromatogr 96:137
78. Lindahl R, Rappe C, Buser HR (1980) Chemosphere 9:351
79. Safe S (1990) CRC Crit Rev Toxicol 21:51
80. McKinney JD (1989) Environ Health Perspec 82:323
81. Chu I, Villeneuve DC, Secours V, Valli VE (1989) J Environ Health B24:493
82. Chu I, Villeneuve DC, Secours V, Valli VE (1990) J Environ Sci Health B25:225
83. Chui YC, Addison RF, Law FCP (1990) Xenobiotica 20:489

84. Metcalfe CD, Metcalfe TL, Cormier JA, Huestis SY, Niimi AJ (1997) Environ Toxicol Chem 16:1749
85. Hornung MW, Zable EW, Peterson RE (1996) Toxicol Appl Pharmacol 140:227
86. Fahrig R, Nilsson C-A, Rappe C (1978) Genetic activity of chlorophenols and chlorophenol impurities. In: Rao KR (ed) Pentachlorophenol. Plenum Press, New York and London, p 325
87. Manson JM (1986) Environ Health Perspec 70:137
88. Howie L, Dickerson R, Davis D, Safe S (1990) Toxicol Appl Pharmacol 105:254
89. Harper N, Howie L, Connor K, Arellane L, Craig A, Dickerson C, Safe S (1993) Fundam Appl Toxicol 20:496
90. Kerkvliet NI, Brauner JA, Mätlock JP (1985) Toxicology 36:307
91. Nevalainen T, Kolehmainen E (1994) Environ Toxicol Chem 13:1699
92. Rosiak K, Li M-H, Degitz SJ, Skalla DW, Chu I, Francis BM (1997) Toxicology 121:191
93. Francis BM (1989) Environ Toxicol Chem 8:681
94. Poland A, Glover E, Kende AS (1976) J Biol Chem 251:4936
95. Carlson GP, Smith EN, Johnson KM (1980) Drug Chem Toxicol 3:293
96. Chui YC, Hansell MM, Addison RF, Law CP (1985) Toxicol Appl Pharmacol 81:287
97. Iverson F, Newsome H, Hierlihy L (1987) Fd Chem Toxic 25:305
98. Koistinen J, Sanderson JT, Giesy JP, Nevalainen T, Paasivirta J (1996) Environ Toxicol Chem 15:2028
99. Kopponen P, Sinkkonen S, Poso A, Gynther J, Kärenlampi S (1994) Environ Toxicol Chem 13:1543
100. Kodavanti PRS, Ward TR, McKinney JD, Waller CL, Tilson HA (1996) Toxicol Appl Pharmacol 138:251
101. McKinney JD, Fawkes J, Jordan S, Chae K, Oately S, Coleman RE, Briner W (1985) Environ Health Perspect 61:41
102. Gray LE, Kavlock RJ (1983) Toxicol Lett 15:231
103. Gillner M, Jakobsson E (1996) Organohalogen Compounds 29:220
104. Rosiak KL, Chu I, Seo B-Y, Francis BM (1997) Environ Sci Health B 32:377
105. Müller AMF, Makropoulos V, Bolt HM (1995) TEN 2:68
106. Navas JM, Segner H (1998) Environ Sci Pollut Res 5:75
107. Poon G, Chui YC, Law CP (1986) Xenobiotica 16:795
108. Newsome WH, Iverson F, Shields JB, Hierlihy SL (1983) Bull Environ Contam Toxicol 31:613
109. Zitko V, Carson WG (1977) Chemosphere 6:293
110. Niimi AJ (1986) Aquatic Toxicology 9:105
111. Clement RE, Tosine HM (1989) Analysis of chorinated dibenzo-p-dioxins and dibenzofurans in the aquatic environment. In: Afgan BK, Chau ASY (eds) Analysis of trace organics in the aquatic environment. CRC, Boca Raton, FL, Chap 5
112. Clement RE, Koester CJ, Gray L (1994) Dioxins and furans. In: Kiceniuk JW, Ray S (eds) Analysis of contaminants in edible aquatic resources, book II. VCH, New York, chap 4
113. Koistinen J, Stenman O, Haahti H, Suonperä M, Paasivirta J (1997) Chemosphere 35:1249
114. Verta M, Ahtiainen J, Hämäläinen H, Jussila H, Kiviranta H, Korhonen M, Kukkonen J, Lehtoranta J, Lyytikäinen M, Malve O, Mikkelson P, Moisio V, Niemi A, Paasivirta J, Palm H, Rantalainen A-L, Salo S, Vartiainen T, Vuori K-M (1999) Organochlorine compounds and heavy metals in the River Kymijoki: occurrence, transport, impacts and health risks. Final report of the KYPRO project. Finnish Environment Institute (in Finnish)
115. Coburn JA, Comba M (1981) Unpublished data. NWRI, Burlington, Ontario, Canada
116. Kuehl DW, Durhan E, Butterworth B, Linn D (1984) Environ Intern 10:45
117. Jaffe R, Stemmler EA, Eitzer BD, Hites RA (1985) J Great Lakes Res 11:156
118. Jaffe R, Hites RA (1986) J Great Lakes Res 12:63
119. Niimi AJ, Metcalfe CD, Huestis SY (1994) Environ Toxicol Chem 13:1133
120. Birkholz DA, Gottschalk R, Boothe D, Ralitsch M (1995) Organohalogen Compounds 23:77
121. Stafford CJ (1983) Chemosphere 12:1487

122. Koistinen J (1990) Chemosphere 20:1043
123. Koistinen J, Mussalo-Rauhamaa H, Paasivirta J (1995) Chemosphere 31:4259
124. Koistinen J, Koivusaari J, Nuuja I, Paasivirta J (1995) Chemosphere 30:1671
125. Koistinen J, Herve S, Paukku R, Lahtiperä M, Paasivirta J (1997) Chemosphere 34:2553
126. Newsome WH, Shields JB (1982) J Chromatogr 247:171
127. Hattula ML (1973) PhD thesis. University of Jyväskylä
128. Lau P-Y, Newsome WH (1982) Proceedings of the annual conference on mass spectrometry and related topics, Honolulu, HI, TOA 7, p 196
129. Huestis SY, Sergeant DB (1992) Chemosphere 24:537
130. Williams DT, LeBel GL (1988) Chemosphere 17:2349
131. Williams DT, Kennedy B, LeBel GL (1991) Chemosphere 23:601
132. Stanley JS, Cramer PH, Ayling RE, Thornburg KR, Remmers JC, Breen JJ, Schwemberger J (1990) Chemosphere 20:981
133. Stanley JS, Cramer PH, Thornburg KR, Remmers JC, Breen JJ, Schwemberger J (1991) Chemosphere 23:1185
134. Ryan JJ, Lizotte R, Newsome WH (1984) J Chromatogr 303:351
135. Stalling DL, Tindle RC, Johnson JL (1972) J Assoc Off Anal Chem 55:32
136. Paasivirta J, Tarhanen J, Juvonen B (1987) Chemosphere 16:1787
137. Jensen S, Sundström G (1974) Ambio 3:70
138. Smith LM, Stalling DL, Johnson JL (1984) Anal Chem 56:1830
139. Vuorinen PJ, Paasivirta J, Keinänen M, Koistinen J, Rantio T, Hyötyläinen T, Welling L (1997) Chemosphere 34:1151
140. Vuori K-M (1996) Joint meeting of the Finnish society of toxicology and the British toxicology society, Tampere, Finland, July 4–6
141. Neely WB, Branson DR, Blau GE (1974) Environ Sci Technol 8:1113
142. Bøwadt S, Hawthorne SB (1995) J Chromat 703:549
143. Calvo R, Obregon MJ, Ruix de Ona C, Escobar del Ray R, Morreale de Escobar G (1990) J Clin Invest 86:588
144. Zabel EW, Cook PM, Peterson RE (1995) Aquatic Toxicol 31:315
145. Humppi T (1985) Chemosphere 4:523

CHAPTER 8

Chlorinated Paraffins

Derek Muir[1] · Gary Stern[2] · Gregg Tomy[2]

D. Muir (e-mail: Derek.Muir@cciw.ca, Tel.: 905-319-6921, Fax: 905-336-6430)
[1] Environment Canada, National Water Research Institute, Burlington ON L7R4A6, Canada
[2] Fisheries and Oceans Canada, Freshwater Institute, Winnipeg MB R3 T 2N6 Canada

Chlorinated paraffins or polychlorinated *n*-alkanes (PCAs) consist of C_{10} to C_{30} *n*-alkanes with chlorine content from 30 to 70% by mass. PCAs are used as high temperature lubricants, plasticizers, flame retardants, and additives in adhesives, paints, rubber, and sealants. This chapter reviews the existing data on the production, uses, reactions, methods of quantitative analysis, and levels of PCAs, and includes an assessment of environmental distribution using a steady state non-equilibrium fugacity-based chemical fate models. Short chain PCAs (C_{10} to C_{13} with 60–70% chlorine) have similar molecular weight and physical properties (octanol-water partition coefficient, water solubility, vapor pressure) to many persistent organochlorines such as PCBs and toxaphene. Medium chain (C_{14} to C_{17}) and long chain (C_{18} to C_{30}) PCAs are extremely hydrophobic and nonvolatile and likely to be associated with particles in aquatic systems. The limited biodegradation data suggests that PCAs are less persistent in water, sediments, or biota than other organochlorines. To date there is limited information on the distribution and fate of PCAs in the environment. There have been recent advances in the analysis of C_{10} to C_{13} PCAs using high resolution negative ion mass spectrometry. Interlab comparisons have shown that both high resolution and low resolution negative ion mass spectrometry can be employed to determine PCAs in environmental samples, although agreement between laboratories is poor compared with the determination quantitation of PCBs and organochlorine pesticides. There are indications that PCAs are widespread environmental contaminants at ng/l levels in surface waters and ng/g (wet weight) levels in biota. However, environmental measurements of PCAs are very limited at the present time in the USA and Canada and are only slightly more detailed in western Europe. Application of the Equilibrium Criterion Model, Level III (steady state, nonequilibrium) to representative short ($C_{12}H_{20}Cl_6$) and medium chain PCAs, ($C_{16}H_{24}Cl_{10}$) using the best available physical-chemical property data and estimated degradation rates, showed that that medium chain compound $C_{16}H_{24}Cl_{10}$ would achieve higher concentrations in sediment and soil than the short chain $C_{12}H_{20}Cl_6$ because of slower degradation rates and lower water solubility. Volatilization from water to air of both compounds was predicted by the model suggesting the possibility of long range transport via atmosphere and oceans. The environmental residence time of $C_{16}H_{24}Cl_{10}$ is estimated to be 350 days compared to 170 days for $C_{12}H_{20}Cl_6$. Future studies will require better analytical methods and reference materials certified for PCA content. Additional data is needed to evaluate exposure to PCAs in the environment, particularly in light of their continued production and usage around the globe.

Keywords. Chlorinated paraffins, Polychlorinated *n*-alkanes, Short chain chlorinated paraffins, Releases, Sources, Environmental fate,, Environmental distribution, Biodegradation, Physical properties, Environmental fate modelling

The Handbook of Environmental Chemistry Vol. 3 Part K
New Types of Persistent Halogenated Compounds
(ed. by J. Paasivirta)
© Springer-Verlag Berlin Heidelberg 2000

1
Introduction

As their name implies, chlorinated paraffins are chlorinated derivatives of paraffinic hydrocarbons. They are referred to in this review as polychlorinated alkanes (PCAs) because they are produced by chlorination of n-alkane feedstocks. Commercial PCA mixtures fall into different categories: C_{10}–C_{13} (short), C_{14}–C_{17} (medium) and C_{20}–C_{30} (long). These mixtures are further subcategorized into their weight content of chlorine: 40–50%, 50–60%, and 60–70% [1, 2]. Knowledge of the environmental chemistry of PCAs is needed because the physical properties of short and medium chain mixtures are similar to those of the persistent organic pollutants or POPs (PCBs, DDT, toxaphene etc). PCAs have received much less attention in terms of exposure and risk assessment than the POPs because of lower mammalian toxicity than most POPs and the lack of environmental measurements for estimating human and animal exposure.

The short chain PCAs are of particular concern because they have the greatest potential for environmental release and the highest aquatic and mammalian toxicity of PCA products [2]. In the United States, short chain PCAs have been placed on the Environmental Protection Agency (EPA) Toxic Release Inventory (TRI), and in Canada they are listed as "Priority Toxic Substances" under the Canadian Environmental Protection Act. In Europe voluntary restrictions for short chain PCA use have been implemented by industry. PCAs represent the largest group of high molecular weight chlorinated hydrocarbons in commercial use in terms of quantities produced globally. With the phasing out of many POPs in the past 20 years, most of which are chlorinated, PCAs are among the last industrially produced high molecular weight organochlorine compounds [3].

The environmental chemistry of chlorinated paraffins was first reviewed in the Handbook of Environmental Chemistry by Zitko [4]. There have been many other reviews, especially for environmental and human exposure assessment, beginning with a report prepared for the US Environmental Protection Agency by Howard et al. [5] and more recently by Mukherjee [6], Environment Canada [2], UK Dept of the Environment [7], the World Health Organization [8], and by Tomy et al. [9]. In addition to environmental chemistry aspects, the above reviews have covered the toxicity and bioaccumulation of PCAs. This chapter will focus on the recent advances in knowledge of the physical properties, degradation, analysis, and environmental levels of PCAs with special emphasis on the C_{10}–C_{13} group.

2
Sources

2.1
Industrial Synthesis and Other Possible Sources

PCAs are produced by chlorination of C_{10}–C_{30} n-alkanes using molecular chlorine, either of the liquid paraffin or in a solvent, typically carbon tetrachloride [5, 10]. For specialized and more limited applications, PCAs have also been produced by addition chlorination of α-olefins [11]. Depending upon the n-alkane feedstock, the reaction takes place at temperatures between 50 and 150 °C, at elevated pressures and/or in the presence of UV light [1, 4]. After chlorination the product is stripped of solvent, residual chlorine, and reaction products (e.g., HCl) by gas sparging. Final products are mixtures that are viscous, colorless or yellowish dense oils, except for C_{20}–C_{30} PCAs of high chlorine content (70%) which are solids [5].

There has been speculation that PCAs could be found as by-products of other industrial syntheses or processes involving chlorine [5]. While the presence of PCAs in chlorinated solvents (e.g., CCl_4, perchloroethylene, methylene chloride) is unlikely because most are distilled or fractionated, PCAs, or compounds resembling them in molecular weight and physical properties, could be by-products of incomplete polymerization of chlorinated ethylene monomers [5]. The possible inadvertent production of compounds resembling commercial PCAs (i.e., >50% Cl by weight) during aqueous chlorination, e.g., from hydrocarbons in waste water or drinking water treatment, is highly unlikely because of the dilute conditions [5].

2.2
Composition and Complexity of PCAs

Because the alkane feedstock consists of n-alkanes with a range of chain lengths, the final preparation contains a mixture of their chlorinated analogues. PCA mixtures may be contaminated by isoparaffins, aromatic compounds, sulfur, metals, and unreacted n-alkanes [4, 5]. As the purity of n-paraffin feedstocks has improved so to has the purity of the PCA products [11, 12]. Commercial products may contain additives added to inhibit decomposition of the PCA, via HCl loss, at elevated temperatures and to increase flame retardancy (e.g., antimony oxide). Common stabilizers include epoxides and organotin compounds [6]. The concentration of these additives, however, is usually below 0.05% [13].

The commercial polychlorinated alkanes are *multi*congeneric mixtures, one or two orders of magnitude more complex than PCB mixtures. When analyzed by high resolution gas chromatography or liquid chromatography, the PCAs generally elute over a wide retention time range, and individual components are not resolved [14, 15] (Fig. 1). Short- and medium chain commercial PCAs products with 50% Cl or more do not contain appreciate amounts (i.e., >0.1%) of mono- to tetrachloro-alkanes (Tomy G, unpublished data). Assuming one chlorine per carbon, since a second chlorine atom does not readily substitute for hydrogen at a carbon already bound to chlorine [16–18], and taking into account that isomers with four chlorines or less are not present, the theoretical number of positional isomers possible for C_{10}–C_{17} PCAs ranges from 327 for penta to decachloro-n-decane to 53,000 for penta- to heptadecachloro-heptadecane [9]. GC-MS analysis of commercial short chain PCA mixtures shows that hexa- and heptachloro- decanes, undecanes, and dodecanes are the predominant isomers [14]. The actual substitution pattern on the carbon chain is unknown. Less chlorinated alkanes are likely to have 1,3,5-type substitution due to steric considerations, but short chain PCAs with high chlorine contents (e.g., 70%) must have numerous congeners with vicinal chlorines [5]. The lack of CCl_2 groups in commercial PCA mixtures with less than 60% Cl has been confirmed by NMR [19].

2.3
Quantities Used and Applications

The first large scale usage of PCAs began in 1932 when they were incorporated as extreme pressure additives in lubricants [5, 6]. Hardie [20] reported world consumption estimates of 38–50 kt/year in 1961, while in 1977, estimates were reported to be about 230 kt/year [12]. Global consumption estimates for 1993

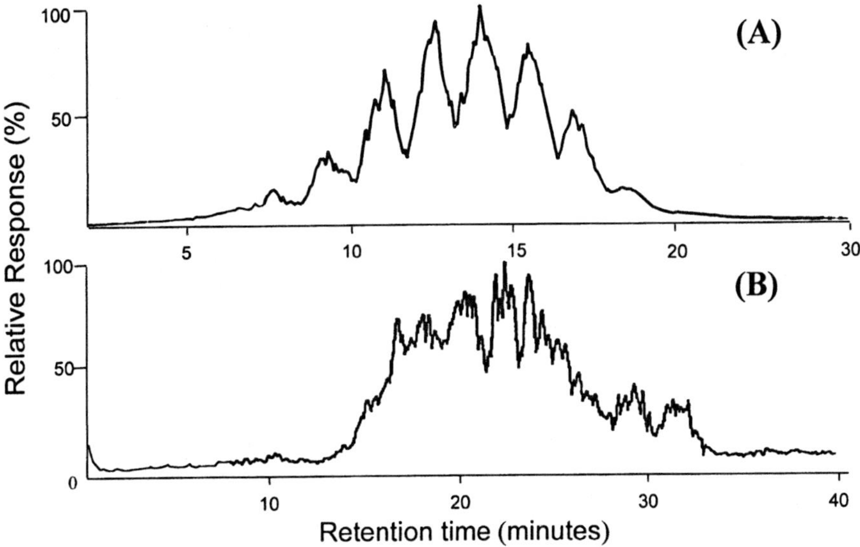

Fig. 1a, b. a RGC-ECNI-MS total ion chromatogram (m/z 65–600) of PCA-60. b HPLC-NPSP-MS total ion chromatogram (m/z 350–650) of PCA-70 [14, 15]

were reported to be 300 kt/year [2]. In the United States and in Canada consumption estimates are reported to be 44–45 kt/year and 3.5–5 kt/year, respectively. In Europe, annual consumption ranges from 100 kt to 200 kt [7]. Production in the US, Europe, and Japan, the major consumers of chlorinated paraffins globally, remained relatively constant throughout the 1990s [11].

Common applications of PCAs include high temperature lubricants, plasticizers, flame retardants, and additives in adhesives, paints, rubber, and sealants [4–6]. Table 1 shows the consumption patterns of PCAs in the United States, Canada, and Western Europe.

Table 1. Consumption patterns of PCAs in USA, Canada, and Western Europe Total consumption estimated to be 300 kT per year in the 1990s [2]

Consumption pattern	Country and % usage		
	USA[a]	Canada[b]	Western Europe[c]
Lubricating additives and fire retardants	45	20	70
Plastics	20	65	4
Rubber	13	8	10
Paints	9	3	8
Adhesives and sealants	6	2	4
Miscellaneous	7	2	4

[a] From [10].
[b] From [2].
[c] From [7].

A major use of PCAs is as extreme temperature additives in metal working fluids (short, medium, and long carbon chain length compounds of 50–60 wt% Cl) for a variety of engineering and metal working operations such as drilling, machining/cutting, drawing, and stamping [6, 7]. The PCA content of the metal working fluid usually ranges from 2% to 10%, but can be up to 80% or more for specialty applications [7].

PCAs (typically medium carbon chain length compounds of 50–60 wt% Cl) are also used as secondary plasticizers for polyvinyl chloride (PVC) and in other plastics (polyesters, polyolefins, polystyrene) and rubbers (neoprene). PCAs can partially replace primary plasticizers such as phthalates and phosphate esters [21]. As plasticizers, PCAs lend flame retardancy to the product, unlike phthalate esters [5]. They are added prior to the processing step for the polymer.

PCAs did not replace PCBs in the late-1970s, despite having some similarities in physical properties and flame retardant characteristics [5]. PCAs were not good PCB replacements for uses requiring high heat stability (e.g., capacitors, transformers) because of their much lower thermal stability.

2.4
Pathways of Release to the Environment

The release of PCAs into the environment could occur during production, storage, transportation, industrial use, and carry-off on manufactured products. Release could also occur due to release from plastics, paints, and sealants in which they are incorporated, leaching, runoff, or volatilization from landfill, sewage sludge amended soils, or other waste disposal sites. Of these, however, the major releases are thought to be from production and from industrial usage [2, 3, 21, 22].

Waterborne releases from production sites may occur either from spills or facility wash-down, in particular, during the cleaning of the reactor vessel [2, 22]. Surveys of two production facilities in the US and Canada detected ug/l concentrations in effluent of manufacturing plants [23, 24]. However, the majority of waterborne release of PCAs into the environment, particularly of C_{10}–C_{13} PCAs, is thought to be via industrial usage [2, 3, 21]. Releases can result from improper disposal of used metal-working lubricants and storage drums or carry-off from work pieces. Used drums, which have been sent to drum reconditioning companies where they are washed and refurbished, may be another potential source for entry into the environment [2].

In Sweden it has been estimated that as much as 55%, i.e., ~227 tons/year, of the C_{10}–C_{13} PCAs used as high-temperature lubricants in industry may be directly discharged as waste into the air and water [3]. A much lower estimate, on a per capita basis, of ~161 tons was made for the United States in 1991 [2, 21].

Disposal and burning of waste containing PCAs may be another potential source of entry of these compounds into the environment. Land filling of products such as plastics, textiles, painted materials, paint cans, and oils containing PCAs may result in slow leaching and/or volatilization from these matrices [2, 13]. Because of their relatively low vapor pressure and water solubility [25, 26], releases from materials in which they are incorporated into the matrix are thought to be low [13]. These releases, however, have not been studied experimentally.

3
Physical-Chemical Properties

The environmentally important physical-chemical properties of PCAs (such as octanol-water partition coefficient (K_{ow}), (water solubility WS), and vapor pressure (VP)) have been determined using the commercial products or with synthetic products resembling components of the commercial mixtures [13, 27]. Drouillard et al. [25, 26] have reported physical-chemical properties of individual PCA congeners that were synthesized by chlorine additions to *n*-alkenes [28].

3.1
Water Solubility, Vapor Pressures, and Henry's Law Constant

Water solubilities of PCAs are generally in the $\mu g\,l^{-1}$ or $ng\,l^{-1}$ level and vary tremendously with carbon chain length and chlorine content (Table 2). Drouillard et al. [26] reported WS of individual (C_{10}-), (C_{11}-), and (C_{12}-) PCA congeners, measured by the generator column technique [29], ranging from 22.4 $\mu g\,l^{-1}$ to 994 $\mu g\,l^{-1}$. The authors also noted that the Cl substitution pattern has significant effects on WS, and that there is a trend of increasing WS with increasing degree of chlorination up to five chlorines. Water solubilities of short chain PCAs are typically 10–100× higher than those of chlorinated aromatic compounds of similar molecular weight such as PCBs [30].

Short chain PCAs have similar vapor pressures to other chlorinated organics of the same molecular weight range such as PCBs [30] and toxaphene [31]. The tetra- to hexachlorodecanes, with VPs ranging from 0.066 to 0.001 Pa, can be classed as semi-volatile organics because a significant proportion will be in the gas phase at ambient temperatures. Vapor pressures of C_{12}–C_{20} PCAs with >50% Cl are much lower, ranging from 1.5×10^{-3} Pa to 1.9×10^{-10} Pa based on direct measurements and estimates (Table 2). Drouillard et al. [25] reported experimentally measured subcooled liquid VPs [32] of individual C_{10}–C_{12} PCA congeners, ranging from 0.00049 Pa to 0.5 Pa. A significant trend of decreasing VPs with increased carbon chain length and degree of chlorination was observed for the C_{10}–C_{12} PCAs [25].

Henry's Law Constants (HLC) of chlorodecanes range from 0.8 Pa m^3 mol^{-1} to 15 Pa m^3 mol^{-1} [25], similar to those of PCBs and some organochlorine pesticides [30]. Some lower chlorinated undecanes and dodecanes also have HLCs in this range. This implies that low molecular weight PCAs may volatilize from water to air in temperate and tropical environments similar to the behavior of PCBs and many pesticides. Medium chain (C_{14}–C_{17}) PCAs with higher molar chlorine have relatively low estimated HLCs (<0.34 Pa m^3 mol^{-1}). Drouillard et al. [25] found a trend of decreasing HLCs with increasing degree of chlorination. However, in general, HLC values for PCAs do not show the large

Table 2. Physical properties of PCA congeners and mixtures of isomers

PCA	%Cl	VP (Pa)[a]	HLC[b] (Pa m^3 mol^{-1})	WS[c] (mg l^{-1})	log K_{ow}[d]	log K_{OA}[e]
$C_{10}H_{18}Cl_4$	50	0.066	14.67	1.26	5.93	8.2
$C_{10}H_{17}Cl_5$	56	0.004 – 0.0054	2.62 – 4.92	0.678 – 0.994	6.04 – 6.20	8.9 – 9.0
$C_{10}H_{17}Cl_5$	56	0.066	14.67	–	–	–
$C_{10}H_{16}Cl_6$	61	0.001 – 0.002	–	–	–	–
$C_{10}H_{13}Cl_9$	70	2.4×10^{-4}	0.83	–	–	–
$^{14}C_{11}$	59	–	–	0.15	–	–
$C_{11}H_{20}Cl_4$	48	0.01	6.32	0.575	5.93	8.5
$C_{11}H_{19}Cl_5$	54	0.001 – 0.002	0.68 – 1.46	0.546 – 0.962	6.20 – 6.40	9.6 – 9.8
$C_{11}H_{18}Cl_6$	58	0.002 – 0.0005	–	–	6.4	–
$^{14}C_{12}H_{21}Cl_5$	51	0.0016 – 0.0019	1.37	–	–	–
$C_{12}H_{20}Cl_6$	56	–	–	0.037	6.40 – 6.77	–
$^{14}C_{12}H_{20}Cl_6$	56	$1.4 – 5.2 \times 10^4$		–	6.8	–
$C_{12}H_{19}Cl_7$	59	–	–	–	7.00	–
$C_{21}H_{18}Cl_8$	63	–	–	–	7.00	–
$^{14}C_{12}H_{16}Cl_{10}$	67	–	–	–	7.3	–
$C_{13}H_{23}Cl_5$	49	3.2×10^{-4}	4.18	0.03	6.61	9.4
$C_{13}H_{22}Cl_6$	53	–	–	–	6.77 – 7.00	–
$C_{13}H_{21}Cl_7$	58	–	–	–	7.14	–
$C_{13}H_{16}Cl_{12}$	70	2.8×10^{-7}	0.34	$4.9 \times 10-4$	–	–
$C_{14}H_{23}Cl_7$	56	1.1×10^{-5}	0.36	$1.4 \times 10-2$	–	–
$C_{14}Cl_7$	52	1.3×10^{-4}	–	–	–	–
$C_{14}Cl_7$	37–70	–	10.9	–	–	–
$^{14}C_{15}$	43	–	–	0.005	–	–
C_{16}	42	–	–	0.01	–	–
$^{14}C_{16}H_{31}Cl_3$	32	–	–	–	6.9	–
$^{14}C_{16}H^{21}Cl_{31}$	68	–	–	–	7.5	–
$C_{17}H_{32}Cl_4$	37	4.0×10^{-6}	51.3	$2.9 \times 10-5$	–	–
$C_{17}H_{27}Cl_9$	58	1.7×10^{-8}	0.01	$6.6 \times 10-4$	–	–
$C_{18}H_{34}Cl_4$	36	7.9×10^{-7}	33	$9.4 \times 10-6$	–	–
$C_{18}H_{30}Cl_8$	54	1.1×10^{-11}	0.07	$8.6 \times 10-5$	–	–
$C_{20}H_{38}Cl_4$	34	4.5×10^{-8}	54.8	–	–	–
$C_{20}H_{33}Cl_9$	54	1.9×10^{-10}	0.02	$5.3 \times 10-6$	–	–
C_{23}	42–52	2.7×10^{-3}	–	–	–	–
$C_{26}H_{44}Cl_{10}$	50	6.3×10^{-15}	0.003	$1.6 \times 10-9$	–	–
$^{14}C_{25}$	51	–	–	< 0.005	–	–
$^{14}C_{25}$	70	–	–	< 0.005	–	–

[a] Vapor pressures from [2, 7, 13, 25].
[b] Henry's Law Constant from [2, 7, 25].
[c] Water solubility from [2, 13, 22, 26, 27].
[d] Octanol-water partition coefficient from [22, 33, 36].
[e] Octanol-air partition coefficient calculated from K_{ow}/K_{AW} using results from [25, 33, 36]. K_{AW} = HLC/RT where R = gas constant 8.319 Pa m^3 mol^{-1} κ^{-1} and T = 293 K.

differences among short, medium, and long chain groups that are observed for WS and VP.

3.2
Octanol-Water (K_{OW}) and Octanol-Air (K_{OA}) Partition Coefficients

On the basis of log K_{OW} values, PCAs are very hydrophobic compounds because of their high Cl content, with all major components of commercial formulations having log K_{OW} values > 5.5. Sijm and Sinnige [33] used a slow-stirring octanol-water technique [34] on a commercial short chain PCA formulation, and reported log K_{OW}s ranging from 5.85 to 7.14. The authors found a parabolic relationship between the total number of carbon and chlorine atoms (N_{tot}) and log K_{OW}. The log K_{OW} of PCAs increased linearly at low N_{tot}, and leveled off at higher N_{tot}. Estimations of log K_{OW} using fragment constant methods [35] yield log K_{OW} values of 5.06–8.12 for C_{10}–C_{13}, 6.83–8.96 for C_{14}–C_{17} and 8.70–12.68 for C_{18}–C_{26} mixtures [22]. Reverse phase HPLC analysis of C_{16} PCAs also yields log K_{OW} values close to the range predicted by fragment constants [36].

Octanol-air partition coefficients of PCAs have not been measured directly but they can be estimated from the ratio of K_{OW}/K_{AW} (where K_{AW} = unitless air-water partition coefficient). Short chain with 50–60% Cl PCAs have log K_{OA}s ranging from 8.2 to 9.8 (Table 2). It should be noted that K_{OA}s calculated from K_{OW} and K_{AW} may be half a log unit lower than actual values due to the fact that the octanol phase represents a phase saturated with water [37]. These log K_{OA} values for PCAs are similar to those for PCBs which range from 9.0 to 10.8 for penta- to heptachlorobiphenyls [37]. More highly chlorinated and longer chain PCAs, with lower HLCs and higher K_{OW}s, would have log K_{OA} values > 10. These values imply a high partitioning to plant and soil surfaces, as well as to airborne particulates, for gas phase PCAs.

3.3
Organic Carbon Partition Coefficient (K_{OC})

K_{OC} values for PCAs are also relatively high, typical of compounds which partition onto particulate organic carbon (POC) and dissolved organic carbon (DOC) in aquatic systems. Drouillard [38] found an average log K_{OC} of a $C_{12}H_{20}Cl_6$ mixture of 4.86 for POC obtained from freshwater sediments and filtered lake water. Fisk et al. [36] reported log K_{OC} values for the $C_{12}H_{20}C_{l6}$ mixture studied by Drouillard [38], as well as another C12 (69% Cl) and two C16 (35% and 70% Cl) of 4.1, 4.7, 5.0, and 5.2, respectively. The lower K_{OC} values observed by Fisk et al. [36] were from a system with a high sediment to water ratio. In general, these measured K_{OC} values are lower than estimated using K_{OW} values equation ($K_{OC} = 0.41 \times K_{OW}$) [39] and may reflect a colloidal or third phase effect and non-equilibrium conditions.

The fraction of PCAs freely dissolved in water (f_W) will vary depending on the mass fraction of POC and DOC (kg/l): $f_W = 1/(1 + POC \times K_{POC} + DOC \times K_{DOC})$ [40]. If the true K_{OC} value for POC and DOC in lake water can be approximated by K_{OW} [40], then f_W for C_{10}–C_{13} PCAs would range from 20% to

80 % at 0.2 mg/l POC (ultraoligotrophic systems) and 0.5 % to 9.0 % at 10 mg/l (moderately eutrophic systems). A substantial fraction of short chain PCAs may therefore be in the dissolved phase in oligotrophic natural waters. However, the more hydrophobic medium chain PCAs would be expected to be mainly sorbed to particles in most environmental situations.

4
Abiotic and Biotic Transformations

4.1
Photolysis, Hydrolyis, Oxidation, and Thermal Reactions

PCAs do not undergo direct photolysis under environmental conditions due to lack of appropriate chromophores absorbing UV light > 290 nm. Friedman and Lombardo [41] demonstrated that PCAs do not absorb high intensity UV light (550 W mercury vapor lamp), and observed no photochemical degradation. PCAs may be subject to attack, via *indirect* photolysis, by oxidizing radicals in the troposphere [2]. Based on Atkinson's OH radical reaction model [42], theoretical half-lives of PCAs in the atmosphere would be inversely proportional to the carbon chain length ranging from 1.2 days to 1.8 days for C_{10}–C_{13}, 0.85 days to 1.1 days for C_{14}–C_{17}, and 0.5 days to 0.8 days for C_{18}–C_{30} [7]. Indirect photolysis reactions in the aquatic environment have not been studied but could involve reaction with free radicals, especially OH radical generated from photolysis of dissolved organic and inorganic compounds in the photic zone.

Rates of hydrolysis and oxidation of PCAs in natural waters are considered negligible at ambient temperatures [2, 6, 7]. However, reactions involving catalysts, known to be present in the aquatic environment, might induce hydrolysis or oxidation reactions, although no studies have been carried out to demonstrate this. Reiger and Ballschmiter [43] have noted, for example, that the interaction of PCAs with activated alumina results in dehydrochlorination. Lahaniatis et al. [44] were able to reductively dechlorinate a commercial formulation using sodium in an ammonia-diethyl ether solution; the resulting products were identified as *n*-alkanes and *n*-alkenes.

The utility of PCAs as flame retardants and cutting fluid additives is due to their ability to release HCl at elevated temperatures. Heating of industrial PCA formulations at 175 °C yields less than 0.5 % HCl [45]; however, heating at 300 °C yields rapid HCl emission [1, 4, 5]. Pyrolysis experiments performed on PCAs by Bergman et al. [46] showed that the decomposition products formed were dependent on the degree of chlorination of the PCA. For a synthesized C_{12} mixture containing 59 % Cl, the major decomposition products were unchlorinated or lower chlorinated aromatics (benzenes, biphenyls, and naphthalenes), while a C_{12} mixture containing 70 % Cl yielded polychlorinated aromatics with up to six chlorines. Pyrolysis of the higher chlorine containing mixture also resulted in the formation of mono- and dichlorodibenzofurans; however, it was unclear whether they were formed directly from the PCAs themselves or by secondary degradation of PCBs [46]. In addition, it was thought that the aromatic compounds containing 12 carbon atoms or fewer were formed by intramolecular reactions of the dehydrochlorinated products, while those with more than 12 carbons (e.g., anthracene and phenanthrene) were formed by intermolecular additions.

4.2
Microbial Biodegradation

Microbial degradation is potentially the most important pathway for removal of PCAs in waste streams, yet much remains to be learned about transformation rates under various conditions as well as the products of degradation. In general, studies suggest that biodegradation of PCAs does occur, and is influenced by chlorine content and carbon chain length.

Zitko and Arsenault [47] examined the aerobic and anaerobic biodegradation of two long carbon chain (C_{20}–C_{30}) PCAs (42 and 70% Cl) in a suspension of sea water and decomposing organic matter at room temperature (19–22 °C) and found that the rate of biodegradation was higher under anaerobic than aerobic conditions, and that the higher chlorinated PCA (70% Cl) was degraded to a greater extent than the lower chlorinated PCA (42% Cl).

In the most extensive examination of the biodegradation of PCAs, Madeley and Birtley [48] used BOD tests to examine the biodegradation of a range of PCAs with different carbon chain lengths and chlorine contents. They concluded that (i) acclimatized microorganisms showed a greater ability to degrade PCAs than did organisms normally used for treating domestic sewage, (ii) increasing chlorination inhibited biodegradation, (iii) short carbon chain PCAs (<60% Cl) appeared to be rapidly and completely degraded, and (iv) medium and long carbon chain PCAs with up to 45% Cl degraded more slowly than shorter carbon chain PCAs. No significant oxygen uptake was observed in tests using the highly chlorinated PCAs, which included two short carbon chain (60% and 70% Cl) PCAs and one medium carbon chain (58% Cl) PCA. Madeley and Birtley [48] also examined the breakdown of a ^{14}C-labeled 42% Cl pentacosane (C_{25}) and found that after 8 weeks with nonacclimatized microorganisms, 11% of the original ^{14}C could be collected as CO_2.

Omori et al. [49] studied the PCA dechlorination potential of a series of soil bacterial strains. They found that different strains pretreated with *n*-hexadecane had different dechlorination abilities. A mixed culture (four bacterial strains) released 15–57% of the Cl of five PCA products, with the amount of Cl released decreasing with increasing carbon chain length and chlorine content. Activated sludge from a sewage treatment plant, acclimated to *n*-hexadecane for 60 days, dechlorinated only 2% of a medium chain PCA ($C_{15.4}Cl_{5.6}$). Fisk et al. [36] found that four ^{14}C-labeled PCAs (two C_{12} (56% and 69% Cl) and two C_{16} (35% and 69% Cl) PCAs) were degraded at 12 °C in aerobic sediments used for a study of bioavailability of PCAs in oligochaetes. Half lives of the C_{12}-PCAs in sediment were 13 days and 30 days for the 56% and 69% Cl products, respectively, while for the C_{16}-PCA, half lives of 12 days and 58 days for 35% and 69% Cl PCAs were observed.

5
Analytical Methods for Determination of PCAs

In this section only the most recent methods (post 1990) for extraction and quantitative determination of PCAs in commercial products and environmental samples are discussed. While extraction and isolation techniques have relied mainly on techniques already developed for POPs, there have been major advances in the quantification of PCAs using gas chromatography mass spectrometry in electron capture negative ionization mode (GC-ECNIMS). No attempt will be made to discuss older methods of analysis such as thin-layer chromatography [13, 50], and neutron activation methods [51].

5.1
Extraction and Isolation

Extraction and cleanup techniques for the analysis of PCAs in environmental samples are similar to the methods used for determination of other persistent organochlorines (OCs). Solvents such as dichloromethane (DCM) [14, 23] and mixtures such as acetone-hexane [52], diethylether-hexane [52], and cyclohexane-isopropanol [43] have all been used to extract PCAs from environmental samples.

Tomy et al. [14] extracted PCAs from fish and sediments by using DCM extraction, lipid removal by SX-3 Biobeads gel permeation chromatography (GPC) [53], and cleanup with Florisil column chromatography. Fractionation on Florisil was achieved by eluting with hexane (F1), which eluted all the PCBs, chlorinated benzenes, 4,4'-DDE, then with (15:85) DCM/hexane (F2), and finally with (1:1) DCM/hexane (F3). Fractions F2 and F3 contained PCAs along with 4,4'-DDT, toxaphene, *cis*- and *trans*-chlordane and other more polar organics, such as heptachlor epoxide and dieldrin. The mean percentage recovery of PCAs for this method was 85%.

Reiger and Ballschmiter [43] described a multistep method for PCA analysis in sewage sludge by cyclohexane-isopropanol extraction and cleanup on silica gel column chromatography. Fractionation on silica gel was achieved by eluting with hexane (F1), which desorbed hexachlorobenzene, 4,4'-DDE, PCB, PCDD, and PCDF. PCAs were then desorbed from the column with (90:10) hexane/diethyl ether. The recovery of PCAs by this method was 86%. These authors also noted that cleanup chromatography on activated alumina should be avoided because PCAs were either totally or partially destroyed by dehydrochlorination during the adsorption process.

Jansson et al. [52] described a multiresidue method for PCA analysis in biological samples by acetone-hexane and diethylether-hexane extraction, oxidation with sulfuric acid to remove lipid, and isolation of PCAs on SX-3 Biobeads GPC. A combination of silica and activated charcoal column chromatography was then used to isolate other OCs. With diethylhexylphthalate (DEHP) as a retention indicator and (1:1) DCM/hexane as the mobile phase for GPC, PCAs were selectively removed from other persistent OCs by collecting an initial fraction that corresponded to a factor times the retention time of DEHP. All other compounds were collected in a second fraction. The mean percentage recovery of PCAs by this method was 97%.

Metcalfe-Smith et al. [23] extracted PCAs with DCM and employed the same alumina column chromatography cleanup step as did Murray et al. [24]. Using this method, PCA recoveries were reported to be greater than 90% [23]. These authors did not encounter problems with degradation of PCAs on alumina as reported by Reiger and Ballschmiter [43].

5.2
Quantitation by Gas Chromatography Mass Spectrometry

Jansson et al. [52] developed a low resolution method for PCA analyses, based on GC/ECNI-MS in the selected ion mode (SIM). In this method, PCAs were selectively removed from other common environmental contaminants by GPC, and quantification was performed by integrating the response of the Cl_2^- (m/z 70) ion, an ion that predominates in the mass spectra of individual PCA congeners at high ion source temperatures [54].

Junk and Meisch [55] developed a low resolution method for PCA analyses based on GC/electron ionization (EI)-MS in the SIM mode. By introducing a commercial formulation directly into the ion source of the MS via a direct insertion probe, and under full scan conditions, they selected the $C_5H_{10}^{35}Cl^+$ (*m/z* 105) ion to be the characteristic ion, i.e., the quantitation ion, of the standard. The integrated area of this ion was compared to that of the same ion from an injection of a known amount of an external standard to calculate the amount of PCAs in the sample.

Rieger and Ballschmiter [43] developed a low resolution method for PCA analyses, based on GC/ECNI-MS in the SIM mode. Their method of quantifying PCAs in environmental samples was based on a triangulation method previously developed in their laboratory for the analysis of toxaphene [56]. For PCAs, typically four ions, known to be prominent in the external standard, were monitored in separate injections of a known amount of standard and in the sample, and the areas of the broad PCA peak were then compared. The choice of external standard used was based on pattern matching, i.e., visually comparing the elution time and signal structure of the sample to those of a number of "in-house" standards [57].

Metcalfe-Smith et al. [23] employed a low resolution method for PCA analyses based on GC/ECNI-MS in the full scan mode. They reported that, although the environmental samples taken from the St. Lawrence River contained interferences, namely PCBs and chlorinated dibenzofurans, their presence was not a problem because their full scanning technique could distinguish them from the PCAs.

Parlar et al. [58] described a novel approach to quantitation of chlorodecanes using low resolution GC-ECNIMS. They synthesized specific C_{10}-PCAs (Cl_4 to Cl_6) and used characteristic ions of each formula group to quantify components in cod liver oil. Because of the difficulty of separating PCAs even on high resolution GC columns, Parlar et al. [58] recommended the use of direct introduction into the mass spectrometer via a short uncoated column. Other authors have recommended similar approaches to address the difficulty of resolving PCA components [55].

The methods outlined above all rely on low resolution MS and inherently lack selectivity [14]. In addition, methods relying on full scan mode or EI mode also lack the sensitivity required to measure trace amounts of PCAs [23, 24, 55] although they may be suitable for analysis of effluents or commercial products.

Procedures based on monitoring the ubiquitous (in organochlorine negative ion mass spectra) m/z 70–73 ions, i.e., Cl_2^- and HCl_2^- [59], present the problem that many other persistent OCs fragment to yield such ions, e.g., p,p'-DDT, p,p'-DDE, lindane, dieldrin, aldrin, and endrin, to name a few [60–62]. Thus, if these contaminants are not selectively removed from the sample matrix during the extraction or clean-up procedures, they would ultimately contribute to the response of the quantitation ion, Cl_2^- (m/z 70), and lead to an overestimation in the level of PCAs in samples.

Methods that monitor ions at nominal mass [23, 43, 55, 63, 64] have interferences from higher PCBs, toxaphene, and chlordane-related compounds, all of which have similar GC-retention times to PCAs and similar molecular masses to PCAs (i.e., 350–500 Daltons) [14]. Reiger and Ballschmiter [43] have applied a simple subtraction technique to compensate for the effects of these interferences, but it is unclear how this procedure compromises the accuracy of their quantitation method.

A recent advance in the analysis of short chain PCAs is the use of electron capture negative ionization high resolution MS to circumvent the problems associated with earlier methods [14]. With the high resolving power no interferences from other POPs, e.g., from toxaphene, PCBs, and chlordane, were detected [14]. By using simpler synthesized mixtures as surrogates for commercial PCA formulations, and at an ion source temperature of 120°C, Tomy et al. [54] were able to selectively increase the abundance of the $[M\text{-}Cl]^-$ ion relative to the ubiquitous and structurally noncharacteristic Cl_2^- and HCl_2^- ions. At a resolving power of 12,000, and in the SIM mode, the MS was tuned to monitor the two biggest peaks in the multipeak $[M\text{-}Cl]^-$ species, one for quantitation and the other for confirmation for formula groups within each alkane species: C_{10} (Cl_5 to Cl_{10}), C_{11} (Cl_5 to Cl_{10}), C_{12} (Cl_6 to Cl_{10}), and C_{13} (Cl_7 to Cl_9). Formula group abundance profiles were then generated by correcting the electronically integrated ion signals for each formula group for isotopic and response factors. Figure 2 shows the formula group abundance profile for PCA-60, i.e., C_{10}–C_{13}, 60% Cl. Visual inspection of the abundance profiles, created for the samples and standards, then allows for: (i) selection of the most abundant ion (corresponding to the most abundant formula group) on which the quantitation is based, and (ii) application of correction factors that account for variations in the abundances of formula groups in the standard and sample.

Generation of formula group abundance profiles allowed PCA concentrations to be reported according to individual formula and homologue groups yielding much more detailed information on the extent of transformation of PCAs in environmental samples as well as previously uncharacterized proportions of these groups in commercial products.

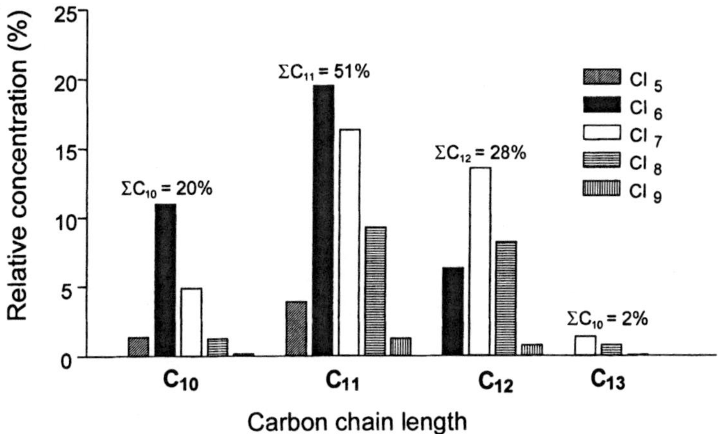

Fig. 2. Formula group abundance profiles of a C_{10}–C_{13} commercial PCA mixture containing 60 mass % Cl (PCA-60) [14]. Each bar represents the relative proportion of each group in total PCAs

5.3
Quality Assurance

There are potential quality assurance problems with the analysis of PCAs at trace levels in environmental samples because of their use in some consumer products like PVC or paints, and in metal cutting oils. Standard precautions used for the analysis of PCBs and other organochlorines such as use of glass distilled solvents, high temperature heating of filters, glassware, and sodium sulfate, and minimal use of plastics, should also minimize contamination by PCAs. This contamination is potentially more difficult to control under field conditions where PCAs may be incorporated into outdoor paints and sealants or may be residues on or in sampling equipment.

Few papers on the analysis of PCAs or their measurement in environmental samples have reported on techniques to minimize contamination. PCAs (C_{10}–C_{13}, 60–70% Cl) levels ranging from 4 ng g^{-1} to 25 ng g^{-1} in sodium sulfate were found in procedural blanks used in sediment extractions [28]. PCAs (C_{10}–C_{13}, 60–70% Cl) were also detected in DCM (0.15 µg l^{-1}) left to evaporate in an open flask overnight; it was unclear, however, whether contamination was a result of airborne PCAs or was from the DCM itself [28]. Similar problems have been encountered with airborne PCB contamination of analytical labs [65]. Significant procedural blanks result in higher method detection limits, i.e., the mean plus three times the standard deviation in the background signals from procedural blanks (sodium sulfate) [14, 66, 67].

Procedures for the analysis of some organochlorines in environmental samples (e.g., chlorinated dioxins – EPA Method 1613) [68] also require the use of surrogate standards (usually ^{13}C-labeled) and certified or standard reference materials (CRMs or SRMs). At present there are no stable isotope labeled reference materials and no reference materials have yet been certified for PCA content. Tomy [28], however, found (C_{10}–C_{13}, 60–70% Cl) PCAs in two SRMs from the National Institute for Standards and Technology (NIST). SRM 1588, a cod liver oil extract, was found to contain 49 ng g^{-1} of PCAs, while SRM 1945, a blubber extract from a stranded pilot whale [69], contained 172 ng g^{-1} of PCAs. These SRMs are, therefore, possible candidates for future certification for PCA content.

5.4
Inter-Laboratory Study on Quantitative Methods

An international inter-laboratory study recently compared quantitative methods used for measuring short chain (C_{10}–C_{13}) PCAs [70]. In this study, the samples to be quantified consisted of a standard solution of a commercially available PCA product containing 70% chlorine by mass (PCA-70), a synthetic PCA mixture consisting of purified products derived from the chlorination of 1,5,9-decatriene (PCA-1) and two fish sample extracts. The fish samples consisted of a cleaned up extract (lipid-free) of muscle tissue from a yellow perch fish, collected at the mouth of the Detroit River at Lake Erie, in 1995. All participating labs were asked to quantify the PCA-1 and PCA-70 mixtures and at least one fish extract against a primary standard solution which contained a known concentration of a second short chain commercial PCA product containing 60 mass % chlorine (PCA-60) and, if desired, by using other commercially available C_{10}–C_{13} PCA mixtures.

The choice of GC conditions employed was left to the discretion of the seven participating laboratories. Instrument types used in this study included three quadrupole systems and four magnetic sector instruments. One of the laboratories performed the analyses using both electron capture and mass spectral detection. Two laboratories performed the analysis at a high resolving power (RP > 10,000) while others used the mass spectrometer operating at nominal resolution. All laboratories performed the analysis under ECNI conditions

The results from the study are shown in Table 3. The accuracy and precision of the analytical methods were based on the measurements reported for the PCA-1 and PCA-70 mixtures of known concentration. The PCA-70 mixture was selected because its elution profile closely resembled that of the supplied external standard (PCA-60). In contrast, PCA-1 had a profile that was very different to that of PCA-60. The rationale for using these two mixtures was to see if the quantitative data would become more accurate for mixtures with profiles more closely resembling the standard. The data derived from the quantitative measurements on the environmental samples provided an estimate on the inter-laboratory precision.

From this study it is clear that there is currently a high degree of variability associated with the quantitative methods used for the determination of short chain PCAs. Results from various laboratories are likely to be comparable only within a factor of two at best. Additional variability is also introduced when different commercial formulations are used as external standards. For the environmental samples, the reported measurements for the first fish sample were more consistent than those made on the second. The authors suggest this could

Table 3. Results of an inter-laboratory study on quantification of PCAs in a synthetic product (PCA-1), a commercial products (PCA 70) and fish tissue sample [70][a]

Lab No	PCA-1[b]			PCA-70[c]			Fish sample 1		Fish sample 2		Detection method
	Value	CV	% error	Value	CV	% error	Value	CV	Value	CV	
1	81 ± 1	0.01	10	309 ± 12	0.04	160	nd	–	58 ± 6	0.01	LRMS/SIM
2	100	–	35	480	–	310	nd	–	42	–	LRMS/SIM
3	nd	–	–	78 ± 8	0.10	30	nd	–	22 ± 6	0.3	ECD
	nd	–	–	267 ± 13[d]	0.05	130	nd	–	36 ± 6[d]	0.2	ECD
	nd	–	–	267 ± 6[e]	0.06	130	nd	–	31 ± 5[e]	0.2	ECD
4	120 ± 20	0.17	63	310 ± 40	0.13	160	nd	–	80 ± 20	0.25	LRMS/SIM
5	75	–	2	352	–	200	36.7	–	35.4	–	LRMS/TIC
	102[f]	–	38	477[f]	–	300	55.7[f]	–	–	–	LRMS/TIC
6	88.9	–	21	236.7	–	100	32.2	–	nd	–	HRMS/SIM
7	128 ± 1	0.007	74	163.6 ± 16	0.01	40	54 ± 2	0.04	24 ± 1	0.04	HRMS/SIM
$\bar{x}$ ± sd[g]	99.3 ± 19.5			297 ± 132			44.6 ± 11.9		41 ± 19		
CV	0.20			0.44			0.27		0.47		
ADM	17.6			98.5			10.2		16.2		

[a] The external standard was PCA-60, except as noted. LRMS = low resolution mass spectrometry; HRMS = high resolution mass spectrometry; ECD = electron capture detector; SIM = selected ion monitoring; TIC = total ion current; CV = coefficient of variation; ADM = average deviation from mean; nd = not determined.
[b] True concentration (by weight) = 74 ng/µl.
[c] True concentration (by weight) = 118 ng/µl.
[d] Triangulation method of quantification.
[e] Integrated area of PCA elution signal.
[f] These based on Celeclor-S52 as the external standard.
[g] Mean ± standard deviation of the laboratory means.

have resulted from the fact that the former measurements were all made by laboratories that corrected for co-eluting interferences, while the latter measurements included laboratories that chose not to do so. As a result, it was recommended that procedures involving use of low resolution mass spectrometry should try to take into account, and correct for, the possible presence of other organochlorine contaminant interferences.

The reported values for PCA-1, and in particular PCA-70, were higher than their respective true values. It is not clear why results for the PCA-70 mixture, whose GC profile and composition are similar to those of the PCA-60 standard, were less accurate then the results for the PCA-1 sample, whose GC profile and composition were quite different to the external standard. One possible explanation could be the amount of additives/stabilizers used by the manufactures, which are not measurable using ECNI or ECD detection. This makes the preparation of standard solutions from commercial products problematic for quantitation of PCAs and suggests that only pure PCA commercial formulations or synthetic mixtures prepared by free-radical chlorination of pure n-alkanes should be used for the preparation of external standards.

6
Environmental Levels

In this section, recent measurements of PCAs in environmental samples, from 1990 to the present will be reviewed. With many different analytical methods being used for estimating residues of PCAs, comparing measured environmental concentrations from different laboratories is tenuous at best. Nevertheless some general comparisons will be attempted because the interlaboratory study shows results may be comparable within a factor of two.

6.1
PCA Levels in Abiotic Matrices

The environmental concentrations of PCAs reported in water, sediment, and air are presented in Table 4. Much more sampling and analysis for PCAs has been done in western Europe than in North America. No results for PCAs have been reported in Japan, the other major use region. In general, levels reported in western European waters and sediment are higher than those detected to date in North America.

In the United Kingdom, Willis et al. [7] reported C_{10}–C_{30} (no information given on the chlorine content) PCA concentrations ranging from 0.6 µg l^{-1} to 4 µg l^{-1}. Ballschmiter [71] reported C_{10}–C_{30} PCA levels ranging from 0.05 µg l^{-1} for the River Lech, Augsburg Germany, to 1.2 µg l^{-1} for the River Danube, both close to industrial centers. In Canada, Tomy reported C_{10}–C_{13} (60–70% Cl) PCA levels ranging from 0.02–0.05 µg l^{-1} in water from the Red River downstream of the city of Winnipeg, a manufacturing center [72].

In sediments from Germany, Ballschmiter [71] reported C_{10}–C_{13} PCA levels ranging from 0.017 µg g^{-1} in Hamburg Harbour to 0.7 µg g^{-1} from the River Lech. In Canada, Muir et al. [73] reported C_{10}–C_{13} (60–70% Cl) PCA concentra-

Table 4. Published concentrations of PCAs in abiotic environmental matrices

PCA measured	Location	Country	Sample type	Conc.	Units	Method of Analysis[d]	Ref.
$C_{20}-C_{30}$	Derwent Reservoir[a]	UK	water	1.5	µg l^{-1}	Tic-ar	[7]
$C_{20}-C_{30}$	River Trent, Newmark[a]	UK	water	0.9	µg l^{-1}	Tic-ar	[7]
$C_{20}-C_{30}$	Trent Mersey[a]	UK	water	0.6	µg l^{-1}	Tic-ar	[7]
$C_{20}-C_{30}$	River Derwent[a]	UK	water	0.6	µg l^{-1}	Tic-ar	[7]
$C_{20}-C_{30}$	River Ouse, Goole[a]	UK	water	0.9	µg l^{-1}	Tic-ar	[7]
$C_{20}-C_{30}$	River Ure[a]	UK	water	1.5	µg l^{-1}	Tic-ar	[7]
$C_{20}-C_{30}$	River Rother[a]	UK	water	2.1	µg l^{-1}	Tic-ar	[7]
$C_{10}-C_{30}$	Walton on Trent[a]	UK	water	1.5	µg l^{-1}	Tic-ar	[7]
$C_{10}-C_{30}$	River Don[a]	UK	water	1.8	µg l^{-1}	Tic-ar	[7]
$C_{10}-C_{30}$	River Aire[a]	UK	water	1.3	µg l^{-1}	Tic-ar	[7]
$C_{10}-C_{30}$	River Ouse, York[a]	UK	water	1.8	µg l^{-1}	Tic-ar	[7]
$ClWC_{30}$	River Cover[a]	UK	water	1.0	µg l^{-1}	Tic-ar	[7]
$C_{10}-C_{30}$	River Trent, Gainsborough[a]	UK	water	3.1	µg l^{-1}	Tic-ar	[7]
$C_{10}-C_{30}$	River Trent, Burton[a]	UK	water	3.9	µg l^{-1}	Tic-ar	[7]
$C_{10}-C_{30}$	River Trent, Humber[a]	UK	water	4.0	µg l^{-1}	Tic-ar	[7]
$C_{10}-C_{30}$	Hall Docks[a]	UK	water	3.4	µg l^{-1}	Tic-ar	[7]
$C_{10}-C_{30}$	River Lech, Augsburg[a]	Germany	water	0.05	µg l^{-1}	HRGC/ECNI/MS	[71]
$C_{10}-C_{30}$	River Lech, Gersthofen[b]	Germany	water	01–0.60	µg l^{-1}	HRGC/ECNI/MS	[71]
$C_{10}-C_{30}$	River Lech, Rain[a]	Germany	water	0.12	µg l^{-1}	HRGC/ECNI/MS	[71]
$C_{10}-C_{30}$	River Danube[b]	Germany	water	0.1–1.2	µg l^{-1}	HRGC/ECNI/MS	[71]
$C_{10}-C_{13}$ 62% Cl	Sewage plant run-off[b]	Germany	water	0.12	µg l^{-1}	HRGC/ECNI/MS	[43]
$C_{10}-C_{13}$ 62% Cl	Upstream Sewage plant[b]	Germany	water	0.08	µg l^{-1}	HRGC/ECNI/MS	[43]
$C_{10}-C_{13}$ 62% Cl	Downstream Sewage plant[b]	Germany	water	0.07	µg l^{-1}	HRGC/ECNI/MS	[43]
$C_{10}-C_{13}$ 62% Cl	River upstream from city (background)[b]	Germany	water	0.03	µg l^{-1}	HRGC/ECNI/MS	[43]
$C_{10}-C_{13}$ 50–70% Cl	Effluents from sewage treatment plant[b]	Canada	water	0.60–4.48	µg l^{-1}	HRGC/ECNI/HRMS	[73]
$C_{14}-C_{17}$ 52% Cl	St. Lawrence River[b]	Canada	water	< 1	µg l^{-1}	HRGC/ECNI/MS	[23]
$C_{10}-C_{13}$ 50–70% Cl	Red River, Selkirk[a]	Canada	water	0.03	µg l^{-1}	HRGC/ECNI/HRMS	[28]
$C_{10}-C_{13}$	River Lech[b]	Germany	Sediment	< 5–700[c]	µg kg^{-1}	HRGC/ECNI/MS	[71]
$C_{10}-C_{13}$	River Elbe, Hamburg[a]	Germany	Sediment	17–25[c]	µg kg^{-1}	HRGC/ECNI/MS	[71]
$C_{10}-C_{13}$	Outer, Hamburg[a]	Germany	Sediment	36[c]	µg kg^{-1}	HRGC/ECNI/MS	[71]

Table 4 (continued)

PCA measured	Location	Country	Sample type	Conc.	Units	Method of Analysis[d]	Ref.
C_{10}–C_{13}	River Main[a]	Germany	Sediment	25–50[c]	µg kg^{-1}	HRGC/ECNI/MS	[71]
C_{10}–C_{13}	River Rhein, Rheinfelden[a]	Germany	Sediment	26–83[c]	µg kg^{-1}	HRGC/ECNI/MS	[71]
C_{10}–C_{13}	Hamburg, Harbour[a]	Germany	Sediment	17[c]	µg kg^{-1}	HRGC/ECNI/MS	[71]
C_{10}–C_{13} 60% Cl	Industrial[b]	S. Germany	Sewer film	0.5–30[c]	µg g^{-1}	HRGC/ECNI/MS	[43]
C_{10}–C_{13} 60% Cl	Industrial/residential[b]	S. Germany	Sewer film	1–17[c]	µg g^{-1}	HRGC/ECNI/MS	[43]
C_{10}–C_{13} 60% Cl	Residential[a]	S. Germany	Sewer film	0.5–15[c]	µg g^{-1}	HRGC/ECNI/MS	[43]
C_{10}–C_{13} 60% Cl	Sewage Sludge[b]	S. Germany	Sludge	47–65[c]	µg g^{-1}	HRGC/ECNI/MS	[43]
C_{14}–C_{17} 52% Cl	Hamilton Harbour (Windemere basin)[b]	Canada	Sediment	290[c]	µg kg^{-1}	HRGC/ECNI/HRMS	[73]
C_{10}–C_{13} 60–70% Cl	Lake Nipigon[a]	Canada	Sediment	18[c]	µg kg^{-1}	HRGC/ECNI/HRMS	[72]
C_{10}–C_{30}	Manufacturing Plant[b]	Germany	Air	30	mg m^{-3}	NA	[6]
C_{10}–C_{30}	Western States[b]	Germany	Air	250	kg	Tic-ar	[7]
C_{10}–C_{13} 60–70% Cl	Canadian arctic, Alerta	Canada	Air	< 1–8.5	Pg m^{-3}	HRGC/ECNI/HRMS	[74]
C_{10}–C_{13} 60–70% Cl	Southern Ontario[b]	Canada	Air	543	Pg m^{-3}	HRGC/ECNI/HRMS	[28]
C_{10}–C_{13} 60–70% Cl	UK, Lancaster[a]	UK	Air	99	Pg m^{-3}	HRGC/ECNI/HRMS	[75]

[a] Sampling site remote from industrialized area.
[b] Sampling site near industrialized area.
[c] Dry weight.
[d] Tic-ar = thin layer chromatography with argentation, HRGC/ECNI/MS (/HRMS) = high resolution gas chromatography/electron capture negative ion/mass spectrometry (high resolution mass spectrometry); NA = method not available.

tions ranging from 0.007 µg g^{-1} to 0.29 µg g^{-1} in surface sediment samples from harbour areas along Lake Ontario. Highest concentration were found at the most industrialised site, Windemere basin in Hamilton Harbour. Tomy et al. reported a C_{10}–C_{13} (60–70% Cl) PCA concentration of 0.018 µg g^{-1} in a surface sediment sample from Lake Nipigon in northwest Ontario. This lake is sparsely populated and would likely have few sources of PCAs except atmospheric inputs [72].

Rieger and Ballschmiter [43] measured C_{10}–C_{13} (60% Cl) PCA concentrations ranging from 0.5 µg g^{-1} to 30 µg g^{-1} and 1 µg g^{-1} to 17 µg g^{-1} (dry weight), respectively, in surface films (formed by deposition of organic matter) from industrial and mixed industrial/residential waste water sewer pipes of a city in southern Germany. In both areas, metal-working plants where PCAs are used in high pressure additives for metal processing were present. Concentrations in sewer film samples collected from residential areas with minimal industrial activity, ranged from 0.5 µg g^{-1} to 15 µg g^{-1}. In sewage sludge sampled near the treatment plant, concentrations ranged from 47 µg g^{-1} to 65 µg g^{-1}. Run-off water from the sewage plant, which is discharged into a nearby river, had a PCA concentration of 0.2 µg l^{-1} while levels measured in the river water, up-stream and down stream from the purification plant, were 0.08 µg l^{-1} and 0.07 µg l^{-1}, respectively. A tributary river upstream of the city had a short chain PCA concentration of 0.03 µg l^{-1} and was used to represent the background level in the region. Muir et al. [73] reported C_{10}–C_{13} (60–70% Cl) PCA concentrations ranging from 0.60 µg l^{-1} to 4.48 µg l^{-1} in final effluent from sewage treatment plants in southern Ontario. Higher levels were observed in samples from treatment plants in industrialized areas such as Hamilton and St. Catherines compared to non-industrialized towns such as Niagara-on-the-Lake and Niagara Falls.

Mukherjee [6] reported atmospheric concentrations of PCA (C_{10}–C_{30}, no information given on Cl content) of 30 mg m^{-3}, measured at a PCA manufacturing plant in Germany in 1988. In 1990, atmospheric emissions of PCAs, either sorbed onto dust particles or as a vapor, in the western states of Germany, were estimated to be 250 kg [7]. In Canada, Stern et al. [74] reported that PCAs (C_{10}–C_{13}, 60–70% Cl) were detected in gas phase samples of air collected at Alert (northern tip of Ellesmere Island in the high Arctic) in the fall and winter of 1992. Concentrations ranged from <1 pg m^{-3} to 8.5 pg m^{-3}. Tomy [28] reported PCA (C_{10}–C_{13}, 60–70% Cl) concentrations ranging from 65 pg m^{-3} to 924 pg m^{-3} in gas phase air samples collected every day over a four month period in the summer of 1990 at Egbert, ON, Canada, and in the UK, Peters et al. [75] reported a mean concentration of 99 pg m^{-3} (C_{10}–C_{13}, 60–70% Cl) in air collected from a semi-rule site in Lancaster. PCA concentrations in air from the latter three sites were in the order, S. Ontario > Lancaster > Alert, and most likely reflect the nearness of these sites to local source areas. Comparison between homologue distribution patterns of the latter three sites showed that the profiles from Alert and Lancaster were similar. This may reflect the fact that organochlorine contaminant content of the atmosphere in the high Arctic during winter is influenced by European sources [76, 77] or that both sites are removed from the local source area. The Southern Ontario homologue pattern, relative to that of the Alert, Lancaster, and the C_{10}–C_{13}, 60% Cl profiles, has higher proportions of the pentachloro-*n*-decanes and -undecanes, and the hexachloro-*n*-dodecanes ($C_{10}H_{17}Cl_5$, $C_{11}H_{19}Cl_5$, and $C_{12}H_{20}Cl_6$, respectively). This may be attributable to the nearness of the Egbert site to local sources and the use of a variety of industrial mixtures, some of which may have a chlorine content of <60%.

6.2
PCA Levels in Biota

The environmental PCA concentrations reported in aquatic biota, terrestrial mammals and human milk are shown in Table 5. There is now a growing body of information on PCAs in biota, especially in Canada. In Europe, measurements of PCAs in biota using GC-ECNIMS are limited to Sweden.

In Sweden, Jansson et al. [59] found PCA concentrations (C_{10}–C_{13}, 60% Cl) ranging from 0.13 µg g^{-1} in ringed seal from Kongsfjorden to 1.6 µg g^{-1} in herring from Skagerrak. PCA concentrations (C_{10}–C_{13}, 60% Cl) in whitefish (Lake Storvindeln), arctic char (Lake Vättern), and grey seal (Baltic Sea) were 1 µg g^{-1}, 0.57 µg g^{-1}, and 0.28 µg g^{-1}, respectively. Herring from the Bothnian Sea and Baltic Proper were found to contain similar PCA concentrations (C_{10}–C_{13}, 60% Cl) of 1.4 µg g^{-1} and 1.5 µg g^{-1}, respectively [59]. In the United States, yellow perch, catfish, and zebra mussels from the Detroit River (MI) had measured PCA concentrations (C_{10}–C_{13}, 60–70% Cl) of 1.1 µg g^{-1}, 0.3 µg g^{-1}, and 1.2 µg g^{-1}, respectively [14]. In Canada, Metcalfe-Smith et al. were unable to detect (<3.5 µg g^{-1}) PCAs in zebra mussels, yellow perch, or white suckers from the St Lawrence River downstream of a chlorinated paraffin manufacturing plant. Tomy [28], however, measured mean C_{10}–C_{13} (60–70% Cl) PCAs concentrations of 0.57 µg g^{-1} and 0.93 µg g^{-1} in blubber samples collected from St Lawrence beluga in Baie-des-Sables and Ste Flavie, respectively. Tomy [28] also reported mean PCA (C_{10}–C_{13}, 60–70% Cl) concentrations of 0.23 µg g^{-1} and 0.16 µg g^{-1}, respectively, in beluga whale blubber from Saqqaq and Nuussuaq (northwestern Greenland). Walruses (Thule, northwestern Greenland) and ringed seal (Eureka, southwestern Ellesmere Island) blubber samples were found to contain mean PCA (C_{10}–C_{13}, 60–70% Cl) levels of 0.43 µg g^{-1} and 0.53 µg g^{-1}, respectively [28]. Stern et al. [74] reported mean ΣPCAs (C_{10}–C_{13}, 60–70% Cl) concentrations of 0.63 µg g^{-1}, 0.20 µg g^{-1}, 0.32 µg g^{-1}, and 0.46 µg g^{-1} in blubber from male beluga collected in Hendrickson Island (Southern Beaufort Sea near the Mackenzie River delta), Arviat (western Hudson Bay), Sanikiluaq (Belcher Island area in southern Hudson Bay), and Pangnirtung (southeastern Baffin Island), respectively. Mean Σ PCA concentrations in the Pangnirtung and Hendrickson Island animals were significantly higher (t-test, p < 0.05) than those from Hudson Bay, while the Sanikiluaq animals had levels significantly higher than those from Arviat [74].

Stern et al. [74] noted that the formula group profiles for PCAs in tissues of Arctic animal (Fig. 3) showed higher proportions of the lower chlorinated congeners (Cl_5–Cl_7), suggesting that the major source of contamination to the Arctic is via long range atmospheric transport. In St Lawrence beluga, the formula group profile more closely resembles that of PCA-60, which implies that local

Table 5. Published concentrations ($\mu g\ kg^{-1}$) of PCAs in aquatic biota, terrestrial mammals and human milk

PCA measured	Location	Country	Sample type	n	Sex	Tissue	Conc. $\mu g\ kg^{-1}$	Method of analysis[i]	Ref.
C_{10}–C_{13} 60% Cl	Lake Storvindeln[a]	Sweden	Whitefish	35[c]	M + F	Muscle	1000[f]	HRGC/ECNI/MS	[59]
C_{10}–C_{13} 60% Cl	Lake Vättern[a]	Sweden	Arctic Char	15[c]	M + F	Muscle	570[f]	HRGC/ECNI/MS	[59]
C_{10}–C_{13} 60% Cl	Bothnian Sea[a]	Sweden	Herring	100[c]	M + F	Muscle	1400[f]	HRGC/ECNI/MS	[59]
C_{10}–C_{13} 60% Cl	Baltic Proper[a]	Sweden	Herring	60[c]	M + F	Muscle	1500[f]	HRGC/ECNI/MS	[59]
C_{10}–C_{13} 60% Cl	Skagerrak[a]	Sweden	Herring	100[c]	M + F	Muscle	1600[f]	HRGC/ECNI/MS	[59]
C_{14}–C_{17} 52% Cl	Hamilton Harbour[b]	Canada	Carp	3[d]	nd	Whole fish	2630[g]	HRGC/ECNI/HRMS	[73]
C_{10}–C_{13} 60–70% Cl	Detroit River[b]	U.S.	Yellow Perch	1[d,e]	nd	Muscle	1148[g]	HRGC/ECNI/HRMS	[14]
C_{10}–C_{13} 60–70% Cl	Detroit River[b]	U.S.	Catfish	1[d,e]	nd	Muscle	305[g]	HRGC/ECNI/HRMS	[14]
C_{10}–C_{13} 60–70% Cl	Middle Sister Ile.[b]	U.S.	Zebra Mussel	1[d,e]	–	Mussel	201[h]	HRGC/ECNI/HRMS	[14]
C_{10}–C_{13} 60–70% Cl	Saqqaq[a]	Greenland	Beluga	1	M	Blubber	253[g]	HRGC/ECNI/HRMS	[28]
C_{10}–C_{13} 60–70% Cl	Saqqaq[a]	Greenland	Beluga	1	F	Blubber	215[g]	HRGC/ECNI/HRMS	[28]
C_{10}–C_{13} 60–70% Cl	Nuussuaq[a]	Greenland	Beluga	2[d]	F	Blubber	164[g]	HRGC/ECNI/HRMS	[28]
C_{10}–C_{13} 60–70% Cl	Hendrickson Isl.[a]	Canada	Beluga	17[d]	M	Blubber	626[g]	HRGC/ECNI/HRMS	[74]
C_{10}–C_{13} 60–70% Cl	Hendrickson Isl.[a]	Canada	Beluga	3[d]	F	Blubber	851[g]	HRGC/ECNI/HRMS	[74]
C_{10}–C_{13} 60–70% Cl	Arviat[a]	Canada	Beluga	5[d]	M	Blubber	204[g]	HRGC/ECNI/HRMS	[74]
C_{10}–C_{13} 60–70% Cl	Arviat[a]	Canada	Beluga	3[d]	F	Blubber	169[g]	HRGC/ECNI/HRMS	[74]
C_{10}–C_{13} 60–70% Ci	Sanikiluaq[a]	Canada	Beluga	5[d]	M	Blubber	323[g]	HRGC/ECNI/HRMS	[74]
C_{10}–C_{13} 60–70% Cl	Sanikiluaq[a]	Canada	Beluga	4[d]	F	Blubber	396[g]	HRGC/ECNI/HRMS	[74]
C_{10}–C_{13} 60–70% Cl	Pangnirtung[a]	Canada	Beluga	31[d]	M	Blubber	457[g]	HRGC/ECNI/HRMS	[74]
C_{10}–C_{13} 60–70% Cl	Baie-des-Sables[a]	Canada	Beluga	1	M	Blubber	370[g]	HRGC/ECNI/HRMS	[28]
C_{10}–C_{13} 60–70% Cl	Baie-des-Sables[a]	Canada	Beluga	1	F	Blubber	773[g]	HRGC/ECNI/HRMS	[28]
C_{10}–C_{13} 60–70% Cl	Ste Flavie[a]	Canada	Beluga	2[d]	M	Blubber	1064[g]	HRGC/ECNI/HRMS	[28]
C_{10}–C_{13} 60–70% Cl	Ste Flavie[a]	Canada	Beluga	1	F	Blubber	665[g]	HRGC/ECNI/HRMS	[28]
C_{10}–C_{13} 60–70% Cl	S.W. Ellesmere Isl.[a]	Canada	Ringed Seal	1	M	Blubber	374[g]	HRGC/ECNI/HRMS	[28]
C_{10}–C_{13} 60–70% Cl	S.W. Ellesmere Isl.[a]	Canada	Ringed Seal	5[d]	F	Blubber	556[g]	HRGC/ECNI/HRMS	[28]
C_{10}–C_{13} 60% Cl	Kongsfjorden[a]	Sweden	Ringed Seal	7[c]	F	Blubber	130[f]	HRGC/ECNI/MS	[59]

Table 5 (continued)

PCA measured	Location	Country	Sample type	n	Sex	Tissue	Conc. $\mu g\ kg^{-1}$	Method of analysis[i]	Ref.
C_{10}–C_{13} 60% Cl	Baltic Sea[a]	Sweden	Grey Seal	8[c]	F	Blubber	280[f]	HRGC/ECNI/MS	[59]
C_{10}–C_{13} 60–70% Cl	Thule[a]	Greenland	Walrus	2[d]	M	Blubber	426[g]	HRGC/ECNI/HRMS	[28]
C_{10}–C_{13} 60% Cl	Revingeshed, Skåna[a]	Sweden	Rabbit	15[c]	nd	Muscle	2900[f]	HRGC/ECNI/MS	[59]
C_{10}–C_{13} 60% Cl	Grimsö, Västmanland[a]	Sweden	Moose	13[c]	M + F	Muscle	4400[f]	HRGC/ECNl/MS	[59]
C_{10}–C_{13} 60% Cl	Ottsjö, Jämtland[a]	Sweden	Reindeer	31[c]	M + F	Suet	140[f]	HRGC/ECNI/MS	[59]
C_{10}–C_{13} 60% Cl	Sweden[a]	Sweden	Osprey	34[c]	M + F	Muscle	530[f]	HRGC/ECNI/MS	[59]
C_{10}–C_{13} 52% Cl	N. Quebec[a]	Canada	Human milk	3[d]	F	Milk	12.8[f]	HRGC/ECNI/HRMS	[28]

[a] Sampling site remote from industrialized area.
[b] Sampling site near industrialized area.
[c] Pooled sample.
[d] Mean value.
[e] Duplicate sub-sample.
[f] Lipid weight.
[g] Wet weight.
[h] Dry weight.
[i] HRGC/ECNI/MS (/HRMS) = high resolution gas chromatography/electron capture negative ion/mass spectrometry (high resolution mass spectrometry).
nd = not determined.

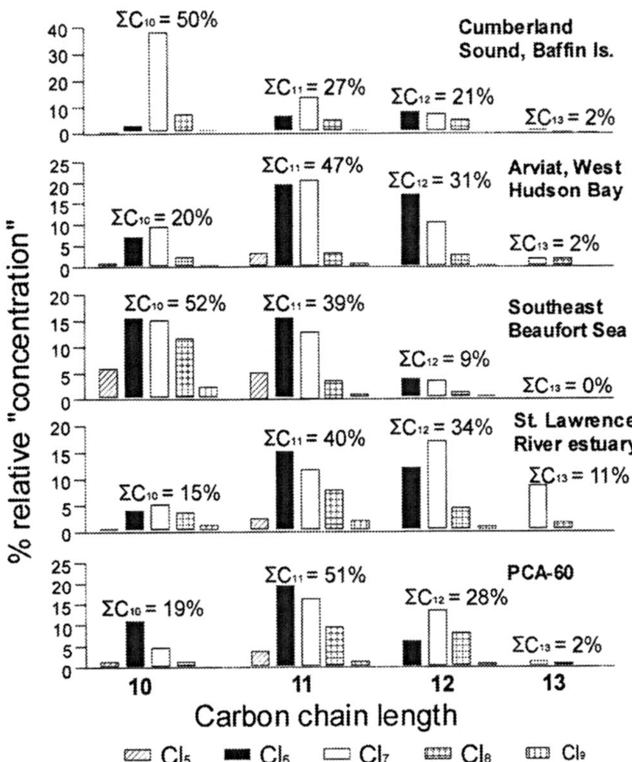

Fig. 3. Chlorinated *n*-alkane formula group profiles for beluga whales (*Delphinapterus leucas*) from Cumberland Sound, western Hudson Bay, the southeastern Beaufort Sea, and the St. Lawrence river estuary are compared with a commercial short chain chlorinated paraffin with 60% chlorine (PCA-60). The arctic beluga show higher proportions of the lower chlorinated congeners (Cl_5–Cl_7) compared with the commercial product and the St. Lawrence River animals. Results from Tomy et al. [78]

sources of PCAs, possibly from the Great Lakes and/or the lower industrialized regions of the St Lawrence River, predominate in the St Lawrence River estuary.

Figure 4 shows the ECNI selected ion chromatograms of the $C_{10}H_{15}Cl_7$ formula group, based on the [M-Cl]⁻ ion, for two Arctic beluga blubber samples, the Alert air sample, and the PCA-60 standard [78]. Reasons for the altered appearance of the elution profile of this species, and its relatively high abundance (see Fig. 3), in the Pangnirtung beluga is unclear. It is postulated that this may be the result of the PCA formulation that these animals were exposed to in their diets or, perhaps, differences in food web structures. With the exception of the Pangnirtung blubber samples, the chromatographic profile shown for the Sanikiluaq animals is typical of those observed for the beluga from the other regions investigated in the study and is also similar to that of the Alert air.

To date, very limited information is available on PCA concentrations in tissues of terrestrial mammals. In Sweden, Jansson et al. [59] reported PCA

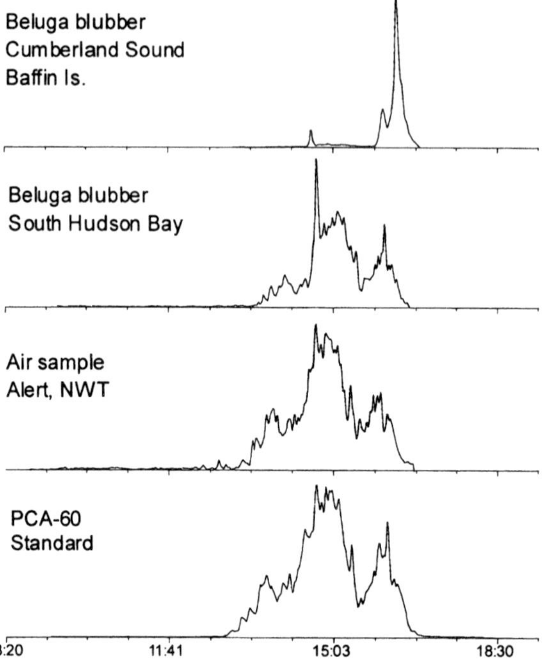

Fig. 4. GC-ECNI high resolution mass spectrometry selected ion chromatograms of the $C_{10}H_{15}Cl_7$ formula group, based on the $[M\text{-}Cl]^-$ ion, for beluga blubber samples from Cumberland Sound and Hudson Bay are compared with an extract of an air sample from Alert (NWT) at the northern tip of Ellesmere Island and in the PCA-60 standard. The air sample shows a strong resemblance to the commercial product while the profile in beluga suggests considerable biotransformation. Results from Tomy et al. [78]

(C_{10}–C_{13}, 60% Cl) concentrations in rabbit (Revingeshed, Skåne), moose (Grimsö, Västmanland), reindeer (Ottsjö, Jaäamtland), and osprey (from various regions in Sweden) to be 2.9 µg g^{-1}, 4.4 µg g^{-1}, 0.14 µg g^{-1}, and 0.53 µg g^{-1}, respectively. Tomy [28] reported a PCA (C_{10}–C_{13}, 60–70% Cl) mean concentration of 0.013 µg g^{-1} (lipid wt) in human breast milk from Inuit women living in communities on Hudson Strait in Northern Québec. Like the Arctic beluga blubber, the PCA formula group profile consists of a higher proportion of the shorter chain lower chlorinated congeners.

7
Environmental Fate Modeling of PCAs

In this section the environmental distribution of PCAs will be estimated using Mackay's Equilibrium Criterion (EQC) level III fugacity model [79]. Level III refers to a steady state, nonequilibrium system among soil, air, and water compartments, with the chemical undergoing reactions or inputs and removal processes (advection, volatilization, deposition, photolysis, hydrolysis, and biodegradation).

Two congeners have been selected for this study, namely $C_{12}H_{20}Cl_6$ and $C_{16}H_{24}Cl_{10}$. Table 6a shows the physical-chemical properties and reaction rates of the two congeners used for the level III calculations. Physical-chemical properties were taken from recent literature values or calculated from relationships based on carbon and chlorine number (see footnote Table 6a). Values of environmental degradation rates were more difficult to find. Degradation rates for air were based on photolysis rates estimated from Atkinson's OH radical reaction model [42]. Half-lives of $C_{12}H_{20}Cl_6$ and $C_{16}H_{24}Cl_{10}$ in water were estimated at 60 days and 100 days, respectively, based on conclusions of Madeley and Birtley [48] for biodegradation of chlorinated paraffins. Degradation rates for soil and sediment were based on inferences from the biodegradation data cited previously which suggests relatively long half-lives, i.e., 3–6 months. Half-lives for PCAs in soils were assumed to be two-fold greater than in sediment based on the likelihood that this compound would undergo aerobic as well as anaerobic degradation in sediments but only aerobic degradation in surface soils. $C_{12}H_{20}Cl_6$ was assigned a half-life of 180 days in soil and 90 days in sediment while $C_{16}H_{24}Cl_{10}$, with high Cl content and long chain length, was assumed to have half-lives of 360 days in soil and 180 days in sediment. Inputs to soil and water of 1000 kg/h into a unit world (air $= 1 \times 10^{14}$ m^3; water $= 2 \times 10^{11}$ m^3; soil =

Table 6a. Physical chemical properties and reaction rates used for EQC level III calculations

	$C_{12}H_{20}Cl_6$		$C_{16}H_{24}Cl_{10}$	
Molecular weight (g mol^{-1})	374		566	
water solubility (g m^{-3})[a]	0.037		2.9×10^{-4}	
vapor pressure (Pa)[b]	2.0×10^{-4}		2.3×10^{-8}	
log K_{ow}[c]	6.4		7.6	
HLC (Pa · m^3 mol^{-1})[d]	1.37		0.044	
Degradation rates	*(d^{-1})*	*half-life (d)*	*(d^{-1})*	*half-life (d)*
air[e]	0.347	2	0.347	2.0
water[f]	0.012	60	0.0069	100
soil[g]	0.0039	180	0.0019	360
sediment[g]	0.0078	90	0.0038	180

[a] Water solubility of $C_{12}H_{20}Cl_6$ and $C_{16}H_{24}Cl_{10}$ was estimated from K_{ow} using the relationships (i.e., log WS $= -1.38 \times$ log $K_{ow} + 1.17$) developed by Isnard and Lambert [82].

[b] Vapor pressure of $C_{12}H_{20}Cl_6$ was from [25] and for $C_{16}H_{24}Cl_{10}$ was estimated from equation based on carbon and chlorine number developed by Drouillard et al. [25].

[c] Log K_{ow} of $C_{12}H_{20}Cl_6$ and $C_{16}H_{24}Cl_{10}$ were estimated from equation based on carbon and chlorine number developed by Sijm and Sinnige [33].

[d] HLC of $C_{12}H_{20}Cl_6$ was from [25] and for $C_{16}H_{24}Cl_{10}$ was derived from the equation HLC = VP/WS.

[e] Degradation rates for air are based on photolysis rates estimated from Atkinson's OH radical reaction model [42].

[f] Degradation rates for water are estimated based on Madeley and Birtley [48] for short chain low chlorinated paraffins.

[g] Degradation rates for sediment and soil are based on inferences from sediment core studies [72].

Table 6b. Concentrations, total amounts and reaction rates of $C_{12}H_{20}Cl_6$ and $C_{16}H_{24}Cl_{10}$ determined from a level III EQC model [79] assuming emissions to soil and water of 1000 kg h^{-1}

Compartment	Emissions (kg d^{-1})	Concentration	Percent in each compartment	Reaction rate	Largest intercompartment transfers (kg d^{-1})	Residence time (d)
$C_{12}H_{20}Cl_6$						
air	0	5 ng m^{-3}	0.0055	160	112 to soil	170
water	2.4×10^3	1170 µg m^{-3}	2.8	2.700	16,230 to sediment	
soil	2.4×10^3	0.23 µg g^{-1}	74.8	24,100	16 to water	
sediment	0	2.9 µg g^{-1}	22.4	14,400	905 to water	
$C_{16}H_{24}Cl_{10}$						
air	0	0.006 ng m^{-3}	3.6×10^{-6}	0.21	0.38 to soil	350
water	2.4×10^3	945 µg m^{-3}	1.1	1310	20,210 to sediment	
soil	2.4×10^3	0.46 µg g^{-1}	73.9	24,000	30 to water	
sediment	0	8.4 µg g^{-1}	24.9	16,180	2020 to water	

1.8×10^{10} m^3) were assumed following a preliminary model run which showed that this rate of emission yielded concentrations within an order of magnitude of reported environmental levels. Default parameters were assumed for intermedia transfer by processes such as wet deposition from the air, sediment deposition in the water, and soil runoff.

Results from the level III EQC model suggest that both of these PCA congeners would achieve their highest concentrations in sediment and soil (Table 6b). Concentrations in water and air are extremely low for both congeners. These results are not surprising, as PCAs are expected to be sorbed to particulate organic matter, and their low vapor pressures and high Koa values will minimize partitioning to air from soils or materials such as plastics in which they are incorporated. The greatest quantity of these congeners will also be found in soil and sediment, which is due to the large size of these compartments; most of the degradation of these PCAs will, therefore, likely occur in sediment and soil. $C_{16}H_{24}Cl_{10}$ would achieve higher concentrations in sediment and soil than $C_{12}H_{20}Cl_6$ because of its slower degradation rates and lower water solubility.

The environmental residence time of $C_{16}H_{24}Cl_{10}$ is estimated to be 350 days compared to 170 days for $C_{12}H_{20}Cl_6$. These results are sensitive to degradation rates in soils and sediment, the two compartments with the largest amounts of each compound. Although no emissions to air were assumed, air concentrations of 5 ng m^{-3} for $C_{12}H_{20}Cl_6$ and 0.006 ng m^{-3} for $C_{16}H_{24}Cl_{10}$ are predicted because of volatilization from water to air as well as from soil to air. This small but significant transfer from water to air of both compounds suggests the possibility of long range transport of PCAs via atmosphere and oceans. These results suggest that characteristics which result in higher relative vapor pressures of PCAs, i.e., lower chlorination and short carbon chain length, may influence degradation, atmospheric transport, and fate of PCAs. However, these results should be viewed with caution because there are no measured photolysis rates, and the presence of PCAs in Arctic sediments and biota [28, 72], suggest that their degradation in air may be slower than the values used for our model calculations. On the other hand, biodegradation in soils and sediments are highly uncertain and may be more rapid than we have assumed.

Willis et al. [7] performed a level I fugacity calculation (all compartments at equilibrium) on $C_{16}H_{28}Cl_6$ using a United Kingdom "model world" and found that the distribution of this PCA was similar to what we observed for the C_{12} and C_{16} PCAs. Willis et al. [7] carried the fugacity model a step further by estimating the release rate of this PCA in England, which allowed a comparison to environmental concentrations. Measured and calculated concentrations were similar, although background levels in water were somewhat higher in the real world.

8
Conclusions

PCAs are a group of high molecular weight organochlorine chemicals with a significant volume of production worldwide. Yet, compared to most other high MW chlorinated organics, information on their current levels in the environment and data on reaction rates (biodegradation, hydrolysis, photolysis) is very limited. Thus a thorough assessment of the exposure and risk to aquatic or terrestrial animals including humans is not yet possible. Development of the appropriate data equivalent, for example, to information available on chlorinated dioxins or PCB congeners, will be difficult given the wide range of products (C_{10}–C_{13} to C_{18}–C_{30} with varying degrees of chlorination) and lack of analytical standards for individual congeners or formula groups. Future studies will require better analytical methods and reference materials certified for PCA content. More attention must also be given to the potential for sample contamination during collection as well as during analysis of samples for trace amounts of PCAs.

The presence of PCAs in top predators such as seals and whales also suggests that bioaccumulation and possibly biomagnification of some PCAs is occurring. The magnitude of biomagnification is not yet known because of the absence of data for food web organisms, water, and sediments at the same sites where information exists for biota. From the survey by Jansson et al. [59] we can surmise that there is much less biomagnification of PCAs in the Baltic sea food web, e.g., herring, seals, guillemot, than for PCBs, DDT, or toxaphene. On the other hand, the same study reported relatively high levels in terrestrial herbivores. Half-lives of PCAs in fish are also generally shorter than those for PCBs, toxaphene, and chlordane components when studied under identical conditions [80, 81]. There is a need for studies of the biomagnification of PCAs in aquatic and terrestrial food webs in order to understand pathways of exposure and the importance of carbon chain length and Cl content.

Of all the data available for assessment of environmental fate of PCAs, perhaps the most uncertain are the rates of reaction in sediment, atmosphere, and soils. This basic information is required to predict fate in environmental compartments and ultimately exposure through food, air, and water. While the EQC model result suggests that PCAs are only moderately persistent in the environment, the results are uncertain because of limited knowledge of reaction rates. Short chain PCAs, with low Cl content, generally appear to have shorter half-lives in BOD tests and other biodegradation experiments than persistent POPs such as PCBs, DDT, and chlordane. Use of other compounds, such as the cyclic aliphatic POPs, as surrogates for understanding the fate of PCAs or empirical structure-activity relationships is problematic because of the unique structures of the chlorinated *n*-alkanes, differences in physical-chemical properties, and very limited knowledge of the effect of Cl content and carbon chain length on their biodegradation. Future studies of PCA persistence and biodegradation should include positive controls with chlorinated organics whose fate is reasonably well studied, e.g., DDT and dieldrin, so that relative rates can be calculated. The use of synthetic PCAs, including individual congeners for physical property and bioaccumulation studies, has enhanced our understanding of the environmental chemistry of PCAs. Future work using single congeners in biodegradation studies, and as standards for quantitation of PCAs in environmental samples, is likely to occur and will advance our knowledge of the environmental chemistry of this complex group of chemicals.

Acknowledgments. We wish to thank the Natural Sciences and Engineering Research Council of Canada (Ottawa ON), Environment Canada (Commercial Chemicals Evaluation Branch) and the Canadian Chlorine Coordinating Committee (Burlington ON) and the Canadian Chemical Producers Assoc. for research grants which enabled the preparation of this review. We thank A. Fisk (Canadian Wildlife Service, Ottawa), P. Ostrowski (Occidental Chemicals, Niagara Falls NY), Bo Jansson (Stockholm University, Stockholm, Sweden), Don Bennie (Environment Canada, National Water Research Institute), and Heidi Karlsson (National Water Research Institute) for information and helpful discussions.

References

1. Kirk-Othmer (1980) Chlorinated paraffins. Kirk-Othmer encyclopedia of chemical technology, 3rd edn. Wiley, New York
2. Environment Canada (1993) Priority substances program, CEPA assessment report, chlorinated paraffins. Commercial chemicals branch, Hull Quebec
3. Swedish National Chemicals Inspectorate (1991) In: Risk reduction of chemicals, a government commission report, chap 9, chlorinated paraffins, pp 167–187; KEMI Rep 1/91
4. Zitko V (1980) In: Handbook of environmental chemistry, vol 3A. Springer, Berlin Heidelberg New York, pp 149–156
5. Howard PH, Santodonato J, Saxena J (1975) Investigation of selected potential environmental contaminants: chlorinated paraffins. EPA-560/2–75–007. NTIS, Springfield, Va 22151
6. Mukherjee AB (1990) The use of chlorinated paraffins and their possible effects in the environment. National Board of Waters and the Environment, Helsinki, Finland, Series A 66
7. Willis B, Crookes MJ, Diment J, Dobson SD (1994) Environmental hazard assessment: chlorinated paraffins. Toxic Substances Division. Dept. of the Environment. London, UK

8. World Health Organization (1996) Environmental Health Criteria 181: chlorinated paraffins. WHO, Geneva, Switzerland

9. Tomy GT, Fisk A, Westmore JB, Muir DCG (1998) Environmental chemistry and toxicology of polychlorinated n-alkanes. Rev Environ Contam Toxicol 158:53–128

10. Windrath OM, Stevenson DR (1985) Chlorination and bromochlorination paraffins as flame retardants. Plastics Compounding, pp 38–52

11. Calder R (1998) Normal paraffins (C9-C17). SRI Consulting, New York

12. Ostrowski PJ (1997) Personal communication. Occidental Chemical Corp, Basic Chemicals Group, Niagara Falls, NY

13. Campbell I, McConnell G (1980) Chlorinated paraffins and the environment. 1. Environmental occurrence. Environ Sci Technol 14:1209–1214

14. Tomy GT, Stern GA, Muir DCG, Fisk AT, Cymbalisty CD, Westmore JB (1997) Quantifying C_{10}–C_{13} polychloroalkanes in environmental samples by high resolution gas chromatography/ electron capture negative ion mass spectrometry. Anal Chem 69:2762–2771

15. Tomy GT, Muir DCG, Westmore JB, Stern GA (1993) Liquid chromatography thermospray mass spectrometry for characterization of commercial chlorinated n-paraffin mixtures. Proceedings of the 41st Annual Conference on Mass Spectrometry and Allied Topics, San Francisco, CA

16. Colebourne N, Stern ES (1965) The chlorination of some n-alkanes and alkyl chlorides. J Chem Soc 3599–3605

17. Fredricks PS, Tedder JMJ (1960) Free-radical substitution in aliphatic compounds. Part II. Halogenation of the n-butyl halides. J Chem Soc 144–150

18. Chambers G, Ubbelohde ARJ (1955) The effect of molecular structure of paraffins on relative chlorination rates. J Chem Soc 285–295

19. Gusev MN, Urman YG, Mochalova OA, Kocharyan LA, Slonim IY (1968) Nuclear magnetic resonance study of the structure of chloroparaffins. Izv Akad Nauk SSR, Ser Khim 7:1549

20. Hardie DWF (1964) Chlorinated paraffins. Kirk-Othmer Encyl Chem Technol, 2nd edn, vol 5

21. Environmental Protection Agency (1991) Office of Toxic Substances. Rm1 decision package. Chlorinated paraffins. Environmental risk. EPA, Washington DC

22. Environment Canada (1993) Health and Welfare Canada, Priority substances list assessment report: chlorinated paraffins. Government of Canada, Catalogue En 40 215/17 E, ISBN 0 662 20515 17 E

23. Metcalfe-Smith JL, Maguire RJ, Batchelor SP, Bennie DT (1995) Occurrence of chlorinated paraffins in the St Lawrence river near a manufacturing plant in Cornwall, ON. National Water Research Institute. Environment Canada, Burlington, ON

24. Murray TM, Frankenberry DH, Steele DH, Heath RG (1988) Chlorinated paraffins: a report on the findings from two filed studies, Sugar Creek, Ohio and Tinkers Creek, Ohio, vol 1, Technical Report, US Environmental Protection Agency, EPA/560/5 87/012

25. Drouillard KG, Tomy GT, Muir DCG, Friesen KJ (1997) volatility of chlorinated n-alkanes. Environ Toxicol Chem 17:1252–1260

26. Drouillard KG, Hiebert T, Tran P, Tomy GT, Muir DCG, Friesen KJ (1997) Water solubilities of chlorinated n-alkanes. Environ Toxicol Chem 17:1261–1267

27. Madeley JR, Gillings E (1983) The determination of the solubility of four chlorinated paraffins in water. ICI Confidential Report BL/B/2301

28. Tomy GT (1997) The mass spectrometric characterization of polychlorinated n-alkanes and the methodology for their analysis in the environment. PhD Thesis, University of Manitoba

29. May WE, Wasik SP, Freeman DH (1978) Determination of the aqueous solubility of polynuclear aromatic hydrocarbons by a coupled column liquid chromatographic technique. Anal Chem 50:175–179

30. Mackay D, Shiu WY, Ma KC (1992) Illustrated handbook of physical-chemical properties and environmental fates for organic chemicals, vol I. Monoaromatic hydrocarbons, chlorobenzenes, and PCBs. Lewis, Chelsea, MI

31. Mackay D, Shiu WY, Ma KC (1997) Illustrated handbook of physical-chemical properties and environmental fates for organic chemicals, vol V. Chlorinated pesticides. Lewis, Chelsea, MI
32. Schwarzenbach RP, Gschwend PM, Imboden DM (1993) Environmental organic chemistry. Wiley, Toronto, ON
33. Sijm DTHM, Sinnige TL (1995) Experimental octanol/water partition coefficients of chlorinated paraffins. Chemosphere 31:4427–4435
34. de Bruijn JF, Busser F, Seinen W, Hermens J (1989) Determination of octanol/water partition coefficients with the 'slow-stirring' method. Environ Toxicol Chem 8:499–512
35. Lyman WJ, Reehl WF, Rosenblatt DH (1990) Handbook of chemical property estimation methods: environmental behavior of organic compounds. American Chemical Society, Washington DC
36. Fisk AT, Weins SC, Bergman A, Muir DCG (1998) Accumulation and depuration of sediment-sorbed C_{12} and C_{16}-polychlorinated alkanes by oligochaetes (*Lumbriculus variegatus*). Environ Toxicol Chem 17:2019–2026
37. Harner T, Bidleman TF (1996) Measurements of octanol-air partition coefficients for polychlorinated biphenyls. J Chem Engin Data 41:895–899
38. Drouillard KG (1996) Physico chemical property determination on chlorinated *n*-alkanes (C10 to C12). Parameters for estimation of the environmental fate of chlorinated *n*-paraffins. MSc Thesis, University of Manitoba
39. Karickhoff SW (1981) Semi-empirical estimation of sorption of hydrophobic pollutants on natural sediments and soils. Chemosphere 10:833–846
40. US Environmental Protection Agency (1995) Great lakes water quality initiative technical support document for the procedure to determine bioaccumulation factors. Office of Water, EPA-820–9-95–005
41. Friedman D, Lombardo P (1975) Photochemical technique for the elimination of chlorinated aromatic interferences in the gas-liquid chromatographic analysis for chlorinated paraffins. J Assoc Offic Anal Chem 58:703–706
42. Atkinson R (1986) Kinetics and mechanisms of gas phase reactions of the hydroxyl radical with organic compounds under atmospheric conditions. Chem Rev 86:69–201
43. Reiger R, Ballschmiter K (1995) Semivolatile organic compounds polychlorinated dibenzo-*p*-dioxins (PCDD), dibenzofurans (PCDF), biphenyls (PCBs), hexachlorobenzene (HCB), 4,4' DDE and chlorinated paraffins (CP) as markers in sewer films. Fres J Anal Chem 352:715–724
44. Lahaniatis ES, Parlar H, Klein W, Korte F (1975) Analysis and photochemistry of high-boiling polychlorinated paraffin mixtures. Chemosphere 4:83–88
45. Camino G, Costa L (1980) Thermal degradation of a highly chlorinated paraffin used as a fire retardant additive for polymers. Polym Deg Stab 2:23–28
46. Bergman Å, Hagman A, Jacobsson S, Jansson B, Åhlman M (1984) Thermal degradation of polychlorinated alkanes. Chemosphere 13:237–250
47. Zitko V, Arsenault E (1977) Fate of high molecular weight chlorinated paraffins in the aquatic environment. Adv Environ Sci Technol 8:409–418
48. Madeley J, Birtley R (1980) Chlorinated paraffins and the environment. 2. Aquatic and avian toxicology. Environ Sci Technol 14:1215–1221
49. Omori T, Kimura T, Kodama T (1987) Bacterial cometabolic degradation of chlorinated paraffins. Appl Microbiol Biotechnol 25:553–557
50. Hollies JI, Pinnington DF, Handley AJ, Baldwin MK, Bennett D (1979) The determination of chlorinated long chain paraffins in water, sediment and biological samples. Anal Chim Acta 111:201–213
51. Svanberg O, Bengtsson BE, Linden E, Lunde G, Baumann Ofstad E (1978) Chlorinated paraffins – a case of accumulation and toxicity to fish. Ambio 7:64–65
52. Jansson B, Andersson R, Asplund L, Bergman Å, Litzén K, Nylund K, Reutergårdh L, Sellström U, Uvemo U-B, Wahhlerg C, Wideqvist U (1991) Multi-residue method for the gas-chromatographic analysis of some polychlorinated and polybrominated pollutants in biological samples. Fresenius J Anal Chem 340:439–445

53. Stalling DL, Tindle RC, Johnson JL (1972) Cleanup of pesticide and polychlorinated biphenyl residues in fish extracts by gel permeation chromatography. J Assoc Offic Anal Chem 55:32–38
54. Tomy GT, Tittlemier SA, Stern GA, Muir DCG, Westmore JB (1998) Effects of temperature and sample amount on the electron capture negative ion mass spectra of polychloro-n-alkanes. Chemosphere 37:1395–1410
55. Junk SA, Meisch H-U (1993) Determination of chlorinated paraffins by GC MS. Fresenius J Anal Chem 347:361–363
56. Walter B, Ballschmiter K (1991) Quantitation of camphechlor/toxaphene in cod-liver oil by integration of the HRGC/ECD-pattern. Fresenius J Anal Chem 340:246–249
57. Mullin MD, Pochini CM, McCrindle S, Romkes M, Safe SH, Safe LM (1984) High-resolution PCB analysis: synthesis and chromatographic properties of all 209 PCB congeners. Environ Sci Tecnnol 18:468–476
58. Parlar H, Coelhan M, Saraçi M, Lahaniatis ES, Lachermeir C, Koske G, Nitz S, Leupold G (1998) Quantification of C10-chloroparaffins with purely synthesized chloroalkanes as standards. Organohalogen Compounds 35:395–398
59. Jansson B, Andersson R, Asplund L, Litzen K, Nylund K, Sellstrom U, Uvemo U, Wahlberg C, Wideqvist U, Odsjo T, Olsson M (1993) Chlorinated and brominated persistent organic compounds in biological samples from the environment. Environ Toxicol Chem 12:1163–1174
60. Ong VS, Hites RA (1994) Electron capture mass spectrometry of organic environmental contaminants. Mass Spectrom Rev 13:259–283
61. Dougherty RC, Dalton J, Biros FJ (1972) Negative chemical ionization mass spectra of polycyclic chlorinated insecticides. Org Mass Spectrom 6:1171–1181
62. Rankin PC (1971) Negative ion mass spectra of some pesticidal compounds. J Assoc Off Anal Chem 54:1340–1348
63. Schmid PP, Muller MD (1985) Trace level detection of chlorinated paraffins in biological and environmental samples, using gas chromatography/mass spectrometry with negative-ion chemical ionization. J Assoc Offic Anal Chem 68:427–430
64. Gjøs N, Gustavsen KO (1982) Determination of chlorinated paraffins by negative ion chemical mass spectrometry. Anal Chem 54:1316–1318
65. Wallace JC, Basu I, Hites RA (1996) Sampling and analysis artifacts caused by elevated indoor air polychlorinated biphenyl concentrations. Environ Sci Technol 30:2730–2734
66. Freeman DH (1980) Guidelines for data acquisition and data quality evaluation in environmental chemistry. Anal Chem 52:2242–2249
67. Long GL, Winefordner JD (1983) Limit of detection. A closer look at the IUPAC definition. Anal Chem 55:712A–724A
68. US Environmental Protection Agency (1989) Office of Water Regulations and Standards, Industrial Technology Division. Method 1613: Tetra-through octa-chlorinated dioxins and furans by isotope dilution HRGC/HRMS
69. Schantz MM, Koster BJ, Oakley LM, Schiller SB, Wise SA (1995) Certification of polychlorinated biphenyl congeners and chlorinated pesticides in a whale blubber standard reference material. 67:901–910
70. Tomy GT, Stern SA, Muir DCG, Fisk AT, Westmore JB (1999) Interlaboratory study on quantitative methods of analysis of C_{10}–C_{13} polychloro-n-alkanes. Anal Chem 71:446–451
71. Ballschmiter K (1994) Determination of short and medium chain length chlorinated paraffins in samples of water and sediment from surface water. Abt Anal Chem und Umweltchem. 10 May 1994. Unpublished data report
72. Tomy GT, Stern GA, Muir DCG, Lockhart L, Westmore JB (1997) Occurrence of polychloro-n-alkanes in Canadian mid-latitude and Arctic lake sediments. Organohalogen Compounds 33:220–226
73. Muir DCG, Bennie D, Tomy GT, Stern GA (1998) SCCPs – an updated assessment based on recent environmental measurements and bioaccumulation results. Unpublished report to Environment Canada, Commercial Chemicals Evaluation Branch. National Water Research Institute Burlington, ON

74. Stern GA, Tomy GT, Muir DCG, Westmore JB, Dewailly E, Rosenberg B (1998) Polychlorinated *n*-alkanes in aquatic biota and human milk. Presented at the American Society for Mass Spectrometry and Allied Topics, 45th Annual Conference, May, Palm Springs, CA
75. Peters PJ, Tomy GT, Stern GA, Jones KC (1998) Polychlorinated alkanes in the atmosphere of the United Kingdom and Canada-analytical methodology and evidence of the potential for long-range transport. Organohalogen Compds 35:439–442
76. Stern GA, Halstall CJ, Barrie LA, Muir DCG, Fellin P, Rosenberg B, Ya Rovinsky F, Ya Kononov E, Pastuhov B (1997). Polychlorinated biphenyls in Arctic air. 1. Temporal and spatial trends 1992–1994. Environ Sci Technol 31:3619–3628
77. Halstall CT, Bailey R, Stern GA, Barrie LA, Fellin P, Muir DCG, Rosenberg B, Rovinski F, Konovov E, Pastukhov B (1998) Multiyear observations of organohalogen pesticides in the Arctic atmosphere. Environ Pollut 102:51–62
78. Tomy GT, Stern GA, Koczanski K, Halldorson T (1998) Polychloro-*n*-alkanes in beluga whales from the arctic and the St. Lawrence river estuary. Organohalogen Compds 35:399–402
79. Mackay D, Di Guardo A, Paterson S, Cowan CE (1996) Evaluating the environmental fate of a variety of types of chemicals using the EQC model. Environ Toxicol Chem 15:1627–1637
80. Fisk AT, Cymbalisty CD, Bergman Å, Muir DCG (1996) Dietary accumulation of C_{12}- and C_{16}-chlorinated alkanes by juvenile rainbow trout (*Oncorhynchus mykiss*). Environ Toxicol Chem 15:1775–1782
81. Fisk AT, Norstrom RJ, Cymbalisty CD, Muir DCG (1998) Dietary accumulation and depuration of hydrophobic organochlorines: bioaccumulation parameters and their relationship with K_{ow}. Environ Toxicol Chem 17:951–961
82. Isnard P, Lambert S (1989) Aqueous solubility and *n*-octanol/water partition coefficient correlations. Chemosphere 18:1837–1853

CHAPTER 9

Toxaphene. Analysis and Environmental Fate of Congeners

Walter Vetter · Michael Oehme[2]

[1] Friedrich-Schiller-University Jena, Department of Food Chemistry,
Dornburger Strasse 25, D-07743 Jena, Germany, (e-mail: b5wave@rz.uni-jena.de)
[2] Organic Analytical Chemistry, University of Basel, Neuhausstrasse 31, CH-4057 Basel,
Switzerland, (e-mail: oehme@ubaclu.unibas.ch)

Toxaphene is an organochlorine pesticide of complex composition. Several hundreds of components consisting mainly of polychlorinated bornanes are present in technical mixtures. This number is high but low in comparison to the theoretically possible number of congeners which exceeds 30,000. In the environment and particularly in higher organisms, most of the components of toxaphene are degraded, and only few are accumulated. A direct synthesis of individual polychlorinated bornanes is not possible. Nevertheless, more than 40 individual compounds were produced by partial synthesis and/or degradation combined with further isolation. Several of these polychlorinated bornanes have been used as external standards for the quantitation of toxaphene in environmental samples. This article summarizes recent results on the chemistry, characterization, and environmental fate of single toxaphene congeners.

Keywords. toxaphene, chlorobornane, nomenclature, identification, separation, toxicity, congener-specific, biota, sediment.

The Handbook of Environmental Chemistry Vol. 3 Part K
New Types of Persistent Halogenated Compounds
(ed. by J. Paasivirta)
© Springer-Verlag Berlin Heidelberg 2000

1
Introduction – What is Toxaphene?

Toxaphene (Camphechlor), introduced in 1945 by Hercules Powder Inc. [1], is one of the world-wide most widely applied pesticides. Table 1 lists some citations from the scientific literature dealing with the environmental impact of toxaphene.

On reading these citations it is astonishing that so little attention was paid to the "toxa-phen(e)-omenon" over a long period. In 1978, shortly after toxaphene application had reached its climax, Pollock and Kilgore stated that "toxaphene remains the most heavily used, and perhaps, the least understood insecticide presently available" [5]. Screening studies carried out at the beginning of the 1990s proved that toxaphene is a major contaminant in the Arctic and the Antarctic [6, 7], and the most uniformly distributed organochlorine compound in Scandinavia [8]. Toxaphene is ranked in position 32 on the priority list of hazardous substances issued in 1993 on the basis of the Comprehensive Environmental Response, Compensation, and Liability Act of 1980, CERCLA [9].

Table 1. Citations documenting the environmental impact of toxaphene

"Recent surveys showed that the pesticide toxaphene is a major organochlorine contaminant of aquatic life in polar regions and lakes all over the world."	[2]
"Toxaphene is among the eleven critical pollutants to be studied for sources, transport, and remedial action by the International Joint Commission of the Great Lakes Water Quality Board."	[3]
"Toxaphene, a complex mixture of polychloroterpenes and a major insecticide, is mutagenic."	[4]

Several countries have limits concerning levels of toxaphene in food. In the USA, the use of toxaphene was restricted in 1982: Distribution or sale of remaining stocks was permitted until 1986. All registered applications of toxaphene were canceled in July 1990 because of sufficient scientific evidence about its harmful health effects for humans and animals [10]. In 1971, The Federal Republic of Germany introduced maximum residue limits for toxaphene in food. However, no violations were registered before 1993. At this point of time, it was reported that 70% of all herring samples exceeded the German toxaphene tolerance level [11]. Similar data were reported from Norway [12]. Furthermore, non-harmonized residue limits between Canada and the USA caused problems with respect to fishing permits in the Great Lakes area [13].

The main reason for the detection of food levels exceeding thresholds values was not a sudden increase of toxaphene levels in the environment but the development of suitable analytical methods. While residues of PCBs and DDT have been analyzed routinely since the 1970s, the determination of toxaphene lagged behind by about two decades. In 1980, Zell and Ballschmiter prognosticated that a quantitation of toxaphene in environmental samples would only be possible with an acceptable accuracy if the compound distribution were similar or identical to that of technical toxaphene [14]. Using technical toxaphene as the standard for quantification, Wideqvist et al. concluded that the observed uncertainties make it difficult if not impossible to compare results from different laboratories [15]. Several investigations found that toxaphene levels were quite different when GC/ECD and GC/MS were employed in parallel [15–18]. Onuska claimed that changes in the pattern between the technical mixture and the environmental samples might be the main reason for the found deviations due to differences in the response factors of individual compounds [19].

A milestone in the analysis of toxaphene was the commercial availability of environmentally relevant toxaphenes as single reference standards. The first standards were presented at the "Workshop on Toxaphene" held at Burlington, Canada, in February 1993 by the group of Parlar. The quantification of toxaphene residues based on single standards often resulted in more exact results and allowed us for the first time to determine toxaphene on a routine basis in environmental samples. In Germany, official food control laboratories have to quantify three (and only these three) selected single toxaphene congeners in fish and fish products, and there is some evidence that this convention will be introduced in other countries as well. It is also strongly desirable that toxicity studies are carried out with individual toxaphene components. Such single standards enabled us also to discover several analytical peculiarities which could not be identified with the complex technical product.

The scientific interest in toxaphene has increased significantly during the past five years as can be seen from the increasing number of papers registered in the Chemical Abstracts Services. The availability of single standard components is probably one of the major reasons for this. Therefore, the following chapters focus mainly on a critical discussion of congener-specific analysis techniques.

1.1
Production, Use and Properties of Toxaphene

Toxaphene (CAS: toxaphene mixture [8001-35-2]) or Camphechlor was introduced in 1945 by Hercules Inc. as a non-systemic pesticide under the product name "Hercules 3956" [5, 20]. The crude product (see below) is a yellow to amber waxy solid that smells like turpentine [21]. The major producer remained Hercules Inc. although other producers also introduced similar products on the market (see Table 2).

Toxaphene has a broad spectrum of pesticidal activity [3]. It was mainly used on cotton, but was also sprayed extensively on soybeans, peanuts, tobacco, vegetables, turf, and many other crops [3, 20, 26]. The use on 168 agricultural commodities has been reported [20]. One special field of application of toxaphene was during flowering due to its presumed moderate toxicity to bees [20]. The product manufactured in the former German Democratic Republic was most likely called Melipax (Fig. 1) due to this property: The name is an acronym of *Apis mellifera* (zoological name of the honey bee) and pax (Latin for peace) [27].

The pesticide has also been used to control *exo*parasites on livestock (cattle, sheep, and goats) [20]. Toxaphene is highly toxic to fish. For a short period in the 1960s, it was used as a piscicide to eliminate undesired fish species in lakes prior to the release of commercially attractive fish species. Saleh reported about 817 registered products containing toxaphene in a concentration of 1% to 80% [3]. Toxaphene was marketed as emulsifiable concentrates, powders, granules, and baits [3].

In the mid-1970s, toxaphene was the most applied pesticide in the USA and other countries. The total global production was estimated at $0.45-1.33 \times 10^6$ tons [28]. The highest production rates were reported for the USA, Brazil,

Table 2. Trade names of commercial toxaphene products [22–25]

Altox; Alltox; Chem
Phene M5055; Chlor Chem T
590; Crestoxo, Delicia
Fribal; Estonox; Fasco
Terpene; Geniphene; Gy
Phene; Hercules 3956;
Huilex; Melipax; Penphene; Phenacide; Phenatox; Polychlorcamphen; Strobane T 90;
Toxakil; Toxaphene; Toxon 63

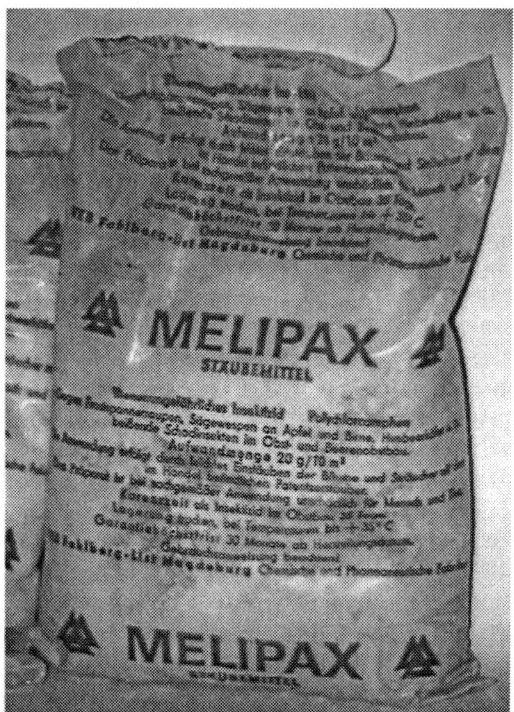

Fig. 1. Typical toxaphene product (here: the East-German Melipax). This formulation was used against insects in fruit-growing and contained about 10% toxaphene. The recommended application amount was 20 g (i.e., 2 g toxaphene) per 10 m². It was distributed in 500 g units at a price of about 1 $/unit

the former USSR and the former German Democratic Republic as well as Central America [28]. Data on the present status of usage do not reveal a clear picture. Most of the application estimates in the literature are presented without source information. The most comprehensive literature search carried out so far by Voldner and Li (1993) indicate a continued usage in Mexico and the former Soviet Union [28]. Other articles (1987, 1997) report applications also on the African continent, in Eastern Europe, and South America [3, 29].

The high water solubility (see below) was another reason for the very frequent use of toxaphene. It was found that the solubility of other organochlorines increased when they were mixed with toxaphene [10]. Consequently, toxaphene was applied in mixtures together with lindane, DDT [30], and methylparathion [31].

Most of the physical properties of toxaphene cover a relatively wide range due to the complex composition which varies between manufacturers and the use of different methodology. Technical products have a typical chlorine content of 67–69%, an average molecular mass of 414 g/mol and a specific gravity of 1.63 g/mL [3]. Toxaphene melts between 65–90°C [3, 20, 25, 32], whereas the melting points of single components are significantly higher (see Sect. 2.1.4).

Pollock and Kilgore found that the amber color of toxaphene arises from the most polar components in the technical product [33]. Parlar et al. reported that recrystallization of toxaphene yielded white crystals which mainly consisted (95%) of polychlorinated bornanes (or dihydrocamphenes) [32]. The mother liquid of the recrystallization contained about 25% unsaturated toxaphene components. A fraction enriched in unsaturated toxaphene components can be obtained in this way.

Literature values of the water solubility of technical toxaphene range between 0.4 and 3 mg/L [3, 20, 25, 32]. There is some evidence that the water solubility of higher chlorinated toxaphene components is lower: Murphy et al. estimated a water solubility of 0.55 mg/L for an average molecular mass of 350 g/mol which is lower than the 414 g/mol of toxaphene [34]. The partitioning of a chemical between air and water is described by the constant of Henry's law. Bidleman et al. estimated it for toxaphene to be 1.7×10^{-6} atm m^3 mol^{-1} [35]. This is within the range of compounds having a moderate tendency to evaporate from water. Therefore, toxaphene can be dispersed to remote areas by the atmosphere [36]. The vapor pressures reported in the literature vary significantly [3]. Based on data collected by Saleh [3] and Wauchope et al. [26], a vapor pressure of 4×10^{-6} mm Hg (20 °C) appears to be realistic for toxaphene. The soil organic carbon sorption coefficient or K_{OC} (calculated by measuring the ratio of sorbed to solution pesticide concentration after equilibrium of a pesticide in a water/soil slurry) of toxaphene is 10^5 mL/g [26].

1.2
Synthesis of Toxaphene

Toxaphene mainly consists of chlorinated components with a bicyclo[2.2.1] heptane (norbornane) skeleton and three additional methyl or methylene substituents (see Fig. 2). The synthesis of toxaphene is based on the chlorination of α-pinene or of camphene which is the isomerization product of the former [20]. First investigations on the chlorination mechanisms of camphene were made by Langlois in 1919 [37]. In 1949, Tishchenko discovered that the primary product of the chlorine addition to camphene is 2-*exo*,10-dichlorobornane [38, 39]. In the 1960s it was found that the reaction mechanism includes a Wagner-Meerwein rearrangement [38]. In a first step, a chloronium ion is added to the primary olefinic carbon of camphene. The positive charge in the six-membered ring is distributed between C3, C4, and C5. The resulting stabilization of the carbocation (usually) results in no charge on C10 (Markovnikov's rule). The tricycline primarily rearranges to bornane by a nucleophilic attack at C4 which results in 2-*exo*,10-dichlorobornane [40]. Further radical chlorination leads to the higher-chlorinated bornanes.

Higher-chlorinated camphenes are the most plausible side products. They may be formed together with other cyclic carbon skeletons if the Wagner-Meerwein rearrangement (to polychlorinated bornanes) is hindered [38, 39]. The ionic addition of chlorine to the primary olefinic carbon is always in competition with radical chlorination. Although the first mechanism is more likely, the second one may also take place. It was found that a gentle chlorination of

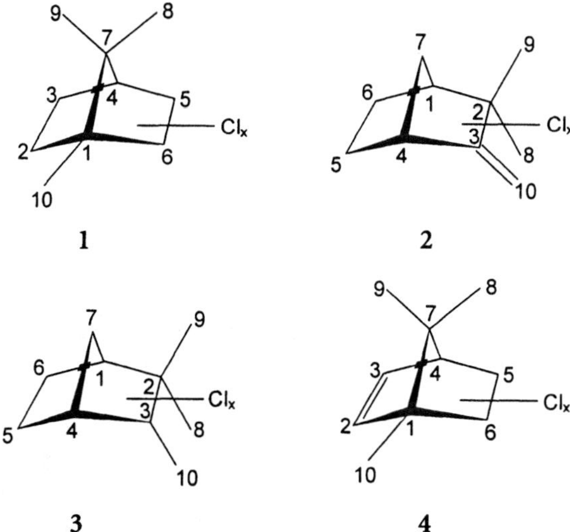

Fig. 2. Synthesis products of toxaphene (1) bornane; (2) camphene; (3) dihydrocamphene; (4) bornene

camphene in the dark resulted in 2-*exo*,10-dichlorobornane as well as 5-chlorocamphene and 10-chlorocamphene [41]. A radical chlorine substitution of these positions on camphene hinders the Wagner-Meerwein rearrangement [40] and all polychlorinated camphenes identified so far have chlorine atoms at C5 and/or C10 [41, 42].

A further class of compounds detected in toxaphene are chlorinated dihydrocamphenes [41, 43, 44]. In the 1970s, Landrum et al. isolated a heptachlorodihydrocamphene from technical toxaphene. Note that dihydrocamphene is not to be confused with camphane. The term camphane has been used earlier as a synonym for bornane, but was abolished by the IUPAC [45]. For example, Kwart described several dichlorocamphanes in 1953, which nowadays have to be called dichlorobornanes according to current nomenclature [46]. A relict of earlier assignments is the trivial name campher given to a compound having a bornane skeleton.

Furthermore, Landrum et al. and other scientists [41, 44, 47, 48] did not apply the correct IUPAC numbering of the carbon skeleton [49, 50]. Therefore, the correct name of the component isolated by Landrum et al. has to be 3-*exo*, 5-*exo*,6-*exo*,8,8,9,10- or 3-*exo*,5-*exo*,6-*exo*,8,9,9,10-heptachlorodihydrocamphene. Until now, the IUPAC rules do not distinguish between carbons C8 and C9 [51] but, due to the alphabetic order of the second letter in *endo* and *exo*, the *endo*-methyl group has been labeled as C8 and the *exo*-methyl group as C9 in this presentation (see Fig. 2). Presently, IUPAC is working on a definite nomenclature of camphene [52].

The formation of polychlorinated dihydrocamphenes is not as straightforward as it might seem. Addition of HCl to the double bond of camphene [44] is unlikely since an ionic attack at the double bond is more favorable due to the stabilization effect of the Wagner-Meerwein rearrangement resulting in the formation of bornane. Chlorination together with a maintenance of the camphene structure occurs only if the Wagner-Meerwein rearrangement is hindered [40]. Stabilization of the non-classic carbocation is more difficult in the case of a chlorine substituent at C5. Substitution at C5 allows a stabilization of the positive charge due to delocalization between C1, C3, and C4 (see Fig. 2). Nevertheless, compared to the bornyl cation, this carbocation is energetically less favored due to a higher ring strain in the neighboring four-membered ring (C1, C2, C3, and C7). Stabilization by addition of chlorine on C4 would lead to chlorinated pinanes which is less favorable than the formation of dihydrocamphenes generated by addition of chlorine at C3. Therefore, all dihydrocamphenes should have a chlorine substitution at C3, C5, and C10 [41, 43, 44].

Polychlorinated bornenes have also been discussed as constituents of toxaphene [3]. However, so far, no single polychlorinated bornene has been isolated directly from the technical mixture. The hexa- and heptachlorobornenes reported in the literature [3] have been produced in laboratory experiments by dehydrochlorination of individual polychlorinated bornanes [53]. Based on mass spectrometric detection only, it was concluded that bornenes are part of technical toxaphene [54]. However, a direct formation of bornenes by synthesis is very unlikely since a rearrangement of camphene to bornene is impossible. Elimination of HCl or Cl_2 from chlorinated bornanes is also not very plausible since high amounts of chlorine are present during toxaphene synthesis. Nevertheless, the group of Parlar recently identified traces of polychlorinated bornenes by FTIR and GC/FTIR in a technical product [55]. The synthetic conditions should have a great influence on the formation of polychlorinated bornenes. Therefore, one might find substantial concentration variations between different batches and products from various manufacturers [40]. Furthermore, colorless HPLC fractions of technical toxaphene turned amber after one year of storage in the dark, indicating the possibility of dehydrochlorination reactions of toxaphene over time [56].

Finally, the presence of tricyclic components has also been discussed [44]. The synthetic pathways discussed above do not exclude the formation of traces of polychlorinated pinanes.

1.3
Gas Chromatographic Analysis of the Composition of Toxaphene

Before the introduction of gas chromatography, the detection of toxaphene was carried out mainly by infrared spectroscopy [57, 58], measurement of total chloride content [59], or colorimetric methods [60]. Colorimetric methods had a detection limit around 100 µg/mL [60]. Coulson et al. published the first gas chromatogram of toxaphene in 1959 which clearly demonstrates the problem of the analysis of toxaphene. Due to the high number of components no signals of single compounds were observable in contrast to the case for other chloropesticides such as DDT, lindane, and chlordane [61].

In the 1970s, the GC/ECD analysis of toxaphene with packed columns resulted in about 25 to 30 separated peaks [36] which is comparable to the resolution by current HPLC techniques. Due to the low resolution achieved on packed co-lumns, it has been mentioned that some signals identified as PCBs in GC/ECD chromatograms could have been toxaphenes. A breakthrough in the gas chro-matographic determination was the introduction of wall-coated open tubular (WCOT) capillary columns resulting in an increase in the separation perfor-mance by nearly two orders of magnitude. This allowed one to separate the technical mixture into many compounds. Low detection limits were still achiev-ed due to the use of sensitive detectors such as the electron-capture detector or the mass spectrometer.

High resolution gas chromatography has been applied to quantify toxaphene compounds both in the environment and to investigate the composition of technical mixtures. Even with the available separation performance, it is not possible to achieve a sufficient separation of all compounds of the complex mixture resulting in a significant rise of the baseline in the center part of the gas chromatogram. About 100 separated signals are present in a GC/ECD chroma-togram of toxaphene. A better resolution can be obtained by applying a second separation technique such as adsorption chromatography on silica [62] or other liquid chromatography [63, 64], normal phase HPLC [65], or multidimensional GC [66–68]. An interesting method was presented by Nikiforov et al. [69]. Toxaphene was separated into 55 fractions by adsorption chromatography, and their compositions were investigated by ^{1}H-NMR spectroscopy. Using these techniques, the number of compounds in technical toxaphene compositions was estimated at ≥ 177 [62], > 200 [67], > 300 [65], 341 [69], ≥ 675 [64].

In addition to separation problems, a partial decomposition of toxaphene in the injection port and on the column have been observed [70]. The thermal sta-bility of toxaphene is relatively low. Above a temperature of 120 °C [25] to 155 °C [3], a significant degradation has been reported (see also Sect. 3.2).

1.4
Theoretically Possible Number of Congeners and Structural Restrictions

In 1993, the number of theoretically possible polychlorinated bornanes was cal-culated as 32,768 congeners (16,128 pairs of enantiomers and 512 achiral struc-tures) [71]. This surprisingly high number is obtained by the product of the possibility to chlorinate nine positions on secondary or tertiary carbons (2^9) and three primary carbon positions (C8, C9, and C10) with a variety of 4 giving

$$2^9 \times 4^3 = 32,768$$

possible compounds [71].

However, the real number in technical toxaphene is restricted due to the syn-thetic mechanisms and steric hindrance as mentioned in [51, 72]: (i) Trichloro-methyl groups do not exist, (ii) the maximum degree of chlorination of gemi-nal methyl groups is 3, (iii) the geminal methylene groups (C2/C3 and C5/C6) only carry ≤ 3 chloro substituents, and (iv) the bridgehead carbon atom C4 is generally not substituted [72]. Sterically, a chlorination of C4 is possible [73] as seen for chlordane structures which have a bridged six-membered ring with two chlorine atoms at both bridgehead carbons.

Table 3. Theoretically possible variety of toxaphene components in comparison to other classes of organochlorines [56, 66, 71]

Class of compounds	No. of pairs of enantiomers *plus* achiral compounds	No. of enantiomers *plus achiral compounds*
Heptachlorbornanes	1946	3840
Octachlorobornanes	2437	4818
Nonachlorobornanes	*2626*	*5192*
Polychlorinated bornanes	16640	32768
Polychlorinated camphenes	12288	25576
Polychlorinated dihydrocamphenes	*32768*	*65536*
Toxaphene, totally	61696	123880
Polychlorinated biphenyls (PCBs)	209	228[a]/287[b]
Polychlorinated dibenzo-*p*-dioxins (PCDDs)	75	75
Polychlorinated dibenzofurans (PCDFs)	135	135
Polychlorinated naphthalenes	75	75
Polychlorinated terphenyls	8149	n.c.[c]
Monomethylsulfonyl-PCBs	836	n.c.
Polychlorinated benzenes	14	14
1,2,3,4,5,6-hexachlorocyclohexanes (HCHs)	8	9

[a] Including the 19 stable atropisomers.
[b] Including all 78 chiral PCBs.
[c] Not calculated.

These restrictions allow a maximum degree of chlorination of only eleven, however, the presence of *dodeca*chlorobornanes in toxaphene has been reported [3]. Recently, the so-called "bridge- and-*exo*" rule was formulated which states that C8 dichloromethyl groups require an *exo*-chloro on C6, and dichloromethyl groups on C9 an *exo*-chloro on C2 [51]. Table 3 summarizes the theoretically possible numbers of isomers and congeners of substance classes present in technical toxaphene and compares them to the variety in other classes of organochlorines.

1.5
Nomenclature of Toxaphene and its Components

The complex composition of toxaphene also creates nomenclature problems. Initially, toxaphene was the trademark of the product manufactured by the Hercules Inc. However, due to the non-restricted use of the trademark, toxaphene has become a general term for this pesticide. Further frequently applied terms were "camphechlor", polychlorinated "bornanes", "camphenes", and "terpenes", as well as "chlorobornanes". The expression "toxaphene" is not the same as the trademark "Toxaphene®", since residues in the environment may also originate from other technical products (see Table 2). "Toxaphene is the reaction product of the chlorination of technical camphene" is a suitable definition of the expression used in the scientific language [27]. Owing to the problems with abbreviations as described below, toxaphene will be used in the following chapters as a synonym for the compounds of technical toxaphene.

Due to the variable composition, no general abbreviation is used in the literature in contrast to the case for, e.g., polychlorinated biphenyls which are normally assigned as PCBs. Nevertheless, the following abbreviations are found in the literature: PCCs (for polychlorinated camphenes) [74], CHBs [75], CBNs [2], and CB (for chlorinated bornanes or chlorobornanes) [76], and CTTs (compounds of *technical* toxaphene) [77]. The latter includes all substance classes of toxaphene but so far it has not found a broad application.

The abbreviations PCCs, CHB, CBNs, and CB are not satisfying since they do not include all substance classes present in, or formed from toxaphene [40, 44]. PCC has the further disadvantage that is also a common abbreviation for biological parameters, resulting in unsatisfactory literature search results.

The nomenclature for single components is also problematic. The correct numbering of the bornane backbone is given in Rules A-72/72.1 and A-74 for cyclic and bicyclic terpenes published by the International Union of Pure and Applied Chemistry (IUPAC) [45]. They can also be found in the well-known "Handbook of Chemistry and Physics" [49, 50].

Chlorinated bornanes can be formally derived from monocyclic *p*-menthane (4-isopropyl-1-methylcyclohexane) by a ring closure between C1 and C8 (see Fig. 3). Therefore, the carbon numbering of bornane follows the same rules as for *p*-menthane. The lower numbered methyl group C9 of the isopropyl function is at the same side as the methylene groups of the cyclohexane ring with the highest carbon numbers C5 and C6. The second methyl group C10 is at the side of C2/C3. It has to be observed that this is a formal convention since the isopropyl group of *p*-menthane can rotate freely, so that a differentiation between C9 and C10 is not possible.

During the past decades, the numbering of C8 and C9 did not always follow these rules. In some cases the methyl groups were not assigned at all resulting in a not exactly defined structure. In the 1980s different labeling systems of the bornane structure were more often published compared to the 1970s when the carbon numbering followed the IUPAC rules in most cases.

In the 1970s the main compounds in the technical mixtures were successfully isolated, for example, Toxicant A, Toxicant B, Toxicant C, and Toxicant A$_c$ [78–81]. Chandurkar et al. elucidated the structure of Toxicant A$_c$ by ^{1}H-NMR spectroscopy in 1978 [80]. Although the correct IUPAC numbering was applied, a wrong structure was given due to the erroneous assignment of the ^{1}H-NMR

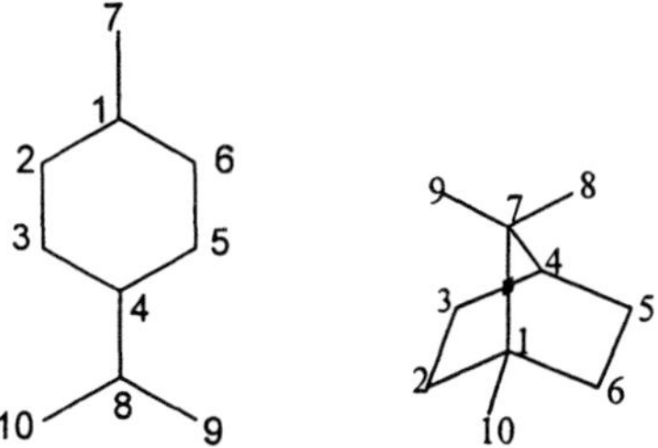

Fig. 3. Chemical structure and carbon backbone numbering of *p*-menthane *(left)* and bornane *(right)*

signals for the protons on C8 and C9, since Chandurkar et al. did not carry out nuclear Overhauser enhancement (NOE) experiments. This technique is, however, mandatory for an unequivocal assignment of the protons at the bridge (see Sect. 2.2.2). This problem was first solved in 1992, when Stern et al. elucidated the structure of a compound isolated from blubber of beluga whales and realized that its ^{1}H-NMR data were identically to those of Toxicant A_c [82]. Unfortunately, Stern et al. reversed the numbering of C8 and C9. The chemical structure and name of Toxicant A_c was given correctly for the first time nearly two decades after its isolation [83]. In 1991 Saleh published the structure and nomenclature of 20 toxaphene components [3], however, the names of at least three compounds were not identical with the assigned structures.

As mentioned before, most toxaphene components are chiral. Polychlorinated bornanes contain a plane of symmetry. The structures of enantiomers are generated by reflection of the substituents at C2, C3, and C9 with those at C6, C5, and C8, respectively (see Fig. 4). Consequently, enantiomers of chlorinated bornanes have a different nomenclature which causes additional troubles with naming [71, 83].

To overcome this problem, a set of rules was published in 1994 [83] by the IUPAC. The following presentation is brief compilation of the original text:

(i)   The carbons of bornane have to be numbered as shown in Fig. 2.
(ii) Chlorine positions should be listed for both enantiomers in increasing order.
(iii) The decision which of the enantiomers obtains the IUPAC name is based on the following rules:
    (1) First priority is given to the enantiomer with the lowest numbering of the chlorine positions. For example, 2,2,6,6,8-pentachlorobornane is selected and not 2,2,6,6,9-pentachlorobornane.
    (2) If the preceding rule does not allow a decision, the enantiomers are distinguished by the first *endo/exo* position which is different; *endo* is preferred due the earlier alphabetical order of the second letter.
        For example, 2-*endo*,6-*exo*-dichlorobornane is favored over 2-*exo*, 6-*endo*-dichlorobornane.
        However, 2-*exo*,6-*endo*,8-trichlorobornane is still preferred over 2-*endo*, 6-*exo*,9-trichlorobornane due to the first rule.

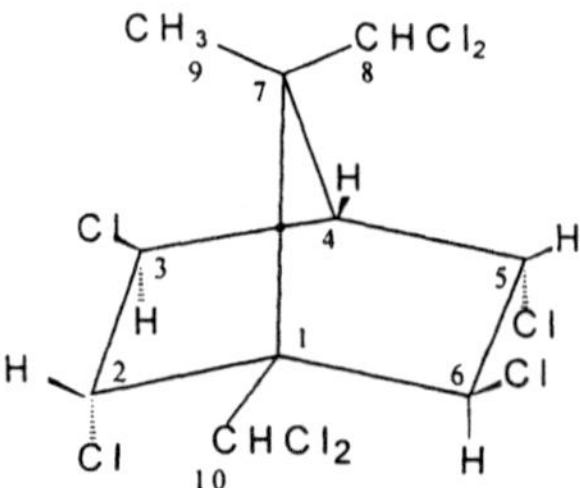

2-*endo*,3-*exo*,5-*endo*,6-*exo*,8,8,10,10-octachlorobornane (B8-1413a)

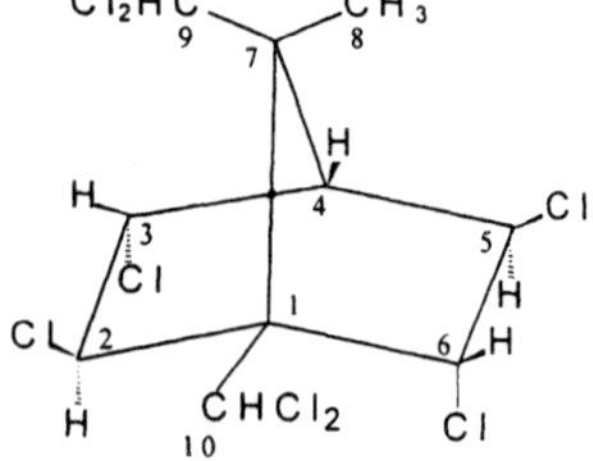

2-*exo*,3-*endo*,5-*exo*,6-*endo*,9,9,10,10-octachlorobornane (B8-1413b)

Fig. 4. Chemical structures of the enantiomers of B8-1413 (P-26)

Due to the enormous number of theoretically possible structures of polychlorinated bornanes (see Sect. 1.4), their numbering is not straightforward. When abbreviations are used for selected congeners, it has to be considered that enantiomers have different chemical names. Several systems of coding have been introduced during the last years. The numbering of the 17 toxaphene congener signals present in the gas chromatograms of beluga blubber extracts was the origin of the assignments T2 and T12 used for the two most abundant peaks [84]. Parlar numbered selected toxaphene signals from 11 to 69 on the basis of the retention time observed for the GC stationary phase 95% methyl-/5% phenylpolysiloxane [85]. So far, these "Parlar numbers" are most frequently used. The major disadvantage of this designation is that newly identified compounds, which elute in between other congeners, cannot be incorporated.

Therefore, new systems were developed allowing a complete assignment of all possible components. The first universal coding was presented by Nikiforov et al. in 1994 and published the year after [86]. 13 of the 18 substitution positions of bornane were selected and the presence of hydrogen was assigned by 0 and of chlorine by 1. The remaining five positions normally always have the same substitution (Cl or H atom). The resulting 13-digit binary code was then converted into a decimal number. This scheme was modified by Oehme and Kallenborn [87], who described the degree of chlorination on the six-membered ring (C1 – C6) by one binary code and that of the methyl groups C8, C9, and C10 by a second one. Both were converted to a four- and a three-digit decimal number which were combined with a hyphen. By converting both numbers into binary information, the position of the chlorine atoms could be deduced.

The following year, Andrews and Vetter introduced a system which allowed a conversion of the structure into a code with a computer program [88]. These so-called AV-codes (Andrews and Vetter codes) start with a number and a letter representing the C-backbone and the degree of chlorination which are connected to a one to four-digit number by a hyphen. The latter is the isomer-specific assignment. So far, this system is restricted to polychlorinated bornanes since the IUPAC nomenclature for camphenes and dihydrocamphenes is still not completely clarified (see Sect. 1.2).

In 1997, Wester et al. [89] proposed a code system which applied binary codes for secondary carbons and tertiary codes for primary carbons. The system allows one also to deduct the structure from the codes but suffers from a complex decoding system and lengthy numbers which may be hard to remember.

To facilitate comparison and conversion, a list of 39 polychlorinated bornane structure is given in Table 4 which includes the coding according to all systems discussed above.

In the following chapters, each congener is presented by its AV-code and the Parlar number in brackets, since these are currently the mostly used codes. In addition, both are the only ones which assign only one code to a pair of enantiomers which is according to IUPAC requirements.

Table 4. Systematic acronyms, chemical names, and chemical abstract numbers (CAS) of polychlorinated bornanes[a]

IUPAC name (enantiomer A)	Enantiomer B	AV-code	Parlar. no.	OK-code[b]	Nikiforov	Wester	CAS
2-exo,3-endo,6-exo,8,9,10	2-exo,5-endo,6-exo,8,9,10	B6-923	-	137-111	HxCB 3124	B[21002]-(11)	57981-29-0
2,2,5,5,9,10,10	3,3,6,6,8,10,10	B7-499	Parlar #21	99-013	HpCB-6533	B[300309-(012)	165820-13-3
2,2,5-endo,6-exo,8,9,10	2-exo,3-endo,6,6,8,9,10	B7-515	Parlar #32	195-111	HpCB-6452	B[30012]-(111)	51775-36-1
2-endo,3-exo,5-endo,6-exo,8,9,10	2-exo,3-endo,5-exo,6-endo,8,9,10	B7-1001	–	198-111	HpCB-4916	B[12012]-(111)	70649-42-2
2-endo,3-exo,6-exo,8,9,10,10	2-exo,5-exo,6-endo,8,9,10,10	B7-1059	–	134-113	HpCB-4661	B[12002]-(112)	
2-exo,3-endo,5-exo,6-exo,8,9,10	2-exo,3-exo,5-endo,6-exo,8,9,10	B7-1440	–	169-111	HpCB-3242	B[21022]-(111)	
2-exo,3-endo,5-exo,8,9,10,10	3-exo,5-endo,6-exo,8,9,10,10	B7-1450	–	41-113	HpCB-3210	B[21020]-(112)	
2-exo,3-endo,5-exo,9,9,10,10	3-exo,5-endo,6-exo,8,8,10,10	B7-1453	–	41-033	HpCB-3196	B[21020]-(022)	177344-52-0
2-exo,3-endo,6-endo,8,9,10,10	2-endo,5-endo,6-exo,8,9,10,10	B7-1462	–	265-113	HpCB-3157	B[21001]-(112)	
2-exo,3-exo,6-endo,8,9,10,10	2-endo,5-exo,6-exo,8,9,10,10	B7-1618	–	261-113	HpCB-2644	B[22001]-(112)	
2-exo,5,5,8,9,10,10	3,3,6-exo,8,9,10,10	B7-1712	–	97-113	HpCB-2445	B[20030]-(112)	
2-exo,5,5,9,9,10,10	3,3,6-exo,8,8,10,10	B7-1715	–	97-033	HpCB-2439	B[20030]-(022)	
2,2,3-exo,5-endo,6-exo,8,9,10	2-exo,3-endo,5-exo,6,6,8,9,10	B8-531	Parlar #39	199-111	OCB-6964	B[32012]-(111)	64618-67-3
2,2,5,5,8,9,10,10	3,3,6,6,8,9,10,10	B8-786	Parlar #51	99-113	OCB-6549	B[30030]-(112)	165820-18-8
2,2,5,5,9,9,10,10	3,3,6,6,8,8,10,10	B8-789	Parlar #38	99-033	OCB-6535	B[30030]-(022)	165820-15-5
2,2,5-endo,6-exo,8,8,9,10	2-exo,3-endo,6,6,8,9,9,10	B8-806	Parlar #42a	195-311	OCB-6460	B[30012]-(211)	58002-19-0
2,2,5-endo,6-exo,8,9,9,10	2-exo,3-endo,6,6,8,8,9,10	B8-809	Parlar #42b	195-131	OCB-6454	B[30012]-(121)	177695-50-0
2,2,5-endo,6-exo,8,9,10,10	2-exo,3-endo,6,6,8,9,10,10	B8-810	Parlar #49a	195-113	OCB-6453	B[30012]-(112)	n.a.
2-endo,3-exo,5-endo,6-exo,8,8,9,10	2-exo,3-endo,5-exo,6-endo,8,9,9,10	B8-1412	–	198-311	OCB-4924	B[12012]-(212)	
2-endo,3-exo,5-endo,6-exo,8,8,10,10	2-exo,3-endo,5-exo,6-endo,9,9,10,10	B8-1413	Parlar #26	198-303	OCB-4921	B[12012]-(202)	142534-71-2
2-endo,3-exo,5-endo,6-exo,8,9,10,10	2-exo,3-endo,5-exo,6-endo,8,9,10,10	B8-1414	Parlar #40	198-113	OCB-4917	B[12012]-(112)	166021-27-8
2-exo,3-endo,5-exo,8,9,9,10,10	3-exo,5-endo,6-exo,8,8,9,10,10	B8-1945	Parlar #41	41-133	OCB-3223	B[21020]-(122)	165820-16-6
2-exo,5,5,8,9,9,10,10	3,3,6-exo,8,8,9,10,10	B8-2229	Parlar #44	97-033	OCB-2455	B[20030]-(122)	165820-17-7
2,2,3-exo,5,5,8,9,10,10	3,3,5-exo,6,6,8,9,10,10	B9-715	Parlar #58	103-113	NCB-7061	B[32030]-(112)	165820-20-2
2,2,3-exo,5,5,9,9,10,10	3,3,5-exo,6,6,8,8,10,10	B9-718	–	103-033	NCB-7047	B[32030]-(022)	
2,2,3-exo,5-endo,6-exo,8,9,9,10	2-exo,3-endo,5-exo,6,6,8,8,9,10	B9-742	–	199-131	NCB-6966	B[32012]-(121)	
2,2,3-exo,5-endo,6-exo,8,9,10,10	2-exo,3-endo,5-exo,6,6,8,9,10,10	B9-743	–	199-113	NCB-6965	B[32012]-(112)	
2,2,5,5,6-exo,8,9,9,10	2-exo,3,3,6,6,8,8,9,10	B9-1011	–	227-131	NCB-6583	B[30032]-(121)	

2,2,5,5,8,9,9,10,10	3,3,6,6,8,8,9,10,10	B9-1025	Parlar #62	99-033	NCB-6551	B[30030]-(122)	154159-06-5
2,2,5-*endo*,6-*exo*,8,8,9,10,10	2-*exo*,3-*endo*,6,6,8,9,9,10,10	B9-1046	Parlar #56	195-313	NCB-6461	B[30012]-(212)	64618-71-9
2,2,5-*endo*,6-*exo*,8,9,9,10,10	2-*exo*,3-*endo*,6,6,8,8,9,10,10	B9-1049	Parlar #59	195-133	NCB-6455	B[30012]-(122)	155750-49-5
2-*endo*,3-*exo*,5-*endo*,6-*exo*, 8,8,9,10,10	2-*exo*,3-*endo*,5-*exo*,6-*endo*, 8,9,9,10,10	B9-1679	Parlar #50	198-313	NCB-4925	B[12012]-(212)	66860-80-8
2-*exo*,3,3,5-*exo*,6-*endo*,8,9,10,10	2-*endo*,3-*exo*,5,5,6-*exo*,8,9,10,10	B9-2006	–	301-113	NCB-3797	B[23021]-(112)	
2-*exo*,3,3,5-*exo*,6-*endo*,9,9,10,10	2-*endo*,3-*exo*,5,5,6-*exo*,8,8,10,10	B9-2009	–	301-033	NCB-3783	B[23021]-(022)	
2-*exo*,3-*endo*,5-*exo*,6-*exo*, 8,8,9,10,10	2-*exo*,3-*exo*,5-*endo*,6-*exo*, 8,9,9,10,10	B9-2206	Parlar #63	169-313	NCB-3261	B[21022]-(212)	182266-92-8
2,2,3-*exo*,5,5,8,9,9,10,10	3,3,5-*exo*,6,6,8,8,9,10,10	B10-831	–	103-133	DCB-7063	B[32030]-(122)	
2,2,3-*exo*,5-*endo*,6-*exo*,8,9,9,10,10	2-*exo*,3-*endo*,5-*exo*,6,6,8,8,9,10,10	B10-860	–	199-133	DCB-6967	B[32012]-(122)	
2,2,5,5,6-*exo*,8,9,9,10,10	2-*exo*,3,3,6,6,8,8,9,10,10	B10-1110	Parlar #69	227-133	DCB-6583	B[30032]-(122)	151183-19-6
2-*exo*,3,3,5-*exo*,6-*endo*,8,9,9,10,10	2-*endo*,3-*exo*,5,5,6-*exo*,8,8,9,10,10	B10-1981	–	301-133	DCB-3799	B[23021]-(122)	

[a] Only compounds with determined structure were considered (some CTTs isolated in the 1970 s were not included in the list). n. a.= not available.
[b] Corrected.

2
Production and Characterization of Single Toxaphene Standards

Unfortunately, no specific synthesis of highly chlorinated toxaphene congeners has been achieved so far. A formation of chlorinated bornenes by a Diels-Alder reaction also failed due to the electron-withdrawing chlorine substituents on the methyl group of the diene preventing the addition of the dienophile. This blocked the possibility to generate chlorinated bornanes by addition of H_2, HCl, or Cl_2. An attempt starting with less withdrawing groups instead of chlorine atoms was not fruitful either [40]. The only successfully synthesized structures reported were 5-*exo*,6-*exo*,8-trichlorocamphene and 5-*exo*,6-*exo*,8,10-tetrachlorodihydrocamphene. Both are not suited for a further reaction to more relevant, higher chlorinated toxaphenes [40]. Another pathway would be the formation of the bicyclic system from selectively chlorinated *p*-menthane (see Fig. 3, above). However, the obvious formation of the C1 → C8-bonds of *p*-menthane leading to bornane has not been achieved. In general, a systematic synthesis of polychlorinated bornanes is strongly complicated by the camphene/ bornane skeleton rearrangement. In addition, the Wagner-Meerwein rearrangement to bornane fails if the camphene has electron-withdrawing substituents [40]. The impossibility to synthesize higher chlorinated bornanes directly means that other strategies are required to obtain single components. These are:

(i) isolation of components from technical toxaphene [43, 54, 78–80, 90–92];
(ii) degradation of technical toxaphene followed by the isolation of the remaining components [93, 94];
(iii) selective chlorination of pure di- and trichlorobornanes and isolation of highly chlorinated reaction products [69, 94–97];
(iv) isolation of persistent toxaphene components from environmental samples [82, 83, 92, 98–100].

In 1973, variant (i) was introduced by Khalifa et al. and the year after three reports were published which described the isolation and structure elucidation of single components from technical toxaphene [54, 78, 91]. These initial studies focused on the isolation and structure elucidation of the major components in the technical mixture. However, it turned out that the isolated components were easily degraded in biota. Nevertheless, this approach revealed interesting results such as the presence of dihydrocamphenes in toxaphene [43, 44] and the isolation of a non-racemic component from the technical product Melipax [27].

Variant (ii) was performed by Lach et al. who treated technical toxaphene photochemically [93]. By partial degradation a toxaphene congener pattern was obtained similar to that found in fish. Therefore, it was suggested to use the resulting mixture as reference standard for the quantification in fish extracts. Hainzl et al. isolated 25 polychlorinated bornanes and camphenes from photochemically degraded toxaphene [94]. Fig. 5 compares the GC/ECD chromatogram of technical toxaphene with that of a standard containing 22 components obtained in this way. Such photochemical reactions were also part of the study of the stability and fate of single toxaphene components [101].

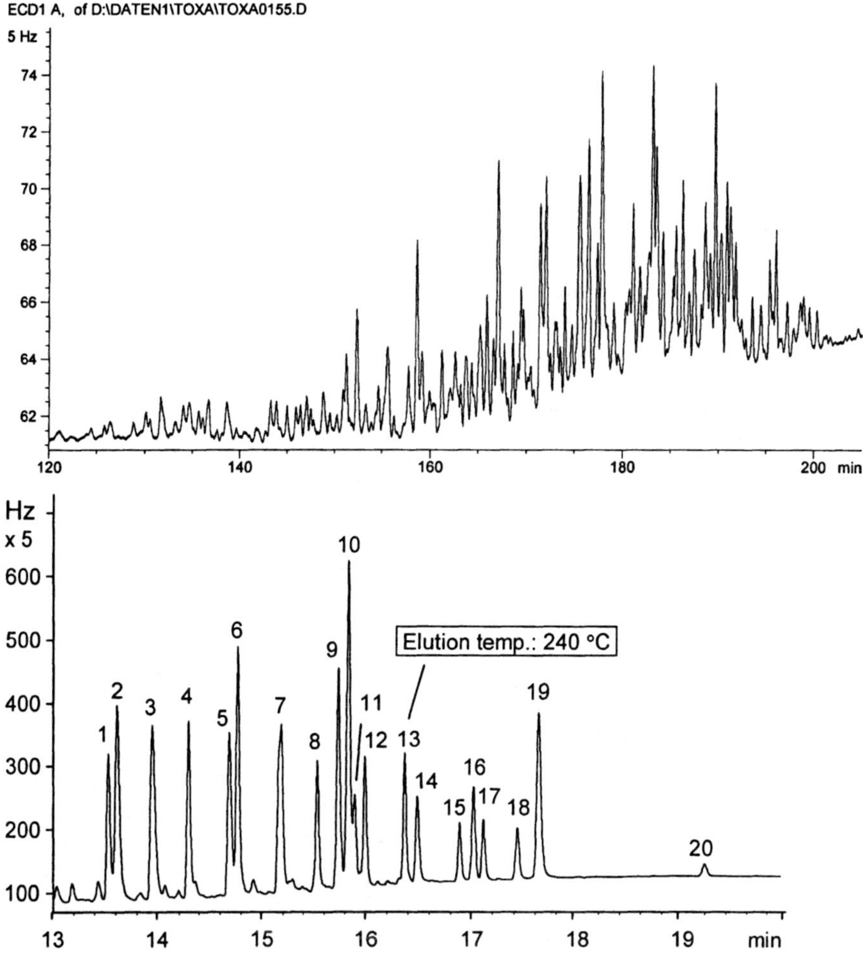

Fig. 5. GC/ECD of technical toxaphene and the 22-compound standard on DB 5 [102]; peaks: 1 = (P-11), 2 = (P-12), 3 = (P-15), 4 = B7-499 (P-21), 5 = (P-25), 6 = B8-1413 (P-26), 7 = (P-31) + B7-515 (P-32), 8 = B8-789 (P-38), 9 = B8-531 (P-39), 10 = B-1414 (P-40) + B8-1945 (P-41), 11 = B8-806/B8-809 (P-42), 12 = B8-229 (P-44), 13 = B9-1679 (P-50), 14 = B9-786 (P-51), 15 = B9-1046 (P-56), 16 = B9-715 (P-58), 17 = B0-1049(P-59), 18 = B9-1025 (P-62), 19 = B9-2206 (P-63), 20 = B10-1110 (P-69)

The aim of variant (iii) is to obtain well-defined low-chlorinated single toxaphene components. Further chlorination of the single components allows one to produce several higher chlorinated components. The idea of this approach goes back to the initial investigators of the mechanisms of camphene chlorination [38, 39, 46]. In 1976, Parlar et al. prepared seven low-chlorinated bornanes by adding small amounts of Cl_2 to camphene [103]. In a similar way, Turner et al. isolated 2-*exo*,10-dichlorobornane (B2-20). Photochlorination of B2-20 resulted in B7-515 (P-32) [53]. Parlar and co-workers applied this method to produce more than 20 pure polychlorinated bornanes [94, 95, 104]. Nikiforov et al. [69] adapted it further and obtained several polychlorinated bornanes relevant in biota starting from 2-*exo*,10,10-trichlorobornane (B3-90).

Variant (i), (ii), and (iii) do not necessarily lead to environmentally relevant components. An unequivocal identification of the structure of a toxaphene compound present in environmental samples is only possible after its isolation (variant (iv)). However, variant (iv) enables one to only isolate low quantities which do not permit performance of a complete toxicological investigation or the preparation of commercially available standard solutions.

The application of variant (ii) and (iii) was the main reason for the breakthrough concerning commercially available toxaphene standards. This was supported by the preceding isolation of two of the most abundant toxaphene components in marine mammals (B8-1413 (P-26) and B9-1679 (P-50)) [105, 106] which allowed one to prove the identity of the isolates with the synthesized products [82, 83]. Despite this success it has to be remembered that one cannot generally conclude that a synthesized compound is identical with one present in environmental samples on the basis of identical GC retention times only as has been published in several cases. Additional confirmation is necessary by applying several stationary phases, multidimensional gas chromatography, preseparation by liquid chromatography or, preferably, a complete structure elucidation according to variant (iv).

2.1
Spectroscopic Characterization of Toxaphene Components

The isolation of single substances has to be combined with a detailed characterization and structure elucidation. For this purpose, the following three techniques were mostly applied in the past: Mass spectrometry, nuclear magnetic resonance spectroscopy, and X-ray crystallographic techniques. They will be discussed in more details in the following subsections.

2.1.1
Mass Spectrometric Analysis

Structural information about toxaphene components can be obtained by gas chromatography coupled to electron ionization mass spectrometry (GC/EI-MS), positive ion chemical ionization mass spectrometry (GC/PICI-MS) or negative ion chemical ionization MS(GC/NICI-MS). Mass separation is performed by low resolution quadrupole or high resolution magnetic field instruments, by ion trap systems (GC/IT-MS), or by tandem mass spectrometry (MS/MS) offering a broad spectrum of possibilities.

2.1.1.1
GC/EI-MS

Compounds with a sufficient volatility and thermal stability can be ionized in the vapor phase by bombardment with high-energy electrons which removes electrons from occupied orbitals [107]. According to our search of the literature, the first mass spectrometric analysis of toxaphene was reported in 1966 [108], and the first GC/EI-MS of a single toxaphene was published in 1974 [62]. Nowadays, numerous references are available containing many mass spectra and detailed descriptions of the EI-MS-fragmentations of polychlorinated bornanes [109–116]. Here, only the basic approach is given for the interpretation of EI-MS. Most EI-MS of polychlorinated bornanes lack the molecular ion in contrast to toxaphene components with one or two double bonds where it is clearly visible [109].

EI-MS of toxaphene components show a strong fragmentation resulting in 20 or more fragment ions formed by elimination of Cl, HCl, Cl_2, and carbon skeleton fragments (see Fig. 6 for an example).

The type and construction of the mass spectrometer as well as instrument parameters such as ionization energy and ion source temperature have an influence on the fragmentation pattern of toxaphene components. Therefore, an EI-MS published in the literature might differ from one's own recording concerning both the presence and relative abundances of fragment ions. However, EI-MS of isomers recorded under identical conditions are highly characteristic for a structure and contain fingerprint information for the respective compo-

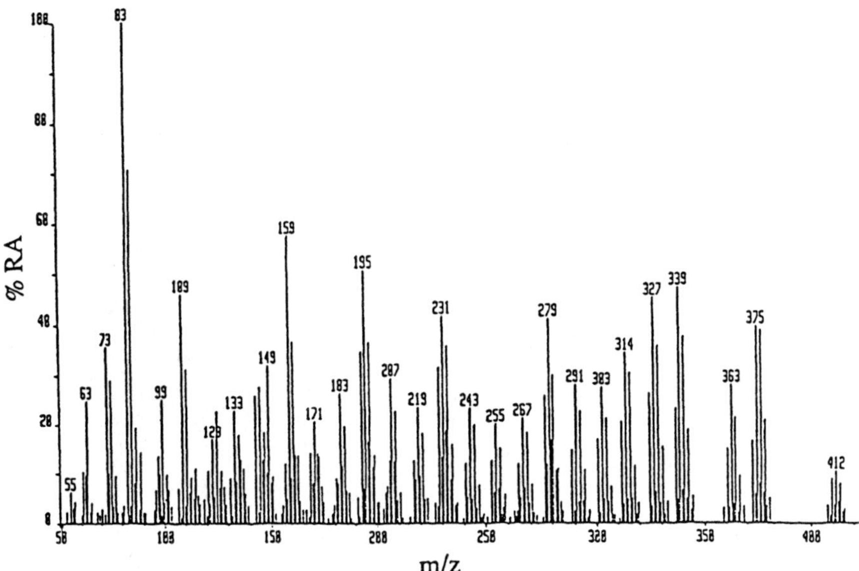

Fig. 6. GC/EI mass spectrum of B9-1679 (P-50) [110]

nent. This allows us to extract important structural information. For this purpose, fragment ions can be divided into four groups delivering information about the C_{10}-, C_9-, C_8-, and C_7-skeletons:

- C_9-fragment ions contain information about the substitution of (eliminated) primary carbons on the bornane molecule.
- C_8-fragment ions include retro-Diels-Alder (RDA)-products, which inform us about the number and distribution of chlorine atoms at the six-membered ring.
- C_7-fragment ions may be formed after the elimination of the bridge (C7 to C9) or by combination of RDA-reactions with elimination of a primary carbon unit (see above).

The latter example clearly demonstrates that a careful evaluation of the fragmentation pathway is required. Particularly, at masses below m/z 150 an unequivocal interpretation of the fragmentation mechanism is often not possible due to overlap of, e.g., fragment ions with the same formal mass but different isotope compositions ($\Delta m = 1, 2$, or 4 u). Furthermore, rearrangements may occur leading to changes of the carbon skeleton. For example, the EI-MS of B7-515 (P-32) published by Holmstead et al. [62] shows an abundant fragment ion at m/z 83 ($[CHCl_2]^+$) although the component does not contain a dichloromethyl group. Therefore, a verification of fragmentation pathways by MS/MS-experiments is often necessary [111, 117]. Furthermore, a low ion source temperature is recommended in order to avoid thermal fragmentation and to increase the abundance of high mass fragment ions as also recommended by Stemmler and Hites [118] for NICI-MS of organochlorines. An illustration of several EI-MS fragmentation pathways is given in Fig. 7.

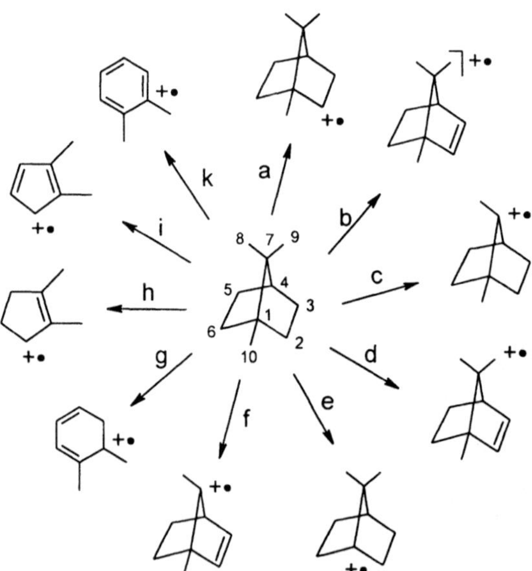

Fig. 7. Postulated EI-MS fragment ions of polychlorinated bornanes according to Parlar [119]

Kantner and Mumma found that m/z 159 is the most abundant fragment ion above 100 u common to all toxaphene components [108]. This was confirmed by Saleh who proposed a formation pathway and the dichlorotropylium cation $C_7H_5^{35}Cl_2]^+$ as structure [109]. An investigation of published GC/EI-MS revealed an interference of the m/z 159 ion cluster by a fragment ion at m/z 161 ($[C_7H^{35}_7Cl_2]^+$) leading to a deviation from the correct isotope ratio of 1.5634 between $^{35}Cl_2$ and $^{35}Cl^{37}Cl$. The abundance ratio between m/z 159 and m/z 161 decreases with the degree of chlorination (nona-, m/z 159/161 > 1; octa-, $\approx$ 1; heptachloro, < 1) [56]. In addition, m/z 193 (trichlorotropylium cation) and m/z 125 (chlorotropylium cation) are also typical fragment masses of toxaphene components [3] (see Sect. 3.3.1.2). Buser and Müller detected some components in toxaphene with an abundant fragment ion at m/z 100. They suggested a RDA formation pathway and concluded that m/z 100 is an indication for dihydrocamphenes [120].

Unlike the case for other chlorinated compounds such as PCBs, double charged ions (at m/z values < MG/2 and with only 1 u between the chlorine isotope peaks) have not been mentioned for toxaphene components. The analysis of the GC/EI-MS fragmentation patterns allows a partial structure elucidation and the confirmation of the presence of a bornane skeleton. This knowledge facilitates a further structure elucidation by ^{1}H-NMR spectroscopy [111].

2.1.1.2
GC/PICI-MS

The introduction of a moderation or reagent gas into the ion source such as methane or ammonia reduces the vacuum significantly and leads to an ionization of this gas which reacts further with analyte molecules by proton transfer or ion-molecule adduct formation. The first GC/PICI-MS results were published in 1974 [62]. PICI-MS are less complex than EI-MS. The softer ionization technique decreases the number of C-C bond cleavages [121]. Holmstead et al. found no molecular ion but instead high mass fragment ions such as $[M–Cl]^+$, $[M–Cl–HCl]^+$, and $[M–Cl–2\ HCl]^+$ when employing methane as reagent gas. The $[M–Cl]^+$ fragment is the base ion of several components [62] and is probably formed by loss of HCl from the protonated molecule [109]. Saleh and Casida described the presence of the protonated molecular ion $[M+1]^+$ of low abundance in the PICI-MS of a hexachlorobornene and a pentachlorotricyclene but not of three hepta- and hexachlorobornanes [122]. If so, this technique may be used to distinguish polychlorinated bornanes from components with another carbon backbone. Unfortunately, this technique has not been followed-up further. Due to lower sensitivity compared to NICI-MS and less achievable structure-relevant information in contrast to EI-MS, the application of PICI-MS for toxaphene has not received much attention.

2.1.1.3
GC/NICI-MS

The introduction of a moderation gas into the ion source allows the generation of electrons with only thermal energy from the primary electron beam. In addition, reagent gas anions can be formed which can react further with the introduced analytes to negative ions. Electron capture of thermal electrons by molecules with high electron affinity is the most frequently observed mechanism for generating anions. This technique has been abbreviated as negative chemical ionization (NCI), or more correctly as electron-capture negative ion (ECNI) or negative ion chemical ionization (NICI) mass spectrometry. To achieve a good yield and correspondingly low detection limit of the electron capture mechanism, a vacant orbital of low energy has to be available. This requirement is fulfilled by non-occupied π-orbitals as well as σ-orbital of non-bonding electron pairs as available in Cl substituents [107]. Therefore, polychlorinated compounds can be detected by NICI-MS with high sensitivity [107, 123]. Ribick et al. reached detection limits for polychlorinated bornanes in the ppb range using GC/NICI-MS [124]. With this technique, molecular ions can hardly ever be observed. However, often an abundant $[M–Cl]^-$ fragment is formed. Depending on the structure, further fragment ions are present such as $[M–HCl]^-$, $[M–HCl–Cl]^-$, and $[M–2\ HCl–Cl]^-$ (see Fig. 8 for examples of NICI mass spectra).

NICI mass spectra do not contain isomer-specific information. Therefore, this technique is most suitable for the determination of molecular mass information and quantification (see Sect. 3.3.1.1).

2.1.2
NMR Investigations

^{1}H-NMR is most commonly applied for structure elucidation of toxaphene components. Since its first use in 1974 [91], the structures of over thirty polychlorinated bornanes have been elucidated by ^{1}H-NMR [41, 44, 53, 69, 79, 80, 82, 83, 92, 94–97, 100, 125]. Due to the high number of resonances and couplings in the range of 2–7 ppm relative to tetramethylsilane (TMS), the use of minimum 300 MHz (7.1 Tesla) instruments and preferably 500 (11.75 Tesla) or 600 MHz (14.1 Tesla) systems is suggested to achieve a sufficient resolution. In general, the stronger the magnetic field, the easier the differentiation is between chemical shifts and couplings. Two coupling systems independent from each other are present in the bornane molecule, the six-membered ring and the carbons C8 and C9 at the bridge. Interactions between hydrogens close in space from two coupling systems can be verified by the induced nuclear Overhauser effect. It can be observed in the bornane structure since the bridge carbons, always substituted with hydrogens, are close to the six-membered ring protons in *exo*-positions. It has to be remembered that the solvent (e.g., $CDCl_3$ and C_6D_6) exerts a large influence on the chemical shift of the protons in ^{1}H-NMR spectra [83]. $CDCl_3$ has been suggested since it usually results in a better dispersion of signals at lower ppm than C_6D_6 [3]. During the past years, several rules concerning chemical shifts and coupling constants of polychlorinated bornanes have been published [3, 83, 94]. The quintessence of this information is summarized in Table 5.

Besides structure elucidation, two more applications of NMR spectroscopy deserve discussion. The first is the purity control of standard components.

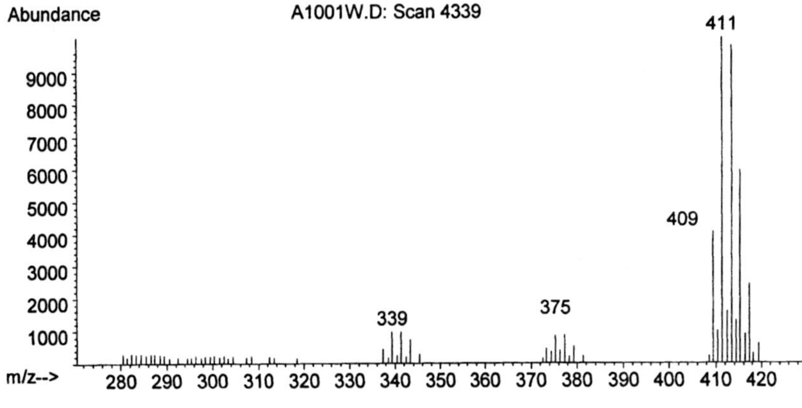

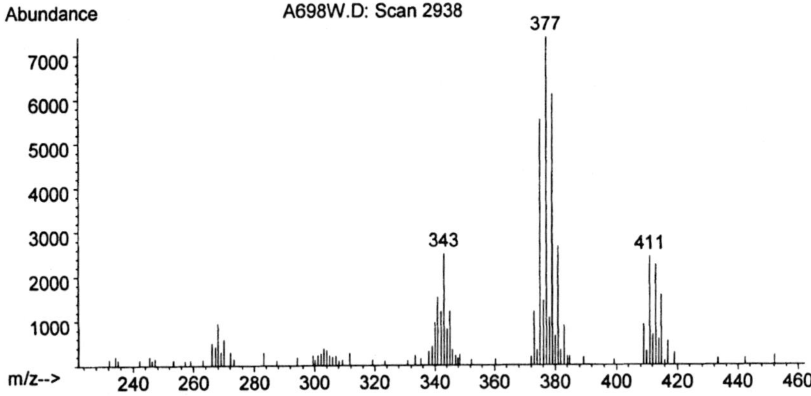

Fig. 8. GC/NICI mass spectra of (top) B9-1679 (P-50) and (bottom) B9-1025 (P-62) [56]

Table 5. Typical ^{1}H-NMR chemical shifts and coupling constants (in $CDCl_3$) of polychlorinated bornanes [3, 83, 94]

Spin coupling	Range of the coupling constant J	Proton position [Hz]	Range of the chemical shift [ppm]
Geminal on secondary carbons	> 15	$-CHCl_2$	6.0–7.2[a]
Geminal on primary carbons	11–14.5	$-CH_2Cl$	3.7–5.1
Vicinal *syn*-coupling (H2-H3)	8.5–9.5	$-CH_3$	1.8–2.0
Vicinal *anti*-coupling (H2-H3)	4.0–6.5	$-CH_2-$	2.7–3.8[b]
Bridgehead H4-H3-*exo*	2.5–5	$-C(H\text{-}endo)Cl-$	4.1–6.0
Bridgehead H4-H3-*endo*	0.3–2.1	$-C(H\text{-}exo)Cl-$	4.6–5.2
Long-range W H8-H9	0.5–3.1	H4	2.6–3.7

[a] Some protons of dichloromethyl groups of decachlorobornanes appear more down field; except for these cases, chemical shifts of $-CHCl_2$ groups on C8 or C9 appear more down field than those on C10.

[b] In the most cases Δppm < 0.5.

Turner et al. used this technique to prove that the unresolved peak in the gas chromatograms of Toxicant A consisted of the two components B8-806 (P-42a) and B8-809 (P-42b) [79]. The second approach is quantitative NMR. This technique can be used to calculate the amount of a component in a solution since the integrals of proton resonances are in the first order independent of the component structure and proton position. Nikiforov et al. studied the composition of toxaphene in this way. Toxaphene components with geminal chlorine atoms on primary carbons have proton signals in the range of 6–7 ppm. Monitoring this ppm range allows the determination of the number of components in a mixture [69].

2.1.3
X-Ray Analysis

In addition to structure elucidation by NMR, X-ray crystallographic investigations have been performed to elucidate the structures of several toxaphene congeners in the solid state. Information about symmetry, space group, bond lengths, bond angles, and preferred orientation of the substituents on primary carbons in the solid state can be obtained [47, 126, 127]. It was found that B7-515 (P-32) crystallizes in orthorhombic form, B8-1413 (P-26) and B9-1025 (P-62) triclinic, and B9-1679 (P-50) monoclinic [126, 127].

It has been suggested that an assignment of the orientation of the chlorine atoms at the primary carbons is necessary for an exact description of the structure [94]. However, other studies claim that this is not necessary due to a lack of stable atropisomers [51, 128]. According to this, the structure of polychlorinated bornanes can be described unequivocally by ^{1}H-NMR investigations [51]. However, the orientation of the chlorine substituents may influence the chemical stability of toxaphene congeners [73] and, therefore, this knowledge could help to interpret the chemical behavior of toxaphene components [51]. Although such information can be obtained by molecular modeling as well [83, 125], more X-ray data of toxaphene components are of high interest.

2.1.4
Miscellaneous Techniques

In 1962, Clark suggested the application of infrared spectroscopy to quantify toxaphene [58]. A characteristic strong band at 1280–1308 cm^{-1} corresponding to C-Cl single bonds was used [58]. This technique was soon substituted by more suitable gas chromatographic detection methods. Shafer et al. recommended gas chromatography coupled with Fourier-transform infrared spectroscopy (GC/FTIR) as a method providing information complementary to GC/MS for the analysis of toxaphene and other pesticides [129]. In 1991, Parlar mentioned that GC/FTIR is able to distinguish between bornenes and camphenes [119]. With this technique he found evidence that toxaphene contains rather camphenes than bornenes [119]. The range between 1652 cm^{-1} and 1595 cm^{-1} is typical for C=C valence vibrations [130]. Suitable wavenumbers of chlorinated bornenes are 1608 and 1597 cm^{-1}, while 1633 and 1616 cm^{-1} were suggested for chlorinated camphenes, respectively [130].

Finally, melting points are suitable to characterize toxaphene components as well as to provide additional data on the purity of the respective component. Melting points of B7-515 (P-32) and B8-806/B8-809 (P-42) of 166–167°C and 134–136°C, respectively, were significantly higher than that of technical toxaphene (see above).

3
Analytical Methods

3.1
Sample Clean-Up Procedures

The current determination of toxaphene is based on quantification by GC/ECD or GC/MS [3]. Both techniques exhibit excellent sensitivity but show different selectivity. While GC/ECD is only highly selective for compounds with a high electron affinity (among them chlorinated compounds), the higher selectivity of the mass spectrometer allows one to identify toxaphene components with different degrees of chlorination and to distinguish them from other compounds when using the selected ion monitoring mode. These differences in selectivity require different sample preparation techniques for ECD and MS.

Since toxaphene is resistant to strong acids, most sample clean-up techniques using acids and originally developed for organochlorines are also suitable for toxaphene. However, the use of strong alkaline conditions (e.g., alkaline saponification) should be avoided since toxaphene is readily degraded and dechlorinated at high pH values resulting in a toxaphene pattern with mainly earlier eluting compounds [131]. The comparably high water solubility of toxaphene has also to be considered (see Sect. 1.1).

If ECD is selected, PCBs and other components which elute in the same retention range should be eliminated during sample clean-up. Several methods based on adsorption chromatography, photodehalogenation, and nitration have been described during the last decades. Silica is the most commonly used adsorption material to separate toxaphene from PCBs. The first methods using deactivated silica were published in the 1970s [132, 133]. On silica, PCBs can be eluted prior to toxaphene components with n-hexane. The fraction containing toxaphene is then collected with a more polar solvent to achieve faster elution. Mixtures of n-hexane with diethyl ether, toluene, ethyl acetate, or dichloromethane have been applied for this purpose. Using technical toxaphene as standard, it was reported that 1 g silica deactivated with 1.5% water is sufficient for a complete PCB/toxaphene group separation [134, 135]. However, Alder et al. could not confirm this with single standards [17]. The most abundant octachlorobornane B8-1413 (P-26) in fish tissue was found almost completely in the PCB fraction. Using single standards, Krock et al. observed that a separation on 4.0 g of activated silica was sufficient for a complete toxaphene separation [136]. However, when high amounts of p,p'-DDE and other organochlorines were present in real samples, the elution of toxaphene was accelerated, and the amount of silica had to be increased to 8.0 g for a secure PCB/toxaphene group separation [136]. B8-1413 (P-26) as an early eluting and B7-515 (P-32) as late eluting toxaphene congeners were suggested as indicators for the elution range of toxaphene on silica columns [136]. Zell and Ballschmiter used 13 g slightly deactivated florisil (1.5% water) and found also some toxaphene components in the PCB fraction [14]. Separations of toxaphene from PCBs were also performed with alumina using a microcolumn with 2 g alumina. However, 5% of toxaphene was found in the PCB fraction [137]. These observations clearly demonstrate that the method validation of a PCB/toxaphene group separation should be carried out with single toxaphene standards. A control with technical toxaphene as reference may result in unobserved losses of single compounds.

HPLC with silica columns has been used to separate PCBs from toxaphene [138–140] as well as for isolating toxaphene components [65, 92, 94]. Marth et al. found that HPLC silica phases with a large surface (here: 650 m^2/g) were suitable for a PCB/toxaphene group separation but not those with a small one (here: 170 m^2/g) [138]. The elution order of toxaphene standards by RP-HPLC on octadecylsilane phases was also determined. However, as for amino-, cyano-, and nitro-modified HPLC phases [138], no complete separation of PCBs from toxaphene could be achieved [94, 141]. Brumley et al. suggested the use of high performance gel-permeation chromatography for the PCB/toxaphene group separation [142].

Alternative methods others than liquid chromatography have also been suggested. Parlar et al. used photodechlorination reactions to eliminate interfering compounds [143]. Irradiation at 254 nm with low-pressure-mercury lamps mineralized quantitatively PCBs and further organochlorines within 30 min, while toxaphene was only partly degraded [8].

Electrophilic nitration of aromatic organochlorines with a 1:1 mixture of concentrated HNO_3 and H_2SO_4 was recommended to eliminate interfering compounds [144]. Both methods have the disadvantage that PCBs are destroyed so that they have to be quantified before. Nitration was also suggested as an additional clean-up step after pre-separation of PCBs; DDT and its metabolites are eliminated from the toxaphene fraction [137].

3.2
Optimization of HRGC

A careful optimization of the complete gas chromatographic system consisting of "injector-column-detector" is an important requirement for the analysis of complex mixtures such as toxaphene.

The injection should transfer the sample quantitatively or at least reproducibly to the capillary column. On-column and splitless injectors are suitable for ultratrace analysis. The capillary column should enable a complete separation of all components by selection of a suitable stationary phase and optimization of other parameters such as column length, diameter, film thickness, carrier gas flow and temperature program. The detector should have a sufficient sensitivity and selectivity. Both requirements are fulfilled by NICI-MS applying the selected ion monitoring mode. However, the ECD is not able to distinguish between toxaphene components and other polychlorinated components with high electron affinity such as PCBs, chlordane, and DDT and their metabolites.

The formation of unsaturated toxaphene components by HCl-elimination from the thermally rather unstable toxaphene was suggested as a reason for the rising baseline during GC analysis [70]. This reaction may happen either in the injector or on the column.

In the following, the optimization of the different parts of the GC system is discussed starting with the most important part, the separation column and conditions.

3.2.1
Selection of HRGC Stationary Phases

The search of the literature carried out for this survey, allows us to estimate that approximately 90% of all toxaphene separation were carried out on non-polar stationary phases similar to 95% methyl-/5% phenylpolysiloxane and commercially available under the trade names DB 5, CP-Sil 8, HP-5, Ultra 2, Rtx 8, SE 52, etc. Two reports in the literature suggest the use of non-polar only. Alder et al. mentioned the decomposition of B8-1413 (P-26) and B9-1025 (P-62) on the polar DX-4 phase [145]. Baycan-Keller and Oehme tested four different stationary phases (capillary length 12 to 30 m) and found a remarkable decomposition of labile toxaphene compounds on 90% dicyanopropyl-/10% phenylcyanopropylpolysiloxane (Rtx-2330) [146]. Therefore, the exclusive use of non-polar stationary phases was recommended for the quantification of toxaphene as well as a second column for confirmation of the results. A very non-polar stationary phase (CP-Sil 2) was suggested for the congener-specific separation of toxaphene [147, 148].

Non-polar stationary phases were also found to show a structure-dependent elution order of toxaphene congeners [147]. For example, the substitution pattern at the bridge can be deducted. 8,8-substituted polychlorinated bornanes eluted much earlier than 8,9-substituted ones [147, 149, 150]. The less polar the stationary phase, the more this rule comes to fruition. However, such structure-dependent selectivities do not necessarily improve the separation performance. For 8,8,9- or 8,9,9-substituted chlorobornanes having otherwise the same substitution pattern (e.g., B8-806 (P-42a) and B8-809 (P-42b)), the difference in retention decreases with decreasing polarity of the stationary phase. These components can only be separated on more polar stationary columns. Resolution of B8-806 (P-42a) and B8-809 (P-42b) was obtained on Optima 17 and Optima 3 stationary phases [151, 152]. When using ECD it has to ensured in addition that no other organochlorine compounds interfere with the toxaphene compounds.

Due to the many toxaphene components present in environmental samples, a sufficient GC separation has to be achieved. For example, a suitable stationary GC has to be able to separate B8-1414 (P-40) from B8-1945 (P-41) which was only realized for some studies while others reported no baseline separation or coelution. 50 or 60 m columns are recommended in this connection. Furthermore, the selection of a proper stationary phase has to be followed up by a proper selection of carrier gas type and flow, and the temperature program. For a better comparability of literature data, it is highly recommended that relative retention times (RRT) or retention indices (RI) be reported. RI have been listed relative to B9-1679 (P-50), Aldrin [76, 94], the sum of B8-1413 (P-26) and B9-1679 (P-50) [147], and on the basis of the M series standards [149] introduced by Manninen et al. [153]. So far, RI and RRT have only been published scarcely.

3.2.2
Injection Systems

Vaporizing injectors in the split/splitless mode have to be used with caution for toxaphene components since some toxaphene congeners may already be degraded at temperatures as low as 160 °C (see above). The use of cold on-column injection avoids this problems and ensures a quantitative transfer of the sample onto the column. However, this technique is less suitable for automation and requires a better sample clean-up than for split/splitless injection where non-volatile matrix residues are deposited on the glass insert and not inside the column. However, such matrix layers may act as active surfaces catalyzing the decomposition of labile components. Furthermore, as demonstrated with single standards, even a clean and deactivated glass liner is able to degrade the most labile toxaphene components [102]. Evaluating an intercalibration exercise, Alder et al. recommended injector temperatures of below 240 °C to minimize decomposition of labile compounds such as B9-1025 (P-62) [18]. However, many studies have been carried out at injector temperatures of 270 to 285 °C (see, e.g., [154–156]).

More recent studies preferred injector temperatures of below 200 °C to avoid decomposition in the splitless injector [146]. However, low injector temperatures increase the risk of an incomplete transfer of the sample onto the GC column. To overcome this problem, a pressure-pulsed injection has been suggested [125, 157–159]. Detailed investigations with single toxaphene components have shown that the response factors of less volatile components can be increased fourfold by this technique using an injector temperature of 230 °C (see Fig. 9 for an example) [158, 159].

Figure 9 also demonstrates clearly that the detector responses of single toxaphene components are different [95]. However, due to thermodegradation they might also be influenced by the injection method [76, 94].

An alternative to the pressure-pulse mode is the use of a temperature-programmable injector (abbreviated as PTV, CIS/KAS) which was initially applied for the injection of large sample volumes [160] or for injection of toxaphene at a low temperature followed by a very fast temperature ramp. In this way toxaphene components can be volatized with a minimum of thermal stress [161]. Although no detailed study has been conducted with toxaphene, this technique should result in less degradation than a splitless injection at a constant high temperature.

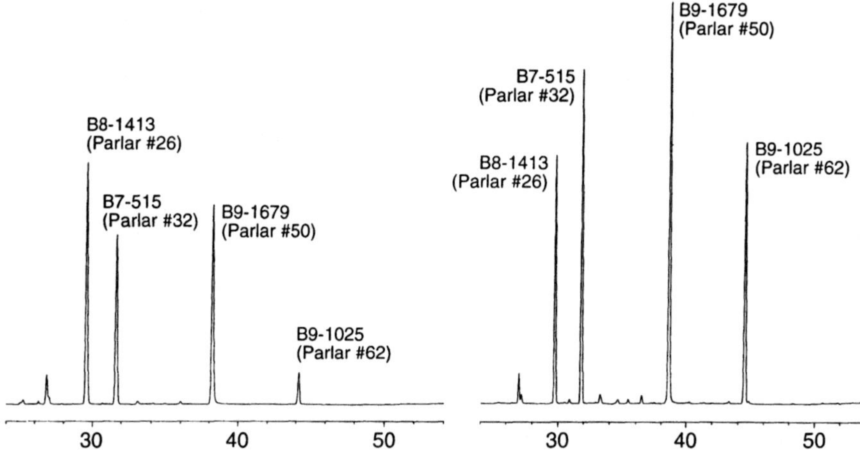

Fig. 9. GC/ECD chromatogram of B8-1413 (P-26), B7-515 (P-32), B9-1679 (P-50), and B9-1025 (P-62) [56]. Constant flow injection at 230 °C *(left)* and pressure pulse injection *(right)*

Another way of sample introduction into the mass spectrometer was performed by Hunt et al. and Onuska et al. who used a direct probe inlet [162, 163]. It was claimed that this technique combined with NICI-MS or GC/HREI-MS detection is more precise and ten times faster than GC/MS [19]. However, this technique does not provide structure-specific information.

3.3
Detection Techniques

3.3.1
Quantitation of Toxaphene by GC/MS

3.3.1.1
NICI-MS

Besides GC/ECD, GC/NICI-MS is the most commonly applied technique for the quantitation of toxaphene. The most important methods were published by Jansson and Wideqvist [63], and Swackhamer et al. [164]. Both groups suggested the use of two isotope signals from the $[M-Cl]^-$ ion cluster for hepta- to dodecachloro compounds and two from the molecular isotope cluster for hexachloro components. However, Fingerling et al. found that only one of two recently isolated hexachlorobornanes formed an intensive molecular ion while the $[M-Cl]^-$ fragment ion was abundant for both [165]. Therefore, hexa- to decachlorinated toxaphene components should be monitored by two ions selected from the $[M-Cl]^-$ isotope cluster [16, 142].

One problem of NICI-MS quantitation of toxaphene can be caused by the presence of traces of oxygen in the ion source [63]. The $[M-Cl+O]^-$ ion and the corresponding fragments formed from PCBs interfere with the SIM-masses for toxaphene selected as described before. An interference by hydroxylated PCBs can be easily discovered, since the isotope ratio between the two selected ions is different from that of toxaphene components [166]. However, such interferences prevent a proper toxaphene quantitation.

Another difficulty is caused by the strongly different NICI response factors for toxaphene isomers which, in addition, deviate from those obtained for ECD. Depending on the type of instrument, response factors of single toxaphene components may vary by more than one order of magnitude [76]. Therefore, a correct quantitation of unknown toxaphene congeners is not possible. B9-1025 (P-62) is an example for a very low NICI response (underestimation of quantity) [167] and B8-1412 for a very high one (overestimation of concentration) [168]. The low NICI response factor of B9-1025 (P-62) is caused by a partial degradation in the hot injector (see above). Furthermore, B9-1025 (P-62) shows a higher degree of fragmentation resulting in a higher abundance of the $[M-HCl-Cl]^-$ and sometimes even the $[M-2\ HCl-Cl]^-$ fragments compared to the $[M-Cl]^-$ fragment. This special behavior was also the reason why several participants of an intercalibration exercise organized by Health Canada in 1995, were not able to detect this abundant congener in fish [13, 169]. Therefore, it has been suggested to quantify B9-1025 (P-62) via the $[M-HCl-Cl]^-$ fragment ion [148, 166]. $[M-HCl-Cl]^-$ fragments of Cl_X bornanes have the same mass as the $[M-Cl]^-$ ions of Cl_{X-1} bornenes and camphenes. Therefore, the presence of $[M-71]^-$ may mimic the presence of polychlorinated bornenes and camphenes in technical toxaphene. An elimination of HCl is present in the NICI mass spectra of selected polychlorinated bornanes such as B8-1413 (P-26) and a corresponding monitoring of mass m/z 374 has been suggested to identify this persistent octachlorbornane [168].

For GC/NICI-MS mostly quadrupole instruments are applied with unity mass resolution. In recent years, several groups have started to employ high-resolution magnet instruments, which can give lower detection limits [170].

At present, the lack of isotope-labeled standards does not allow application of the isotope dilution for quantification which would result in a better reliability. As a substitute, several compounds without a toxaphene structure have been used as internal standards, among them a synthetic chlordane [171]. However, their suitability has not been tested in detail.

3.3.1.2
EI-MS

Due to extensive fragmentation, no ion of dominant abundance is present in the EI mass spectra of toxaphene congeners. Furthermore, this results in a lower sensitivity of low resolution MS (LRMS) compared to NICI-MS and ECD. Therefore, EI-LRMS has been used only for a few applications.

The dichlorotropylium cation $C_7H_5Cl_2$ at m/z 159 was suggested for quantitation [109]. The fragment m/z 125 is also formed by most toxaphene components. However, it is also produced by PCBs and is therefore less suitable [161]. Although m/z 159 is relatively specific for toxaphene, several other components such as endosulfan also generate this fragment mass, which may lead to interferences and an overestimation of toxaphene levels [166]. The selection of a suitable ion for confirmation is problematic. The isotope signal of $C_7H_5Cl_2$ at m/z 161 is disturbed since also $C_7H_7Cl_2$ fragments are formed. The $^{35}Cl_2$ isotope ion of $C_7H_7Cl_2$ and the $^{35}Cl^{37}Cl$ isotope signal of $C_7H_5Cl_2$ have nearly the same mass ($\Delta m = 0.02059$) and can only be resolved by high resolution mass spectrometry (HRMS). Several studies used EI-HRMS for toxaphene quantification [142, 172, 173]. Lau et al. measured both m/z 158.9769 and 160.9739 and found a rather good agreement of the isotope ratio for the Parlar 22 component standard [166].

3.3.1.3
Miscellaneous MS Techniques

In the past years, several studies have applied MS/MS techniques using quadrupoles [111, 117, 120] or ion traps [161]. In this way Buser et al. [111] verified fragmentation pathways identified by normal EI-MS [112]. For quantification, two common fragment ions in the mass spectra of toxaphene components (of interest) are monitored, and the one (daughter) must be produced from the other (parent). This gives a very high selectivity. Chan described two common parent $\rightarrow$ daughter ion fragmentations for toxaphene components such as m/z 159 $\rightarrow$ m/z 125 and m/z 125 $\rightarrow$ m/z 89, whereas the first one was less selective for toxaphene [161]. Santos et al. studied parent $\rightarrow$ daughter transitions from $[M-Cl]^-$ and $[M-HCl]^-$ to $[M-HCl-Cl]^-$ by high resolution MS/MS for each degree of chlorination [117].

3.3.2
GC/ECD

GC/ECD is the most applied method for the determination of toxaphene congeners in fish and other matrices. Due to the limited selectivity of the ECD, co-elutions of toxaphene components with other polychlorinated compounds such as PCBs cannot be identified. However, since B8-1413 (P-26), B9-1679 (P-50), and B9-1025 (P-62) are very abundant in fish, a significant interference (>30%) by other toxaphene congeners can be excluded. The risk of interferences can be reduced by analyzing the same sample on two different stationary phases. Since identical disturbances on both systems are not very likely, comparable concentrations increases the confidence in the obtained results. De Boer et al. and Shoelb et al. used multidimensional gas chromatography with ECD as an alternative to avoid interferences [67, 68].

3.4
Other Techniques for Quantitation of Toxaphene

Techniques without a preceding separation step such as direct inlet probe MS [163] or ^{1}H-NMR [69] allow one to estimate the sum concentration of toxaphene. In connection with GC/ECD, Walter and Ballschmiter used the so-called triangle method for this purpose. It estimates the area under the raised baseline caused by not resolved toxaphene peaks [174]. Although this method has to be considered as semiquantitative, the results for samples showing a high degree of metabolization such as cod liver oil, were surprisingly close to those obtained by an exact quantitation (see Sect. 4.1).

4
Environmental Fate and Toxicology of Single Toxaphene Components

The question about the fate of toxaphene in the environment cannot be answered in a simple way. In the environment, only some toxaphene components remain unmetabolized and are accumulated in biota. This leads to a simpler toxaphene residue pattern compared to the technical mixture. In 1979, Saleh et al. studied the metabolism of B7-515 (P-32) and technical toxaphene with seven test animals. Both were metabolized the least in chicken and the most in monkeys [175]. Furthermore, the major toxaphene congeners present in the technical mixture are not the same as those found in environmental samples due to selective bioaccumulation/metabolization and abiotic degradation mechanisms (see Table 6).

Table 6. Major toxaphene components in different matrices [73]

Technical mixtures [79, 78]	Marine mammals [82, 98]	Lake sediment [125]
B7-515 (P-32)	B8-1413 (P-26)	B6-923 (–)
B8-806/B8-809 (P-42)	B9-1679 (P-50)	B7-1001 (–)

Lahaniatis et al. found that a thermal degradation of toxaphene at 400 °C to 800 °C leads to the formation of aromatic compounds, among them tri- to hexachlorobenzenes, low-chlorinated naphthalenes, biphenyls, and even dibenzofurans [176]. The major processes of degradation seem to be dechlorination and dehydrochlorination [54]. However, Chandurkar and Matsumura reported that at least some hydroxylated toxaphene metabolites are formed [81]. So far, only very few oxidation products of toxaphene have been identified [177, 178, 179] which seem to be rather unstable [177].

The bulk of the toxaphene components detected in environmental samples are polychlorinated bornanes and a few camphenes or bornenes. Since reductive dechlorination of polychlorinated bornanes leads to bornanes with one less chlorine atom, at present it is not possible to distinguish unequivocally between components originating from technical toxaphene and dechlorinated metabolites. This question can only be solved with fate studies of single components in the laboratory [122, 165].

One major problem with the comparability of toxaphene levels reported from different studies is that different numbers of single toxaphene congeners or technical toxaphene standards have been used as reference for quantitation. In addition, Carlin Jr. et al. found significant differences in the composition of technical toxaphene standards obtained from different suppliers [180]. This causes additional quantification problems. The quantification techniques used at the moment, can be divided into five main approaches:

(i)   The use of a three toxaphene congener standard (BgVV 3) as suggested by Alder and Vieth [17] and commercially available from Dr. Ehrenstorfer and Promochem. BgVV 3 contains B8-1413 (P-26), B9-1679 (P-50), and B9-1025 (P-62).

(ii) Application of a six-component standard available from Dr. Ehrenstorfer.

(iii) The use of the 22-component standard available from Dr. Ehrenstorfer.

(iv) Employment of isolated, not commercially available components such as B6-923 [125], B7-1001 [125], B7-1453 [92], and B8-1412 [100].

(v)   Semiquantitative determination of further, still unknown components, for example TC1-TC10 [120], 7-1, 8-2, 8-6, 8-7 [148], AES 1 [68], and others.

Frequently quantified reference components are summarized in Table 7 and compared with their presence in biota.

Most toxaphene levels presented in publications were based on the quantification of a few selected toxaphenes. Therefore, no reported result for a certain toxaphene congener does not necessarily mean that this component was not present in the respective samples.

Depending on the number of toxaphene congeners quantified in a certain matrix, the sum of the levels of single congeners correspond to 10% to 85% of the total toxaphene burden. Therefore, data about toxaphene levels reported in the literature are hardly comparable, and the present state of toxaphene quantification needs further improvement (see next section).

4.1
Intercalibrations and Toxaphene Levels in Reference Materials

As early as 1982, Ribick et al. noted that methods for the identification and quantification of toxaphene residues in environmental samples are poorly documented and vary greatly among laboratories participating in intercalibrations [124]. This problem seems to be less pronounced when only a few toxaphene standards are determined. Following the suggestions of Alder and Vieth [17], the most prominent polychlorinated bornanes B8-1413 (P-26), B9-1025 (P-62), and B9-1679 (P-50) were quantified at two intercalibration exercises, one organized by Health Canada and one by the German BgVV. They were also quantified in standard reference material with certified levels of PCB congeners and a number of pesticides but not toxaphene (see Table 8).

The reported levels in SRM 1588 deviated surprisingly much, and the intra- and interlaboratory deviations were significantly higher than those acceptable for a certification. Since the applied analytical and quantification methods were very different, a final conclusion cannot be drawn concerning the reasons for these deviations. Some of the results were reported shortly after the introduction of single toxaphene standards. In the meantime, more experience has been gained concerning parameter optimization for the quantification of single toxaphenes, and it is reasonable to assume that the quality standard is higher now allowing a better precision.

Table 7. Chlorobornanes in different standard mixtures and environmental samples

AV-code	Parlar no.	22-comp. std.	6-comp. std.	5-comp. std.	Bg VV 3	fish	seal, wahle, dolphin	human milk	sediment	toxaphene
B6-923						–	–	–	H	l
B7-499	P-21	×				–	–	–	–	
B7-515	P-32	×		×		–	–	–	l	H
B7-1001						–	–	–	H	l
B7-1059										
B7-1440										
B7-1450										
B7-1453						L	L			L
B7-1462										
B7-1618										
B7-1712										
B8-531	P-39	×								
B8-786	P-51	×					m			
B8-789	P-38	×								
B8-806	P-42a	×								H
B8-809	P-42b	×					m			H
B8-810	P-49						m			
B8-1412	-					M	M			
B8-1413	P-26	×	×	×	×	H	H	H		
B8-1414	P-40	×	×			L-M	L			
B8-1945	P-41	×	×			L-M	M			
B8-2229	P-44	×	×			M	M	H		l
B9-715	P-58	×								
B9-718	–									
B9-742	–									
B9-743	–									
B9-1011	–									
B9-1025	P-62	×	×	×	×	H	L			
B9-1046	P-56	×								
B9-1049	P-59	×								
B9-1679	P-50	×	×	×	×	H	H	H		
B9-2006	–									
B9-2009	–									
B9-2206	P-63	×								
B10-831	–									
B10-860	–									
B10-1110	P-69	×		×						
B10-1981	–									

L = low, M = medium, H = high abundant. Small letters: result is based on few reports or component is only detected in selected samples.

Table 8. Toxaphene levels reported for the reference material SRM 1588 (cod liver oil). All concentrations in µg/kg

Research group	Alder and Vieth	Chan et al.	Andrews et al.	Klobes et al.	Walter and Ballschmiter	Wade et al.	Muir and Grift	Fowler et al.
Reference	[17]	[161]	[169]	[181]	[174]	[173]	[182]	[183]
Detector type	ECD/NICI-MS	Ion trap	NICI[a]/ECD[a]	ECD	ECD	HREI-MS	ECD	NICI
Injector temp. [°C]	230	135 to 300	various	230 and 250	250	280	unknown	270
GC phase (length m)	DB 5/DB 5 (60)	DB 5 (30)	various	CP-Sil 2 (50)/ CP-Sil 8/20 % C18 (50)[b]	DB 5 (60)	DB 5 (60)	DB 5 (60)	DB 5 (60)
B8-1413 (P-26)	298/375	220	260/244	244/251	n.a.	n.a.	n.a.	
B7-515 (P-32)	n.a./n.a.	< 40	11/31	n.a.	n.a.	n.a.	n.a.	
B9-1679 (P-50)	434/448	280	388/360	372/414	n.a.	n.a.	n.a.	
B9-1025 (P-62)	220/264	140	179/187	281/interfered	n.a.	n.a.	n.a.	
Σ Toxaphene	n.a./n.a.	3430	4100/4270	n.a.	2230	4810	2510	5410
Percentage of single congeners quantified		18.7	30.6/25.7					

[a] Median values as suggested by Chan et al. [161] which were significantly lower and more reliable than mean values.
[b] CP-Sil 8/20 % C18 contains 80 % CP-Sil 8 (= DB 5) and 20 % methyl octadecyl polysiloxane; n.a. = not analyzed.

4.2
Congener-Specific Levels of Toxaphene in Environmental Samples

Media exposed to toxaphene contamination in the environment can be differentiated according to their metabolization properties as follows:

(i)   Marine mammals and marine biota (fish and birds) as well as other mammals at high trophic levels.
(ii) Media with mainly anaerobic degradation such as soil and sediment, and anaerobic sewage sludge.
(iii) Media with poorly metabolized toxaphene patterns such as water and air.

The first category is most suited for a congener-specific analysis since many of the major components present in these matrices are available as single standards. However, the two most abundant toxaphene congeners in media with anaerobic degradation are not commercially available as a single standard at present. Finally, in compartments such as water and air the toxaphene pattern is often rather comparable to the technical mixture. Therefore, single standards are more important for structural assignment and identification of degradation pathways than for quantitative determination.

The following subsections are not a review about toxaphene levels reported so far in different matrices. They are intended as a help for scientists to select the important toxaphene congeners which should be detected depending on the sample type.

4.2.1
Toxaphene in Mammals and Marine Species

So far, fish is the most common matrix selected for a congener-specific toxaphene determination. Several studies have been conducted employing reference standards with 3, 5, 6, 8, 11, or 22 toxaphene single congeners. However, not all standard components could be detected in samples. B8-1413 (P-26), B9-1679 (P-50), and B9-1025 (P-62) were both present in all quantitation standards *and* found in all samples. In addition, these three toxaphene congeners did not show significant interferences on the stationary phases DB-5, Rtx1701, and Rtx2330 while other toxaphene components were significantly disturbed by co-elutions [184]. Most of the studies applied GC columns coated with 95% methyl-/5% phenylpolysiloxane or very similar phases. The quantification of toxaphene in SRM 1588 by different groups indicates a precision of ±50% (Table 8). However, a recent intercalibration using a cleaned-up fish extract showed that the results for B8-1413 (P-26), B9-1679 (P-50), and B9-1025 (P-62) were within ±40% [185].

The concentration ratios of the three main components in fish (B8-1413 (P-26), B9-1679 (P-50), and B9-1025 (P-62)) differed between species and sites. However, B9-1679 (P-50) was most abundant in more than 80% of more than 200 investigated samples. Delorme et al. estimated the half life of B8-1413 (P-26), B9-1679 (P-50), and technical toxaphene in lake trout (*Salvelinus namaycush*) and white suckers (*Catastomus commersoni*). The half life time of the toxaphene compounds in white suckers was about twice that of lake trout. Delorme et al.

concluded that fish species show different elimination rates for technical toxaphene and single components under natural living conditions [186].

In addition to the three congeners mentioned before, B8-2229 (P-44), B8-1414 (P-40), B8-1945 (P-41), B8-1412, and B7-1453 have been quantified in fish [136, 187]. B8-806/B8-809 (P-42) was also present in selected samples [187], but B7-515 (P-32) was below the detection limits in most samples analyzed so far. Several organochlorines may coelute with B7-515 (P-32) and pretend its presence in GC/ECD chromatograms. However, at a site which was heavily contaminated with toxaphene [188], Maruya et al. found significant B7-515 (P-32) levels in the respective fish samples [189]. Tribulovich et al. identified in addition B7-1059 as a major toxaphene component in fish as well as B7-1584, B7-1715, B8-789 (P-38), and B9-718 [190].

Kimmel et al. compared toxaphene levels quantified with the 22-component standard, the six-, and the five-component standard and found the six-component standard to be the most useful [187]. Quantitation of these six toxaphene components and additionally B7-1453 and B8-1412 in cod liver (*Gadus morhua*) and other marine species was suggested by the group of Vetter [103, 136].

The percentage of total toxaphene burden determined by selected congeners is difficult to estimate due to variations in the response factors of the congeners. Dependent on the applied method, the three indicator congeners accounted for 19 to 31% of the total toxaphene level in SRM 1588 (see Table 8).

B8-1413 (P-26) and B9-1679 (P-50) are also the dominating toxaphene congeners in blubber from marine mammals representing up to 80% of the total toxaphene levels [6, 7, 82, 148, 191]. Next to B8-1413 (P-26) and B9-1679 (P-50), abundant toxaphene congeners in Antarctic seal species were B8-2229 (P-44), B8-1414 (P-40), B8-1945 (P-41), B8-1412, and B7-1453 [148]. B9-1025 (P-62), a major congener in fish is metabolized in seals as confirmed by, e.g., seals from the North Sea [192]. Ringed seals (*Phoca hispida*) from the Arctic accumulated B8-1414 (P-40), B8-531 (P-39), and B8-806/B8-809 (P-42) [193]. Belugas (*Delphinapterus leucas*) from the Western Arctic accumulated toxaphene levels comparable to PCBs in blubber [194]. B8-1413 (P-26) and B9-1679 (P-50) dominated, the latter being 2- to 2.5-fold higher than the first one [195]. The two most abundant congeners represented about 50 to 65% of total toxaphene in beluga blubber [196]. A significant correlation between B9-1679 (P-50) levels and age was found for adult females but not for males or for B8-1413 (P-26) [196]. Seal blubber contained less toxaphene components than blubber from cetaceans. Seals have a more specialized enzyme system with a higher capacity to metabolize toxaphene components [192].

Few data about toxaphene levels in human adipose tissue were found in a search of the literature carried out in 1991 [197], i.e., before the introduction of single toxaphene standards. Gill et al. carried out two studies about levels in human blood and serum [198, 199]. The congener-specific analysis was performed with on-column injection on a 25 m long DB-5 capillary and NICI-MS detection. In the first study nine octachloro- or nonachlorobornanes were detected. B8-1413 (P-26) and B9-1679 (P-50) were the most abundant congeners [198]. Using the 22-component standard, B8-1414 (P-40) or B8-1945 (P-41) and B8-2229 (P-44) were also identified as abundant peaks [199]. B8-1413 (P-26) and B9-1679 (P-50) are also dominant in human milk [82, 200], while B9-1025 (P-62) is present at a much lower level [67]. De Boer and Wester found significantly different toxaphene patterns in human milk samples from different countries [201].

In 1979, Saleh et al. studied the metabolism of B7-515 (P-32) and technical toxaphene in six mammals and chicken. Both were metabolized the least in chicken and the most in monkeys [175]; B7-515 (P-32) was readily degraded, and three metabolites were formed. Two of them were B6-923 and B6-913, which were excreted with the feces [175]. Also, the octachlorobornanes B8-806/B8-809 (P-42) were metabolized by all test species, and highest degradation rates were found for monkeys [175]. A second feeding study of primates was conducted by Andrews et al. [202]. B8-1413 (P-26), B9-1679 (P-50), B8-2229 (P-44), and B9-1025 (P-62) were identified as the most abundant toxaphene congeners in both blood and adipose tissue [202]. In blood an equilibrium level of approximately 40 ppb was reached after 10 weeks and in adipose tissue levels of ca. 4000 ppb after 15–20 weeks [202]. Different abundance ratios were found for GC/ECD and GC/NICI-MS, and GC/NICI-MS was considered as less suitable for this study [202].

Recently, a feeding test was performed with chickens [151]. With a daily dose of 5 ppm toxaphene mixed into the diet from the 22nd to the 60th week of life, 9 of the congeners in the 22 Parlar component standard as well as unknown toxaphenes were detected in different tissues [151]. Next to congeners detected in fish, B8-806/B8-809 (P-42), B9-715 (P-58), and B10-1110 (P-69) were found. The results of Saleh et al. were confirmed, the latter found degradation of B7-515 (P-32) and partial degradation of B8-806/B8-809 (P-42) in chicken [175]. The half lives of seven components ranged from 25 to 51 days, and only small differences between kidney and fatty tissue were observed [151].

B8-1413 (P-26) and B9-1679 (P-50) were also the most abundant toxaphene congeners in adipose tissue from $3^1/_2$ year-old and mature polar bear (*Ursus maritimus*) shot in Iceland [203]. On the other hand, Zhu and Norstrom found that the quantity of B8-1413 (P-26) and B9-1679 (P-50) contributed only about 10% to the overall level of the 15 toxaphene components detected in adipose tissue of a 3 year-old female polar bear from the Northwest Territories of Canada [204].

Buser and Müller found B8-1413 (P-26), B9-1679 (P-50), and six further chlorobornanes (TC1, TC2, TC5–TC8) in penguin tissue [120]. TC8, tentatively identified as B8-806/B8-809 (P-42), was present in low amounts [120]. Some of the other components may be B8-1412, B8-1414 (P-40), B8-1945 (P-41), B8-2229 (P-44), which have been identified in the same matrix [100]. Usually, in biota from the top level of the food chain B8-1413 (P-26) and B9-1679 (P-50) were most abundant. In addition, the number of toxaphene peaks decreased with increasing trophic level. The results obtained so far can be summarized as follows. Most of the toxaphene components are degradable, and only a few persistent congeners are accumulated in higher organisms.

4.2.2
Toxaphene in Anaerobic Milieu

Nash and Woolson studied the fate of nine chloropesticides in Congaree sandy loam and Evesboro loamy sand. Toxaphene was the most persistent pesticide after

15 years [205]. The toxaphene pattern in anaerobic milieu (sediment, soil, and sewage sludge) is different to that of technical toxaphene and of marine biota. In sediment with a high microbial activity, the GC elution profile is changed to components with lower retention times than that of technical toxaphene [206]. The structures of the dominating B6-923 (Hx-SED) and B7-1001 (Hp-Sed) [207] were elucidated by Stern et al. [125]. However, the toxaphene pattern seems to vary with the microbial activity. Stern et al. presented the toxaphene patterns in sediments of five Canadian lakes all of which were different [208]. It was proposed that some of the variances were due to local versus atmospheric input [208, 209]. Swackhamer [210] made the controversial proposal that paper mills might be toxaphene sources. However, Rantio et al. found no toxaphene in pulp mill effluents [16]. Rappe et al. confirmed that river sediments close to paper mills had no increased toxaphene content in contrast to PCDDs and PCDFs where a significant dependence between level and distance from the factory was observed [211]. However, chlorinated monoterpene alcohols and monoterpene hydrocarbons formed by kraft pulp bleaching were detected in the chlorination stage effluent [212, 213, 214]. The components had similar retention times and responses as toxaphene components [214]. Stuthridge concluded that the log K_{OW} values of chlorinated monoterpenes (bornanes and pinanes) indicate persistency and bioaccumulation [213].

Fingerling et at. studied the fate of six toxaphene single components in soil and observed that two hexachlorobornanes were formed with the structure of B6-923 and B6-913 [165]. These two components were also generated by reductive dechlorination of B7-515 (P-32) [122]. Sewage sludge was used to study the anaerobic degradation of toxaphene [215]. A breakdown of particularly octa- and nonachlorobornanes was observed after two weeks leading to B6-923, B7-1001, B8-1945 (P-41), B8-2229 (P-44), and B9-2206 (P-63) as the major abundant isomers in GC/NICI-MS chromatograms [215].

Reductive dechlorination at positions with geminal chloro substituents is assumed to be the major way of degradation [83, 122, 215]. The fact that C1, C4, and C7 on bornane are never chlorinated leaves seven positions available. Therefore, dead-end metabolites in an anaerobic milieu should be chlorobornanes with a maximum of seven chlorine atoms [216]. Chlorobornanes which have two hydrogens at one carbon such as B7-515 (P-32) and B8-806/B8-809 (P-42) should be degraded to hexachlorobornanes. Consequently, pentachlorobornanes were predicted as the final anaerobic metabolic products of B8-2229 (P-44) and B9-1025 (P-62) [216], and their presence could be recently confirmed in sediment samples [217].

Therefore, analysis of microbial-controlled samples on the basis of the most frequently determined toxaphene compounds in fish may result in not detectable toxaphene quantities though B6-923 and B7-1001 are abundant. This clearly demonstrates the restriction of a not matrix optimized congener-specific analysis.

4.2.3
Toxaphene in Air and Water

The degradation of toxaphene is less pronounced in water and air than in biota. Due to the many more components present in air and water, co-elutions may occur, and an unequivocal structure assignment of the signals in the chromatogram is difficult [75]. B8-1413 (P-26) and B9-1679 (P-50) have been detected in air and water [75], and another major octachlorobornane was recently identified as B8-1412 [218].

At the moment it is not certain if a congener-specific analysis will give a more precise quantitation than one based on technical toxaphene as external standard. Anyhow, only a congener-specific analysis will provide information about the structure of the most abundant compounds and thus information on the structure-dependent fate of toxaphene.

Three important properties which describe the behavior of a contaminant in aqueous systems are log K_{OW}, water solubility, and vapor pressure [219]. Due to the complex composition of toxaphene and analytical uncertainty in the determination of these parameters (see Sect. 1.1), an exact evaluation of the congener-specific fate of toxaphene in water and air is presently not possible. However, Shoelb et al. identified B7-1001, B8-1413 (P-26), B8-806/B8-809 (P-42), B8-531 (P-39), B9-1679 (P-50), and further unknown toxaphene congeners in air from the north of Lake Ontario [68]. Interestingly, toxaphene levels in deep water (225 m) were significantly higher than in pelagic water (10 m) [220].

Volatilization followed by atmospheric transport is thought to be the major pathway of toxaphene dispersion from source areas [221]. The time until half of the toxaphene amount disappeared from cotton fields was < 14 days [221]. Nevertheless, at present toxaphene is probably still more emitted by volatilization from old residues than by fresh application [222]. Equilibration of toxaphene with air and water was estimated to be > 90 % gas phase-controlled and an efficient wet deposition was expected [34]. Recently, several studies have been performed to distinguish between long atmospheric transport input and local toxaphene sources [129, 156, 223, 224].

4.3
Bioaccumulation of Toxaphene

In general, only highly hydrophobic compounds have a potential to bioaccumulate [225]. Based on fugacity modelling, components with log K_{OW} < 6 are not expected to bioaccumulate [226]. Using food chain models, Suedel et al. claimed that compounds with log K_{OW} > 4.0 may have a potential to bioaccumulate in aquatic food webs [227]. One study estimated log K_{OW} of technical toxaphene as 6.44 [3], while another study found a value < 5 [227]. Therefore, a categorization of toxaphene in terms of bioaccumulation is difficult.

The fragment constant method [228] allowed the calculation for B7-1001, B8-1413 (P-26), and B9-1679 (P-50) of log K_{OW} values of 6.2, 6.5, and 6.6, respectively [229]. Geyer et al. [230] estimated log K_{OW} and bioconcentration factors (BCFs) for seven persistent toxaphene components. The log K_{OW} ranged from 5.8 to 7.9, those for B8-1413 (P-26) and B9-1679 (P-50) were similar to those reported by Fisk et al. [229]. Log K_{OW} values of individual toxaphene components which exceed the average log K_{OW} of 6.44 for technical toxaphene should be particularly considered as candidates for bioaccumulation.

Information exists about bioaccumulation and food chain biomagnification of toxaphene in aquatic and freshwater ecosystems [227]. Paasivirta found decreasing toxaphene levels from fish (*Clupea harengus* and *Salmo salar*) to seal (*Halichoerus grypus*) and whitetailed eagle (*Haliaetus albicilla*) in a Baltic food chain [231]. Bioaccumulation has also been studied in an Arctic food web: The toxaphene levels increased up to the trophic level of Arctic cod, but declined again from Beluga whale to ringed seal and polar bear [220]. This was attributed to the higher metabolization potential of seals for toxaphene compared to cetaceans and whales [220] which was also mentioned in other studies [178, 192, 232]. On the whole, the major toxaphene components present in marine mammals, B8-1413 (P-26) and B9-1679 (P-50), are also bioaccumulated in ringed seals and beluga whales from the Arctic Ocean/Beaufort Sea [233]. No bioconcentration was, however, found for the major toxaphene congeners in sediment, B6-923 (Hx-Sed) and B7-1001 (Hp-Sed) [233]. This was confirmed by Fisk et al. who found a BCF of < 1 for B7-1001 and of > 1 for B8-1413 (P-26) and B9-1679 (P-50) in an aquatic food chain [229]. The authors concluded that only (selected) octa- and nonachlorobornanes do bioaccumulate [229] due to the decreased water solubility of higher chlorinated bornanes (see Sect. 1.1). In a food chain from Lake Baikal, the bioaccumulation of technical toxaphene in seals was inferior than for DDT and PCBs, most likely due to more pronounced metabolism of toxaphene [234].

4.4
Enantioselective Determination of Toxaphene Components

Chromatographic enantioseparation of chiral xenobiotics and their metabolites is a versatile tool for process studies in marine and terrestrial ecosystems [235]. In 1994, three papers focused on the enantioselective determination of toxaphene components [120, 236, 237]. Buser and Müller found that technical toxaphene mixtures are not necessarily racemic [237]. This observation was supported after isolation of non-racemic B7-1453 from the product Melipax which had an excess of ca. 25% of the dextrorotary enantiomer [27, 238]. The enantioselective separation of toxaphene components is almost restricted to chiral stationary phases (CSPs) based on randomly derivatized *tert*-butyldimethyl-silylated *β*-cyclodextrin (commercially available from BGB Analytik, Adliswil, Switzerland). So far, only a few toxaphene components were enantioseparated on other CSPs [239, 240]. Some of these CSPs are not well defined as well, and for this reason a test mixture called CHIROTEST X was suggested for initial column testing [241].

It was found that the most abundant congeners B8-1413 (P-26) and B9-1679 (P-50) were nearly racemic, even in tissue from the top level of aquatic food webs such as marine mammals and birds. A few studies, however, mentioned that enantiomer ratios of less persistent toxaphene components significantly deviated from 1.0 [239, 242, 243]. However, the enantioseparation of less abundant components in sample extracts is more difficult than that of B8-1413 (P-26) and B9-1679 (P-50). Suitable strategies to avoid compound interferences are preseparation by liquid chromatography, multidimensional GC, and MS/MS.

4.5
Toxicological Studies of Single Standards Compared to Technical Toxaphene

First toxicological studies with single toxaphene components were carried out in the 1970s to find important relations between structure and toxicity [244]. At that time, the toxic potential of components isolated from technical toxaphene was investigated. The compounds B7-515 (P-32 or Toxicant B) and B8-806/9 (P-42 or Toxicant A) were tested concerning their acute toxicity for fish, rats and insects.

B9-1327, isolated by Anagnostopoulos et al. [71] and later assigned as Toxicant C, was four times more toxic in the *Musca domestica* test than technical toxaphene [71]. Chandurkar et al. found B9-1679 (P-50) to be four times more toxic to fish and as harmful to mosquito larvae as the technical mixture [80]. Furthermore, B8-806/B8-809 (P-42) was more potent than B9-1679 (P-50) [80]. Olson et al. found a difference in the toxicity of B7-515 (P-32) and B8-806/9 (P-42) for lake trout compared to technical toxaphene by studying behavioral parameters such as inferior swimming ability and delayed righting reflex [245].

Recently, Calciu et al. investigated the effect of technical toxaphene, B8-1413 (P-26), B9-1679 (P-50), and a 1:1 mixture of B8-1413 (P-26) and B9-1679 (P-50) on cultivated rat embryos [246]. Exposure of embryos to B8-1413 (P-26) and B9-1679 (P-50) alone had a less significant effect on growth than technical toxaphene and the 1:1 mixture using identical levels. This suggests synergistic effects of toxaphene components on growth retardation. On the other hand, the single components had a significantly stronger effect on the biotic system.

Furthermore, both components differed in their toxic properties. B8-1413 (P-26) caused limb and flexion abnormalities which were not observed with B9-1679 (P-50). It is noteworthy, that these two congeners are most abundant in human blood. Calciu et al. concluded that environmentally predominant toxaphene congeners can have organ-specific embryotoxic effects not predicted by studies on technical toxaphene [246]. Consequently, there is a need for further studies concerning the health implications of toxaphene congeners, particularly among the highly exposed Inuit populations [246].

Steinberg et al. tested the mutagenic activity of technical toxaphene as well as B7-515 (P-32), B8-1413 (P-26), B9-1025 (P-62), and B9-1679 (P-50) in the Amos-test with the *Salmonella typhimurium* strains TA98 and TA100 [247]. In contrast to the technical mixture, the single congeners were not mutagenic [247]. This partly confirms results of Hooper et al. who reported that the most insecticidal components such as B7-515 (P-32) were less mutagenic for TA100 than the complete toxaphene mixture [4, 10]. Furthermore, recrystallized toxaphene was less mutagenic than the original toxaphene [4].

The mutagenic potential (genotoxicity or DNA-damaging potential) was also studied with the Mutatox assay by Boon et al. [192]. Only the technical mixture and B7-515 (P-32) were effective in the direct assay. As expected, the S9 fraction from rats [248] as well as microsomes from harbor seals (*Phoca vitulina*) and albatross (*Diomedea immutabilis*) decreased the test-assay response [192]. No mutagenic potential was found for technical toxaphene using the Mutatox assay together with rat S9 activation [191]. Although no enzyme could be identified, it was concluded that species with low ability to metabolize toxaphene are those most likely to be affected by the carcinogenic properties of toxaphene [192]. The mammalian metabolism of toxaphene may reduce the overall genotoxic effect of technical toxaphene [10].

The estrogenic activity of toxaphene and synergetic transactivation caused by other organochlorines according to Arnold et al. [249], have been discussed controversially [250–252]. The answer to this question is of central importance, since toxaphene itself is a multicomponent mixture: If there is a synergetic potential, it is important to know which congeners have endocrine effects and whether or not there are synergetic effects between the compounds present in technical toxaphene. Because of these unsolved problems, a detailed discussion of this phenomenon is not possible at the moment.

Although the toxicological aspects of toxaphene congeners have not been investigated in detail, the tolerable daily intake (TDI) of toxaphene was estimated by Health Canada to be 0.2 $\mu g \cdot kg^{-1} \cdot day^{-1}$ [253]. This TDI is exceeded in some indigenous populations in the Western Northwest Territories of Canada [253].

5
Structure-Related Observations

Along with the identification of an increasing number of toxaphene components persistent in the environment, the elucidation of the mechanisms leading to biodegradation or bioaccumulation of the components has gained in interest. As mentioned before, the dominating toxaphene components in sediment are not identical with those in marine biota.

In marine mammals, B8-1413 (P-26) and B9-1679 (P-50) are the most abundant congeners. The stability of the two components was attributed to an alternating *endo-exo-endo-exo* conformation of the chlorine atoms in the six-membered ring [65]. Furthermore, these persistent compounds are the least polar and therefore elute earlier from silica. Consequently, polarity was considered as another important factor to explain persistency and bioaccumulation [136]. Geminal chlorine atoms at the secondary carbon atoms of the six-membered ring seem to be most labile, and polychlorinated bornanes without them appear to be more stable [83, 244]. Parlar divided polychlorinated bornanes into three groups with respect to their photostability [254]. The presence of geminal chlorine atoms at the six-membered ring was suggested to be the reason for pronounced photodegradation [254]. Vetter and Scherer reported a low persistency of polychlorinated bornanes with two *exo*-chlorine atoms vicinal to C1 (i.e., 2-*exo* and 6-*exo*) [51]. Using molecular modelling they showed that steric hindrance was the major reason for this behavior [73]. Molecular modelling of toxaphene components was introduced in 1994 [83] to support structure elucidation of toxaphene components based on spectroscopic data [83, 100, 125, 193]. The thermodynamic stability of several polychlorinated bornanes was calculated with semi-empirical methods. The heat of formation (ΔH°) was found to be a determining factor for the stability of polychlorinated bornanes [73].

6
Further Summarizing Literature

The last comprehensive review on toxaphene was published in 1991 by Saleh [3]. Although it was written before the introduction of toxaphene standards, most of the data and conclusions are still valid. Earlier helpful reviews are those of Pollock and Kilgore [5], Korte et al. [20], WHO [25], Parlar [70], as well as the more recent ones of Muir and de Boer [255] and de Geus et al. [256]. Toxicological profiles of toxaphene were summarized by Roper [21] and in a recent update [10]. The enantioselective analysis of organochlorines and toxaphene was evaluated by Vetter and Schurig [257] and Hühnerfuss [235]. About 70 reports from the US EPA and several patents deal also with toxaphene.

7
Conclusions and Future Research Topics

The availability of single standards for quantification of toxaphene resulted in the following results and conclusions:

- The scientific interests in toxaphene increased.
- MS and ECD response factors of toxaphene components can vary significantly.
- Decomposition of labile toxaphene components may occur in the injector.
- Degradation of labile toxaphene components may occur on polar GC phases.
- Congener-specific toxaphene analysis generated more data for fish and marine mammals.
- Structure elucidation of polychlorinated bornanes has become a routine method.
- First toxicological studies with single toxaphene components were performed.
- The dependency between structure and persistency could be evaluated.

However, the following problems remained:

- More standard components are needed for a more complete study of the fate of toxaphene.
- Not all standards relevant for anaerobic media are available.
- Single standards normally quantified in biota are hardly present in sediments leading to far too low levels being determined.

These points clearly show the advantages and disadvantages of congener-specific investigations. Furthermore, the validation of analytical methods (sample clean-up and GC quantification) with reference materials is highly advisable. SRM 1588 cod liver oil and the three congeners B8-1413 (P-26), B9-1025 (P-62), and B9-1679 (P-50) should be used for method development. Furthermore, certification of toxaphene levels in more reference materials has to be an important task in the future.

B8-1413 (P-26) is an early eluting congener from silica and is therefore most suitable for the control of a complete PCB/toxaphene group separation during sample clean-up. B9-1025 (P-62) is the most problematic congener at gas chromatographic separations, and B9-1679 (P-50) appears to be the most abundant toxaphene component in most sample matrices.

Depending on the number of congeners quantified in a sample, different toxaphene levels will result. This has to be considered when literature data are compared. Nevertheless, a congener-specific quantification has to be state-of-the-art when investigating the fate of toxaphene in the environment. Toxicity studies should also be toxaphene congener as well as enantioner-specific.

References

1. US Patents 2,565,471 and 2,657,164, Hercules Inc.
2. Bidleman TF, Muir DCG (ed) (1993): Analytical and environmental chemistry of toxaphene. Workshop held at Burlington Ontario, Canada, 4-6 February 1993. Chemosphere 27, No 10 (ISBN 0045-6535)
3. Saleh MA (1991) Rev Environ Cont Toxicol 118:1
4. Hooper NK, Ames BN, Saleh MA, Casida JE (1979) Science 205:591
5. Pollock GA, Kilgore WW (1978) Res Rev 69:87
6. Bidleman TF, Walla MD, Muir DCG, Stern GA (1993) Environ Toxicol Chem 12:701
7. Luckas B, Vetter W, Fischer P, Heidemann G, Plötz J (1990) Chemosphere 21:13
8. Paasivirta J, Rantio T (1991) Chemosphere 22:47
9. Agency for Toxic Substances and Disease Registry, ATSDR (1994) 1993 CERCLA Priority List of Hazardous Substances that will be Subject for Toxicological Profiles and Support Document. US Department of Health and Human Services, Public Health Service, Atlanta, GA, USA
10. Research Triangle Institute (1994) Toxicological profile for toxaphene. Draft, 1994, prepared for the US Department of Health and Human Services.
11. Alder L, Karl H (1995) lecture at Bremerhaven, January 18, 1995
12. Oehme M (1995) personal communication, April 21, 1995
13. Andrews P (1995) Health Canada, personal communication to W Vetter
14. Zell M, Ballschmiter K (1980) Fresenius Z Anal Chem 300:387
15. Wideqvist U, Jansson B, Reutergårdh L, Olsson M, Odsjö T, Uvemo U-B (1993) Chemosphere 27:1987
16. Rantio T, Paasivirta J, Lahtiperä M (1993) Chemosphere 27:2003
17. Alder L, Vieth B (1996) Fresenius J Anal Chem 354:81
18. Alder L, Bache K, Beck H, Parlar H (1997) Chemosphere 35:1391
19. Onuska FI, Maguire RJ (1997) Environmental applications of mass spectrometry: toxaphene analysis. In: Caprioli RM et al. (eds) Selected topics and mass spectrometry in the biomolecular sciences, pp 533–558
20. Korte F, Scheunert I, Parlar H (1979) Pure Appl Chem 51:1583
21. Roper, WH (1990) Toxicological profile for toxaphene. US Department of Health and Human Services, Public Health Service, Agency for Toxic Substances and Disease Registry, TP-90-26
22. Wideqvist U, Jansson B, Reutergårdh L, Sundström G (1984) Chemosphere 13:367
23. Maier-Bode H (1965) Pflanzenschutzmittel-Rückstände. Eugen Ulmer Verlag, Stuttgart
24. US Environmental Protection Agency (1988) Toxaphene. Rev Environ Cont Toxicol 104:203
25. WHO Geneve (1984) Camphechlor. Environ Health Crit 45 (ISBN 92-4-154185-7)
26. Wauchope RD, Buttler TM, Hornsby AG, Augustijn-Beckers PWN, Burt JP (1992) Rev Environ Cont Toxicol 123:1
27. Vetter W, Krock B, Klobes U, Luckas B (1997) J Agric Food Chem 45:4866
28. Voldner EC, Li YF (1993) Chemosphere 27:2073
29. Fishbein L (1997) Global Environ 1:481
30. Liebmann R, Hempel D, Heinisch E (1971) Arch Pflanzensch 7:131

31. Casida JE, Holmstead RE, Khalifa S, Knox JR, Ohsawa T (1974) Toxaphene composition and metabolism in rats. In: Koivistoinen P (Ed.), Pesticides. Lectures held at the IUPAC 3. International Congress of Pesticide Chemistry, Helsinki July 3–9. Supplement Volume III. Thieme, Stuttgart, pp 365-369
32. Parlar H, Nitz S, Michna A, Korte F (1978) Z Naturforsch 33b:915
33. Pollock GA, Kilgore WW (1980) J Toxicol Environ Health 6:115
34. Murphy TJ, Mullin MD, Meyer JA (1987) Environ Sci Technol 21:155
35. Bidleman TF, Wideqvist U, Jansson B, Soderlund R (1987) Atmos Environ 15:619
36. Bidleman TF, Olney CE (1975) Nature 257:475
37. Langlois G (1919) Ann Chim (Paris) [IX]/12:193
38. Jennings BH, Herschbach GB (1965) J Org Chem 30:3902
39. Richey Jr HG, Grant JE, Garbacik TJ, Dull DL (1965) J Org Chem 30:3909
40. Krock B, Vetter W, Luckas B, Scherer G (1999) Chemosphere 39:133
41. Tribulovich VG, Nikiforov VA, Bolshakov VS (1996) Organohalogen Compd 28:385
42. Coelhan M, Parlar H (1996) Chemosphere 32:217
43. Seiber JN, Landrum PF, Madden SC, Nugent KD, Winterlin WL (1975) J Chromatogr 114:361
44. Landrum PF, Pollock GA, Seiber JN, Hope H, Swanson KL (1976) Chemosphere 5:63
45. Rigaudy J, Klesney SP (1979) Nomenclature of Organic Chemistry Sections A, B, C, D, E, F and H. Pergamon Press, Oxford, pp 49–52, and 498
46. Kwart H (1953) J Am Chem Soc 75:5942
47. Hainzl D (1995) J Agric Food Chem 43:277
48. Wester PG, Geus H-J de, Boer J de, Brinkman UAT (1997) Chemosphere 35:2857
49. Weast RC (ed) (19988/89) Handbook of Chemistry and Physics, 69th edn. CRC Press, Cleveland, pp C-24
50. Lide DR (Editor-in-Chief) (1993/1994) Handbook of Chemistry and Physics, 74th edn. CRC Press, Boca Raton (special student edition) pp 2–66 and 2–67
51. Vetter W, Scherer G (1998) Chemosphere 37:2525
52. Dr. NcNaught, IUPAC (1995) personal communication to W Vetter
53. Turner WV, Engel JL, Casida JE (1977) J Agric Food Chem 25:1394
54. Casida JE, Holmstead RL, Khalifa S, Knox JR, Ohsawa T, Palmer KJ, Wong RY (1974) Science 183:520
55. Parlar H (1998) unpublished results
56. Vetter W (1998) unpublished results
57. Kenyon WC (1952) Anal Chem 24:1197
58. Clark WH (1962) J Agric Food Chem 10:214
59. Phillips WF, DeBenedictus ME (1954) J Agric Food Chem 2:1226
60. Hornstein I (1957) J Agric Food Chem 5:446
61. Coulson DM, Caranagh LA, Stuart J (1959) J Agric Food Chem 7:250
62. Holmstead RL, Khalifa S, Casida JE (1974) J Agric Food Chem 22:939
63. Jansson B, Wideqvist U (1983) Intern J Environ Anal Chem 13:309
64. Jansson B (1982) Separation av toxafenkomponenter (in Swedish). Statens naturvaardsverk, Rapport 82–04, Specialanalytiska Laboratoriet, Wallenberglaboratoriet, 10691 Stockholm, Sweden
65. Zhu J, Mulvihill MJ, Norstrom RJ (1994) J Chromatogr A 669:103
66. Vetter W, Luckas B (1995) Sci Total Environ 160/161:505
67. Boer J de, Geus H-J de, Brinkman UAT (1997) Environ Sci Technol 31:873
68. Shoelb M, Brice KA, Hoff RM (1997) Organohalogen Compd 33:68
69. Nikiforov, VA, Tribulovich VG, Karavan VS (1995) Organohalogen Compd 26:379
70. Parlar H (1985) Intern J Environ Anal Chem 20:141
71. Vetter W (1993) Chemosphere 26:1079
72. Hainzl D, Burhenne J, Parlar H (1994) Chemosphere 28:245
73. Vetter W, Scherer G (1998) Organohalogen Compd 35:235
74. Ballschmiter K, Zell M (1980) Intern J Environ Anal Chem 8:15
75. Bidleman T, Falconer RL, Walla MD (1995) Sci Total Environ 160/161:55

76. Alawi M, Barlas H, Hainzl D, Burhenne J, Coelhan M, Parlar H (1994) Fresenius Environ Bull 3:350
77. Vetter W, Luckas B, Heidemann G, Skírnisson K (1996) Sci Total Environ 186:29
78. Khalifa S, Mon TR, Engel JL, Casida JE (1974) J Agric Food Chem 22:653
79. Turner WV, Khalifa S, Casida JE (1975) J Agric Food Chem 23:991
80. Chandurkar PS, Matsumura F, Ikeda T (1978) Chemosphere 7:123
81. Chandurkar PS, Matsumura F (1979) Arch Environ Cont Toxicol 8:1
82. Stern GA, Muir DCG, Ford CA, Grift NP, Dewailly E, Bidleman TF, Walla MD (1992) Environ Sci Technol 26:1838
83. Vetter W, Scherer G, Schlabach M, Luckas B, Oehme M (1994) Fresenius J Anal Chem 349:552
84. Muir DCG, Ford CA, Stewart REA, Smith TG, Addison RF, Zinck ME, Béland P (1990) Can Bull Fish Aquat Sci 224:165
85. Parlar H, Angerhöfer D, Coelhan M, Kimmel L (1995) Organohalogen Compd 26:357
86. Nikiforov, VA, Tribulovich VG, Karavan VS (1995) Organohalogen Compd 26:393
87. Oehme M, Kallenborn R (1995) Chemosphere 30:1739
88. Andrews P, Vetter W (1995) Chemosphere 31:3879
89. Wester PG, de Geus H-J, de Boer J, Brinkman UT (1997) Organohalogen Compd 33:47
90. Black DK (1974) Division of Pesticide Chemistry, 168th ACS national meeting, Atlantic City, NY, September 1974
91. Anagnostopoulos ML, Parlar H, Korte F (1974) Chemosphere 3:65
92. Krock B, Vetter W, Luckas B, Scherer G (1996) Chemosphere 33:1005
93. Lach G, Parlar H (1990) Chemosphere 21:29
94. Hainzl D, Burhenne J, Barlas H, Parlar H (1995) Fresenius J Anal Chem 351:271
95. Burhenne J, Hainzl D, Xu L, Vieth B, Alder L, Parlar H (1993) Fresenius J Anal Chem 346:779
96. Malaiyandi M, Buchanan GW, Nikiforov V, William DT (1993) Chemosphere 27:1849
97. Tribulovich VG, Nikiforov VA, Karavan VS, Miltsov SA, Bolshakov S (1994) Organo-halogen Compd 19:97
98. Vetter W, Luckas B, Oehme M (1992) Chemosphere 25:1643
99. Stern GA, Loewen MD, Kidd KA, Miskimmin M, Muir DCG, Schindler DW, Westmore JB (1995) Organohalogen Compd 26:363
100. Vetter W, Klobes U, Krock B, Luckas B, Glotz D, Scherer G (1997) Environ Sci Technol 31:3023
101. Fingerling G, Maurer M, Coelhan M, Parlar H (1998) Fresenius Environ Bull 7:610
102. Baycan-Keller R, Oehme M (1998) unpublished results
103. Parlar H, Gäb S, Michna A, Korte F (1976) Chemosphere 5:217
104. Hainzl D, Burhenne J, Parlar H (1993) Chemosphere 27:1857
105. Vetter W, Luckas B, Oehme M (1992) Organohalogen Compd 8:177
106. Stern GA, Muir DCG, Ford CA, Grift NP, Dewailly E, Bidleman TF, Walla MD (1992) Organohalogen Compd 8:381
107. Dougherty RC (1981) Anal Chem 53:625A
108. Kantner TR, Mumma RO (1966) Res Rev 16:138
109. Saleh MA (1983) J Agric Food Chem 31:748
110. Stern GA, Muir DCG, Westmore JB, Buchannon WD (1993) Biol Mass Spectrom 22:19
111. Buser H-R, Oehme M, Vetter W, Luckas B (1993) Fresenius J Anal Chem 347:502
112. Vetter W, Oehme M, Luckas B (1993) Chemosphere 27:597
113. Parlar H, Becker F, Müller R, Lach G (1988) Fresenius Z Anal Chem 331:804
114. Parlar H, Gäb S, Nitz S, Korte F (1976) Chemosphere 5:333
115. Parlar H, Nitz S, Gäb S, Korte F (1977) J Agric Food Chem 25:68
116. Parlar H, Michna A (1983) Chemosphere 12:913
117. Santos FJ, Galceran MT, Caixach J, Rivera J, Huguet X (1997) Rapid Commun Mass Spectrom 11:341
118. Stemmler EA, Hites RA (1988) Electron capture negative ion mass spectra of environ-mental contaminants and related compounds. VCH, Weinheim (ISBN 0-89573-708–6)

119. Parlar H (1991) Nachr Chem Tech Lab 39/1:26
120. Buser H-R, Müller MD (1994) Environ Sci Technol 28:119
121. Vetter W (1993) doctoral thesis, University of Hohenheim
122. Saleh MA, Casida JE (1978) J Agric Food Chem 26:583
123. Dougherty RC, Roberts JD, Biros FJ (1975) Anal Chem 47:54
124. Ribick MA, Dubay GR, Petty JD, Stalling DL, Schmitt CJ (1982) Environ Sci Technol 16:310
125. Stern GA, Loewen MD, Miskimmin BM, Muir DCG, Westmore JB (1996) Environ Sci Technol 30:2251
126. Palmer KJ, Wong RY, Lundin RE, Khalifa S, Casida JE (1975) J Am Chem Soc 97:408
127. Frenzen G, Hainzl D, Burhenne J, Parlar H (1994) Chemosphere 28:2067
128. Nikiforov VA, Tribulovich VG, Karavan VS, Miltsov SA (1997) Organohalogen Compd 33:53
129. Shafer HH, Cooke M, Deroos F (1981) Appl Spectroscop 35:469
130. Parlar H, Angerhöfer D (1998) personal communication to W Vetter
131. Archer TE, Crosby DG (1966) Bull Environ Cont Toxicol 1:70
132. Jansson B, Vaz R, Blomkvist G, Jensen S, Olsson M (1979) Chemosphere 8:181
133. Stalling DL (1975) International Conference on Environmental Sensing and Assessment, September 14–19, Las Vegas
134. Specht W, Tillkes M (1985) Fresenius Z Anal Chem 322:443
135. Fürst P, Fürst Ch, Groebel W (1989) Deutsch Lebensm Rdschau 9:273
136. Krock B, Vetter W, Luckas B (1997) Chemosphere 35:1519
137. Atuma SS, Jensen S, Mowrer S, Örn U (1986) Intern J Environ Anal Chem 24:213
138. Marth P, Schramm K-W, Lattmann L, Oxynos K, Kettrup A (1995) Organohalogen Compd 26:351
139. Petrick G, Schulz DE, Duinker JC (1988) J Chromatogr 435:241
140. Haglund P, Andersson R, Buser H-R, Wågman N, Rappe C (1998) Organohalogen Compd 35:275
141. Bartha R, Klobes U, Vetter W, Luckas B (1998) Organohalogen Compd 35:247
142. Brumley WC, Brownrigg CM, Grange AH (1993) J Chromatogr 633:177
143. Parlar H, Becker F, Müller R, Lach G (1988) Fresenius Z Anal Chem 331:804
144. Musial CJ, Uthe JF (1983) Intern J Environ Anal Chem 14:117
145. Alder L, Beck H, Khandker S, Karl H, Lehmann I (1996) Chemosphere 34:1389
146. Baycan-Keller R, Oehme M (1998) J High Resol Chromatogr 21:298
147. Vetter W, Klobes U, Krock B, Luckas B (1997) J Microcol Sep 9:29
148. Vetter W, Krock B, Luckas B (1997) Chromatographia 44:65
149. Nikiforov VA, Tribulovich VG, Bolshakov S (1995) Organohalogen Compd 28:355
150. Vetter W, Müller U, Krock B, Luckas B (1996) Organohalogen Compd 28:359
151. Kaltenecker M, Schwind K-H, Ueberschär K-H, Hecht H, Petz M (1998) Organohalogen Compd 35:281
152. Baycan-Keller R, Oehme M, Galliker B (1998) Organohalogen Compd 35:229
153. Manninen A, Kuitunen ML, Julin L (1987) J Chromatogr 394:465
154. de Boer J, de Geus H-J (1995) Organohalogen Compd 26:345
155. Wade TL, Chambers L, Gardinali PR, Sericano JL, Jackson TJ (1996) Organohalogen Compd 28:399
156. Glassmeyer ST, De Vault DS, Myers TR, Hites RA (1997) Environ Sci Technol 31:84
157. Onuska FI, Terry KA, Seech A, Antonio M (1994) J Chromatogr A 665:125
158. Bartha R, Vetter W, Luckas B (1997) Fresenius J Anal Chem 358:812
159. Vetter W, Bartha R, Luckas B (1997) Organohalogen Compd 33:13
160. Stottmeister E, Hermenau H, Hendel P, Welsch T, Engewald W (1991) Fresenius J Anal Chem 340:31
161. Chan HM, Zhu J, Yeboah F (1998) Chemosphere 36:2135
162. Hunt DF, Shabanowitz J, Harvey TM, Coates M (1985) Anal Chem 57:525
163. Onuska FI, Terry KA (1989) J Chromatogr 471:161
164. Swackhamer DL, Charles MJ, Hites RA (1987) Anal Chem 59:913
165. Fingerling G, Hertkorn N, Parlar H (1996) Environ Sci Technol 30:2984

166. Lau B, Weber, D, Andrews P (1996) Chemosphere 32:1021
167. Lau BP-Y, Weber D, Andrews P (1994) Rapid Commun Mass Spectrom 8:849
168. Vetter W, Luckas B (1998) Rapid Commun Mass Spectrom 12:312
169. Andrews P, Intercalibration Study on Toxaphene, Organized by Health Canada
170. Barrie L, Bidleman T, Dougherty D, Fellin P, Grift N, Muir D, Rosenberg B, Stern G, Toom D (1993) Chemosphere 27:2037
171. Karlson H, Oehme M (1997) Chemosphere 34:951
172. Andrews P, Newsome WH, Boyle M, Collins P (1993) Chemosphere 27:1865
173. Wade TL, Chambers L, Gardinali PR, Sericano JL, Jackson TJ, McDonald TJ, Tarpley RJ, Suydam R (1995) Organohalogen Compd 26:339
174. Walter B, Ballschmiter K (1991) Fresenius J Anal Chem 340:245
175. Saleh MA, Skinner RF, Casida JF (1979) J Agric Food Chem 27:731
176. Lahaniatis ES, Bergheim W, Kettrup A (1991/1992) In: Proceedings of the "International symposium on ecological approaches of environmental chemicals", April 15–17, 1991, in Debrecen/Hungary. GSF-Berichte 4/92 (ISSN 0721-1694)
177. Parlar H (1988) Chemosphere 17:2141
178. Boon JP, Helle M, Dekker M, Sleiderink HW, Leeuw JW de, Klamer HJ, Govers B, Wester P, Boer J de (1996) Organohalogen Compd 28:416
179. Fingerling GM, Parlar H (1997) J Agric Food Chem 45:4116
180. Carlin FJ Jr, Hoffman JM (1997) Organohalogen Compd 33:70–75
181. Klobes U, Vetter W, Luckas B, Weichbrodt M (1999) Organonalogen Compd, in press
182. Muir DCG, Grift N, unpublished results, cited in Schantz MM, Parris RM, Wise SA (1993) Chemosphere 27:1915
183. Fowler B, Hoover D, Hamilton MC (1993) Chemosphere 27:1891
184. Karlsson H, Oehme M (1996) Organohalogen Compd 28:369
185. QUASIMEME: Laboratory performance studies, Round 14, DE-2 Exercise 379. Toxaphene Development Exercise, Aberdeen, UK, January 1999
186. Delorme PD, Muir DCG, Lockhart WL, Mills KH, Ward FJ (1993) Chemosphere 27:1965
187. Kimmel L, Angerhöfer D, Gill U, Coelhan M, Parlar H (1998) Chemosphere 37:549
188. Maruya KA, Lee RF (1998) Environ Sci Technol 32:1069
189. Maruya KA, Wakeham SG, Francendese L (1998) in press
190. Tribulovich VG, Nikiforov VA, Karavan VS (1995) Organohalogen Compd 26:389
191. Johnson BT, Long ER (1998) Environ Toxicol Chem 17:1099
192. Boon JP, Sleiderink HW, Helle MS, Dekker M, van Schranke A, Roex E, Hillebrand MTJ, Klamer HJC, Govers B, Pastor D, Morse D, Wester PG, Boer J de (1998) Comp Biochem Physiol Part C 121:385
193. Loewen MD, Stern GA, Westmore JB, Muir DCG, Parlar H (1998) Chemosphere 36:3119
194. Becker PR, Mackey EA, Demiralp R, Schantz MM, Koster BJ, Wise SA (1997) Chemosphere 34:2067
195. Muir DCG, Koczanski K, Rosenberg B, Béland P (1996) Environ Pollut 93:235
196. Muir DCG, Ford CA, Rosenberg B, Norstrom RJ, Simon M, Béland P (1996) Environ Pollut 93:219
197. Wessel JR, Yess NJ (1991) Rev Environ Cont Toxicol 120:83
198. Gill US, Schwartz HM, Wheatley B (1995) Intern J Environ Anal Chem 60:153
199. Gill US, Schwartz HM, Wheatley B, Parlar H (1996) Chemosphere 33:1021
200. Alder L, Palavinskas R, Andrews P (1996) Organohalogen Compd 28:410
201. Boer J de, Wester PG (1993) Chemosphere 27:1879
202. Andrews P, Headrick K, Pilon J-C, Bryce F, Iverson F (1996) Chemosphere 32:1043
203. Klobes U, Vetter W, Glotz D, Luckas B, Skírnisson K, Hersteinsson P (1998) Intern J Environ Anal Chem 69:67
204. Zhu J, Norstrom RJ (1993) Chemosphere 27:1923
205. Nash RG, Woolson EA (1969) Science 157:924
206. Williams RR, Bidleman TF (1978) J Agric Food Chem 26:280
207. Miskimmin BM, Muir DCG, Schindler DW, Stern GA, Grift NP (1995) Environ Sci Technol 29:2490

208. Stern GA, Billeck B, Lockhart WL, Muir DCG, Tomy GT, Wilkinson P (1998) Organo-halogen Compd 35:299
209. Swackhamer DL, Pearson RF, Schottler S, Symonik DG (1998) Organohalogen Compd 35:303
210. Swackhamer DL, Pearson RF, Eisenreich SJ, Long DT (1995) Organohalogen Compd 26:319
211. Rappe C, Haglund P, Andersson R, Buser H-R (1998) Organohalogen Compd 35:287
212. Stuthridge TR, Wilkins AL, Langdon AG, McFarlane PN, Mackie KL (1990) Environ Sci Technol 24:903–908
213. Stuthridge TR (1993) Arch Environ Contam Toxicol 24:113
214. Pyysalo H, Antervo K (1985) Chemosphere 14:1723
215. Buser H-R, Haglund P, Müller MD, Rappe C (1998) Organohalogen Compd 35:239
216. Vetter W, Bartha R, Stern G, Tomy G (1998) Organohalogen Compd 35:343
217. Maruya KA, Wakeham SG, Vetter W, Francendese L (1998) in press
218. Vetter W, Bidleman T (1998) correspondence, November 1998
219. Ford J, Young TC (1994) In: Steinberg CEW, Wright RF (eds) Acidification of Freshwater Ecosystems: Implications for the Future, Wiley
220. Norstrom RJ, Muir DCG (1994) Sci Total Environ 154:107
221. Willis GH, McDowell LL, Harper LA, Southwick LM, Smith S (1983) J Environ Qual 12:80
222. Bidleman TF, Alegria H, Ngabe B, Green C (1998) Atmos Environ 32:1849
223. Donald DB, Stern GA, Muir DCG, Fowler BR, Miskimmin BM, Bailey R (1998) Environ Sci Technol 32:1391
224. Pearson RF, Swackhamer DL, Eisenreich SJ, Long DT (1997) Environ Sci Technol 31:3523
225. Biddinger GR, Gloss SP (1984) Res Rev 91:103
226. Clark KE, Mackay D (1991) Environ Toxicol Chem 10:1205
227. Suedel BC, Boraczek JA, Peddicord RK, Clifford PA, Dillon TM (1994) Rev Environ Contam Toxicol 136:21
228. Lyman WS, Reehl WF, Rosenblatt RH (1982) Handbook of Chemical-Physical Property Estimation Methods. McGraw-Hill, New York
229. Fisk AT, Norstrom RJ, Cymbalisty CD, Muir DCG (1998) Environ Toxicol Chem 17:951
230. Geyer HJ, Kaune A, Schramm K-W, Rimkus G, Vetter W, Kettrup A, Muir DCG (1998) Organohalogen Compd 35:263
231. Paasivirta J (1998) Toxicol Environ Chem 66:71
232. Vetter W, Luckas B, Heidemann G, Skírnisson K (1996) Sci Total Environ 186:29
233. Muir D, Kidd K, Koczanski K, Stern G, Alaee M, Jantunen L, Bidleman T (1997) Organohalogen Compd 33:34
234. Kucklick JR, Harvey HR, Ostrom PH, Ostrom NE, Baker JE (1996) Environ Toxicol Chem 15:1388
235. Hühnerfuss H (1998) Organohalogen Compd 35:319
236. Kallenborn R, Oehme M, Vetter W, Parlar H (1994) Chemosphere 28:89
237. Buser H-R, Müller MD (1994) J Agric Food Chem 42:393
238. Vetter W, Klobes U, Luckas B, Hottinger G, Schmidt G (1998) J Assoc Off Anal Chem 81:1245
239. Klobes U, Vetter W, Luckas B, Hottinger G (1998) Chromatographia 47:565
240. Vetter W, Klobes U, Luckas B, Hottinger G (1998) Organohalogen Compd 35:305
241. Jaus A, Oehme M, Skopp S, Karlsson H (1998) Organohalogen Compd 35:325
242. Vetter W, Klobes U, Luckas B, Hottinger G (1997) Organohalogen Compd 33:63
243. Geus H-J de, Baycan-Keller R, Oehme M, Boer J de, Brinkman UAT (1998) J High Resol Chromatogr 21:39
244. Saleh MA, Turner WV, Casida JE (1977) Science 198:1256
245. Olson KL, Matsumura F, Boush GM (1980) Arch Environ Contam Toxicol 9:247
246. Calciu C, Chan HM, Kubow S (1997) Toxicology 124:153
247. Steinberg M, Kinoshita FK, Ballantyne M (1998) Organohalogen Compd 35:243
248. Mortelmans K, Haworth S, Lawlor T, Specht W, Tainer B, Zeiger E (1986): Environ Mutagen 8:1

249. Arnold SF, Klotz DM, Collins BM, Vonier PM, Guillette LJ Jr, McLachlan JA (1996) Science 272:1489
250. Ramamoorthy K, Wang F, Chen I-C, Safe S, Norris JD, McDonnell DP, Gaido KW, Boccinfuso WP, Korach KS (1997) Science 275:405
251. McLachlan JA, Arnold SF, Klotz DM, Collins BM, Vonier PM, Guillette LJ Jr (1997) Science 275:405
252. Soto AM, Sonnenschein C, Chung KL, Fernandez MF, Olea N, Olea Serrano F (1995) Environ Health Perspect 103:113
253. Berti PR, Receveur O, Chan HM, Kuhnlein HV (1998) Environ Research A 76:131
254. Parlar H, Fingerling G, Angerhöfer D, Christ G, Coelhan M (1996) ACS Symposium Series 671:348
255. Muir DCG, de Boer J (1995) Trends Anal Chem 14:56
256. de Geus HJ, Besseling H, Brouwer A, Klingsøyr J, McHugh B, Nixon E, Rimkus GG, Wester PG, de Boer J (1999) Environ Health Perspect 107:115
257. Vetter V, Schurig V (1997) J Chromatogr A 774:143

Polychlorinated Dibenzothiophenes (PCDTs), Thianthrenes (PCTAs) and Their Alkylated Derivatives

Seija Sinkkonen

Department of Chemistry, University of Jyväskylä, P.O. Box 35, FIN-40351 Jyväskylä, Finland
(e-mail: ssinkkon@cc.jyu.fi)

Polychlorinated dibenzothiophenes (PCDTs) and polychlorinated thianthrenes (PCTAs) are interesting compounds due to their structural similarity with polychlorinated dibenzofurans (PCDFs) and polychlorinated dibenzo-*p*-dioxins (PCDDs). PCDTs are sulfur analogues of PCDFs and PCTAs are sulfur analogues of PCDDs. PCDTs and PCTAs are probably formed by processes similar to the formation of PCDFs and PCDDs. PCDTs were first detected in the environment in 1986. Since then PCDTs have been analyzed in samples from combustion and metallurgy, sediments, pulp mill effluents, and in aquatic biota. Combustion and metallurgy seem to be the major sources of PCDTs in the environment. PCDTs and PCTAs are analyzed by different gas chromatographic/mass spectrometric (GC/MS) methods. Several model compounds of PCDTs and PCTAs have been prepared for environmental and toxicological purposes. However, due to lack of labeled or/and pure PCDT and PCTA standards, the environmental concentrations, until now, are generally calculated using labeled 2378-TeCDD as an internal standard. Then it is supposed that the MS response of PCDTs and PCDDs is equal. Toxicological studies have shown that the replacement of oxygen atoms by sulfur greatly decreases the AHH/EROD inducing potency of PCDTs and PCTAs compared to that of PCDFs and PCDDs. Anyhow, in vivo studies have shown that PCDTs and PCTAs possess some kind of dioxin type biological activity.

The Handbook of Environmental Chemistry Vol. 3 Part K
New Types of Persistent Halogenated Compounds
(ed. by J. Paasivirta)
© Springer-Verlag Berlin Heidelberg 2000

1
Introduction

Polychlorinated dibenzo-*p*-dioxins (PCDDs), polychlorinated dibenzofurans (PCDFs), polychlorobiphenyls (PCBs), and polychlorophenols (PCPs) are well-known artifacts ubiquitous in the present environment. Due to their lipophilicity and resistance to metabolism or chemical degradation these have the potential to accumulate in the food chain and cause toxic effects [1, 2]. Polycyclic aromatic sulfur heterocycles (PASHs) along with different kinds of polycyclic aromatic compounds (PACs) also occur widely in the environment [3, 4]. In particular, alkylated dibenzothiophenes have previously been found to be persistent residues in the marine environment after oil spills. Dibenzothiophene and its alkylated derivatives have been found to accumulate in fish and other marine organisms [5–7].

Polychlorinated dibenzothiophenes (PCDTs) are sulfur analogues of polychlorinated dibenzofurans (PCDFs) and polychlorinated thianthrenes (PCTAs) are sulfur analogues of polychlorinated dibenzo-*p*-dioxins (PCDDs). PCDTs and PCTAs are environmentally and toxicologically interesting compounds due to their structural similarity with polychlorinated dibenzo-*p*-dioxins (PCDDs) and polychlorinated dibenzofurans (PCDFs). The compound 2378-TeCTA, the sulfur analogue of 2378-TeCDD, has been reported to show some dioxin-like activity in a bioassay which measures the ability of 2378-TeCDD and related compounds to inhibit cell divisions in a mouse epithelial cell culture [8]. These compounds 2378-TeCDT and 2378-TeCTA have been found to be inducers of AHH and EROD [9].

The occurrence of polychlorinated aromatic sulfur compounds in the environment has been reported during the last decade. PCDDs and PCDFs are formed by different chemical, photochemical and enzymatic reactions [10]. It is possible that PCDTs and PCTAs are formed by chemical processes similar to those of the formation of PCDFs and PCDDs. Until now, PCDTs have been analyzed in stack gas and fly ash samples, in sediments, pulp mill effluents, and in some aquatic organisms. Some PCTA congeners have been observed in pulp mill effluents, stack gas, and soil and compost samples. The compound 2468-TeCDT has been found to accumulate from environmentally contaminated sediments to sandworms, clams, and grass shrimp. Accumulation factors have been calculated [11].

In environmental chemical analytical procedures such as extraction, purification, and fractionation with basic aluminum oxide and activated carbon,

PCDTs and PCTAs behave similarly to PCDDs and PCDFs. The exact value of the M^+ ion of TeCDTs is 319.8788 and that of TeCDDs 319.8965. An MS response of about 20,000 in high resolution gas chromatography/high resolution mass spectrometry (HRGC/HRMS) is needed for a unique distinction between them. Separation can be achieved based on the longer retention times of the congeneric PCDTs compared to PCDDs in non-polar SE-30 and DB-5 type columns.

This paper briefly reviews what is known on formation and sources of PCDTs and PCTAs in the environment. Synthesis of model compounds of PCDTs and PCTAs and the structure determination of these is described in more detail, as well as analytical methods and environmental fate of these. Finally, data on biological effects of PCDTs and PCTAs based on some preliminary toxicological investigations is given.

2
Formation and Sources of PCDTs and PCTAs

PCDTs can be formed in chemical processes analogous to those that lead to the formation of PCDFs. Condensation of chlorothiophenols instead of chlorophenols could lead to the formation of PCDTs. Other possibilities are reactions of elemental sulfur in the presence of some metal catalysts or reactions of some reactive sulfur compounds with polychlorinated biphenyls (PCBs).

The major known sources of PCDTs in environment are combustion and metallurgy. PCDTs are known to be formed in waste incineration [12, 13], incineration of PCB, and in metal recycling processes [14]. Still, PCDTs and also alkylated PCDTs are found, though in very low concentrations, in pulp mill effluents and sediments [15, 16]. Other potential sources of PCDTs are automobile exhaust, wood combustion, oil/gas heating, chemical production of PCBs and trichlorobenzene sulfonates, and sewage sludge [17]. One possible source is the ancient use of PCB/sulfur formulations to impart physical and chemical properties such as moisture resistance and flame resistance, adhesion properties, etc. of the products in the preparation of cloth, paper, and wood [17, 18]. Rappe et al. have suggested that PCDTs in the environment could be used as indicators of production and manufacture of iron and steel [19] and have proposed that 2468-TeCDT could be a unique chemical marker from a former 2,4,5-T manufacturing facility [20]. However, sulfur-containing chemicals are not used in the synthesis of 2,4,5-T [10, 21].

Organochlorine compounds such as chlorobenzenes, chlorophenols, PCBs, PCDDs, and PCDFs are formed as unwanted byproducts in metal reclamation. PVC and other chlorine-containing compounds in the raw materials and chlorine-containing chemicals in the reclamation process are possible precursors as well as the so-called de novo synthesis [22]. Gas phase samples from an aluminum smelter and a car shredder and ash have been found to contain PCDTs [14]. The ash was from different processes where the temperatures varied from 350 to 850°C. Raw materials in the aluminum smelter are floated aluminum, dried turnings, Al-sheets and Al-dross. NaCl/KCl is used in the flux. Temperature in the aluminum smelting is from 800 to 1200°C. The gases from the smelting process are cooled and cleaned by a baghouse filter. In the car shredder the cars are crushed and different materials are separated in a screening system. The highest temperatures during the shredding process are 600–700 °C. The gases from the car shredder are cleaned by a cyclone and a scrubber.

The isomer profiles of TeCDTs in the selected-ion monitoring chromatograms of stack gas samples from waste incineration and from pulp mill effluents are quite different. Still, the concentrations of TeCDTs in the stack gas samples were quite high, while the concentrations in pulp mill effluents were very low. The stack emissions contain more isomers which have shorter retention times in the HP-5 column than the pulp mill effluents. This may indicate that the TeCDTs are formed in a different way in waste combustion and in pulp bleaching [15]. SIM chromatograms with the exact value of the $(M+2)^+$ ion of TeCDTs (321.8759) from a stack gas sample and from a pulp mill effluent sample are shown in Fig. 1.

Due to good correlation between the concentration levels of 2378-TeCDD and 2468-TeCDT in sediment and crab samples from Passaic River, a common source of these was suggested [23]. It has been suggested that the elevated levels of 2468-TeCDT in the sediments from the Passaic River in New Jersey originate from a former 2,4,5-trichlorophenoxyacetic acid (2,4,5-T) manufacturing facility [11, 24]. Anyhow, it has been estimated that there are more than 300 PCDD and PCDF sources, which are also potential PCDT sources, within the area of Newark Bay estuary [25]. Still, Cai et al. suggest that 2468-TeCDT is not formed by a route analogous to that of the formation of 2378-TeCDD. The compound 2468-TeCDT might be formed as a by-product in chemical synthesis at the chemical plant or by some environmental processes from the materials discharged from the plant. No TCTA isomers were detected in the crab tissue samples [23].

Small amounts of TriCDTs and TeCDTs are formed in the combustion of mixtures of peat, wood chips, refuse derived fuel (RDF), and liquid packaging boards [26]. RDF was prepared from municipal waste. After pretreatment the waste was crushed and glass and metals were removed. The wood chips were obtained from a sawmill. The temperature used in the combustion experiments was at least 850 °C. The concentrations of TriCDTs and TeCDTs in fly ash were highest when mixtures of wood chips and refuse derived fuel were burned. The concentration levels of TriCDTs were found to be higher than those of TeCDTs. The total number of TeCDT isomers observed was about 15–25, while the number of TeCDD isomers was only 3–4. The fly ash samples from the combustion of pure peat did not contain any PCDTs. Particles formed in the combustion of peat are effectively removed except for the very small particles which pass the cleaning devices. In the atmosphere these are distributed in the surroundings.

OCTA and HeptaCTAs are formed, along with other polychlorinated aromatic sulfur compounds, as by products in the synthesis of pentachlorothiophenol from hexachlorobenzene. Pentachlorothiophenol is used as an additive for the improvement of the vulcanization process of rubber in the tire industry. The formation of OCTA is believed to occur by dimerization of pentachlorothiophenate and by intramolecular cyclization and the formation of HeptaCTAs by dechlorination of higher chlorinated thianthrenes [27].

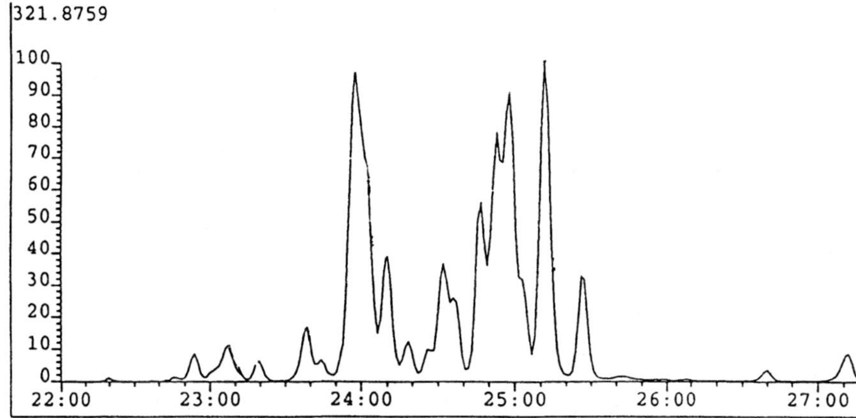

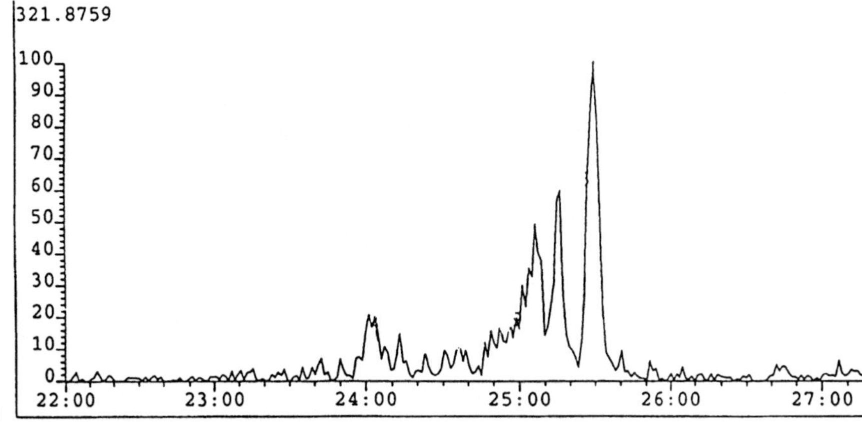

Fig. 1a, b. SIM chromatograms with the exact value of the $(M+2)^+$ ion of TeCDTs (321.8759): **a** from a stack gas sample; **b** from a pulp mill effluent sample. From [15]

3
Preparation of Model Compounds of PCDTs and PCTAs for Environmental Analysis

In the laboratory PCDTs have been prepared by reactions between PCBs, sulfur and aluminum chloride and by direct chlorination of dibenzothiophene [28–30]. The temperature strongly affects the formation of TeCDTs from tetrachlorobiphenyls and sulfur. Temperatures used were 120, 160, 200, and 240°C. Reaction times of 5–50 h were tried. The chlorinated biphenyls (50–200 mg) with sulfur (20–100 mg) and $AlCl_3$ (a few mg) were heated in closed tubes in an oven constructed for this purpose. After cooling the mixtures were extracted with toluene. More chlorine cleaving occurred at higher temperatures. In the case of a mixture of 3,4,3',4'-tetrachlorobiphenyl and 2,3,3',4'-tetrachlorobi phenyl the temperature of 120 °C was too low for the formation of any PCDTs from PCBs. At 240 °C abundant cleavage of chlorine occurred and TriCDTs were produced in great yields. GC/MS showed that the main products were two TriCDTs and three TeCDTs. If it is supposed that the ortho chlorines are preferably cleaved the main products would be 238-TriCDT, 267-TriCDT, 2378-TeCDT, 3467-TeCDT, and 2367-TeCDT. At 160 °C five TeCDTs were produced. Increase in the reaction time from 5 to 50 h did not significantly increase the yields which remained very low [30]. Figure 2 represents the synthesis of TeCDTs and TriCDTs from a mixture of 3,4,3′,4′-tetrachlorobiphenyl and 2,3,3′,4′-tetrachlorobiphenyl.

Buser and Rappe prepared PCDTs from a commercial PCB mixture (Aroclor 1254) which contains tri-, tetra-, penta-, and hexachlorobiphenyls as major compounds, and sulfur. A complex mixture of mono-, di-, tri-, tetra-, and penta-CDTs were obtained. The compound 2378-TeCDT which was found among the TeCDT isomers was supposed to originate from 2,4,5,2′,4′,5′-hexachlorobiphenyl which is abundant in Aroclor 1254 mixture. The yield of PCDTs was estimated to be about 1–2% when a temperature of 540 °C was used. At higher temperatures the yields were even lower [31].

Several TeCDTs have been synthesized by Buser and Rappe at high temperatures (540 °C) from series of individual hexachlorinated biphenyls and sulfur

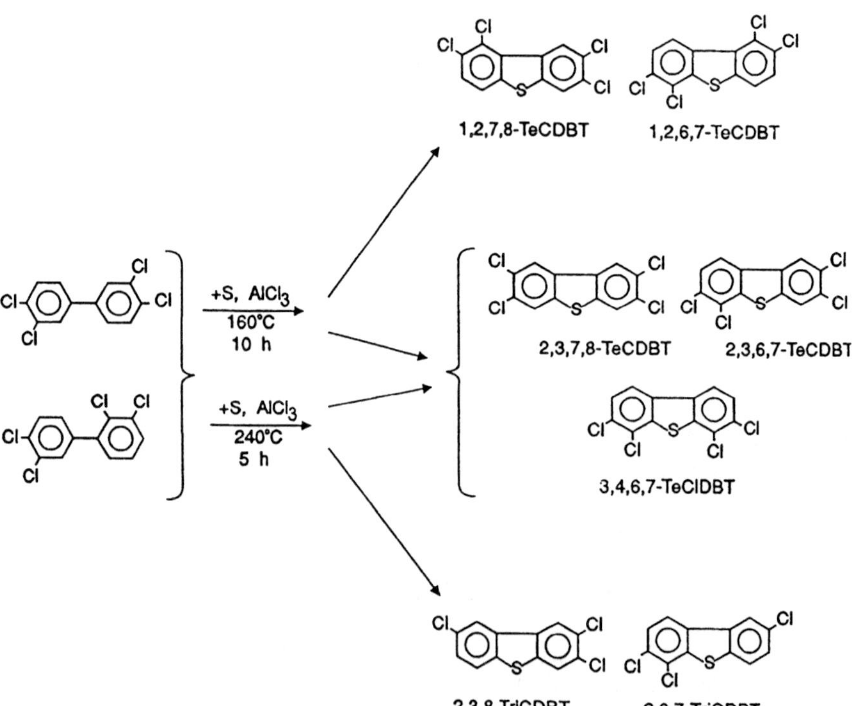

Fig. 2. Synthesis of TeCDTs and TriCDTs from a mixture of 3,4,3′,4′-tetrachlorobiphenyl and 2,3,3′,4′-tetrachlorobiphenyl. From [30]

[31]. Three isomers, 1469-TeCDT, 1269-TeCDT, and 1289-TeCDT, were observed among the products when 2,3,6,2',3',6'-hexachlorobiphenyl was used. TeCDTs were formed through the loss of ortho-substituted chlorine atoms. An inter-molecular cyclization pathway analogous to the reaction pathway which leads to the formation of TeCDFs from PCBs with oxygen could be confirmed [31].

Chlorination of dibenzothiophene produced a complex mixture of PCDTs with one to eight chlorine substituents. Mono- to tri- substituted congeners were most abundant. PCDTs have been found to elute in nonpolar columns in order of the number of chlorine substituents present and to have increased retention times compared to PCDDs [28, 31].

Large aluminum oxide (100–200 g Al_2O_3, Merck Aluminiumoxid 90, 70–230 mesh ASTM) columns and carbon (Alltech, SK 4, 80/100) columns with hexane/dichloromethane and toluene as eluents have been used to fractionate the mixtures of model compounds [29, 30]. Activated carbon chromatography managed to remove most starting compounds and some sulfur from the synthesis mixtures. The mixtures of PCDTs and PCTAs obtained were further fractionated and purified by reversed-phase HPLC with a Silasorb C8 SPH (ELSICO) and a Spherisorb S5 ODS-2 column. Acetonitrile:water (65:35) at a flow rate of 1 ml/min was used as eluent and the UV detector adjusted to 254 nm [30, 32].

Miltsov et al. [33] have tried different synthetic methods to prepare PCDTs. Altogether eight congeners were obtained. The main product of chlorosulfonation of DBT at ambient temperature was 2,8-disulfochloride, which was converted to 28-DiCDT and at elevated temperature 2,4,6,8-tetrasulfochloride, which was converted to 2468-TeCDT. PCDT sulfones were prepared from PCBs by heating with an excess of chlorosulfonic acid. Reduction of 37-DiCDT and 2378-TeCDT sulfones with $LiAlH_4$ gave good yields of 37-DiCDT and 2378-TeCDT.

PCTAs can be prepared by chlorination of thianthrene. The method of Buckholtz uses sulfuryl chloride as a chlorination agent in a mixture of *o*- and *p*-chlorotoluene [34]. For environmental studies small amounts of 2378-TeCTA and some other PCTAs have been prepared by a modified method of Buckholtz [32]. TriCTAs and TeCTAs were prepared by stepwise addition of sulfuryl chloride over 4 h at 60 °C. The degree of chlorination was found to be three to four (only tri- and tetrachlorinated thianthrenes were observed as reaction products) when all of the parent compound was consumed. One TriCTA and one TeCTA were obtained as main products. In addition, two other TriCTAs, four TeCTAs, and some PeCTAs were observed in minor concentrations. Because of the ortho- and para-directing properties of sulfur in electrophilic aromatic substitution reactions, 237-TriCTA and 2378-TeCTA, the thio analogue of 2378-TeCDD, were obtained as the main products. Mass spectrometry and ^{1}H NMR were used in the structure verification.

4
Structure Determination of the Model Compounds

4.1
Mass Spectroscopy

Mass spectroscopy alone does not provide any reliable way for an isomer-specific structure determination owing to the similarity of the mass spectra of PCDT and PCTA isomers. The EI mass spectra of PCDTs and PCTAs all have an intense molecular ion (M^+). The molecular ion and the fragment ions show the typical expected clustering due to chlorine isotopes. The EI mass spectra of PCDTs show a relatively small M^+-2Cl fragment and a smaller M^+-Cl fragment. A GC/MS total ion chromatogram (m/z 50–500) of a mixture of products from direct chlorination of dibenzothiophene by sulfuryl chloride and full scan EI mass spectra of one DiCDT, one TriCDT, and one TeCDT isomer are shown in Fig. 3. Tri- and TeCTAs show strong fragments due to the formation of M^+-Cl and M^+-2Cl ions [32].

4.2
^{1}H NMR Spectroscopy

^{1}H NMR spectroscopy offers a good method to differentiate polysubstituted DTs and TAs. The greatest drawback of ^{1}H NMR spectroscopy when compared to GC/MS is its low sensitivity. This can be partly compensated by the selection of proper solvent and measuring conditions. Sinkkonen et al. have used ^{1}H NMR spectroscopy in the identification of different PCDT and PCTA congeners [29, 30, 32]. All ^{1}H NMR spectra were measured with a Jeol GSX 270 FT NMR spectrometer working at 270.17 MHz. The spectrometer was equipped with a standard C/H dual probe at 30 °C. The following spectral settings were used: spectral width, 2800 Hz; number of data points, 32,000 giving a resolution of 0.17 Hz; flip angle, 8.4 µs (90 °C); acquisition time, 4 s; pulse delay, 1 s; and number of scans, > 1000. All FIDs were exponentially windowed by a line broadening factor of digital resolution prior to Fourier transformation to improve the S/N in the frequency spectra.

A method originally developed for terpenoid type off-flavor compounds was used [35]. By this method a ^{1}H NMR spectrum could be measured reliably from ca. 5 µg of monoterpenoid compound collected directly in the NMR solvent by preparative GC. A thick-wall (5 mm o.d., 2 mm i.d.) sample tube instead of a standard NMR tube was selected for the measurements in order to decrease the solvent volume from 700 to 100 µl. CH_2Cl_2 was found to be a suitable medium, because its ^{1}H NMR signal is locked at 5.30 ppm from tetramethylsilane (TMS). The ^{1}H NMR spectral interpretation was mainly based on the synthetic procedure used, symmetry considerations of possible products, and characteristic intra-aromatic couplings. More details on the NMR spectroscopy of PCDTs and PCTAs are reported elsewhere [29, 30, 32].

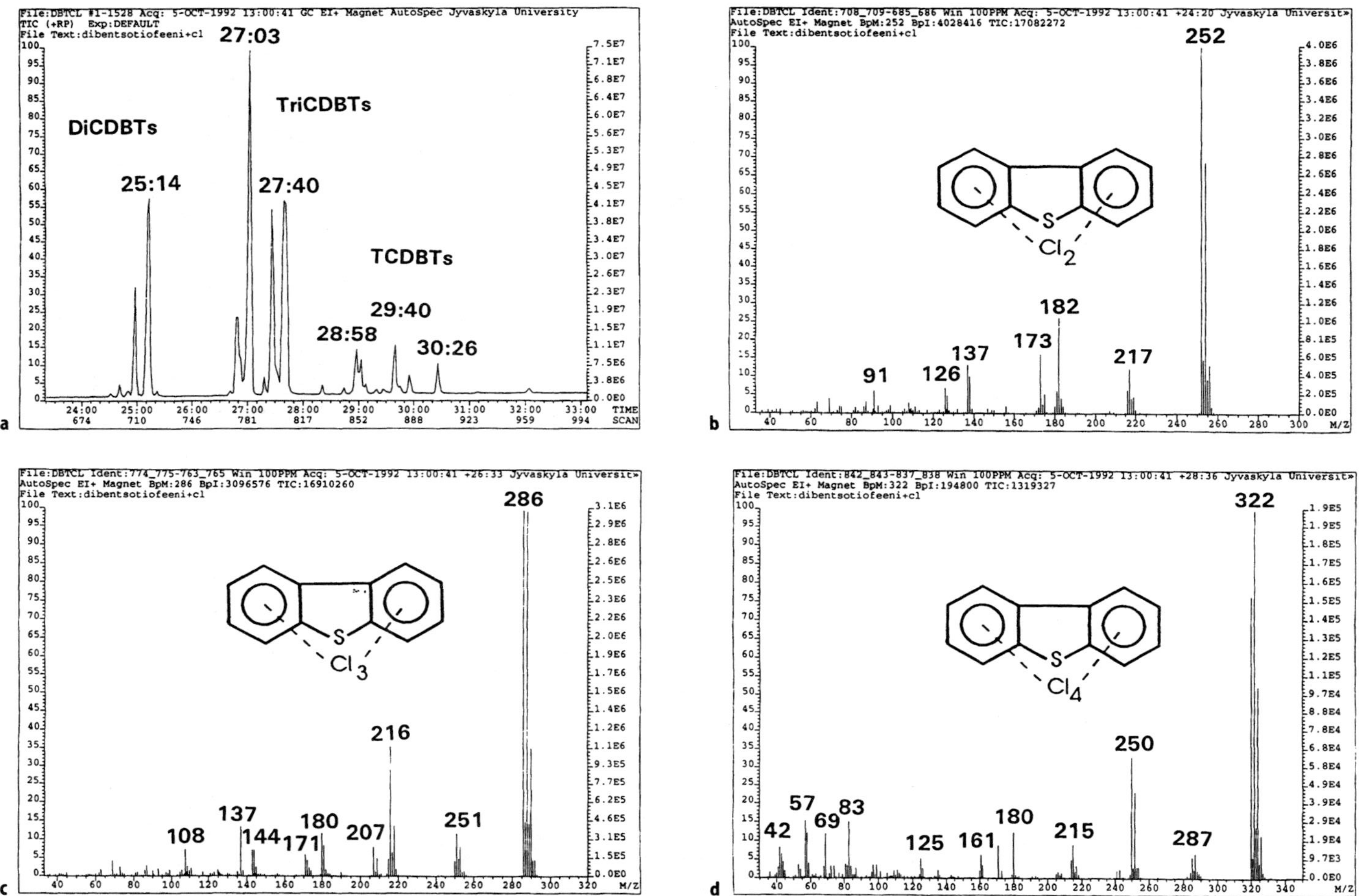

Fig. 3a–d. **a** GC/MS total ion chromatogram (*m/z* 50–500) of mixture of products from direct chlorination of dibenzothiophene by sulfuryl chloride. **b** Full scan EI mass spectra of one DiCDT isomer. **c** Full scan EI mass spectra of one TriCDT isomer. **d** Full scan EI mass spectra of one TeCDT isomer. From [32]

5
Physical and Chemical Properties

The total number of possible PCDT congeners, from mono- to octachlorinated, is 135, including 38 TeCDTs and 28 PeCDTs, and the total number of possible PCTA congeners is 75, including 22 TeCTAs and 14 PeCTAs. As to the 2,3,7,8-substituted congeners, there exist ten possible structures of 2,3,7,8-substituted PCDTs and seven possible 2,3,7,8-substituted PCTAs. The laterally substituted ten PCDT congeners (Fig. 4), like other planar aromatic dioxin-related substances, are assumed to be most persistent and toxic.

Data on the physical and chemical properties of PCDTs and PCTAs are scarce. Due to their structural similarity to PCDFs and PCDDs they are also supposed to possess some likeness in their physical and chemical properties. Sulfur and oxygen are both Group VI elements with two outer shell electrons available for covalent bonding. Structures of thiophene and furan with benzene carbon-sulfur (C_b-S) and carbon-oxygen bond (C_b-O), in PCDTs and PCDFs respectively, suggest similar chemical behavior. The bond dissociation energies (ΔH) show that less energy is required to break the C_b-S bond than the C_b-O bond [17, 36, 37].

Dibenzothiophenes show increasing bioconcentration factors and octanol/water partition coefficients with increasing alkyl substitution [38]. It is not known how persistent PCDTs and PCTAs are in the environment. However, alkylated dibenzothiophenes have previously been found to be persistent oil residue compounds in marine organisms and especially in sediments. DBT derivatives, which are abundant in crude oil, represent a group of most persistent oil residues in the marine environment. Alkyl substituents retard microbial degradation. However, some microbes are known which degrade DBT derivatives [7, 39–41]. Oxidation of the derivatives of dibenzothiophene to the respective sulfoxides and sulfones is possible [42].

In their investigations Pruell et al. calculated accumulation factors for PCB-153, 2378-TeCDD, 2378-TeCDF, and 2468-TeCDT from sediments to biota. The equation

$$AF = [C^i_{organism}/LIPID]/[C^i_{sediment}/TOC]$$

applies where $C^i_{organism}$ is the concentration of compound i in the organism (ng/g dry weight), $C^i_{sediment}$ is the concentration of compound i in sediment (ng/g dry weight), LIPID is the lipid concentration (g/g dry weight), and TOC is the total organic carbon concentration (g/g dry weight) [11, 43]. The accumulation factors in sandworms (*Nereis virens*), clams (*Macoma nasuta*), and shrimp (*Palaemonetes*) were found to be highest for PCB-153 (1.4–2.08). The accumulation factors calculated for 2378-TeCDD, 2378-TeCDF, and 2468-TeCDT were significantly lower than those of PCB-153, in the range of 0.24–1.01. The accumulation factors calculated for 2468-TeCDT in shrimp were significantly higher than the accumulation factors in sandworms and clams. Only trace amounts of 2468-TeCDT could be observed in striped bass (*Morone saxalis*) collected from the same area [11].

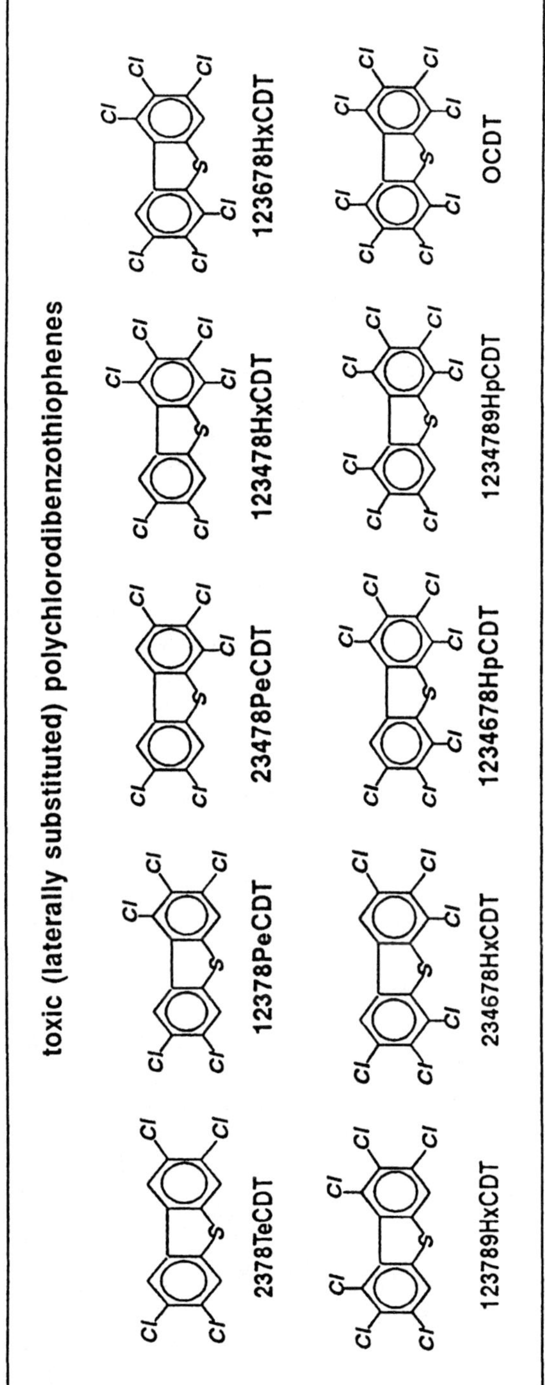

Fig. 4. Structures of the laterally substituted 10 PCDT congeners

6
Analytical Methods

6.1
Extraction and Separation

In chemical analytics PCDTs and PCTAs behave very similar to PCDDs and PCDFs. The chromatographic behavior of PCDTs is similar to that of PCDDs. Purification and fractionation with aluminum oxide and activated carbon originally developed for the analysis of PCDDs and PCDFs can be used [28, 44].

Reversed-phase (RP) TLC with RP-18 plates and acetonitrile-water eluents can be used to separate PCDTs from many interfering compounds [45]. In normal-phase NH_2-TLC with hexane, PCDTs elute faster than many non-chlorinated interfering compounds. If PCDTs are oxidized to the corresponding sulfones, they can be separated by RP-TLC or by RP-HPLC because of the difference in their polarity, but this makes their further analysis by GC more difficult [46, 47].

6.2
Gas, Ash, and Slag Samples

Standard procedures developed for the analysis of PCDDs and PCDFs have been used as follows: 15–20 g of ash or slag sample was Soxhlet-extracted with 220 ml of toluene for 48 h. Toluene was evaporated and the residues were dissolved in 3 ml of hexane, which was then washed three times with concentrated sulfuric acid. Activated carbon (SK-4, 80/100 mesh, Alltech) and basic aluminum oxide (Alumina B, Activity I, ICN Biomedicals) column chromatography were used for the separation of the planar aromatic compounds (PAC) fraction. The first fraction from the carbon column was eluted with 10 ml of hexane-methylene chloride (1:1; v/v) and the second fraction (PACs) with 10 ml of toluene after the column was turned upside down [26].

6.3
Pulp Mill Effluents and Sediments

Alkaline and acid effluents from softwood and hardwood plant from different chlorination stages and input and output effluents from two separate production lines of a modern pulp mill have been investigated [16]. Chlorine, chlorine dioxide, and oxygen were used in the bleaching process. The effluent water samples were filtered twice with Whatman Glass Microfibre Papers GF/D (2.7 μm) and then with Whatman Microfibre GF/F (0.7 μm). The filters with the particles and the water filtrates were analyzed separately. The filters were dried, weighed, and Soxhlet-extracted with toluene for 48 h. The water filtrates were extracted twice with hexane. The $^{13}C_{12}$-labeled 2378-TeCDD was used as an internal standard. The clean-up by activated carbon and basic alumina column chromatography was done as described above. In the case of sediments, the freeze-dried sediment samples were weighed and Soxhlet-extracted for 48 h with toluene. The analysis was continued as with the water samples.

6.4
Biota

Sediment and biota samples have both been analyzed by Pruell et al. by the same method. About 10 g of sediment or tissue sample were mixed with sodium sulfate. The samples were Soxhlet-extracted with an acetone:pentane (30:70, v/v) solvent mixture. Four subsequent column chromatographic purifications with series of layers of activated silica, potassium silicate, sulfuric acid treated silica, sodium sulfate, silver nitrate treated silica, neutral alumina, and activated carbon as adsorbents were used [11].

Cai et al. digested about 10 g of crab tissue with a 30% KOH solution in a mixture of water and ethanol (3:1) at room temperature. Sulfuric acid washing and column chromatography with nonactivated silica, neutral alumina, and activated carbon/silica were used in the purification of the samples [48].

6.5
Coniferous Needles

Coniferous needles were separated due to different age class and dried for 48 h at room temperature. For the analyses of the compounds in the surface wax layer, the needles were extracted twice for 0.5 min with dichloromethane by ultrasonic treatment [49, 50]. The dichloromethane was evaporated and the residue dissolved in hexane. After extraction with dichloromethane the needles were dried and crushed. The crushed needles were extracted twice with hexane by ultrasonic treatment for 1 h. The hexane extracts were evaporated and the analysis continued as above.

7
GC and GC/MS of Environmental Samples

TeCDTs and TeCDDs have in low-resolution MS the same values of M^+, $(M+2)^+$ and $(M+4)^+$ ions. A resolution of about 20,000 is needed for the mass spectrometric separation of TeCDTs and TeCDDs, when the monitoring is done with the exact values of the M^+ ions (m/z 319.8788 for TeCDTs and 319.8965 for TeCDDs). In non-polar GC columns such as SE-30 and HP-5 the retention times of congeneric PCDTs are longer than those of PCDDs. Anyhow, in the case of environmental samples with complex sample matrix and low concentrations, identification and analysis by low resolution GC/MS is not possible due to several interfering peaks eluting in the same GC/MS window as the PCDTs [28]. It is possible that in some previous investigations with low resolution GC/MS some PCDTs erroneously have been analyzed as slowly in GC eluting PCDDs.

A VG AutoSpec high resolution mass spectrometer connected to an HP 5890 Series II gas chromatograph has been used in the determination of PCDTs and PCTAs [14, 15, 32]. A 25 m HP-5 (0.2 mm, 0.11 µm) column was used. The temperature of injector was 260 °C, transfer line 280 °C, and ion source 260 °C. The electron ionization potential was 36 eV. The GC temperature program was 100 °C (1 min) to 20 °C/min to 180 °C to 5 °C/min to 280 °C (15 min). Selected ion monitoring (SIM) mode with the exact values of the M^+, $(M+2)^+$ and $(M+4)^+$ ions was applied. Simultaneously, PCDDs and PCDFs were often analyzed in the samples. A resolution of about 20,000 was used in the analysis of environmental samples to achieve the best possible separation of PCDTs from PCDDs and other interfering compounds. The authentic intensity ratios of the peaks, $M^+/(M+2)^+$ and $(M+2)^+/(M+4)^+$, were used to confirm the analyses of proper compounds.

Buser and co-workers have used different GC/MS techniques in the investigation of PCDTs in fly ash samples and in aquatic organisms [12, 31]. GC/MS and MS/MS were used for detection and to differentiate PCDTs from PCDDs. Various ionization [electron ionization (EI) and negative-ion chemical ionization (NCI)] techniques and MS/MS experiments monitoring different daughter ions on a hybrid instrument were used to distinguish PCDTs from PCDDs [31].

In electron impact (EI) mass spectra PCDTs show a strong molecular M^+ ion and the expected clustering due to chlorine and sulfur isotopes. The major difference in the EI mass spectra of the PCDDs and PCDTs is the formation of a strong M^+-COCl in the former and a strong M^+-2Cl in the latter compounds. PCDDs and PCDFs could be detected with good sensitivity via the reaction $M^+ \rightarrow (M\text{-}COCl)^+$ by QUAD MS/MS. Mass-analyzed ion kinetic energy (MIKE) MS/MS was found suitable for detecting the PCDTs via the reaction $M^+ \rightarrow (M\text{-}2Cl)^+$, whereas PCDDs and PCDFs were monitored by the reaction $M^+ \rightarrow (M\text{-}COCl)^+$ [31].

The major fragments in the low resolution EI mass spectrum of 2468-TeCDT were m/z 285 for [M–Cl], m/z 287 for [M+2Cl], m/z 250 for [M–2Cl], and m/z 252 for [M+2–2Cl] [23].

Cai et al. used the mass-profile monitoring technique that was applied in a GC/HRMS-SIM analysis in the identification of 2468-TeCDT in the crab tissue samples [23, 51]. Confirmation was done by high-resolution peak matching and by a full spectrum by GC/MS. The high-resolution peak matching was done by switching from two perfluorokerosene (PFK) reference ions to the theoretical masses of three TeCDT molecular ions during the elution time window of the unknown compound. The mass peaks of the unknown were within 3 ppm of the theoretical values. The accurate masses of three molecular ions, the full electron ionization mass spectrum, and the chromatographic retention time coincided with those of the 2468-TeCDT standard. The retention time of 1289-TeCDD was very close to that of 2468-TeCDT. In a similar way, a PeCDT isomer was studied. With mass-profile monitoring a mass difference as small as 0.0032 – 0.0046 mass unit could be separated. In the case of PeCDDs and PeCDTs the mass difference was 0.0177 mass unit. In the conventional peak-top mode a mass resolving power of 100,000 would be needed for complete removal of interference between PeCDDs and PeCDTs [23].

8
Environmental Fate; Concentrations in Emissions and in Environmental Samples

The fate of PCDTs and PCTAs released into the environment is quite unknown. For example, the environmental persistence of PCDTs due to photolysis is not known. The water solubility of these compounds is very low so it can be supposed that most PCDTs and PCTAs in water and also in air can be found to be adsorbed onto particles. In the environment PCDTs, until now, have been analyzed in sediments and aquatic organisms. However, the environmental concentrations of PCDTs reported in literature to date are not fully reliable, because these are usually based on the assumption of quite similar mass spectrometric responses of PCDDs and PCDTs in the HRGC/HRMS analyses of these. In several investigations ^{13}C-labeled 2378-TeCDD has been used as an internal standard. In the case of 2468-TeCDT it has been reported that it coelutes with 1289-TeCDD in GC with the SP-2331 column [12]. The SP-2331 column has been generally used in PCDD/PCDF analyses during the 1980s. It is possible that in some previous investigations 2468-TeCDT has been analyzed as 1289-TeCDD. In the modern DB-5 columns, nowadays generally used in PCDD/PCDF analyses, no coelution of 2468-TeCDT and 1289-TeCDD occurs [17].

The concentrations of TriCDTs, TeCDTs, and PeCDTs in soil and sediment samples, gas samples from waste incineration and aluminum smelting, ash from an aluminum smelting plant, a car shredder, and from combustion of wood chips, peat and refuse derived fuel, different effluents from a pulp and paper mill, and crab, carp, and lobster tissue samples, are presented in Table 1.

Peterman was probably the first to detect PCDTs in the environment. Peterman et al. have calculated PCDT concentrations as high as 67,000 ng/kg in soil samples from a waste pit [52, 53]. Electrical capacitors which contained PCBs (Aroclors 1242 and 1254) had been incinerated at this waste pit located in Crab Orchard National Wildlife Refuge, IL.Cl$_2$–Cl$_7$ dibenzothiophenes were tentatively identified by quadrupole full-scan GC/MS. Although the PCDTs have the same nominal molecular masses as the PCDDs, they elute 1–4 min later from a 30 m DB-5 capillary column. The concentrations of the individual PCDT congeners were in the range 0.05–10.0 ng/g. High resolution accurate mass measurements were performed. Definite accurate mass measurements (AMMs) were obtained for 19 separate Cl$_2$–Cl$_5$ PCDTs. AMMs were obtained for at least two ions of each PCDT molecular ion cluster to compute its elemental composition. The AMMs of the M$^+$ and (M+2)$^+$ ions of the 19 PCDT congeners (except one) were within 10 ppm of the exact masses of the PCDTs' elemental compositions [53]. The full-scan mass spectra of the PCDTs showed very abundant molecular ion clusters and the typical (M–70)$^+$ fragment ions.

Hilker et al. have reported one of the very few observations of 2378-TeCTA in the environment [8]. The compound 2378-TeCTA was analyzed in sediment samples from a sanitary sewer near a chemical plant in Niagara Falls, New York. Surprisingly high dioxin-like activity was detected in purified sample extracts. The compound 2378-TeCTA was identified and confirmed by high-resolution mass spectrometry. Full high-resolution mass spectra were acquired. The exact mass measurements of several peaks in these spectra indicated the presence of sulfur and possible molecular formulas of $C_{12}H_5S_2Cl_3$ and $C_{12}H_4S_2Cl_4$. The latter compound is patented by a chemical company located quite near to the sewer where the samples were taken. Confirmation was obtained by synthesizing an authentic 2378-TeCTA. The full-scan EI mass spectra of 2378-TeCTA showed an intense molecular-ion cluster from *m/z* 352 to 358, with the ratio intensities

Table 1. Concentrations of TriCDTs, TeCDTs, and PeCDTs in soil and sediment samples, gas samples from incineration and aluminum smelting, ash from different combustions, pulp mill effluents, and some biological samples

Sample	TriCDTs	TeCDTs	PeCDTs	References
Soil samples from a waste pit where electical PCB capacitors have been incinerated		PCDTs = 67,000 ng/g		Peterman et al. (1986) [52]
Crab, Elisabeth, Newark Bay Estuary	140 pg/g	2468-TeCDT = 8300 pg/g TeCDTs = 8800 pg/g	1300 pg/g 240 pg/g	Buser and Rappe (1991) [31]
Lobster, Elisabeth, Newark Bay Estuary	Nd	2468-TeCDT = 1000 pg/g TeCDTs = 1000 pg/g	60 pg/g 25 pg/g	Buser and Rappe (1991) [31]
Crab, Värö, Sweden	Nd	2468-TeCDT = 45 pg/g TeCDTs = 75 pg/g	40 pg/g 80 pg/g	Buser and Rappe (1991) [31]
Fly ash from electrofilters from four incinerators	Na	< 2–25 ng/g	2–30 ng/g	Buser et al. (1991) [12]
Crab tissue samples, muscle, Newark/Raritan Bay system, September 1991	Na	2468-TeCDT = 480–610 ppt	na	Cai et al. (1994) [23]
Crab tissue samples, hepatopancrease, Newark/Raritan Bay system, September 1991	Na	2468-TeCDT = 290–10,000 ppt	na	Cai et al. (1994) [23]
Crab tissue samples, muscle, Newark/Raritan Bay system, June 1992	Na	2468-TeCDT = 260–270 ppt	na	Cai et al. (1994) [23]
Crab tissue samples, hepatopancrease, Newark/Raritan Bay system, June 1992	Na	2468-TeCDT = 310–4600 ppt	na	Cai et al. (1994) [23]
Acid effluent, softwood kraft, particles > 2.7 μm	Na	3467-TeCDT = 21 pg/l 2378-TeCDT = 55 pg/l Unknown isomer = 17 pg/l	nd	Sinkkonen et al. (1992) [15]
Alkaline effluent, softwood kraft, particles > 2.7 μm	Na	3467-TeCDT = 5 pg/l 2378-TeCDT = 8 pg/l Unknown isomer = 2 pg/l	nd	Sinkkonen et al. (1992) [15]

Acid effluent, hardwood kraft, particles >2.7 μm > 2.7 μm	Na	3467-TeCDT = 1 pg/l 2378-TeCDT = 3 pg/l Unknown isomer < 1 pg/l	nd	Sinkkonen et al. (1992) [15]
Alkaline effluent, harwood plant, particles > 2.7 μm	Na	3467-TeCDT < 1 pg/l 2378-TeCDT < 1 pg/l Unknown isomer <1 pg/l	nd	Sinkkonen et al. (1992) [15]
Input effluent, activated sludge treatment plant, particles > 2.7 μm	Na	3467-TeCDT < 1 pg/l 2378-TeCDT = 3 pg/l Unknown isomer <1 pg/l	nd	Sinkkonen et al. (1992) [15]
Output effluent, activated sludge treatment plant, particles >2.7 μm	Na	3467-TeCDT < 1 pg/l 2378-TeCDT < 1 pg/l Unknown isomer < 1 pg/l	nd	Sinkkonen et al. (1992) [15]
Passaic River sediment	Na	2468-TeCDT = 3 680 pg/g, mean (n = 6), SD = 1 380	na	Pruell et al. (1993) [11]
Hepatopancreas, male blue carbs, Newark Bay	Na	14,800 ng/kg	na	Cooper (1993) [24]
Acid and alkaline effluents, pulp mill, water + particles	210 – 2560 pg/l	< 610 pg/l	na	Sinkkonen et al. (1994) [16]
Input effluent, activated sludge treatment plant, water + particles	1490 pg/l	1740 pg/l	na	Sinkkonen et al. (1994) [16]
Output effluent, activated sludge treatment plant, water + particles	360 pg/l	nd	na	Sinkkonen et al. (1994) [16]
Discharge to sea, water+particles	170 pg/l	40 pg/l	na	Sinkkonen et al. (1994) [16]
Surface sediment, 0.5 km from a pulp mill	20 pg/g dry weight	2 pg/g dry weight	nd	Sinkkonen et al. (1994) [16]

Table 1 (continued)

Sample	TriCDTs	TeCDTs	PeCDTs	References
Sediment, depth 9–12 cm, 0.5 km from a pulp mill	10 pg/g dry weight	nd	nd	Sinkkonen et al. (1994) [16]
Surface sediment, 9 km from a pulp mill	10 pg/g dry weight	nd	nd	Sinkkonen et al. (1994) [16]
Gas sample from aluminum smelting plant, before baghouse filter	98.69–846.44 ng/Nm^3	135.52–367.36 ng/Nm^3	66.41–154.69 ng/Nm^3	Sinkkonen et al. (1994) [14]
Gas sample from aluminium smelting plant, after baghouse filter	41.61–54.97 ng/Nm^3	9.04–34.33 ng/Nm^3	1.28–9.45 ng/Nm^3	Sinkkonen et al. (1994) [14]
Ash from filter in the aluminum smelting palnt	0.53 ng/g	0.39 ng/g	0.10 ng/g	Sinkkonen et al. (1994) [14]
Ash from cyclone in the car shredder plant	8.94 ng/g	5.83 ng/g	1.43 ng/g	Sinkkonen et al. (1994) [14]
Fly ash from combustion of wood chips	430 pg/g	40 pg/g	na	Sinkkonen et al. (1995) [26]
Fly ash from combustion of 85% of wood chips and 15% of RDF	560 pg/g	< 30 pg/g	na	Sinkkonen et al. (1995) [26]
Fly ash from combustion of 70% of wood chips and 30% of RDF	60 pg/g	< 30 pg/g	na	Sinkkonen et al. (1995) [26]
Fly ash from combustion of 55% of wood chips and 45% of RDF	40 pg/g	< 30 pg/g	na	Sinkkonen et al. (1995) [26]
Fly ash from combustion of milled peat	60 pg/g	< 30 pg/g	na	Sinkkonen et al. (1995) [26]
Fly ash from combustion of 85% of peat and 15% of RDF	50–90 pg/g	30–40 pg/g	na	Sinkkonen et al. (1995) [26]
Fly ash from combustion of 70% of peat and 30% of RDF	40 pg/g	< 30 pg/g	na	Sinkkonen et al. (1995) [26]

na = not analyzed, nd = not detected, RDF = refuse derived fuel.

characteristic for tetrachlorinated compounds. The major clusters of fragment ions were from m/z 317 to 323 due to loss of one chlorine atom and from m/z 320 to 326 due to loss of one sulfur atom. The second major ion cluster is between m/z 282 and 288 due to loss of molecular chlorine from the molecular ion and loss of chlorine atom from $C_{12}H_4SCl_4^+$ [8].

Later, in 1992, Benz et al. measured OCTA concentrations in soil, compost, and sewage sludge samples. The concentrations ranged from 1.0 ng/kg to 45 ng/kg [27]. The OCTA concentrations were very low as compared to the OCDD concentrations (< 0.5%). All samples were found to contain OCTA. The concentrations were lowest in the soil samples and highest in the sewage sludge samples.

In 1991 Buser et al. identified PCDTs in fly ash from two municipal solid waste incinerators and from an electric-arc furnace of a car-shredding facility [12]. TeCDTs with three to four major and eight to nine minor isomers were detected. The 2378-TeCDT isomer was included in the TeCDTs and PeCDTs detected in the fly ash samples. The major TeCDT isomers identified in the fly ash samples were 2367-TeCDT, 2378-TeCDT, and 2468-TeCDT. Buser et al. have estimated that the concentrations of PCDTs in fly ash were up to 55 ng/g, which was one order of magnitude below the concentrations the PCDDs and PCDFs in these samples [12].

In Finland in 1991 two different samples from waste combustion were found to contain TeCDTs and PeCDTs [13]. Several TeCDTs and PeCDTs were detected in both samples. In one sample the concentrations seemed to be quite high. Quantitative determinations could not be made due to the unavailability of pure model compounds of TeCDTs and PeCDTS and due to the scarce information about the samples. Selected ion monitoring chromatograms with a resolution of 20,000 with the values 319.8788 and 321.8758 for TeCDTs [exact values of the M^+ and $(M+2)^+$ ions of TeCDTs] and 353.8398 and 355.8369 for PeCDTs [exact values of the M^+ and $(M+2)^+$ ions of PeCDTs] from the other waste incineration sample and with the exact values M^+ of some model compounds are shown in Fig. 5. The sample seemed to contain at least 14 different TeCDT isomers and at least 12 different PeCDT isomers, which eluted within the retention window of the prepared model substances and which had the proper intensity ratios of the M^+ and $(M+2)^+$ ions in the SIM chromatograms. As the mass spectrometer response for PCDTs did not decrease when the MS resolution was increased from 10,000 to 20,000, this still confirmed these compounds to be PCDTs without any interference from other compounds. This sample consisted of the gas phase only.

Later, additional stack gas samples from two different waste incinerators were found to contain TeCDTs and PeCDTs. Two samples contained the gas phase only, and two samples contained both the gas phase and particles. One of the stack gas samples which did not contain any TeCDTs and PeCDTs was strongly suspected to contain TriCTAs and TeCTAs [32]. These samples had previously been found to contain some tri- and tetrachlorinated diphenylsulfides. The detailed analysis of these samples for the chlorinated sulfides have been reported elsewhere [54].

The formation of organochlorine compounds in the bleaching process in pulp and paper mills has been cut down by replacement of chlorine by chlorine

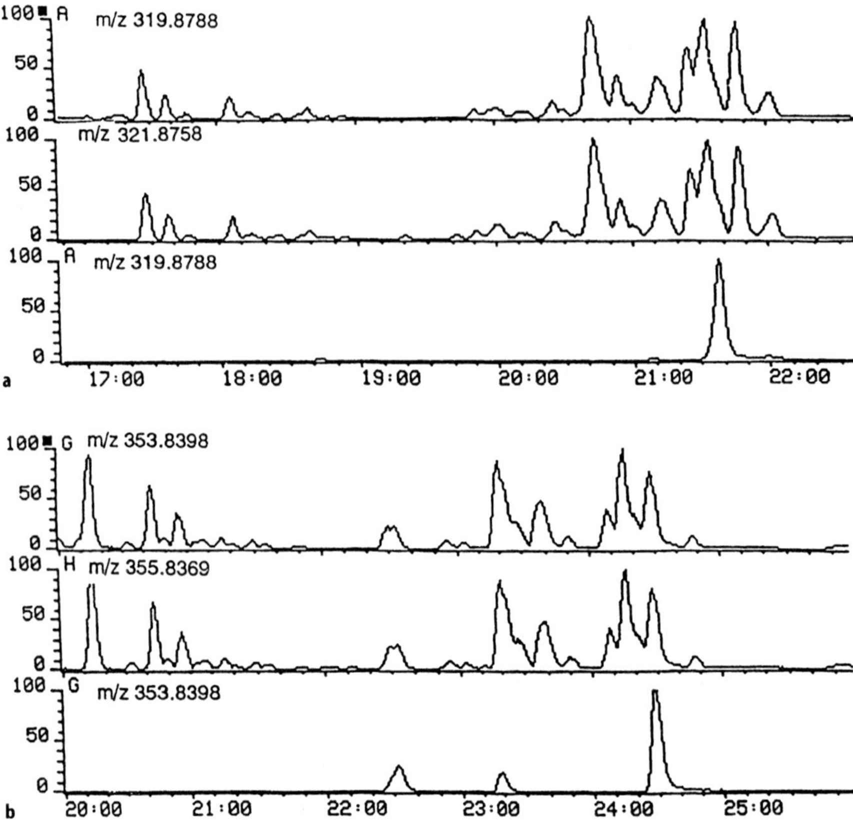

Fig. 5a, b. Selected ion monitoring chromatograms with a resolution of 20,000 with the values: **a** 319.8788 and 321.8758 for TeCDTs [exact values of the M^+ and $(M+2)^+$ ions of TeCDTs]; **b** 353.8398 and 355.8369 for PeCDTs [exact values of the M^+ and $(M+2)^+$ ions of PeCDTs] from a waste incineration sample and with the exact values of M^+ from some model compounds. From [13]

dioxide. Emissions have been reduced by installation of modern active sludge treatment plants. The concentrations of PCDDs, PCDFs, and PCBs in the pulp mill effluents and in recipient waters have been dramatically reduced. The concentrations of on particles adsorbed PCDTs in pulp mill effluents have also been found to be very low. Mainly TeCDTs have been analyzed [15]. Three TeCDT isomers were observed. The concentrations of these were estimated to be below 60 pg/l. The compound 2378-TeCDT was found to be the dominating TeCDT isomer in the effluents from a modern pulp mill in Finland [15]. It is parallel to the occurrence of the oxygen analog 2378-TeCDF as the major toxic dioxin/furan congener formed in chlorobleaching [55]. PeCDTs or any higher chlorinated PCDT congeners were not observed.

Later, TriCDTs were found to be the most abundant PCDT congeners in pulp mill effluents and in the sediments from the nearby recipient watercourse [16]. Both, the water filtrates and the particles in the effluents were analyzed. TriCDT concentrations ranged from 170 to 2560 pg/l in the effluents. The compounds 148-TriCDT and 238-TriCDT seemed to be the dominant TriCDT isomers in the effluents. Total ion GC/MS chromatograms with the exact value of $(M+2)^+$ ion of TriCDTs from three pulp mill effluents and the chlorination mixture of dibenzothiophene are shown in Fig. 6.

TeCDT concentrations (water+particles) were < 600 pg/l. The concentrations of TriCTAs in the effluents ranged from 90 to 420 pg/l and those of TeCTAs were < 180 pg/l. The concentrations of PCDTs and alkylated PCDTs in the sediments were very low. The highest concentrations of TriCDTs were 20 pg/g dry weight and the highest concentrations of TeCDTs were 2 ng/g dry weight in surface sediment samples 0.5 km from the pulp and paper mill.

PCDTs have been analyzed in some sediments and aquatic organisms in Passaic River. The concentrations of 2468-TeCDT in surface sediments [11] and in aquatic organisms [24] exceeded the concentrations of 2378-TeCDD and 2378-TeCDF. Concentrations of 2468-TeCDT as high as 3680 ng/kg have been found in the surface sediments in Passaic River, New Jersey [11]. The concentrations of 2378-TeCDD and 2378-TeCDF were substantially lower.

Cai et al. have identified TeCDTs and PeCDTs in tissues of crabs from Newark/Raritan Bay system [23].The concentration levels of 2468-TeCDT were found to be more than five- to tenfold compared to the concentrations of 2378-TeCDD. The compound 2378-TeCDD was the major dioxin congener found in the crab tissues. Cooper also observed a similar trend in concentrations in the hepatopancreas of male blue crabs from Newark Bay, Raritan Bay, and the Arthur Kill [24]. Concentrations as high as 14,800 ng/kg were observed in the hepatopancreas of male blue crabs [24].The total concentrations of TeCDTs in the male blue crabs were highest in Newark Bay, 14,800 ng/kg, and much lower, 300 and 490 ng/kg, in Sandy Hook and Raritan Bay, respectively. The concentrations of TeCDDs were in the range of 50–940 ng/kg.

The concentration levels of TriCDTs, TeCDTs, and PeCDTs in emissions and waste from an aluminum smelter and a car shredder have been found to be comparable to those of PCDDs and PCDFs. The number of individual TeCDT isomers was larger than the number of TeCDD isomers. PCDTs were effectively removed by the baghouse filter, more effectively than the PCDDs. Different TeCDT isomers had different adsorption tendencies on the filter system. Probably some samples from the car shredder contained some TriCTAs and possibly also some TeCTAs in low concentrations. A very large temporal variation can exist in the concentrations of organochlorine compounds in the emissions from car shredder plants because the raw material used is not very homogeneous [14].

TriCDTs and TeCDTs have been determined in aerosol, particle, and gas phase samples from an aluminum smelting plant. The concentrations of TriCDTs ranged from 40 to 900 ng/Nm^3, the concentrations of TeCDTs from 10 to 400 ng/Nm^3, and those of PeCDTs from 1 to 160 ng/Nm^3 [14]. Due to the scarce number of pure TeCDT model compounds available, the isomerism of most of the TeCDTs in the samples could not be determined. In any case, the concentrations of 2378-TeCDT seemed to be very low. The number of individual TeCDTs in the samples from the aluminum smelter was much higher than the

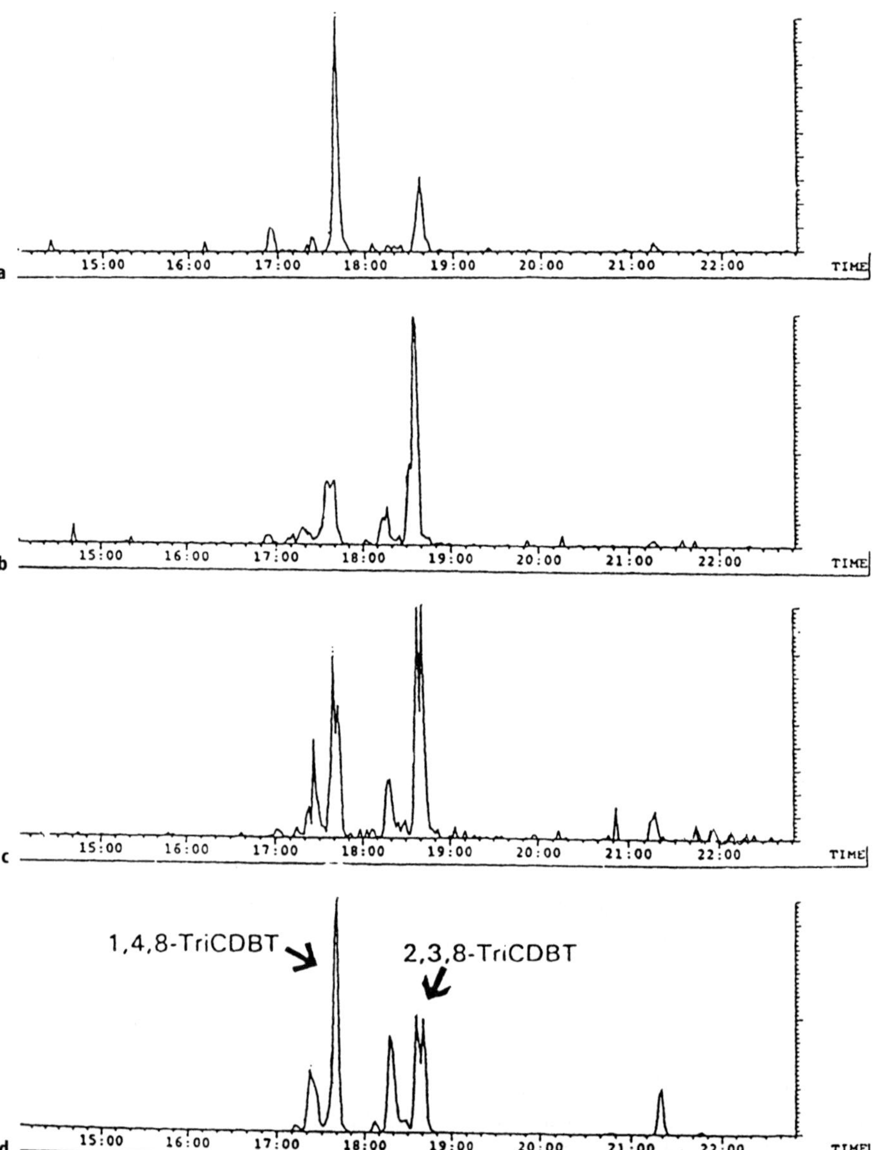

Fig. 6a–d. Total ion GC/MS chromatograms with the exact value of $(M+2)^+$ ion of TriCDTs from three pulp mill effluents; **a** input to activated sludge treatment plant; **b** discharge to sea; **c** alkaline effluent from hardwood kraft; **d** the chlorination mixture of dibenzothiophene. From [16]

number of TeCDDs. The total concentrations of TeCDTs often approached that of TeCDDs in the samples from the smelter. The increased ratio of PCDDs/ PCDTs in the samples taken after the filters indicated the better capacity of the filters to remove PCDTs than PCDDs.

Pine needles from the surroundings of metal reclamation plant probably contained PCDTs in low concentrations. Quantitative determinations were not made due to much interference by other compounds of similar type [49]. The concentrations of TeCDDs were in the range of 5–50 pg/g per dry weight of needles and the concentrations of TeCDFs in the range of 4–15 pg/g. Quite high concentrations of alkylated dibenzothiophenes in pine needles in a recipient area of the emissions from a pulp and paper mill have been measured, which might suggest that PCDTs could also be taken up by needles [56].

9
Biological Effects

From the ecotoxicological and toxicological point of view, PCDTs and PCTAs are interesting compounds due to their structural resemblance with PCDFs and PCDDs. Until now toxicological knowledge of PCDTs and PCTAs has been quite scarce.

The CYP1A1-inducing potencies of 2378-TeCDT, 2378-TeCTA, and 2378-TeCDD have been analyzed, measured as aryl hydrocarbon hydroxylase (AHH) and 7-ethoxyresorufin O-deethylase (EROD) activities in mouse hepatoma cell culture Hepa-1 [9]. Marked differences in the induction potencies were found. The estimated EC50 values for 2378-TeCDD, 2378-TeCTA, and 2378-TeCDT were about 8 pmol/l, 700 pmol/l and 7.5 nmol/l, respectively. It could be concluded that replacement of oxygen atoms by sulfur greatly decreases the potency of a compound to induce AHH/EROD activities. The compound 2378-TeCTA seemed to be slightly cytotoxic. The toxic equivalency factors (TEFs) for PCDDs and PCDFs related to 2378-TeCDD, which has been given a TEF of 1.0, have been calculated by Safe and coworkers [1, 57]. The TEF of 2378-TeCTA was estimated to be 0.01 and the TEF of 2378-TeCDT close to or lower than 0.001 [9]. This indicates that the toxic potencies of these compounds could be close to those of 2,3,7,8-heptachlorinated dioxins and octachlorinated dibenzodioxin/furan. The lower AHH/EROD induction potencies of the sulfur analogues may also be due to metabolism of the compounds. Anyhow, the effect of metabolism may be of minor importance because the electronic structures of these compounds are stable due to uniform distribution of the frontier orbitals and partial charges over the whole molecules.

The biological potencies of 2378-TeCTA and 2378-TeCDT could not be predicted by comparative molecular field analysis (CoMFA, [58]) with a model designed by Poso et al. [59]. The positive point charge of sulfur in 2378-TeCTA and 2378-TeCDT is possibly the reason for the poor performance of the model. As point charges of oxygen in 2378-TeCDD and in its analogues are strongly negative, the sulfur atoms in 2378-TeCTA and 2378-TeCDT are clearly positive.

Preliminary toxicological evaluation of three PCDTs – 3467-TeCDT, 2378-TeCDT, and 134-TriCDT – has been carried out in AH-responsive mice [60].

PCDTs were shown to be considerably less toxic than 2378-TeCDD and the most toxic PCDFs or PCBs. The compound 2378-TeCTA, the sulfur analogue of 2378-TeCDD, has been suggested to have dioxin-like activity. Bioassays which detect the ability of 2378-TeCDD and related compounds to inhibit cell divisions in a mouse epithelial cell culture were used [8]. In in vivo studies 2378-TeCTA has been shown to be less toxic than 2378-TeCDD [61].

A recent paper of Weber et al. reports on rapid elimination of 2378-TeCTA both in vitro and in vivo [62]. 2378-TeCTA was rapidly eliminated in mouse liver and whole body. However, a weekly dosage of 1 mg 2378-TeCTA per kg body weight over 6 weeks did not cause weight loss or other signs of overt toxicity in male mice.

References

1. Safe S (1990) Crit Rev Toxicol 21:51
2. Rappe C, Buser H-R (1989) In: Kimbrough RD, Jensen AA (eds) Halogenated biphenyls, terphenyls, naphthalenes, dibenzodioxins and related products. Elsevier, Amsterdam, p 71
3. Karcher W, Nelen A, Depaus D, Van Eijk J, Glaude P, Jacob J (1981) In: Cooke M, Dennis R (eds) Polynuclear aromatic hydrocarbons: chemical analysis and biological fate. Battelle Press, Columbus, OH, p 317
4. Eastmond A, Booth GM, Lee LM (1984) Arch Environ Contam Toxicol 13:105
5. Ogata M, Miyake Y, Fujisawa K, Kira S, Yoshida Y (1980) Bull Environ Contam Toxicol 25:130
6. Vassilaros DL, Eastmond DA, West RW, Booth GM, Lee LM (1981) In: Cooke M, Denius AJ, Fischer GL (eds) Polynuclear aromatic hydrocarbons: physical and biological chemistry, 6th International Symposium. Battelle Press, Columbus, OH, p 845
7. Friocourt MP, Berthou F, Picart D (1982) Toxicol Environ Chem 5:205
8. Hilker DR, Aldous KM, Smith RM, O'Keefe PW, Gierthy JF, Jurusik J, Hibbins SW, Spink D, Parillo RJ (1985) Chemosphere 14:1275
9. Kopponen P, Sinkkonen S, Poso A, Gynther J, Kärenlampi S (1994) Environ Toxicol Chem 13:1543
10. Rappe C (1994) Fresenius J Anal Chem 348:63
11. Pruell RJ, Rubenstein NI, Taplin BK, LiVolski JA, Bowen RD (1993) Arch Environ Contam Toxicol 24:290
12. Buser H-R, Dolezal IS, Wolfensberger M, Rappe C (1991) Environ Sci Technol 25:1637
13. Sinkkonen S, Paasivirta J, Koistinen J, Tarhanen J (1991) Chemosphere 23:583
14. Sinkkonen S, Vattulainen A, Aittola J-P, Paasivirta J, Tarhanen J, Lahtiperä M (1994) Chemosphere 28:1279
15. Sinkkonen S, Paasivirta J, Koistinen J, Lahtiperä M, Lammi R (1992) Chemosphere 24:1755
16. Sinkkonen S, Kolehmainen E, Paasivirta J, Koistinen J, Lahtiperä M, Lammi R (1994) Chemosphere 28:2049
17. Huntley SL, Wenning RJ, Paustenbach DJ, Wong AS, Luksemburg WJ (1994) Chemosphere 29:257
18. Monsanto Chemical Company (1971) The aroclors: physical properties and suggested applications. Technical Bulletin O-P-115
19. Rappe C, Gara A, Buser HR (1978) Chemosphere 12:981
20. Rappe C, Bergqvist P-A, Swanson S, Belton T, Ruppel B, Lockwood K, Kahn PC (1991) Chemosphere 22:239
21. EPA (1980) Dioxins. US Environmental Protection Agency, Office of Research and Development, November (EPA/6000/2–20–197)

22. Aittola J-P, Paasivirta J, Vattulainen A (1993) Chemosphere 27:65
23. Cai Z, Giblin DE, Ramanujam VMS, Gross ML, Cristini A (1994) Environ Sci Technol 28:1535
24. Cooper, K (1993) Dioxin levels in mollusks and other invertebrates in Newark Bay and Arthur Kill. Symposium on dioxin contaminated sediments in Newark Bay, Society of Environmental Toxicology and Chemistry. Hudson/Delaware Chapter, Environmental and Occupational Health Sciences Institute, Rutgers University, Feb 26
25. Paustenbach DJ (1993) Dioxin levels in sediments from Newark Bay. Symposium on dioxin contaminated sediments in Newark Bay, Society of Environmental Toxicology and Chemistry. Hudson/Delaware Chapter, Environmental and Occupational Health Sciences Institute, Rutgers University, Feb 26
26. Sinkkonen S, Mäkelä R, Vesterinen R, Lahtiperä M (1995) Chemosphere 31:2629
27. Benz T, Hagenmaier H, Lindig C, She J (1992) Fresenius J Anal Chem 344:286
28. Sinkkonen S, Koistinen J (1990) Chemosphere 21:1161
29. Sinkkonen S, Kolehmainen E, Koistinen J (1992) Int J Environ Anal Chem 47:7
30. Sinkkonen S, Kolehmainen E, Laihia K, Koistinen J (1993) Int J Environ Anal Chem 50:117
31. Buser H-R, Rappe C (1991) Anal Chem 63:1210
32. Sinkkonen S, Kolehmainen E, Koistinen J, Lahtiperä M (1993) J Chromatogr 641:309
33. Miltsov SA, Nikiforov VA, Karavan VS, Tribulovich VG, Bolshakov S (1994) Organohalogen Compounds 19:133
34. Buckholtz HE (1976) US Pat 3,989,715
35. Veijanen A, Kolehmainen E, Kauppinen R, Lahtiperä M, Paasivirta J (1992) Wat Sci Tech 25:165
36. Wease RC, Lide DR, Beyer WH (1989) CRC Handbook of chemistry and physics. CRC Press, Boca Raton
37. Cox JD, Pilcher G (1970) In: Thermochemistry of organic and organometallic compounds. Academic Press, London, chap 7
38. Ogata M, Fujisawa K, Ogino Y, Mano E (1984) Bull Environ Contam Toxicol 33:561
39. Fedorak PM, Westlike DWS (1983) Can J Microbiol 29:291
40. Fedorak PM, Westlake DWS (1984) Water, Air, and Soil Pollut 21:225
41. Bayona JM, Albaiges J, Solanas AM, Pares R, Garrigues P, Ewald M (1986) Int J Environ Anal Chem 23:289
42. Vignier V, Berthou F, Dreano Y, Floch HH (1985) Xenobiotica 15:991
43. Lake JL, Rubinstein NI, Lee HII, Lake CA, Heltshe J, Pavignano S (1990) Environ Toxicol Chem 9:1095
44. Tarhanen J, Koistinen J, Paasivirta J, Vuorinen PJ, Koivusaari J, Nuuja I, Kananen N, Tatsukawa R (1989) Chemosphere 18:1067
45. Sinkkonen S (1991) J Chromatogr 553:453
46. Sinkkonen S (1989) J Chromatogr 475:421
47. Sinkkonen S (1989) In Henschel P, Laubereau PG (eds) Water Pollution Research Reports, HPTLC Applied to the Analysis of the Aquatic Environment, Commission of the European Communities, Directorate-General for Science, Research and Development, Environmental and Waste Recycling, Brussels, p 89
48. Cai Z, Ramanujam VMS, Gross ML, Cristini A, Tucker RK (1994) Environ Sci Technol 28:1528
49. Sinkkonen S, Rantio T, Vattulainen A, Aittola J-P, Paasivirta J, Lahtiperä M (1995) Chemosphere 30:2227
50. Reischl A, Reissinger M, Hutzinger O (1987) Chemosphere 16:2647
51. Tong HY, Giblin DE, Lapp RL, Monson SJ, Gross ML (1991) Anal Chem 63:1772
52. Peterman RJ, Smith LM, Stalling DL, Petty JD (1986) Identification of chlorinated biphenylenes and other polycyclic aromatic compounds formed from the incineration of PCB-dielectric fluids at a capacitor plant's disposal site. Proc of 34th Annual Conference on Mass Spectrometry and Allied Topics, Cincinnati, OH, June 8–13, p 486
53. Peterman PH, Lebo JA, Major HJ (1988) Accurate mass determination of polychlorinated dibenzothiophenes in soil from a capacitor plants incineration site, Proc of the 36th ASMS Conference on Mass Spectrometry and Allied Topics, San Francisco, p 240

54. Sinkkonen S, Kolehmainen E, Laihia K, Koistinen J, Rantio T (1993) Environ Sci Technol 27:1319
55. Rappe C, Andersson R, Bergqvist P-A, Brohede C, Hansson M, Kjeller L-O, Lindstöm G, Marklund S, Nygren M, Swanson SE, Tysklind M, Wilberg K (1987) Chemosphere 16:1603
56. Sinkkonen S, Kämäräinen N, Paasivirta J, Lahtiperä M, Lammi R (1998) Chemosphere 36:2475
57. Sawyer TW, Vatcher AD, Safe S (1984) Chemosphere 13:695
58. Cramer RD, Patterson DE, Bunce JD (1988) J Am Chem Soc 110:5959
59. Poso A, Tuppurainen K, Ruuskanen J, Gynther J (1993) J Mol Struct (Theochem) 282:259
60. Mäntylä E, Ahotupa M, Nieminen L, Paasivirta J, Sinkkonen S (1992) Organohalogen Compounds 10:161
61. Olson JR, Tai HL, McGarrigle BP, Barfknecht TR, Smith LW (1991) Toxicologist 11:264
62. Weber R, Hagenmaier H, Schrenk D (1998) Chemosphere 36:2635

CHAPTER 11

Methyl Sulfone and Hydroxylated Metabolites of Polychlorinated Biphenyls

Robert J. Letcher[1] · Eva Klasson-Wehler* · Åke Bergman[2]

[1] Environmental Research Group, Research Institute of Toxicology, Utrecht University, P.O. Box 80176, Yalelaan 2, 3508 TD, Utrecht, The Netherlands, (e-mail: R.Letcher@ritox. vet.uu.nl)
[2] Department of Environmental Chemistry, Stockholm University, SE-10691 Stockholm, Sweden, (e-mail: Ake.Bergman@mk.su.se)

* *Present address*: E. Klasson-Wehler, Department of Pharmacokinetics and Biopharmaceuticals, AstraZeneca R&D, Södertälje, SE-151 85 Södertälje, Sweden

Methyl sulfone (MeSO$_2$-) and hydroxylated (OH-) metabolites of polychlorinated biphenyls (PCBs) have emerged as important classes of environmental contaminants in wildlife and humans. The detection of persistent MeSO$_2$-PCBs was first shown in tissues of Baltic grey seal in the mid-1970s. In the last decade the detection and quantification of these metabolites in biota has gained momentum. MeSO$_2$-PCBs are one of the major classes of organochlorine contaminants in humans and several marine and a few terrestrial mammal species. A number of studies have demonstrated the toxicological potential of MeSO$_2$-PCBs including tissue selective retention via non-covalent protein binding, induction of cytochrome P450 enzymes, and endocrine-related effects. More recently OH-PCBs have gained greater scientific notoriety in environmental toxicology as a consequence of the capability of certain OH-PCB congeners to bind with the thyroxine transport protein, transthyretin, and their interaction with thyroid and estrogen hormone receptors. Research on environmentally persistent MeSO$_2$-PCBs and OH-PCB metabolites reported in the last two decades is presented, discussed and summarized. Topics include the relative importance and mechanisms of biochemical formation in the context of PCB biotransformation, physico-chemical properties, and chemical synthesis, nomenclature and analysis. The chemical analysis summary encompasses tissue extraction, compound separation and methods of detection. MeSO$_2$-PCB and OH-PCB toxicokinetics are addressed such as species- and congener-specific formation and clearance, persistence in biota and tissue specific retention. The known biological and toxicological activities of these PCB metabolites are also summarized.

Keywords. Methylsulfonyl-PCBs, Hydroxylated PCBs, Metabolites, Review

The Handbook of Environmental Chemistry Vol. 3 Part K
New Types of Persistent Halogenated Compounds
(ed. by J. Paasivirta)
© Springer-Verlag Berlin Heidelberg 2000

Abbreviations

MeSO₂-PCB	polychlorinated biphenyl methyl sulfone
OH-PCB	hydroxylated polychlorinated biphenyl
OHS	organhalogen substances
DDT	1,1-bis(chlorophenyl)-2,2,2-trichloroethane
DDE	1,1-bis(chlorophenyl)-2,2-dichloroethene
HCB	hexachlorobenzene
MeO-PCB	polychlorinated biphenyl methyl ether
CYP	cytochrome P450
CB	chlorinated biphenyl
NIH	national health institute
GSH	glutathione
MAP	mercapturic acid pathway
HS	thiol
MeS	methylthio
SAM	S-adenosylmethionine
NMR	nuclear magnetic resonance
UV	ultraviolet
NAS	nucleophilic aromatic substitution
EI	electron ionization
ECNI	electron capture negative ionization
MS	mass spectrometry
ECD	electron capture detection
AED	atomic emission detection

FID	flame ionization detection
DCM	dichloromethane
TTR	transthyretin
I.S.	internal standard
GC	gas chromatography
GPC	gel permeation chromatography
HPLC	high pressure liquid chromatography
DMSO	dimethyl sulfoxide
TIC	total ion current
SIM	selected ion monitoring
MTBE	methyl *t*-butyl ether
FABP	fatty acid binding protein
AHH	aryl hydrocarbon hydroxylase
EROD	ethoxyresorufin-*O*-deethylase
AhR	aryl hydrocarbon receptor

1
Introduction

The metabolism of xenobiotics often generates polar metabolites by enzyme-mediated insertion of oxygen or sulfur into the molecule. Neutral metabolites of xenobiotics containing sulfur were first identified in the late 1960s. The metabolism of e.g. pentachloronitrobenzene and Propachlor (2-chloro-N-isopropylacetanilide), were shown to generate methylsulfonyl-($MeSO_2$-) containing metabolites [1, 2]. Polychlorinated biphenyls (PCBs) were detected as environmental contaminants in 1966 [3], and soon thereafter biotransformation of PCBs was shown to result in hydroxyl-(OH-) and $MeSO_2$-substituted PCB metabolites [4–6]. Since this time $MeSO_2$-PCB congeners have been shown to be present in a growing number of species. Moreover, $MeSO_2$-PCBs are retained in lipid-containing tissues with greater preference to liver, lung and kidney [7–9]. The levels of persistent $MeSO_2$-PCB congeners in some wild species and humans are comparable to other major classes of organohalogen substances (OHS) including PCBs [10–14]. $MeSO_2$-PCB bioaccumulation within marine food chains and temporal studies in humans has emphasized their appreciable biological half-lives [11, 15–17]. The identification of $MeSO_2$-PCBs as neutral, bioaccumulating PCB metabolites has led to the understanding that PCB metabolites can in fact be retained in biota due to their physico-chemical properties. In some cases the retention of $MeSO_2$-PCBs in tissue is also related to their capacity to bind to proteins in the body [18–21].

The DDT metabolites, 4,4'-DDE (1,1-bis(4-chlorophenyl)-2,2-dichloroethene) and 2,4'-DDE are OHS that form $MeSO_2$-containing metabolites that persist in biota and humans [4, 11, 12, 15, 22, 23]. Of particular concern is 3-$MeSO_2$-4,4'-DDE which is bioactivated to a reactive metabolite that irreversibly binds in the adrenal cortex of some species [24–27]. $CYP11\beta$ hydroxylase activity is perturbed in several species, and thus interference with glucocorticoid synthesis *in vivo* is possible. Hexachlorobenzene (HCB) is a major OHS that forms HCB methyl sulfide metabolites in HCB-exposed rats [28, 29]. However, the corresponding MeSO$_2$-HCB metabolites have not been detected in wildlife or humans. Methyl sulfone metabolites of polychlorinated dibenzo-*p*-dioxins and dibenzofurans have also been reported [30, 31].

Hydroxylated aromatic compounds of OHS, phenolic OHS, are not bioaccumulated in lipids. Phenolic OHS are excreted unaltered or as a conjugated entity, unless the phenolic OHS demonstrates a protein binding interaction. Polychlorobiphenylol (OH-PCB) metabolites have been detected in biota, and also as conjugated products in excreta [32–35]. Early *in vivo* studies suggested the potential for OH-PCB selective retention, such as in the intraluminal uterine fluid of mice [36]. A large number of phenolic OHS have been detected in the blood of salmon and in humans [37, 38]. In the mid-1990s, OH-PCB metabolites with specific localization in blood were identified in humans at relatively high concentrations in comparison to their PCB precursors [39–41]. OH-PCBs have now been detected in the blood of fish, birds and mammals [37, 39, 42, 43]. The retention of OH-PCBs in blood has been shown to be due to binding with plasma proteins of the thyroid hormone system [40, 44].

The biotransformation, and thus the recalcitrancy of PCBs, are an integral part of the toxicokinetics, which ultimately defines the levels and congener patterns in biota. The existence in biota of persistent MeSO$_2$-PCBs and retained OH-PCBs in tissues has changed the perception of the toxicokinetics, and potentially the toxicodynamics of PCBs. OH- and MeSO$_2$-PCB possess physico-chemical properties distinct from their parent precursors. Therefore, the potential toxicological risk to exposed organisms is different relative to their parent compounds. In light of MeSO$_2$-PCBs and OH-PCBs in biota, PCB biotransformation can be viewed as a process that generates a second level of persistent OHS. The present chapter focuses on the state of the scientific knowledge for the formation of OH- and MeSO$_2$-PCBs, which are retained in tissues of humans and wildlife. Only brief mention is made of other PCB congeners that form hydroxylated and sulfur-containing metabolites, which are primarily excreted. Topics include physico-chemical properties, methods of chemical synthesis and analysis, environmental levels and congener patterns, species- and congener-specific formation and clearance capacities, and biological and toxicological effects.

2
Methyl Sulfone and Hydroxylated Polychlorinated Biphenyls

2.1
Nomenclature

Abbreviated chemical names applied to individual MeSO$_2$- and OH-PCB metabolites have been suggested as the standard nomenclature (Fig. 1). These abbreviations are an extension of the numbering system applied to PCB congeners formulated by Ballschmiter et al. [45]. The number of a MeSO$_2$-PCB or an OH-PCB metabolite is first determined by omitting any MeSO$_2$- or OH-substituents on the PCB structure. The position of the MeSO$_2$- or OH-functional group is then identified, and primed or unprimed depending on the biphenyl carbon attachment of the functional group (Fig. 1).

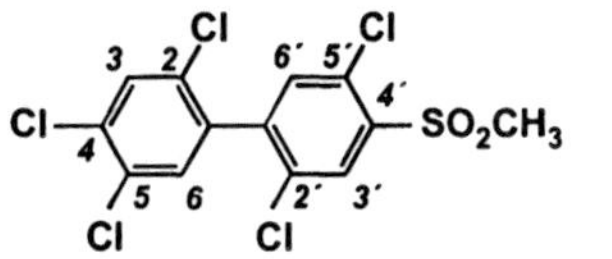

4′-methylsulfonyl-2,2′,4,5,5′-pentachlorobiphenyl
(4′-MeSO₂-CB101)

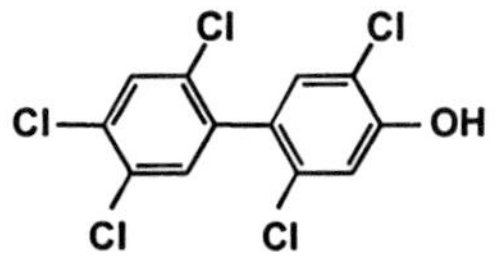

2,2′,4,5,5′-pentachloro-4′-biphenylol
(4′-OH-CB101)

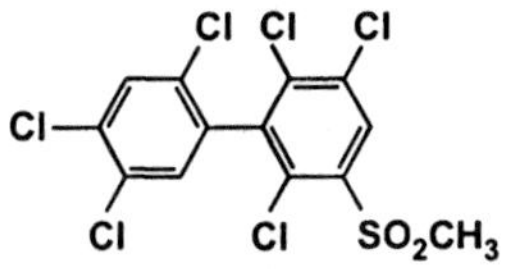

3-methylsulfonyl-2,2′,4′,5,5′,6-hexachlorobiphenyl
(3-MeSO₂-CB149)

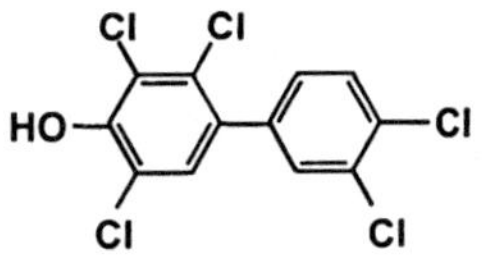

2,3,3′,4′,5-pentachloro-4-biphenylol
(4-OH-CB107)

Fig. 1. The chemical numbering system for individual MeSO$_2$-PCB and OH-PCB congeners. The PCB congener number is based on the PCB numbering system of Ballschmiter et al. [45]

MeSO$_2$-PCB congeners identified in biota so far are derived from the PCB congener of the same number, as exemplified by 4′-MeSO$_2$-CB101 and 3-MeSO$_2$-CB149 (Fig. 1)[1]. General structure-formation relationships with respect to the naming of metabolite compounds do not exist for OH-PCBs. Biotransformation processes such as 1,2-shift of chlorine alters the original chlorine substitution pattern of the precursor PCB congener. Such a process is important for the formation of many of the OH-PCBs retained in blood [46]. Examples of abbreviated OH-PCB chemical names are shown in Fig. 1. OH-PCB metabolites present in tissue samples are often analyzed as the methyl ether (MeO-) derivatives, and their abbreviated names are the same as the OH-PCB from which they are derived [39, 47].

2.2
Metabolic Formation

The following discussion of PCB biotransformation focuses on aspects that relate to formation of MeSO$_2$-PCBs and OH-PCBs known to be retained in biota. The metabolism of PCBs has been extensively reviewed elsewhere [5, 48, 49]. Diet, body condition, age, sex and reproductive status (lactation and gestation) influence the toxicokinetics of PCBs. However, metabolic processes are the major toxicokinetic factors defining not only the biological half-life of a PCB congener, but also the potential bioactivation to a more toxic product (Fig. 2) [50, 51]. The metabolism of xenobiotics, including PCBs, generally results in the formation of polar metabolites, which are subsequently cleared from an organism. Regardless, several MeSO$_2$- and OH-PCB congeners are retained in wildlife spe-

---

[1] Chlorinated biphenyl is abbreviated to CB and corresponds to one single compound of a particular structure.

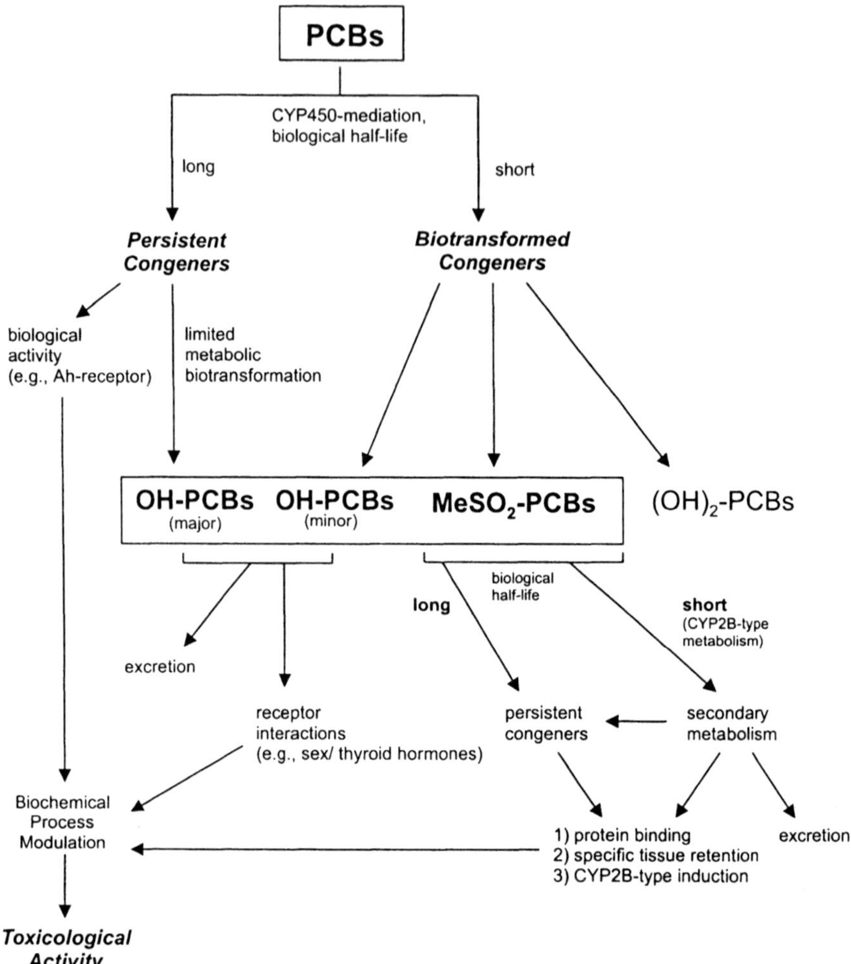

Fig. 2. A schematic diagram illustrating the general factors affecting the biological fate and activity of PCBs and the major classes of metabolites containing OH- and MeSO₂-functional groups

cies and humans, demonstrating the biotransformation of certain PCB congeners to a secondary class of environmental contaminants.

The biological half-live of PCB congeners is determined by their susceptibility to metabolism, which in turn is determined by structural factors related to the number and position of chlorine atoms. Although large species variations exist, common to all species is the greater metabolic resistance of PCB congeners possessing more than five chlorines and *para*-chlorine atoms, such as 2,2′,4,4′,5-pentachlorobiphenyl (CB-99), 2,3′,4,4′,5-pentaCB (CB-118) and 2,2′,3,4,4′,5′-hexaCB (CB-138) [52]. 2,2′,4,4′,5,5′-HexaCB (CB-153) has been shown to possess the greatest biological half-life, estimated to be 338 days in

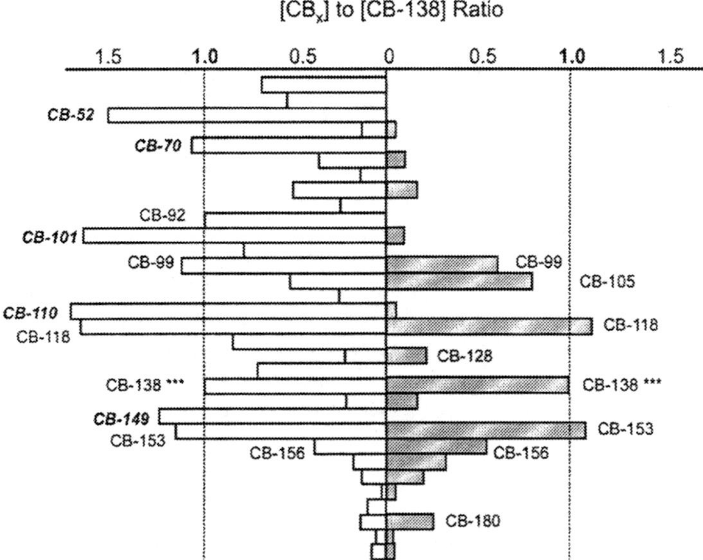

Fig. 3. The relative composition of individual PCBs in the PCB commercial mixture, Clophen A50, before and after an 18 month exposure to mink via the diet [55, 56]. The bars are the concentrations of individual CBs relative to 2,2′,3,4,4′,5′-hexaCB (CB-138). PCBs that are precursors to persistent MeSO$_2$-PCBs, and frequently detected in wildlife and humans are in bold

man [53]. The high resistance of CB-153 to metabolism is due to the lack of adjacent hydrogen atoms on the biphenyl system [52].

PCB congeners with chlorine-unsubstituted *meta-para*-carbons are susceptible to metabolism. Congeners with *meta-para*-chlorines and having more than 5 chlorine atoms are only slowly transformed metabolically [54]. Figure 3 illustrates the disappearance of PCB congeners with chlorine-unsubstituted *meta-para*-carbons in mink (*Mustela vison*). The PCB congener pattern of the technical PCB product, Clophen A50, was shown to undergo extensive metabolic changes in the mink after continuous exposure under experimental conditions [55, 56].

Generally, an initial step in PCB metabolism is the generation of an arene oxide intermediate. With respect to rodents, the initial PCB oxidation is mediated by cytochrome P450s (CYPs) *e.g.*, CYP1A, CYP2B, and possibly CYP2C and CYP3A enzymes [57, 58]. PCB congeners with chlorine-unsubstituted *meta-para*-carbons, and preferably having at least one *ortho*-chlorine, are the favored substrates of CYP2B1 enzymes in rodents and CYP3A4 in humans [57, 58]. PCB congeners having one or less *ortho*-chlorine atoms, and thus having phenyl rings with a more planar orientation, are preferred substrates for CYP1A enzymes. CYP1A, CYP2B isozymes, and possibly CYP2B-, CYP2C- and CYP3A-like enzymes catalyze PCB biotransformation in other species. The CYP enzyme capacity is also species-dependent and responsive to the level of exposure to CYP-inducing OHS in biota [59, 60].

The mechanism determining the PCB product of oxidation varies depending on the CYP enzyme mediating the reaction. For PCBs, oxidation can yield a hydroxy-group by direct insertion in a *meta*-position, or via formation of an arene oxide that subsequently rearranges to a hydroxy-group (Fig. 4). These different oxidation mechanisms have been studied for 2,2',5,5'-tetraCB (CB-52) [61, 62]. At least in rodents, the major pathway is direct insertion in the *meta*-position and is catalyzed by phenobarbital-inducible CYP2B enzymes [61, 63]. The formation of the *para*-hydroxylated CB-52 metabolite is catalyzed by methylcholanthrene inducible enzymes, *i.e.* the CYP1A family, indicating an alternative route of formation [63]. The formation of an intermediate arene oxide in the metabolism of CB-52 was shown by Stadnicki and Preston [64]. The arene oxide of CB-52 was synthesized and shown to be stable for several weeks at room temperature [65]. The stability of the arene oxide may be due to stabilizing properties of the two chlorine atoms on each side of the arene oxide.

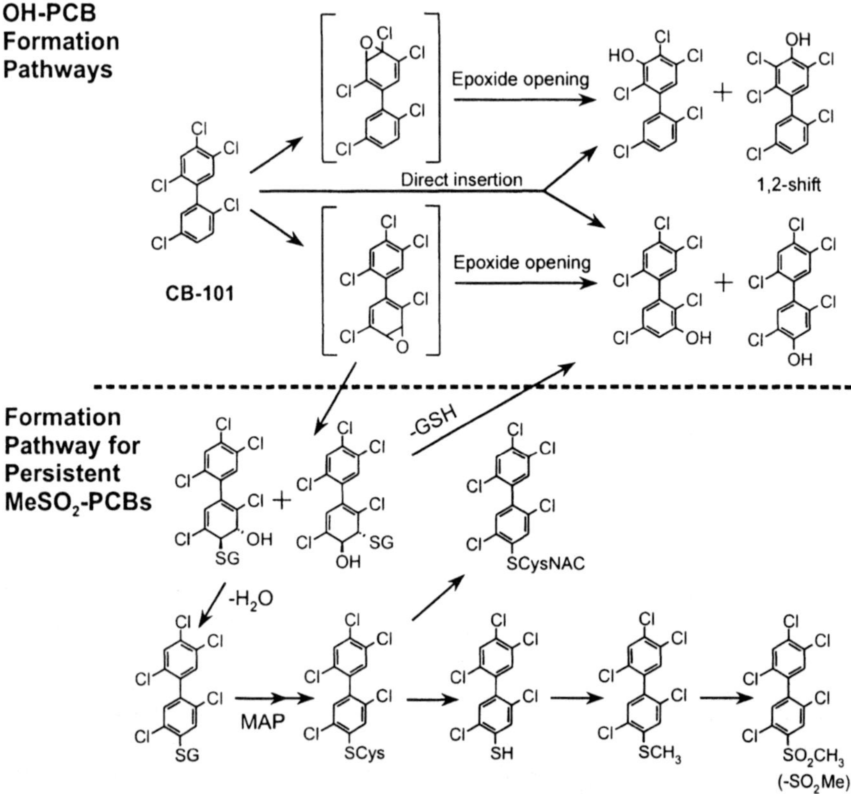

Fig. 4. Metabolism scheme for PCBs as exemplified by 2,2',4,5,5'-pentachlorobipenyl (CB-101). The formation of MeSO₂-PCB and OH-PCB metabolites are shown. CB-101 is one of several PCBs known to form persistent MeSO₂-PCB metabolites in biota. These precursor PCBs possess at least one chlorine-unsubstituted *meta-para* position, facilitating the formation of a 3,4-arene oxide

Arene oxide formation in the *meta-para* position seems to be a common metabolic route, but is species-variable [62, 66, 67]. However, dog, and to a minor extent guinea pig, also form arene oxides in the *ortho-meta* position, which has been demonstrated for CB-153 [68]. Intermediates of PCBs containing a *meta-para*-epoxide with a *meta*-hydrogen and a *para*-chlorine can result in both *meta*- and *para*-OH-PCB metabolites (Fig. 4) as a consequence of a 1,2-shift (NIH-shift) [69]. The formation of a 1,2-shifted hydroxy metabolite is strong evidence of arene oxide formation. The occurrence of 1,2-shifts in PCB arene oxides via phenobarbital inducible CYP enzymes, *i.e.* the CYP2B family, indicates that arene oxides may be formed by similar types of CYP enzymes [46, 57]. OH-PCB metabolites resulting from a 1,2-shift have been reported for 3,3′,4,4′-tetraCB (CB-77), 2,3,3′,4,4′-pentaCB (CB-105) and 2,2′,4,5,5′-pentaCB (CB-101) [70–73], and also 2,3′,4,4′,5-pentaCB (CB-118), CB-153, CB-138 and 2,3,3′,4,4′,5-hexaCB (CB-157) [46]. The 1,2-shift pathway is toxicologically important for the biotansformation of many of the most persistent PCB congeners since they form OH-PCB metabolites that are retained in blood [39, 43, 46].

PCB arene oxides can also generate dihydrodiols via a hydrolytic pathway mediated by microsomal epoxide hydrolase, although metabolism to monohydroxy-metabolites is more commonly observed [74, 75]. OH-PCBs are susceptible to further metabolism, i. e. conjugation reaction with glucuronic acid or sulfate, which increases the water solubility and facilitates excretion. Glucuronic acid and sulfate conjugates of several PCB congeners have been determined in bile and urine from experimental animals exposed to the PCBs [50, 76, 77]. Biliary excretion is the preferred pathway for PCB metabolites, whereas only a small portion is excreted via the urine [77].

Rather than forming OH-PCBs, PCB arene oxide intermediates may react and conjugate with the endogenous tripeptide glutathione (GSH) (Fig. 4) [78]. GSH conjugation occurs spontaneously and/or via glutathione-S-transferase mediation [79]. The formation of isomeric metabolites resulting from the GSH conjugation is (indirect) evidence for the presence of arene oxide as an intermediate. After dehydration, the glutathionyl-PCB (PCB-SG) conjugate undergoes peptidase hydrolysis via the mercapturic acid pathway (MAP) to form a cysteine conjugate, which can be excreted either before or after acetylation to a mercapturic acid [76, 80].

The PCB-cysteine or -mercapturic acid conjugate is subject to biliary excretion into the gastrointestinal tract and is converted to a thiol (-SH) via C-S β-lyase activity in the intestinal microflora [81]. The SH-PCB conjugate can be excreted as a free thiol or after conjugation with glucuronic acid to form S-glucuronides [78, 82]. For $MeSO_2$-PCB formation, the SH-PCB undergoes an S-methyltransferase-catalyzed reaction with S-adenosylmethionine (SAM), mainly in the intestinal mucosa, to form a methylthio- (MeS-)PCB [50, 78, 83]. The importance of S-methyltransferase in the intestinal mucosa was demonstrated with germ-free rats, which lack a viable microflora culture [84]. A two-step CYP-mediated oxidation converts the MeS-PCB to a PCB methyl sulfoxide (MeSO-) and then to a $MeSO_2$-PCB. Oxidation can take place in the liver, which highly expresses CYP enzymes, but has also been shown to occur in the intestinal mucosa [80, 85]. The mechanism for formation of $MeSO_2$-PCBs is described elsewhere in further detail [78].

The relative importance of the MAP-metabolism pathway versus alternate degradation pathways for PCB arene oxides depends on the chlorine substitution pattern of the PCB congener. PCBs with 2,5-dichloro- or 2,3,6-trichlorophenyl rings undergo MAP-directed metabolism to form persistent $MeSO_2$-PCB metabolites, in addition to phenolic metabolites that are not bioaccumulated (Table 1) [76, 78]. CB-52, a PCB containing two chlorine-unsubstituted *meta-para*-positions, likely forms $MeSO_2$-metabolites in both rings [86]. Bis-$(MeSO_2)$-tetrachloroCBs have also been detected in wildlife [11]. In laboratory mice 25% to 30% of CB-101 and 2,4',5-triCB (CB-31) were shown to form MAP-metabolites, whereas the remaining portion formed hydroxylated metabolites [84]. In Canadian ringed seals (*Phoca hispida*) < 10% of the precursor CB congeners metabolized were biotransformed to $MeSO_2$-PCB, assuming that the clearance of the metabolites were at a similarly slow rate as for CB-153 [11]. MAP-metabolites have also been reported for CB-77, CB-105, CB-118 and 2,3,3',4,4',5-pentaCB (CB-156), which do not contain 2,5-dichloro- or 2,3,6-trichloro-substitution [87, 88]. PCB metabolites containing both OH- and $MeSO_2$-groups were determined in human tissues from Japanese „Yusho" patients [89]. A comparative study of the metabolism of the 2,5-dichloro-containing CB-31, –101 and –141 showed that increased chlorination corresponded to an increased proportion of OH-PCB relative to MAP metabolites, but a decrease in the total proportion of metabolites [90]. The increase in hydroxylation was mainly at the *meta*-position, indicating that the metabolic pathway was likely via direct insertion of an OH-group.

In summary, the number and the positional substitution of chlorine atoms determines the pathway and the susceptibility to PCB metabolism. PCB congeners with 2,5-dichloro- or 2,3,6-trichloro-substituted phenyl rings form persistent $MeSO_2$-substituted metabolites, but also hydroxylated metabolites. These $MeSO_2$-PCBs are accumulated in adipose tissue and/or selectively retained in tissues. PCB congeners that are less susceptible to metabolism, possess five or more chlorine atoms and no adjacent, chlorine-unsubstituted *meta-para*-positions, can be metabolized to hydroxylated metabolites that may be retained in blood.

3
Methyl Sulfone Polychlorinated Biphenyls

3.1
Physico-chemical Properties

The $MeSO_2$-PCBs are hydrophobic compounds with log K_{ow} values only slightly lower than the parent PCB congener [11]. $MeSO_2$-CBs with 3 to 6 chlorine atoms possess log K_{ow} values of ca. 4 to 6.5 and are thus sufficiently hydrophobic to accumulate in lipid-bearing tissues. $MeSO_2$-CBs are crystalline, non-volatile substances with high melting points generally greater than 130°C [91–93]. $MeSO_2$-PCBs are much more polar than their parent PCB compounds as a consequence of the $MeSO_2$-functional group. $MeSO_2$-CBs dissolve more readily in chlorinated organic solvents than in non-polar aliphatic solvents such as n-hexane. The higher polarity of MeSO$_2$-PCBs relative to many neutral OHS is useful for separation using liquid chromatography techniques [16, 94–96]. In contrast to other OHS, MeSO$_2$-PCBs are Lewis bases and highly stable to strong acid and base conditions. The unique acid stability of MeSO$_2$-PCBs permits easy separation from neutral OHS by chemical partitioning [4, 15]. The addition of sufficient quantity of water to prepare the sulfuric acid dihydrate regenerates the neutral MeSO$_2$-PCBs for subsequent back-extraction into a neutral organic phase. The addition of water to concentrated acid is not a recommended chemical procedure, and requires that both liquids are first cooled on ice. The cold water is then added slowly (dropwise) with mixing until the desired water content is achieved.

Several of the MeSO$_2$-PCBs synthesized and isolated in their pure form have been characterized by their ^{1}H NMR and UV spectra [91, 97]. The UV spectra are characterized by two adsorption bands at 250–290 nm ($\varepsilon = 3$–4) and at 210–220 nm, respectively.

MeSO$_2$-CB congeners exhibit axial chirality if both of the phenyl rings have an asymmetric chlorine substitution pattern. Of the 28 MeSO$_2$-PCB congeners most frequently identified in wildlife and humans, the tri- and tetra-*ortho*-chlorine substituted congeners, 3- and 4-MeSO$_2$-CB91, -CB95, -CB132, -CB149, and -CB174 (Table 1), possess hindered rotation about the phenyl-phenyl σ-bond at physiological and ambient temperatures, and temperatures used for GC separations. These rotational enantiomers, or atropisomers, were shown to constitute as much as 17% of the sum concentration (Σ-) of individual MeSO$_2$-PCB congeners in the fat tissue of ringed seal and polar bear (*Ursus maritimus*) and human liver [98, 99].

3.2
Chemical Synthesis and Reactivity

Somewhat more than 100 MeSO$_2$-CBs have been synthesized as pure, individual, standard and/or test compounds. The methods used for their preparation have been described in detail in the original articles. Only a brief summary of the methods applied for congener-specific synthesis is presently outlined. The synthesis methods refer to the MeSO$_2$-PCB congeners most frequently determined in biota (Table 1).

Diaryl coupling reactions have been used for the preparation of a large number of MeSO$_2$-CBs. The reactions differ slightly from each other. i) Polychlorinated aniline is diazotized *in situ* by an alkyl nitrite and the resulting diazoaniline is allowed to react with methylthiopolychlorobenzene, acting as solvent, to produce one or more isomers of a methylthiopolychlorobiphenyl [92, 97]. The MeS-PCB is normally purified and thereafter oxidized to the corresponding sulfone by *e.g.*, hydrogen peroxide [91]. ii) A polychloromethylsulfonylaniline or a methylthioaniline may be reacted in a similar way as in i), but with a polychlorobenzene as solvent and reactant in the coupling reaction [92, 97, 100]. Isomers of MeSO$_2$-PCBs are produced directly or after oxidation of the corresponding sulfide. iii) More recently a palladium catalyzed, diaryl coupling between a polychloroiodomethylsulfonylbenzene and a polychlorophenyltrimethyl stanane was described [93]. Occasionally it may be possible to synthe-

Table 1. Polychlorinated biphenyl methyl sulfones (MeSO$_2$-PCBs) identified in tissues of humans and wildlife

Number [a]	Chemical Structure	Abbreviation
1	3-CH$_3$SO$_2$-2,4′,5-trichlorobiphenyl	3-MeSO$_2$-CB31
2	4-CH$_3$SO$_2$-2,4′,5-trichlorobiphenyl	4-MeSO$_2$-CB31
3	3-CH$_3$SO$_2$-2,2′,5,5′-tetrachlorobiphenyl	3-MeSO$_2$-CB52
4	3′-CH$_3$SO$_2$-2,2′,4,5′-tetrachlorobiphenyl	3′-MeSO$_2$-CB49
5	4-CH$_3$SO$_2$-2,2′,5,5′-tetrachlorobiphenyl	4-MeSO$_2$-CB52
6	4′-CH$_3$SO$_2$-2,2′,4,5′-tetrachlorobiphenyl	4′-MeSO$_2$-CB49
7	3-CH$_3$SO$_2$-2,4′,5,6-tetrachlorobiphenyl	3-MeSO$_2$-CB64
8	3′-CH$_3$SO$_2$-2,2′,5,5′,6-pentachlorobiphenyl*	3′-MeSO$_2$-CB95
9	4-CH$_3$SO$_2$-2,4′,5,6-tetrachlorobiphenyl	4-MeSO$_2$-CB64
10	3-CH$_3$SO$_2$-2,2′,4′,5,6-pentachlorobiphenyl*	3-MeSO$_2$-CB91
11	4′-CH$_3$SO$_2$-2,2′,5,5′,6-pentachlorobiphenyl*	4′-MeSO$_2$-CB95
12	4-CH$_3$SO$_2$-2,2′,4′,5,6-pentachlorobiphenyl*	4-MeSO$_2$-CB91
13	3-CH$_3$SO$_2$-2,3′,4′,5-tetrachlorobiphenyl	3-MeSO$_2$-CB70
14	3′-CH$_3$SO$_2$-2,2′,4,5,5′-pentachlorobiphenyl	3′-MeSO$_2$-CB101
15	4-CH$_3$SO$_2$-2,3′,4′,5-tetrachlorobiphenyl	4-MeSO$_2$-CB70
16	4′-CH$_3$SO$_2$-2,2′,4,5,5′-pentachlorobiphenyl	4′-MeSO$_2$-CB101
17	3′-CH$_3$SO$_2$-2,2′,3,4,5′-pentachlorobiphenyl	3′-MeSO$_2$-CB87
18	3-CH$_3$SO$_2$-2,3′,4′,5,6-pentachlorobiphenyl	3-MeSO$_2$-CB110
19	3-CH$_3$SO$_2$-2,2′,4′,5,5′,6-hexachlorobiphenyl*	3-MeSO$_2$-CB149
20	4-CH$_3$SO$_2$-2,3′,4′,5,6-pentachlorobiphenyl	4-MeSO$_2$-CB110
21	4′-CH$_3$SO$_2$-2,2′,3,4,5′-pentachlorobiphenyl	4′-MeSO$_2$-CB87
22	4-CH$_3$SO$_2$-2,2′,4′,5,5′,6-hexachlorobiphenyl*	4-MeSO$_2$-CB149
23	3′-CH$_3$SO$_2$-2,2′,3,4,5′,6′-hexachlorobiphenyl*	3′-MeSO$_2$-CB132
24	4′-CH$_3$SO$_2$-2,2′,3,4,5′,6′-hexachlorobiphenyl*	4′-MeSO$_2$-CB132
25	3′-CH$_3$SO$_2$-2,2′,3,4,5,5′-hexachlorobiphenyl	3′-MeSO$_2$-CB141
26	4′-CH$_3$SO$_2$-2,2′,3,4,5,5′-hexachlorobiphenyl	4′-MeSO$_2$-CB141
27	3′-CH$_3$SO$_2$-2,2′,3,4,5,5′,6′-heptachlorobiphenyl*	3-MeSO$_2$-CB174
28	4′-CH$_3$SO$_2$-2,2′,3,4,5,5′,6′-heptachlorobiphenyl*	4-MeSO$_2$-CB174

[a] The numbers correspond to those labeled on the chromatograms in Fig. 5. The ascending order of the congeners corresponds to the order of elusion from a 95% dimethyl-5% diphenyl-polysiloxane GC column (Sect. 5.1). Details of synthesis procedures are described elsewhere [92, 97, 100]. Chiral MeSO$_2$-CBs are marked with an *asterisk*.

size MeSO$_2$-PCBs from PCB congeners via nucleophilic aromatic substitution (NAS) of a chlorine atom by methane thiolate, and subsequent oxidation of the resulting PCB methyl sulfide [91, 97, 100]. Mixtures of MeSO$_2$-PCBs have been prepared by γ-irradiation of a PCB mixture in dimethyl sulfide that results in the displacement of a Cl-substituent by a MeS-group [101]. The MeS-PCBs were subsequently oxidized to the corresponding sulfones [102]. However, it may be difficult to isolate the products. More than 10 radio-labelled MeSO$_2$-CBs have been synthesized to contain radioactive atoms in the phenyl ring(s) (^{14}C) or in the methyl group (^{14}C or ^{3}H) [91, 103, 104]. A large number of the synthesized MeSO$_2$-PCBs have been characterized by GC/MS (Sect. 3.3).

MeSO$_2$-CBs are resistant to degradation by strong acids and bases, and are thus stable compounds. The sulfone group is not easily reduced by chemical methods without the risk of a concurrent dechlorination reaction. MeSO$_2$-PCBs present in biota with a lower number of chlorine substituents and/or chlorine-unsubstituted *meta-para*-carbons may be metabolically oxidized to subsequently form bis(MeSO$_2$)-PCBs. MeSO$_2$-CBs may also be transformed to secondary metabolites containing a sulfone and a hydroxy group [89]. Theoretically, MeSO$_2$-CBs with a large number of chlorine atoms on the non-MeSO$_2$-containing phenyl ring may undergo NAS with appropriate nucleophiles [91]. Under physiological conditions NAS may occur naturally with good nucleophiles such as the thiol group in GSH [28, 29].

3.3
Detection and Identification

GC-coupled mass spectrometry with electron ionization (EI-MS) or with electron capture, negative ionization (ECNI-MS), and GC with electron capture detection (ECD) have been the major techniques for the analysis of aryl methyl sulfone fractions isolated from tissue (Sect. 5.1). GC/ECD detection and identification of MeSO$_2$-PCBs relies on a comparison of GC retention times relative to authentic standards, which is dependent on the absence of co-eluting interferents. GC/MS techniques have provided important structural information for MeSO$_2$-PCBs, especially in the absence of authentic standards.

MeSO$_2$-PCB isomers with a MeSO$_2$-group in the *ortho*-position have been distinguished by their EI fragmentation pattern from isomers containing a *para*- or *meta*-MeSO$_2$-group. *ortho*-MeSO$_2$-PCBs show a characteristic ion at [M-98]$^+$ due to formation of a dibenzofuran-type fragment [M-(SOCH$_3$+Cl)]$^+$ [100]. MeSO$_2$-PCBs with an *ortho*-Cl atom in one ring and an *ortho*-MeSO$_2$ in the other phenyl ring have been distinguished by the loss of a chlorine atom in EI to generate a dominating [M-Cl]$^+$ ion [92, 100, 102]. Thus, *ortho*-substituted MeSO$_2$-PCBs are easily distinguished from those with *meta*- or *para*-substituted MeSO$_2$-groups. *Meta*- and *para*-MeSO$_2$-PCBs generally have abundant [M-SOCH$_3$]$^+$, [M-SOCH$_2$]$^+$ and [M-(SO$_2$CH$_3$+Cl)]$^+$ ions in EI. Ion trap MS/MS methods in the EI mode can also be used in the characterization of MeSO$_2$-CBs [98]. Fullscan (TIC) mass spectra yielded mainly [M]$^+$, with [M-63]$^+$ as the major daughter ion in MS/MS. [M-63]$^+$ results from a rearrangement reaction of the corresponding aryl methanesulfinate, which looses the fragment CH$_3$SO [105]. High resolution EI-MS using single ion monitoring (SIM) has been used for the determination and quantification of MeSO$_2$-PCBs by measuring [M]$^+$ and [M+2]$^+$ [16].

Besides giving additional structural information, ECNI-MS has been shown to be 500- to 1000-fold more sensitive than EI-MS towards MeSO$_2$-PCBs [106]. The ECNI-MS fragmentation patterns have been investigated for 94 MeSO$_2$-PCB substituted with 3 to 7 chlorine atoms, and with the methyl sulfone group in *ortho*-, *meta*- or *para*-positions [106]. However, it was not possible to make extensive structure-fragmentation relationships since ion-molecule reactions between MeSO$_2$-PCB and oxygen present in the ion source generated a dominant [M-Cl+O]$^-$ ion. Fragment ions generated by ECNI-MS include the dominant [M-CH$_3$]$^-$ and [M-Cl+H]$^-$, and the lesser abundant [M-CH$_3$-Cl+H]$^-$, [M-2Cl]$^-$, [M-2Cl+2H]$^-$ and [M-SO$_2$CH$_3$+H]$^-$ ions [102, 106, 107]. In the absence of oxygen in the ion source, ECNI-MS mass spectra illustrated the existence of structure-fragmentation relationships among 28 3- and 4-MeSO$_2$-PCBs identified as persistent congeners in biota (Table 1) [107]. Compounds with the 4-MeSO$_2$-2,5-dichloro-moiety possessed significantly lower [M-CH$_3$]$^-$ relative to [M]$^-$ ion, where fragment ion abundance could be increased with higher ion source temperatures. Higher chlorinated MeSO$_2$-CB congeners also tended to fragment more significantly to [M-Cl+H]$^-$. Other MeSO$_2$-PCB detection techniques recently reported are GC-atomic emission (AED) [108, 109] and GC-flame ionization detection (FID) [96]. GC/AED was shown to have a similar sensitivity to GC/ECD towards MeSO$_2$-PCBs in the aryl sulfone fraction isolated from Baltic grey seal (*Halichoerus grypus*) tissues when selected for a sulfur or chlorine response.

4
Hydroxylated Polychlorinated Biphenyls

4.1
Physico-Chemical Properties

Hydroxylated PCBs are weak acids. The pK$_a$ of some OH-PCBs have been reported [110], but are in general not available. Congeners with two chlorine atoms substituted on carbons next to the hydroxy group have pK$_a$ values as low as 6.8. The pK$_a$ values up to and over 10 for some OH-PCBs are known to be structure dependent. OH-PCBs with pK$_a$ values in the range of 7 to 10 are only partially ionized, if at all, under neutral conditions (pH = 7), but are present as the corresponding protonated compounds. Despite the ability of the OH-group to hydrogen bond, the hydrophobic, non-OH-containing phenyl ring imparts a lipophilic character to the OH-PCB molecule. OH-PCBs are non-volatile substances. OH-PCBs with 3 or 4 chlorine atoms in the *ortho*-positions exist as enantiomeric pairs similar to PCB and MeSO$_2$-PCB congeners [111]. Among the OH-PCBs so far identified in biota (Table 2), it may be worth mentioning that the 4-OH-CB187 is one of the OH-PCBs that theoretically exists as an enantiomeric pair.

4.2
Chemical Synthesis and Reactivity

A large number of methyl derivatives of OH-PCBs (MeO-PCBs) have been synthesized, and their mass spectrometric characteristics have been studied [47, 112, 113]. The standard synthetic approach for the MeO-PCBs is usually via a Cadogan diaryl coupling reaction between polychlorinated aniline and an anisole. For MeO-PCBs not readily synthesized using the Cadogan diaryl coupling reaction, preparation may be possible via an Ullman reaction [47]. Subsequent chlorination of the reactant generates the desired MeO-PCB. MeO-PCBs are demethylated by boron tribromide in dichloromethane (DCM) to yield the corresponding OH-PCB [114]. Several dihydroxy-PCB congeners have been synthesized via a palladium catalyzed coupling reaction [115].

Table 2. Polychlorinated biphenylols (OH-CBs) identified in biota together with their chemical names and abbreviations

GC-peak no[a]	OH-PCB	Structure[b]
1	4'-OH-CB120[c]	2,3',4,5,5'-pentachloro-4'-biphenylol
2	4-OH-CB107[d]	2,3,3',4',5-pentachloro-4-biphenylol
	4'-OH-CB108[d]	2,3,3',4,5'-pentachloro-4'-biphenylol
3	3-OH-CB153[d]	2,2',4,4',5,5'-hexachloro-3-biphenylol
4	4-OH-CB146[d]	2,2',3,4',5,5'-hexachloro-4-biphenylol
5	3'-OH-CB138[c]	2,2',3,4,4',5'-hexachloro-3'-biphenylol
6	4-OH-CB187[d]	2,2',3,4',5,5',6-heptachloro-4-biphenylol
7	4'-OH-CB159[d]	2,3,3',4,5,5'-hexachloro-4'-biphenylol
8	4-OH-CB162[c]	2,3,3',4',5,5'-hexachloro-4-biphenylol
9	3'-OH-CB180[c]	2,2',3,4,4',5,5'-heptachloro-3'-biphenylol
10	4'-OH-CB172[d]	2,2',3,3',4,5,5'-heptachloro-4'-biphenylol
11	4'-OH-CB201[c]	2,2',3,3',4,5,5',6'-octachloro-4'-biphenylol
12	4,4'-(OH)$_2$-CB202[c]	2,2',3,3',5,5',6,6'-octachloro-4,4'-biphenyldiol
13	4'-OH-CB208[c]	2,2',3,3',4,5,5',6,6'-nonachloro-4'-biphenylol

[a] The numbers correspond to those labeled on the chromatograms shown in Fig. 6. The ascending order of OH-PCB congeners corresponds to the order of elution from a DB-5 GC fused silica column (30 m × 0.25 mm I.D., 25 µm film thickness). Details for analysis are given elsewhere [39]. Methods for synthesis of the compounds shown and their methyl ethers are described elsewhere [47].

[b] The chemical names are adjusted to PCB numbers [45].

[c] Identified by comparison to the authentic reference standards on two different GC columns (unpublished data).

[d] Identified according to Bergman *et al.* [39].

OH-PCB compounds are much more labile than the MeSO$_2$-PCBs. OH-PCBs may undergo oxidation if exposed to air for a prolonged period of time. Most oxidation products are as yet unidentified, although catechols ((OH)$_2$-PCB) and quinone-type products are known to be formed [116]. A number of OH-PCBs, analyzed as their corresponding MeO-PCB derivatives, have so far been identified by comparison to the authentic standards listed in Table 2.

4.3
Detection and Identification

EI mass spectra can be used for the structural determination of MeO-PCBs. Similar to the MeSO$_2$-PCBs, the sensitivity of EI-MS towards MeO-PCBs have been shown to be one to two orders of magnitude lower than for ECNI-MS [47]. MeO-PCBs with the methoxy group in *ortho*-position imparts a characteristic, abundant [M-CH$_3$Cl]$^+$ ion in the EI mass spectra corresponding to the formation of a dibenzofuran-type structure [47, 112, 113]. *Meta*- and *para*-substituted MeO-PCBs are differentiated by i) the abundance of [M-CH$_3$]$^+$, ii) the relative intensity ratio of [M-15]/[M-43] of approximately 1.0 or greater for 4-MeO-PCBs, and iii) a [M-15]/[M-43] ratio less than 0.2 for 3-MeO-PCBs (calculated from [47]). The [M-43] ion corresponds to [M-(CH$_3$+CO)]$^+$. The molecular ion is generally the base peak in both EI and ECNI mass spectra [47].

5
Methyl Sulfone and Hydroxylated Polychlorinated Biphenyls in Wildlife and Humans

$MeSO_2$- and OH-PCB metabolites are retained in wildlife and humans, however, the two metabolite classes are analyzed in tissue in different ways. Depending on the compound, $MeSO_2$-PCBs persistent in tissue as a consequence of i) possessing lipophilic/hydrophobic properties similar to other OHS (including PCBs), ii) being more or less resistant to further metabolic degradation, and iii) protein binding characteristics. $MeSO_2$-PCBs have been detected in a number of lipid-containing tissues, especially fat and liver. In contrast, many OH-PCBs are highly susceptible to conjugation reactions resulting in excretion. So far OH-PCBs have only been reported to be present in blood to any significant degree even though their presence in other tissue compartments of an organism cannot be excluded. The OH-PCBs retained in plasma possess common structural elements. These include an OH-group in the *para*-position, or occasionally in the *meta*-position, with chlorine atoms on the adjacent carbon atoms, and at least one chlorine atom on the non-OH-containing phenyl ring in the *para*-position. These structural elements resemble thyroxine, the natural substrate which binds to transthyretin (TTR) – a thyroxine transport protein. The chlorine substitution pattern of OH-CBs retained in blood resemblance the iodine substitution pattern on thyroxine.

5.1
Methodology for Determination in Tissues

$MeSO_2$- and OH-PCBs require different analytical approaches for isolation from tissue. Lipophilic $MeSO_2$-PCBs are commonly co-extracted with biogenic material and other lipophilic xenobiotics from lipid-containing tissues including blood and milk. $MeSO_2$-PCBs are Lewis bases, but considered to be neutral compounds. Existing analytical methodologies capitalize on the unique physico-chemical properties of $MeSO_2$-PCBs to facilitate the separation of aryl methyl sulfones from co-extracted lipids, mainly triglycerides, and other OHS. Regardless of the methodology used and the cleanliness of the aryl methyl sulfone fraction, studies have consistently reported the percent recovery of $MeSO_2$-PCBs to be 75% or greater.

MeSO_2-PCBs are initially co-extracted from tissue with other OHS and lipids. Lipid-rich tissues such as adipose, liver and lung have been homogenized with dichloromethane (DCM) and *n*-hexane/2-propanol (3:2) to co-extract lipid and OHS [95, 117, 118]. Tissues have also been homogenized with acetone/n-hexane and extracted with *n*-hexane/diethyl ether or methyl *tert*-butyl ether (MTBE) [119]. Mammalian adipose and liver and whole fish have been dehydrated with sodium sulfate prior to extraction with *n*-hexane/DCM (1:1) [11, 96]. Human liver and feces have been extracted with benzene and refluxing by Soxhlet [120]. Formic acid has been added to milk samples in preparation for further clean-up [16]. After extraction, OHS and aryl methyl sulfones have been separated from lipids by various techniques. Lipid destructive strategies include saponification with alcoholic potassium or sodium hydroxide at refluxing temperatures [121–124], and passing the OHS/lipid extract in DCM/n-hexane (1:1) through a column of silica gel impregnated with 33% potassium hydroxide [96].

Lipid removal techniques that are non-destructive to OHS have incorporated size exclusion based gel permeation chromatography (GPC) [96,119], dialysis with semi-permeable membranes [125], and high-pressure liquid chromatography (HPLC) with normal or reversed phase columns. OHS, including the aryl methyl sulfone variety, have a smaller molecular size relative to most long chain lipids. For adipose, liver, lung and milk extracts, GPC with Bio-Beads S-X3 has been used in combination with relatively non-polar mobile phases such as n-hexane/DCM (1:1) and cyclohexane/DCM (50:50) [12, 15, 16, 95, 96, 108, 117]. GPC with Lipidex 5000 in combination with chloroform/methanol/n-hexane (1:1:1) has been used for adipose and liver extracts, milk and plasma [12, 16, 126]. HPLC techniques utilizing Bondpack CN, nitrophenylpropyl silica and polystyrene-divinylbenzene polymer stationary phases have been used to isolate $MeSO_2$-PCBs from other OHS [95, 108, 109, 125].

Liquid-liquid partitioning strategies take advantage of the polarity, Lewis base and acid stability of $MeSO_2$-PCBs for separation from less polar, neutral OHS (*e.g.*, PCBs and chlorinated insecticides) and other xenobiotics, following the separation from bulk lipids. DMSO, acetonitrile and concentrated sulfuric acid in combination with essentially non-polar solvents such DCM and n-hexane have been used [4, 15, 22, 55, 95, 117, 121, 123, 124, 127, 128]. OHS/aryl methyl sulfone separation and the removal of residual biogenic interferents from the aryl sulfone fraction prior to the quantification of the $MeSO_2$-PCBs has also been achieved by several liquid-solid, adsorption chromatography approaches. Heated activated neutral and basic aluminum oxide deactivated up to 5% with water has been used with n-hexane or n-hexane/DCM mobile phases [12, 15, 16, 22, 95, 96, 117, 129]. Adsorption chromatography with heated activated magnesium silicate (Florisil) followed by deactivation with 1.2% water [11, 96, 102, 130] and C_{18}-bonded silica as solid phases [95, 117] have also been reported.

As with other OHS, GC-ECD and -MS techniques are the standard approaches to separating, identifying and quantifying $MeSO_2$-PCBs in aryl methyl sulfone fractions isolated from biological matrices. Until the mid-1980s, GC separation of $MeSO_2$-PCBs incorporated glass-packed columns of about 2 meters in length and containing non-polar solid phases [4, 120, 121, 123, 131]. Coupled with ^{63}Ni ECD detectors and the limited availability of authentic $MeSO_2$-PCB standards, accurate congener identification and quantification was not possible. About 10 years ago capillary GC columns were first used for $MeSO_2$-PCB analysis [122]. GC capillary columns used up to the present time are essentially of a non-polar nature, composed of 95% dimethyl-5% diphenyl polysiloxane or similar type phases [11, 15, 16, 22, 42, 95, 96, 98, 102, 108, 109, 118, 124, 126, 129, 132]. On non-polar, GC capillary columns, the elution envelop of polar $MeSO_2$-PCBs typically appears at least 10 min later than PCBs under similar temperature ramping conditions.

GC/MS as well as GC/ECD are useful for the analysis of $MeSO_2$-PCBs at levels approaching low picograms. ECD or MS quantification of GC separated $MeSO_2$-

PCB congeners have been based on peak height, but mostly peak area. GC/EI-MS has been used to produce reconstructed GC peaks from extracted $[M]^+$ and $[M+2]^+$ ions generated in a TIC mode (Sect. 3.3). GC peaks generated by SIM of $[M]^+$ and $[M+2]^+$ ions increase the sensitivity for quantification [95, 102, 122, 124, 126, 132]. SIM detection limits as low as 0.01 ng/g (lipid weight) have been reported for $MeSO_2$-PCBs isolated from human mother's milk [16].

Since the early 1990s GC/ECNI-MS has been used for $MeSO_2$-PCB quantification, in the TIC and SIM ($[M]^-$ and $[M+2]^-$ ions) modes [15, 16, 22, 23, 42, 96, 98, 118]. Apart from the 500- to 1000-fold greater sensitivity of GC/MS(ECNI) over GC/MS(EI) towards $MeSO_2$-PCBs [106], the former technique in either the TIC or SIM mode was shown to have similar low picogram sensitivity [96]. Further, the response linearities for ECNI-MS are up to the nanogram level, and comparable to GC/ECD [96]. GC/AED has been used in the quantification of $MeSO_2$-PCBs isolated from tissue of Baltic grey seal [108].

The use of capillary GC columns coincided with the application of internal standard (I.S.), rather than external standard techniques for $MeSO_2$-PCB quantification. The analogues 3-$MeSO_2$-4-Me-2',3',4',5,5'-pentaCB and 3-$MeSO_2$-2-Me-2',3',4',5,5'-pentaCB are ideal internal standards since they possess analogous physico-chemical properties to the $MeSO_2$-PCBs, and elute later than the $MeSO_2$-PCB congener envelop found in biological tissues [92]. By the early 1990s, the ECD, EI-MS and ECNI-MS quantification of $MeSO_2$-PCBs in fractions isolated from biological matrices became increasingly more accurate with the use of the I.S. approach in combination with a greater number of commercially available $MeSO_2$-PCB authentic standards [11, 15, 22, 23, 96, 98, 102, 108, 109, 118, 126, 132]. $MeSO_2$-PCB synthesis has been targeted to the environmentally relevant congeners as a consequence of $MeSO_2$-PCB congener identification in biota [92, 102, 103, 106]. Major $MeSO_2$-PCB congeners identified in wildlife and humans are listed in Table 1.

The OH-PCBs are lipophilic and possess weak acidic properties relative to the neutral characteristics of most OHS that are retained in tissue. Tissues where OH-PCBs have been detected include liver, lung, kidney and adipose tissue, but plasma (or blood) is by far the preferred tissue for specific localization [34, 38, 39, 42, 43, 133]. The extraction of OH-PCBs require an acidic pH since the pKa can be below 7; for example *para*-OH-3,5-diCBs [110]. Lipid and OHS extracts (including OH-PCBs) have been isolated from aqueous plasma using methanol and acidification with sulfuric acid and hydrochloric acid for protein denaturation and the rupture of strong OH-PCB-protein interaction [134]. In a more recent report, proteins were denatured with 2-propanol and hydrogen chloride [135, 136]. OHS and lipids were extracted in higher yields than previously reported using the same *n*-hexane/MTBE (1:1) extraction solvent. The initial extraction of liver, lung, adipose, kidney and excreta [70, 77, 137], has been similar to the previous description for $MeSO_2$-PCBs; including the use of Florisil Q, GPC and various types of silica gel preparations [72, 133]. From the lipid extracts, phenolic compounds have been selectively partitioned into ethanolic potassium hydroxide solutions, followed by acidification and re-extraction into *n*-hexane/MTBE. The extraction and clean-up of OH-PCBs based on HPLC separation are also known for liver microsomes, which form the basis of *in vitro* assays to determine the metabolic depletion of PCBs [138, 139]. After OH-PCB derivatization, often via methylation, an adsorption chromatography step with basic alumina [39], silica gel mixed with sulfuric acid (30%) [135, 136] or sulfuric acid partitioning [42] have been used as a final clean-up prior to GC separation and quantification.

GC separation of the isolated phenolic compounds occurs prior to identification and quantification using ECD- and MS-coupled detectors (Sect. 4.3). However, phenolic compounds are polar, and derivatization gives better GC peak shapes in the most commonly used types of GC columns. Hitherto the standard method of derivatization of OH-PCBs retained in blood has been methylation by diazomethane to form the corresponding methoxy-PCB (MeO-PCB) [39, 43, 135]. OH-PCB methylation by ion-pair alkylation with methyl iodide is an alternative to the diazomethane technique [43, 140]. Acetylation has been shown to give comparable recoveries of OH-PCB derivatization as the methyl iodide and diazomethane approaches to OH-PCB methylation [139]. Silylation reactions have also been applied for the derivatization of various OH-PCBs [137].

GC separations have mainly been performed on columns such as 5%-phenylmethyl-derivatized column (e.g., DB-5 or XTI-5) [39,42]. However, co-elution of MeO-PCBs occurs, which may require separation on two columns of different polarity. A cyano-derivatized column (e.g., SP-2331) has been used for this reason to facilitate full identification/separation of the congeners present in the OH-PCB envelope of elution [39]. GC(ECD) and GC/MS(EI) [39, 42], and GC/MS(ECNI) [43] have been used to detect and quantify the MeO-PCBs. The OH-PCB congeners so far identified in wildlife and human blood are listed in Table 2.

5.2
MeSO$_2$-PCBs

5.2.1
Levels and Congener Patterns

In 1976 persistent MeSO$_2$-PCBs and -DDEs were first discovered in the blubber of Baltic grey seals at the low-ppm level [4]. The identification of specific MeSO$_2$-PCB congeners was not possible, however eighteen distinct MeSO$_2$-PCBs were detected by EI-MS. MeSO$_2$-4,4'-DDE was identified shortly after by comparison to an authentic reference standard [141]. The identification of MeSO$_2$-PCBs were reported at ppb levels in Japanese mother's milk a little after the reports in 1976 [131, 142]. Further studies in the mid-1980s on MeSO$_2$-PCBs in the tissues of wildlife and humans coincided with the use of capillary GC for contaminant analysis. Several reports appeared during the mid- to late-1980s on MeSO$_2$-PCBs in autopsied victims of the Japanese "Yusho" incident [95, 117, 121, 122, 124, 143]. Thirteen years after their initial discovery in Baltic grey seal, MeSO$_2$-PCBs in biota were reported in the blubber of blue whale (*Balaenoptera musculus*), fish and mollusks from the Japanese marine environment [95]. Sulfone metabolites present in biota have escaped detection in standard analysis procedures used to determine more common and relatively less polar OHS.

Table 3. Tissues of humans and wildlife species containing persistent polychlorinated biphenyl methyl sulfones[a]

Species	Location	Sampling year	Tissue	#	Σ-MeSO$_2$-PCB, µg/g[b]	MSF Congeners[c] Dect.	MSF Congeners[c] Ident.	Σ-MeSO$_2$-PCB/ Σ-PCB[b]	References
Marine Mammals[d]									
Ringed seal	Canadian Arctic	1993	blubber	11	0.007 – 0.02	41	28	0.02 – 0.04	[11, 98]
Ringed seal	Baltic Sea	1988	blubber	2	4.4/12	> 17	6	0.17/0.04	[4, 15]
Grey seal	Baltic Sea	1988 /91 /98	blubber	17	0.3 – 110	33	23	0.03 – 0.25	[4, 15, 22, 106, 108, 129, 144]
Grey seal	Baltic Sea	1988 /91 /98	liver	7	8.7 – 28	> 25	23	0.91 – 1.10	[4, 22, 106, 108, 144]
Grey seal	Baltic Sea	1998	lung	3	0.7 – 15	> 25	14	–	[108]
Harbour seal	Baltic Sea	1988	blubber	5	0.9 – 12	> 25	6	0.02 – 0.17	[15]
Beluga whale	St. Lawrence	1988 /94–96	blubber	31	0.02 – 1.01	> 25	25	0.01 – 0.21	[22, 23]
Beluga whale	Hudson Bay	1994	blubber	7	0.12 – 0.18	> 26	26	0.02 – 0.05	[23]
Common dolphin	Irish Sea	1987–89	blubber	1	0.03	> 6	6	0.01	[118]
Harbour porpoise	Irish Sea	1987–89	blubber	1	0.58	> 6	6	0.09	[118]
White-sided dolphin	Irish Sea	1987–89	blubber	1	0.31	> 6	6	0.02	[118]
Risso's dolphin	Irish Sea	1987–89	blubber	1	0.07	> 6	6	0.01	[118]
Pilot whale	Irish Sea	1987–89	blubber	1	0.21	> 6	6	0.02	[118]
Striped dolphin	Aegean Sea	1991	blubber	4	0.14 – 0.26	> 6	6	0.01 – 0.01	[118]
False killer whale	Canada	1988	blubber	1	–	5	–	–	[15]
Blue whale	Japan	1989	blubber	1	0.01	≈ 37	–	–	[95]
Finback whale	Atlantic	1997	blubber	1	0.41	> 22	22	0.02	[145]
Terrestrial Mammals									
Human[e]	Japan	1978 (Yusho)	liver	7	0.003 – 0.02	≈ 60	50	0.01 – 0.02	[95, 117, 121, 122, 124]
Human	Japan	1978 (Yusho)	lung	7	0.008 – 0.06	≈ 60	50	0.04 – 0.08	[95, 117, 121, 122, 124]
Human	Japan	1978 (Yusho)	adipose	7	0.02 – 0.03	≈ 60	50	0.001 – 0.002	[95, 117, 121, 122, 124]
Human	Japan	1989 (Yusho)	blood	5	–	–	–	–	[95]
Human	Japan	1976 /89	milk	6	0.004	> 20	–	–	[95]
Human	Sweden	1996	adipose	7	< 0.01 – 0.01	> 15	15	0.002 – 0.12	[12]
Human	Sweden	1996	liver	7	0.01 – 0.36	> 17	17	0.02 – 0.18	[12]

Human	Sweden	1996	plasma	11	0.0008–0.06	23	16	0.001–0.009	[126]
Human	Sweden	1972–92	milk	10–20	0.002–0.009	23	23	0.004–0.009	[16]
Otter	Sweden	1990	muscle	2	0.2/0.6	≈ 30	14	0.05/0.16	[22]
Otter	Sweden	1990	liver	3	14/13/2.1	≈ 30	14	0.19/–/–	[22, 129]
Otter	Sweden	1984	adipose	1	0.4	> 25	8	–	[144]
Wild mink	Sweden	1988	muscle	2	0.1/0.3	≈ 30	14	0.004/0.02	[22]
Wild mink	Sweden	1988	liver	2	0.7/1.9	≈ 30	14	0.04/0.1	[22]
Polar Bear[f]	Canadian Arctic	1985/93–94	liver	18	1.2–4.0	31	25	0.09–0.13	[17, 22, 96, 155]
Polar Bear	Canadian Arctic	1985/93–94	fat	> 20	0.1–0.8	28	24	0.04–0.07	[10, 11, 17, 22, 96, 98]
Birds[g]									
Sea eagle	Sweden	1970s/80s	egg	23	< 1.0	10	10	< 0.01	[94, 129]
Laysan albatross	Pacific ocean	1994	liver	10	0.01–0.08	> 14	14	0.003–0.01	[42]
Cormorant	The Netherlands	1994	adipose	8	0.02–0.13	11	11	0.001–0.01	[127]
Herring gull	Lake Ontario	1989	egg	Pool (118)	≈ 0.01	–	–	–	[96, 148]
Fish[h]									
Sardine	Japan	1989	whole	Pool	0.0009	≈ 50	–	–	[95]
Rainbow trout	Japan	1989	whole	pool	0.004	≈ 28	–	–	[95]

[a] Animals and humans may be male and/or female, but are not distinguished as such.

[b] The maximum and minimum concentrations (lipid weight basis) and ratios are given. Values separated by a slash denote individuals.

[c] Listed are the maximum number of congeners detected and identified in all studies for each tissue.

[d] Recent reports of MeSO$_2$-PCBs exist for Swedish harbour porpoise [125], harbour, hood, harp and ringed seal, and several porpoise and whale species from Atlantic Ocean [145], and grey seals from the Isle of May, Scotland [147].

[e] The "Yusho" incident has been extensively reviewed [143].

[f] Σ-MeSO$_2$-PCB concentrations of <0.2 µg/g (lipid weight) were also found in testes, lung and brain [17].

[g] MeSO$_2$-PCBs have been detected Swedish cormorant, guillemot and eagle owl egg and muscle [216].

[h] MeSO$_2$-PCBs were below detection (<0.00001 µg/g) in Canadian arctic cod [11] and smelt from Lake Huron [96]. Japanese clams and oysters were reported to contain Σ-MeSO$_2$-PCBs <0.001 µg/g (lipid weight), where about 35 congeners were detected [95].

However, the number of reports of MeSO$_2$-PCBs in various tissues of wildlife and humans has grown considerably since the beginning of the 1990s (Table 3). The highlights and trends in the published data are discussed here. These studies can be examined in detail in their original reports.

Reports of persistent MeSO$_2$-PCBs in tissue are predominantly in the fat of marine mammals, and to a lesser extent in liver and lung tissue (Table 3). Sulfones have also been determined in adrenal, prostate, large intestine, testes, brain and kidney in a lone Baltic grey seal [144], and very recently in the blubber, nuchal fat, muscle and brain of Swedish harbour porpoise (*Phocoena phocoena*) [125]. A sum concentration (Σ-MeSO$_2$-PCB) of < 0.2 (µg/g (lipid weight) for 17 individually quantified MeSO$_2$-PCB congeners have been reported in testes, lung and brain of polar bear [17]. The number of individual MeSO$_2$-PCB congeners quantified varies from study to study. Seals and toothed whales from Swedish and Canadian waters have been studied to the greatest extent, and to a lesser extent from the Mediterranean and Japanese marine environments. MeSO$_2$-PCBs also persist in harbour (*Phoca vitulina*), hooded (*Cystophora cristata*), harp (*Phoca groenlandica*) and ringed seals, striped dolphin (*Stenella coerueoalba*), finback (*Balaenoptera physalus*), sperm (*Physeter macrocephalus*) and pygmy sperm (*Kogia breviceps*) whales from the North Sea and the Atlantic waters [145, 146], and grey seals from Isle of May, Scotland [147]. Fat, liver and muscle of polar bear, mink and otter also contain MeSO$_2$-PCBs (Table 3). These terrestrial species are top predators in their respective marine ecosystems. MeSO$_2$-PCBs have been extensively studied in the liver, lung, adipose, milk and blood of Japanese and Swedish human subjects. Over the last 5 years sulfones have been detected in several species of bird, including the liver of Pacific Laysan albatross (*Diomedea immutabilis*) [42], adipose tissue of great cormorant (*Phalacrocorax carbo*) chicks from The Netherlands [127], and partial reports on the eggs of Swedish white-tailed sea eagle (*Haliaetus albicilla*) and Great Lakes herring gull (*Larus argentatus*) (Table 3) [94, 96, 129, 148]. The ages and sex of all species studied tend to be random as a consequence of opportunistic sampling.

In Baltic seal species and otter, and Canadian polar bear Σ-MeSO$_2$-PCB concentrations are among the most abundant classes of anthropogenic OHS [11, 22]. In these species Σ-MeSO$_2$-PCB concentrations can reach the ppm level (lipid weight basis) in the fat tissue. Apart from humans, Σ-MeSO$_2$-PCB concentrations in biota are generally in the order of terrestrial mammals $\approx$ seals > toothed whales > baleen whales > birds > fish based on the present data (Table 3), and reflect among other factors, the species metabolic capacity and the kinetics of MeSO$_2$-PCB formation. The induction of sulfone formation capacity is also influential. For example, Baltic seals have shown a wide range of Σ-MeSO$_2$-PCB concentration, which are proportional to the Σ-PCB concentrations [15]. The apparently low CYP2B-like activities in cetaceans judged from the lack of catalytic and immunochemical evidence [60, 149] has clearly underestimated the capacity for *meta-para*-PCB metabolism and subsequent MeSO$_2$-PCB formation (Table 3) in these species. Studies in western Hudson Bay beluga whale (*Delphinapterus leucas*) versus the more highly exposed (to OHS) St. Lawrence population have indicated that MeSO$_2$-PCB formation and clearance is inducible is this species of cetacean [23].

Σ-MeSO$_2$-PCB concentrations in whole Canadian arctic cod (*Boreogadus saida*), Great Lakes smelt (*Osmerus mordax*) and Japanese sardine (*Sardinops sagax*) and rainbow trout (*Salmo gairdneri*) were at the ppt level or not detectable [11, 95, 96]. Fish and invertebrates are not expected to possess any appreciable sulfone forming capacity. These organisms generally lack significant CYP2B-like activity towards PCBs [150]. CYP2B-like enzyme capacity is necessary to generate the prerequisite 3,4-arene oxide intermediates, which undergo subsequent metabolism, including pathways leading to the formation of persistent MeSO$_2$-PCBs [78].

Σ-MeSO$_2$-PCB concentrations can vary in the adipose tissue of marine mammals over orders of magnitude, whereas the Σ-MeSO$_2$-PCB to Σ-PCB ratios are considerably smaller, and generally range between 0.01 and 0.10 (Table 3). In extreme cases, Baltic ringed and grey seals exposed to high levels of CYP enzyme-inducing OHS possessed Σ-MeSO$_2$-PCB to Σ-PCB ratios of up to 0.25 [15]. The Σ-MeSO$_2$-PCB to Σ-PCB ratio is relatively constant in the adipose tissue of marine mammals, regardless of the capacity for MeSO$_2$-PCB formation. This is most likely related to a species capacity to metabolically eliminate as well as form MeSO$_2$-PCBs. However, for *meta-para* PCBs, competition between metabolic pathways other than those leading to MeSO$_2$-PCB may be a function of the level of OHS exposure. OHS-mediated induction of CYP1A isoforms via the aryl hydrocarbon receptor can sometimes lead to induction of UDP glucosyltransferases required for UDP glucoronic acid conjugation, and glutathione-S-transferases required for GSH conjugation [151].

MeSO$_2$-PCB bioaccumulation within a food chain can also be a factor influencing the overall MeSO$_2$-PCB concentration. The apparent bioaccumulation factor for Σ-MeSO$_2$-PCB concentrations from ringed seal to polar bear was found to be about 30 as compared to 47 for the highly persistent CB-153 [11]. Polar bear was shown to accumulate at least a portion of 3- and 4-MeSO$_2$-CB91, 4-MeSO$_2$-CB149, and 3'- and 4'-MeSO$_2$-CB87 and -CB141 (Table 1) from the diet of ringed seal blubber [11]. In the case of 3'- and 4'-MeSO$_2$-CB132, in bear complete accumulation from seal was observed.

MeSO$_2$-PCBs in the milk of humans [16, 95] and polar bear [17] indicates the ability of sulfones to be mobilized from adipose tissue. Early offspring can therefore be exposed to MeSO$_2$-PCBs via lactational biotransfer. Unlike males, mammalian females appear able to clear a portion of their MeSO$_2$-PCB body burden via lactational clearance. Adipose biopsies of female beluga from the St. Lawrence tended to have lower Σ-MeSO$_2$-PCB concentrations and Σ-MeSO$_2$-PCB to Σ-PCB ratios than in males [23]. In Swedish mother's milk, Σ-MeSO$_2$-PCB concentrations have been shown to decline temporally over a 20-year period at approximately the same rate as Σ-PCB level [16]. This remains the only temporal trend study on MeSO$_2$-PCBs reported in biota.

As many as 60 MeSO$_2$-PCB congeners have been reported in human tissues, and structures have been determined for 50 of these congeners (Table 3) [95, 117, 121, 122, 124]. However, in the tissues of wild mammals between 6 and 41 MeSO$_2$-CBs have been detected, and 6 to 28 congeners identified and quantified. The congeners identified in biota are similar in structure and independent of the species or tissues studied, diet, geographic location, sex, age or reproductive status.

Persistent MeSO$_2$-PCBs are i) 3- and 4-MeSO$_2$-substituted and occur in isomer pairs derived from the same *meta-para*-PCB precursor, ii) trichloro- to heptachloro-substituted, and iii) 2,5-dichloro- or 2,5,6-trichloro-substituted in the MeSO$_2$-containing phenyl ring (Table 1). The majority of these persistent MeSO$_2$-PCBs possess a chlorine atom in the *para*-position of the non-MeSO$_2$-containing phenyl ring. Exceptions for persistent congeners may exist since at least 9 minor 2-, 3- and 4-MeSO$_2$-diCBs lacking 2,5-dichloro- or 2,3,6-trichloro-substitution where identified in the liver, lung and adipose of Japanese Yusho patients [117].

The MeSO$_2$-PCBs present in the adipose tissue of Baltic grey seal, and Canadian ringed seal and polar bear illustrate the dominance of the congeners conforming to i), ii) and iii) (Fig. 5). The preference of 2,5-dichloro- and 2,5,6-trichloro-substitution on the MeSO$_2$-containing ring may be related to the stability of the 3,4-epoxide precursor. Reich et al. [65] showed that the 3,4-epoxide of CB-52 was stable for several weeks, presumably due to the presence of chlorines on either side of the epoxide. Based on the relative peak intensities, roughly 90% to 100% of the Σ-MeSO$_2$-PCB concentrations in biota are comprised of congeners conforming to i), ii) and iii).

Aroclors 1242, 1254 and 1260 and their equivalents manufactured in Europe and Japan are PCB commercial mixtures that have been produced in the highest quantities [152, 153]. If the present rules (*i.e.*, i), ii) and iii)) are accepted for the metabolic formation of the majority of persistent MeSO$_2$-PCBs, only 12 to 17 PCB precursors are likely to form MeSO$_2$-PCBs. There is an additional maximum of 7 PCBs that could form 3- and 4-MeSO$_2$-PCBs if PCBs with 2,5-dichloro- and 2,3,6-trichloro-, and not 4-chloro-substitution on the other phenyl ring are considered.

Species with increasing Σ-MeSO$_2$-PCB levels generally possess a correspondingly simpler MeSO$_2$-PCB congener pattern. CYP2B-like enzyme capacity and the phase II and III enzymatic pathways leading to MeSO$_2$-PCBs are species variable, and dependent on the levels of exposure to enzyme-inducing OHS (*e.g.*, PCBs and Chlordanes). Compared to Baltic seals, ringed seals from the Canadian arctic are exposed to considerably lower OHS levels [15, 154]. The adipose of grey seal from the Baltic Sea and polar bear have been shown to be dominated by 3'- and 4'-MeSO$_2$-CB49, -CB87 and -CB101 (Fig. 5A and C) [11, 22, 98]. Ringed seal adipose tissue, which contains MeSO$_2$-PCB concentration orders of magnitude lower (Table 3) contains many equally dominant conge-

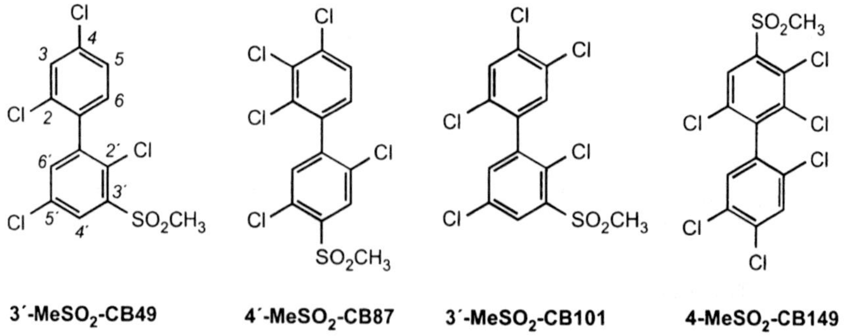

3'-MeSO$_2$-CB49   4'-MeSO$_2$-CB87   3'-MeSO$_2$-CB101   4-MeSO$_2$-CB149

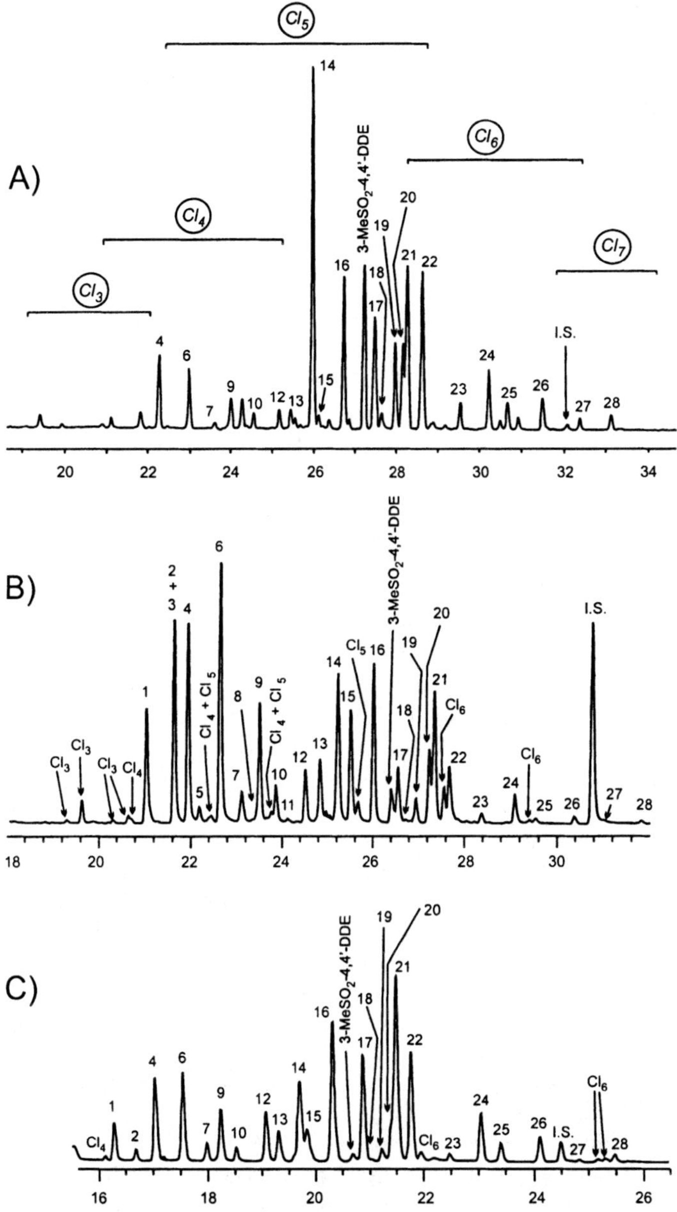

Fig. 5A–C. Chromatograms of the aryl methyl sulfone fraction obtained from, A the blubber of grey seal from the Baltic Sea using GC/ECNI-MS(SIM) [15], B the blubber of ringed seal from the Canandian high Arctic using GC/ECNI-MS(TIC), and C the fat of polar bear using GC/ECD [11]. The number labels refer to the corresponding $MeSO_2$-PCB congeners in Table 1. The Cl_x denotes the number of chlorines in unidentified $MeSO_2$-PCB isomers as indicated by GC/ECNI-MS(TIC). Peaks identified as $MeSO_2$-DDE congeners are also labeled. The I.S. is the $MeSO_2$-PCB internal standard, 3-$MeSO_2$-2-Me-2′,3′,4′,5,5′-pentachlorobiphenyl

ners (Fig. 5B). The adipose tissue of highly contaminated beluga whale from the St. Lawrence was also found to contain mainly 3'- and 4'-MeSO$_2$-CB49, -CB87 and -CB101 in contrast the more complex congener pattern of relatively less contaminated western Hudson Bay beluga [23]. Other species and tissues containing 8 or less dominant MeSO$_2$-PCBs including at least 3'- and 4'-MeSO$_2$-CB87 and -CB101 are Baltic otter (*Lutra lutra*) fat [22,144], harbour seal blubber [15], white-tailed sea eagle egg [94]. Adipose, milk and blood from Swedish individuals were shown to contain dominant levels of 4'-MeSO$_2$-CB87 and 4-MeSO$_2$-CB149 [12, 16, 126].

The congener patterns of MeSO$_2$-PCBs are dependent on the sulfone clearance and formation capacity of a given species. The most persistent MeSO$_2$-PCBs contain at least a 4-chlorine on the non-MeSO$_2$-containing phenyl ring (Table 1). Formation of specific sulfone congeners is dependent on the species. For example, beluga whale appears to have difficulty forming persistent MeSO$_2$-PCBs from PCB precursors with 6 or greater chlorines, regardless of the induced level of enzymes that lead to the formation of MeSO$_2$-PCBs [23, 145], whereas harbour porpoise does not [125]. From a clearance perspective, MeSO$_2$-PCB lacking adjacent, chlorine-unsubstituted carbons, and/or >5 chlorine atoms (Table 1), generally appear to be less susceptible to secondary biotransformation processes. It has been suggested that hindrance of MeSO$_2$-PCB metabolism may also be aided by 3,5-dichloro- or 3,4-dichloro- on the non-MeSO$_2$-substituted phenyl ring [11, 22, 23].

MeSO$_2$-PCBs of low chlorination and possessing *meta-para* and/or *ortho-meta* hydrogens are susceptible to metabolic clearance in several species, via mediation by CYP1A and CYP2B-like enzymes. Trichloro- to pentachloro-MeSO$_2$-PCBs of CB-31, -52, -64, -70 and -95 are effectively reduced relative to the higher chlorinated MeSO$_2$-CB congeners in Baltic grey seal, Canadian polar bears (Fig. 5A and C) and St. Lawrence beluga whale [23], which possess greater CYP activity. CB-52 has been shown to be metabolized to bis-MeSO$_2$-metabolites, and likely via a mono-MeSO$_2$-CB52 intermediate [86]. So far, structurally unidentified bis(MeSO$_2$)-PCBs have also been detected in wildlife [11].

In the adipose tissue of human [12], grey seal [22], ringed seal [98], beluga whale [22, 23], polar bear [98], North Sea long-finned pilot whale (*Globicephalus melas*), white-sided dolphin (*Lagenorhynchus acutus*), common dolphin (*Delphinus delphis*) and Risso's dolphin (*Grampus griseus*), striped dolphin from the Aegean Sea [118], and the plasma [126] and milk of humans [16], the 4-MeSO$_2$-PCB congeners were more dominant than their corresponding 3-MeSO$_2$-PCB isomer. The liver is a special MeSO$_2$-PCB storage tissue in this regard since the 3-MeSO$_2$-PCB as opposed to the 4-MeSO$_2$-PCB pair tends to dominate, as shown in studies on beluga, false killer whale (*Pseudorca crassidens*), grey seal [22], polar bear [17, 155], human [12], Laysan albatross and harbor porpoise [125]. In general MeSO$_2$-PCBs tend to accumulate preferentially in liver and lung, and is the subject of the following section (Sect. 5.2.2).

In summary, MeSO$_2$-PCB metabolites are persistent in biota due to their low reactivity, and lipophilic and bioaccumulative properties. Unlike PCBs, MeSO$_2$-PCB levels and congener patterns are dependent on the capacity of a species to metabolically form as well as clear the compounds. The susceptibility of a PCB congener to form a persistent $MeSO_2$-PCB metabolite, and the biological half-life of the $MeSO_2$-PCB formed are strongly dependent on the chemical structure even though other factors are influential.

5.2.2
Specific Tissue Retention and Protein Binding

$MeSO_2$-PCBs in biota have been detected in a number of different tissues (Table 3). Of the tissues analyzed, the trend among mammals and humans is a preference for the persistent $MeSO_2$-PCB congeners to localize in the liver and lung [7–9, 156]. Swedish grey seal, otter, mink, harbour porpoise and humans, and Canadian polar bear and beluga whale have been shown to contain Σ-$MeSO_2$-PCB concentrations at least 10 times higher in the liver (and in some cases in the lung) than in tissues such as fat, muscle, kidney and testes (lipid weight basis) [12, 15, 17, 22, 125, 144, 156]. Yusho patients were found to have 4- to 5-times higher Σ-$MeSO_2$-PCB levels in liver and lung relative to adipose tissue [122, 124].

Reports detailing studies on $MeSO_2$-PCB protein binding and tissue and fluid retention have appeared regularly in the literature since the early 1980s. Trends exist for the preferential retention of 3-$MeSO_2$-PCBs versus their 4-$MeSO_2$-PCB pairs, although there are inter-species and inter-tissue variations. The 4-$MeSO_2$-CBs generally dominate over their corresponding 3-$MeSO_2$-CB pairs in the adipose tissue of polar bears and arctic ringed seals, and the adipose and milk of humans [11, 12, 16, 98]. In Canadian beluga whale and Swedish harbour porpoise the 3-$MeSO_2$-CBs tended to dominate over their corresponding 4-$MeSO_2$-CB pair [23,125]. Several mechanisms, including protein binding, may underlie the preferential retention of $MeSO_2$-PCBs in the liver, especially for the 3-$MeSO_2$-PCBs. In biota, the number of chlorines does not appear to be a major factor in the preferential storage of $MeSO_2$-PCBs in the liver [12, 17, 22, 42, 125, 145, 155]. However, the 3-$MeSO_2$-CBs invariably dominated over the 4-$MeSO_2$-CBs in the liver of polar bear, humans, Canadian beluga whale, Swedish harbour porpoise and grey seal (Table 3). In a recent feeding study, Lund et al. [157] observed a 2- to 65-fold higher accumulation of 3-$MeSO_2$-PCBs in liver relative to muscle and lung in mink dams exposed to a mixture of 16 *meta*- and *para*-$MeSO_2$-PCBs. Enantioselective biotransformation processes further complicate the accumulation dynamics for some $MeSO_2$-PCBs. High enantiomeric excess was recently shown for 3′-$MeSO_2$-CB132 and -CB149 in human liver [99], and 3- and 4-$MeSO_2$-CB91, 4′-$MeSO_2$-CB132 and 3-$MeSO_2$-CB149 in polar fat [98]. A very high enantiomer selectivity was also shown for each of the metabolite pairs of $MeSO_2$-CB91, -CB132 and -CB149, in rat liver and adipose tissue in PCB (Clophen A50) exposed animals [158].

The tissue-specific retention of $MeSO_2$-PCBs was first observed in experimental studies with mice [7, 9, 159–161]. The tissue-specificity for the retention of $MeSO_2$-PCBs can often be explained by protein binding. The 4,4′-bis ($MeSO_2$)-CB52, a secondary metabolite of 4-$MeSO_2$-CB52, has demonstrated high affinity and reversible protein binding in lung, liver and kidney tissues and the intestine of various laboratory species. A bis($MeSO_2$)-tetrachloroCB detected in the liver of polar bear may in fact be 4,4′-bis($MeSO_2$)-CB52 [11]. [^{3}H]-4,4′-bis($MeSO_2$)-CB52 were shown to form complexes with $\alpha_{2\mu}$-globin in the kidney of rats, and complexes with the major urinary protein in mice [20]. The 4,4'-bis(MeSO$_2$)-CB52 also forms complexes with fatty acid binding protein (FABP) isolated from rat intestinal mucosa [19], and FABP isolated from the liver and intestinal mucosa of chicken [162]. Localization of sulfur containing PCB metabolites was also shown in the feces, bile, liver and lung of rats and mice [61, 137, 163, 164].

The kidney, and especially the lung deserve special attention, with respect to the tissue retention and binding of 4,4'-bis(MeSO$_2$)-CB52 and several mono-MeSO$_2$-PCBs. CB-49 and CB-101 were among the 17 [^{14}C]-labelled PCB that formed isomeric MeSO$_2$-PCB, which localized in the lung of mice. The 4-MeSO$_2$-CB52 was found to concentrate in the bronchi, trachea, larynx and nasal mucosa of mice [165]. Methyl-[^{35}S]sulfonyl-trichlorobiphenyl formed *in vivo* in conventional mice injected with [^{35}S]-cysteine or [^{35}S]-methionine and CB-31 was shown to be localized in lung and kidney, while germ-free mice did not possess this ability [166]. This demonstrates the involvement of the intestinal microflora in the pathway for MeSO$_2$-PCB formation. Evidence for the MAP pathway for the formation of MeSO$_2$-PCBs was given in a series of articles by Bakke and co-workers in the early 1980s [76, 137, 163]. Methyl-[^{35}S]sulfonyl-tetrachlorobiphenyl was also shown to be formed *in vivo* in mice injected with [^{35}S]-cysteine and CB-64 [133]. The radioactive metabolite was shown to be localized in liver, kidney and fat, but mainly in the lung [133]. The MAP pathway for the formation of MeSO$_2$-PCBs is shown in Fig. 4.

Specific binding of 4,4'-bis(MeSO$_2$)-CB52 occurs in the tracheo-bronchial mucosa and bronchoalveolar lavage of the lung and the kidney cortex in mice and rats [21, 167–169]. The accumulation of 4,4'-bis(MeSO$_2$)-CB52 in the bronchi has been shown to be localized to Clara and globet-like cells, which are non-ciliated cells in the lung epithelium [86, 165]. Clara cells secrete material containing MeSO$_2$-PCBs into the airway. Therefore, MeSO$_2$-PCBs can undergo enteropulmonary circulation as a consequence of being swallowed and again re-absorbed from the intestine.

MeSO$_2$-CBs can selectively bind to endocrine-associated proteins, tissues and fluids. The 4,4'-bis(MeSO$_2$)-CB52 binds to steroid binding protein, uteroglobin, expressed in *E. coli* from rat [18]. Recently, the 4,4'-bis(MeSO$_2$)-CB52 was shown to form a complex with purified rat uteroglobin, also called PCB-binding protein [170]. Uteroglobin is a steroid binding protein which complexes with the hormone progesterone. Other MeSO$_2$-PCBs are also known to interact with progesterone from rat lung [18]. MeSO$_2$-CB metabolites of CB-31 are enriched in the uterine fluid of pregnant mice [36]. Johansson et al. [171] recently reported that out of 24 tetrachloro- to hexachloro-3- and 4-MeSO$_2$-CBs found in biota, congeners with 3 chlorines in *ortho*-positions and an MeSO$_2$-group in either the 4- or 4'-position increased the binding affinity to gluco-corticoid receptor (GR) from mouse liver cytosol.

5.2.3
Biological and Toxicological Effects

Deleterious biological and toxicological effects observed in many species, including humans, have been linked to OHS exposure. This linkage may also involve persistent aryl methyl sulfone metabolites in the case of PCBs. It is beyond the scope of this review to cover all the different toxicity aspects of the $MeSO_2$-PCBs. A summarization of the known toxicity studies is listed in Table 4. Mechanistic details can be found in the original research articles cited. Some studies are highlighted below to illustrate the toxicological potential of $MeSO_2$-PCB congeners persisting in biota.

The presence of $MeSO_2$-PCBs in lung tissue was first linked to the pulmonary distress observed in Yusho victims in the 1970s [172], after initial reports on specific tissue localization of 4-$MeSO_2$-PCBs in the lung of mice [9, 159–161]. More regular reports of the biological and toxicological effects of $MeSO_2$-PCB began to appear in the literature in the early 1990s. Congeners such as 3'- and 4'-$MeSO_2$-CB87 and -CB101, which consistently dominate in tissues of wildlife species and humans (Sect. 5.2.1), are frequently included in these studies. The longer half-life and higher biomagification potential of these congeners, which lack adjacent, chlorine unsubstituted carbons, and/or >5 chlorines, are suggestive of a greater potential for toxicological risk from exposure. For example, 3'- and 4'-$MeSO_2$-CB101 can account for 25% to 50% of the Σ-$MeSO_2$-PCB concentrations in some species. Persistent $MeSO_2$-PCBs have demonstrated a limited but diverse range of biological and toxicological activities *in vitro* and *in vivo* in mammal species, but mostly in laboratory rodents (Table 4).

$MeSO_2$-PCBs are capable of interacting with enzymes and perturbing their catalytic activity. In human lymphoblastoid cells 3-$MeSO_2$-3',4,4',5-tetraCB was shown to inhibit aryl hydrocarbon hydroxylase (AHH) activity, whereas the environmentally relevant 3-$MeSO_2$-CB70 did not [173]. Other persistent $MeSO_2$-PCBs and congeners not identified in biota have shown no significant effect on AHH activity [174–176]. Japanese researchers demonstrated that 3-$MeSO_2$-PCBs, and not 4-$MeSO_2$-PCBs, including the sulfones of CB-87 and –101, are inductive of hepatic CYP2B1, CYP2B2, CYP3A2 and CYP2C6 protein levels and related catalytic activities in rats [177–180]. In mice treated with 4-$MeSO_2$-CB52, pulmonary drug metabolism was decreased [181]. The 3- and 4-$MeSO_2$-CB49, -CB101 and -CB149 have been shown to inhibit cell-cell communication between IAR 20 rat liver epithelial cells [182].

Several recent reports have demonstrated the potential of persistent $MeSO_2$-PCB to influence endocrine-related processes. The 3-$MeSO_2$-CB132, -CB141 and -CB149, and 4-$MeSO_2$-CB149 reduced thyroid hormone levels in blood and increased thyroid weight and hepatic CYP protein levels in rats [183]. Interestingly, the sulfone isomers of both CB-132 and CB-149 have specific liver retention properties, and each congener exists as two enantiomers [98, 99]. The 3- and 4-$MeSO_2$-CB52, -CB70, -CB87 and -CB101 are known to be cytotoxic to human placental JEG-3 and JAR choriocarcinoma cells [184, 185]. Chronic and reproductive toxicity was observed in mink exposed *in vivo* to 16 environmentally relevant $MeSO_2$-PCBs [157]. Similar to 3-$MeSO_2$-4,4'-DDE, two $MeSO_2$-PCBs, 4-$MeSO_2$-CB64 and 4-$MeSO_2$-CB110, have been shown to be competitive inhibitors of CYP11B1 in the mouse Y1 adenocortical cell line [186]. As a consequence, the glucocorticoid synthesis was inhibited. $MeSO_2$-PCBs are known to inhibit adrenal CYP11β enzyme activity *in vitro* in grey seals, and may also modulate glucocorticoid synthesis *in vivo* [27, 187]. Thus, it cannot be excluded

Table 4. Biological and Toxicological Activities of Environmentally Persistent Methyl Sulfone Polychlorinated Biphenyls[a]

MeSO$_2$-PCB(s)[b]	Species / Cell/Microsomes	Parameter(s) Measured	References
In Vivo			
3–132, -141, -149; 4–149	Sprague-Dawley rats (male)	Thyroid hormone reduction (T$_3$, T$_4$) in blood; Increased thyroid gland weight; hepatic CYP protein levels	[183]
3- + 4–31, -70, -101, -141; 3–31, -49, -52, -87, -110, -132, -149	Wistar rats (male)	Hepatic CYP2B1/2, CYP3A2, CYP2C6 and b$_5$ induction	[178–180]
5-, 6–77[c]	Wistar rats (male)	CB-77 metabolism; parent + metabolite levels in faeces, liver, blood, lung, kidney, adipose	[87]
3- + 4–31, -70, -101, -141	Wistar rats (male); DDY mice (female)	PCB metabolism; parent + metabolite levels in liver, blood, lung, kidney, adipose, adrenal, brain, heart, testis, spleen	[90, 164]
3–70; 4–52, -87	C57BL/6 N, DBA/2 J mice	Suppression of benzo[*a*]pyrene mutagenicity; inhibition of aryl hydrocarbon hydroxylase (AHH) activity	[174]
4–64; 3- + 4–31; [^{14}C]- 3- + 4–52	C57BL (female) mice; SD rats (male)	MeSO$_2$-metabolite binding; tracheo-bronchial mucosa/kidney cortex	[133, 163, 167]
[^{14}C]MeSO$_2$-31, -77[c]	C57BL mice (preg. female)	Uterine fluid localization in females, and in late gestatinal fetuses (primitive gut)	[36, 191]
3- + 4–52	C57BL / SD mice (male, female)	Binding to protein complex in kidney and lung (Clara cells)	[165]
[^{14}C] / [^{35}S] - 4–52; 3- + 4–31	Wistar/SD rats (male); NMRI mice	MeSO$_2$-metabolite localization in feces, liver and lung; metabolite formation in the intestinal flora	[61, 137, 166]

In Vitro

3- + 4 – 52, -70, -87, -101	human placental JEG-3 and JAR cancer cells	Aromatase (CYP19) activity; cytotoxicity (%LDH leakage, DNA content)	[184]
3- + 4 – 49, -101, -149	IAR 20 rat liver epithelial cells	Cell communication, cytotoxicity	[217]
4 – 101	rat, human	Binding antagonism	[18, 170, 218]
3 – 77 [c]	various mice strains (e.g., C57BL/6 J, etc.) (male)	inhibition of AHH activity, liver microsomes from TCDD- and MC-induced mice	[176]
3 – 70 + -77 [c]; 4 – 52 + -87	human lymphoblastoid cells	inhibition of AHH activity, MC- or TCDD-induced cells	[173, 175]
4 – 52 + -31	adrenal tissue fractions, grey seal	No effect on mitochondrial binding or CYP11β-hydroxylase activity	[71]

[a] The reports are listed in chronological order starting with the most recent.

[b] The MeSO$_2$-PCB congener designation is based on an established PCB numbering system [45]. The congeners listed are those identified in the tissues of wildlife and humans (Table 1). Additional MeSO$_2$-PCBs were investigated in some of the studies.

[c] MeSO$_2$-PCB metabolites of CB-77 have not been shown to be relevant in biota.

that 4-MeSO$_2$-CB64 and 4-MeSO$_2$-CB110 may also have an influence on the enlargement of the adrenal cortex as has been observed in grey seals from the Baltic [188]. The reported affinity of several persistent MeSO$_2$-PCBs to the glucocorticoid receptor of human and mouse also have endocrine-related implications *in vivo* [171]. The 3-MeSO$_2$-CB149 and 7 other MeSO$_2$-CBs were shown to be functional antagonists of ^{3}H-dexamethasone binding with the human GR at dose concentrations at IC$_{25}$ between 5 and 20 µM.

5.3
OH-PCBs

5.3.1
Levels and Congener Patterns

The present review on patterns and levels of OH-PCBs in biota concentrates on those present in blood since the majority of OH-PCBs formed *in vivo* are readily excreted. Most OH-PCBs are therefore anticipated to be present at very low and transient levels. A number of OH-PCBs have been identified or at least indicated in the blood of humans and biota during the last 5 to 10 years (Table 2, Table 5, Fig. 6). However, reports on the identity, presence and levels of OH-PCBs in biota remain very limited in comparison to the MeSO$_2$-PCBs. This situation is expected to change in the near future as a consequence of more scientific interest in this area.

The environmental occurrence of OH-PCBs was first demonstrated in seal feces and guillemot (*Uria aalge*) droppings in the mid-1970s [35]. Twenty years later, several OH-PCB congeners were shown to be selectively retained in human plasma and blood from Baltic grey seals [39]. In the last five years OH-PCBs have also been determined in blood from two albatross species (Fig. 6b) [42], white-tailed sea eagle chicks, polar bears, salmon (*Salmo salar*) and human milk [37, 39, 42, 43]. OH-PCB congeners identified so far in biota are summarized in Table 6.

4-OH-CB107   **4-OH-CB146**   **3-OH-CB153**   **4-OH-CB187**

Up to approximately 30 OH-PCB and a few (OH)$_2$-PCB congeners have been detected in human blood (Fig. 6). As co-elution of OH-PCB congeners is known to occur, the number of OH-PCB congeners present in the samples may still be

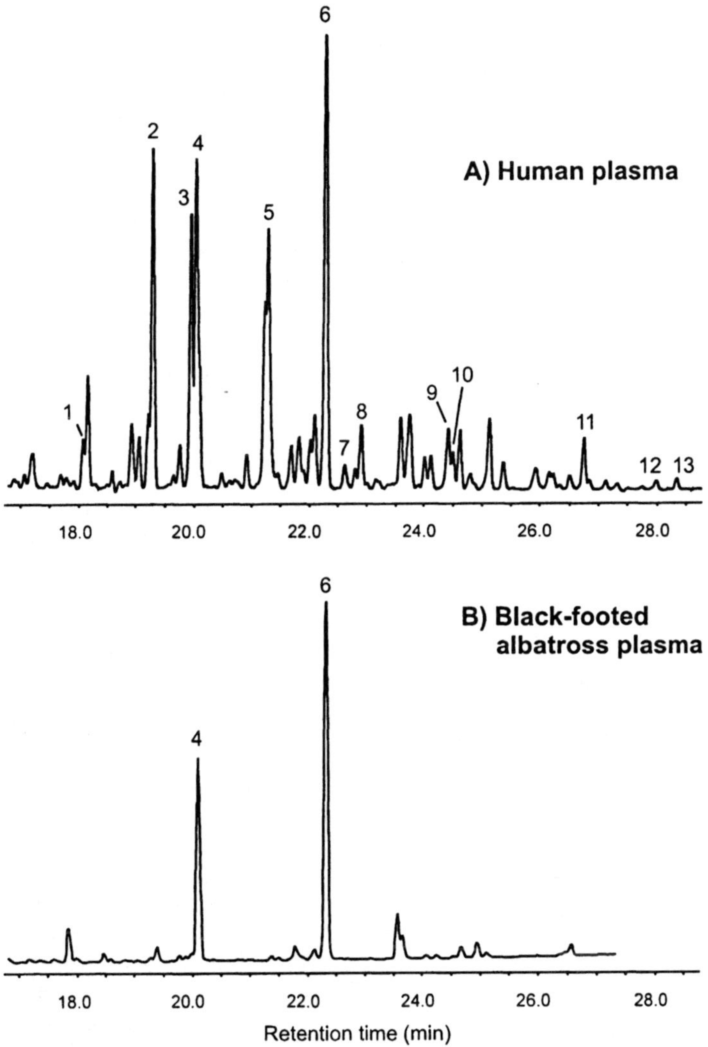

Fig. 6A, B. GC/ECD chromatograms of the phenolic fraction obtained from A a women from the Faeroe islands with a relatively high OHS contamination level in her blood (Å. Bergman et al., unpublished data), and B albatross [42] plasma. The structures of the identified congeners are labeled on the peaks. The number labels refer to the corresponding OH-PCB congeners listed in Table 2. The I.S. is 2,3,3′,4′,5,5′,6-heptachloro-4-biphenylol (4-OH-CB193)

underestimated [38, 39, 46]. The major OH-PCBs present in e.g., human blood are limited to 5 to 10 single compounds. It is notable that the majority of OH-PCBs in blood or plasma are formed from the more persistent PCB congeners with 5 to 7 chlorine atoms [38, 43]. OH-PCBs with 8 and 9 chlorine atoms have also been indicated by GC/MS analysis. Among species the specific OH-PCB congeners retained in blood are essentially the same even though some differ-

Table 5. Polychlorobiphenylols identified in wildlife and humans. The potential and known structures of the parent chlorinated biphenyls are given for the identified OH-CBs

OH-PCB [a]	Environmental occurrence		Origin of OH-PCB metabolites		
	Species	Ref.	Parent PCB	Identity Status[b]	Ref.
4-OH-CB107/	Human	[39, 189, 219]	CB-105	Shown	[46, 71]
4'-OH-CB108[b]	Grey seal	[39]	CB-118	Shown	[46]
	White-tailed sea eagle	[190]			
3-OH-CB153	Human	[39]	CB-153	Shown	[46, 220]
			CB-128	Suggested	
4-OH-CB146	Human	[39, 189, 219]	CB-153	Shown	[46]
	White-tailed sea eagle	[190]	CB-138	Shown	[46]
	Laysan albatross	[42]			
	Black-footed albatross	[42]			
4-OH-CB187	Human	[39, 189, 219]	CB-187	Shown	[46]
	Grey seal	[39]	CB-183	Possible	
	White-tailed sea eagle	[190]			
	Polar bear	[43]			
	Laysan albatross	[42]			
	Black-foted albatross	[42]			
4'-OH-CB159	Grey seal	[39]	CB-156	Suggested	
4'-OH-CB172	Human	[39]	CB-170	Suggested	
	Grey seal	[39]			

[a] The structures of the OH-CBs are given in Table 2.
[b] These OH-PCBs co-elute and are therefore presented together.

ences in the relative levels of different OH-PCB congeners can be observed between individuals. This has been shown for persons from different geographical areas, such as between humans from Latvia, Sweden [189] and the Faeroe Islands (Fängström, pers. commun.). The relative abundance of OH-PCB congeners varies significantly depending on the species. The species-variable OH-PCB patterns in blood are dominated by a few congeners, although the number of congeners detected is generally high [42]. Five to 7 GC OH-PCB peaks dominate the chromatograms of phenolic OHS fractions isolated from human and grey seal blood (Fig. 6A) [39], whereas only two peaks dominate in *e.g.*, albatross (Fig. 6B).

The 4-OH-CB146 and 4-OH-CB187 congeners comprised 90% of the total OH-PCBs quantified in plasma from Laysan and the Black-footed (*Diomedea nigripes*) albatrosses (Fig. 6b) [42]. These two OH-PCBs were also the major congeners in white-tailed sea eagle chicks [190]. Polar bear plasma contained mainly 4-OH-CB187, a minor peak identified as 4-OH-CB193, and an unidentified peak, which likely is 4-OH-CB146 [43]. The OH-PCBs detected in salmon from the Baltic Sea have yet to be identified [37]. Other identified OH-PCB congeners and their probable parent PCB congeners identified in the species stu-

Table 6. Concentrations of Hydroxylated Biphenyls in Wildlife and Humans

Species (number of samples)	Location/ year	Body fluid	No. of OH-PCB	4-OH-CB107 (ng/g, l.w.)[a]	4-OH-CB107 (ng/g, l.w.)[a]	Σ-OH-PCB (ng/g, l.w.)[a]	CB-153 (ng/g, l.w.)[a]	Σ-PCB (ng/g, l.w.)[a]	Ref.
Humans									
Male blood donors (n = 6)	Sweden/ 1994	plasma	14	360 (200–430)	n.a.	n.a.	2200 (740–2800)	3600 (2000–5200)	[39]
Male moderate fish consumers[b] (n = 11)	Sweden/ 1991	plasma	5	71 (46–110)	100 (73–140)	340 (240–480)	450 (340–600)	1600 (1200–2200)	[189]
Male moderate fish consumption[b] (n = 22)	Latvia/ 1991	plasma	5	98 (63–150)	56 (42–75)	270 (190–380)	210 (150–290)	750 (550–1000)	[189]
Wildlife									
Grey seal (n = 5)	Baltic Sea/ 1992	blood coagulate	13	1500 (90–1700)	n.a.	n.a.	7400 (4400– 45000)	30000 (7900– 83000)	[39]
Albatross, Black footed (n = 10)	Midway I./ 1994	plasma	7	n.a.	n.a.	27.1 (20–33) f.w.[c]	35.2 (16.7–66.9) f.w.	92.5 (45–180) f.w.	[42]
Laysan (n = 5)	Midway I./ 1994	plasma	7	n.a.	n.a.	11.5 (5.9–18.5) f.w.[c]	6.1 (4.8–7.8) f.w.	17.6 (15–20) f.w.	[42]
White-tailed sea eagle	Sweden/ 1996–8	whole blood	> 20	18 2.3–590	410 120–1100	– 1100–7800	2900 4700–27000	13000	[190]

[a] The majority of the concentrations listed were normalized for the lipid weight (l.w.). In some cases the concentrations were reported on a fresh weight (f.w.) basis. The mean concentrations and the minimum and maximum concentrations are listed.

[b] The data are from a much larger study, which includes males with high and no fish consumption [189].

[c] The 4-OH-CB187 and 4-OH-CB146 comprised 70% to 90% of the sum concentration (Σ–) for the individually quantified OH-PCB congeners.

n. a. = data not available.

died so far are listed in Table 5. The individual OH-PCB congener and Σ-OH-PCB concentrations are given in Table 6. Based on the work performed to date, it is evident that some of the OH-PCB congeners that are analyzed for, after methylation, may co-elute as has been observed for, *e.g.*, 4-OH-CB107/4'-OH-CB108 [39]. Hence, other co-eluting MeO-PCB congeners cannot be excluded.

OH-PCBs are retained in blood due to an affinity for a plasma protein as discussed below. Their presence in blood is therefore not primarily related to the lipophilic properties of OH-PCBs or to lipid content in the blood. At present it is difficult to make a direct comparison of the existing OH-PCB concentration data reported in biota, since the levels have been determined in different ways, either reporting on the sum of congeners, on congener-specific concentrations, on a lipid weight basis or on a fresh weight basis. Regardless, in the species studied the Σ-OH-PCB concentrations are high relative to Σ-PCB levels, and the concentrations of the OH-PCB and PCB change proportionally. The percent ratios of the Σ-OH-PCB to Σ-PCB concentrations in blood ranges from about 10% in humans to 90% in some albatross samples (Table 6) [39,42].

The parent PCBs of most of the retained OH-PCBs have been identified and are listed in Table 5. The majority of retained OH-PCBs are formed via an arene oxide followed by a 1,2-shift. However, direct insertion of a OH-group may also occur (see Sect. 2.2).

5.3.2
Specific Tissue Retention and Protein Binding

In 1982 selective retention of phenolic PCB metabolites was first observed in the intraluminal uterine fluid of pregnant mice after administration of CB-31, using whole body autoradiography and subsequent chemical analysis [36]. Administration of CB-77 to pregnant mice resulted in a dramatic accumulation of a phenolic metabolite in fetal soft tissue [191]. The retained metabolite was shown to be 4-OH-3,3',4',5-tetraCB and localized in the blood in both the fetus and in adult mice [73]. A metabolism study of the structurally related CB-105 in mice showed significant retention of the *para*-substituted hydroxylated metabolite 4-OH-CB-107 in blood [71]. The 4-OH-CB107 was also shown to be retained in blood from rats administered Aroclor 1254 [39], and was observed at high levels in blood and brain tissue from rat fetuses exposed *in utero* [192].

In the first report on phenolic PCB metabolites in blood from rats after exposure to Aroclor 1254, and from environmentally exposed Baltic grey seals and humans only a few OH-PCBs were structurally identified [39]. Since then, additional metabolites have been identified and are shown in the chromatogram of a human blood sample taken from a Faeroe Island woman (Fig. 6A). All the identified metabolites have chlorine atoms attached to the carbons *ortho* to the oxidized carbon. With a few exceptions, the hydroxy-group is attached to one of the two *para*-positions of the PCB molecule. These structural elements are also found in 3,3',4',5-tetraiodo-L-thyronine, or thyroxine, the natural substrate of TTR [44]. The affinities of the retained OH-PCB congeners are up to 10 times greater than for thyroxine [44, 134]. The structural similarity to thyroxine was suggested as an explanation for OH-PCB retention in fetal rat brain [192]. TTR is the principal transport protein for thyroxine to the fetal rat brain.

As mentioned above, the congener patterns of retained OH-PCBs in blood vary depending on the species, but also, as experimentally observed, with time after a single dose. At 1, 7 and 14 days after Aroclor 1254 administration to rat, the OH-PCBs in blood were shown to be dominated by 4-OH-CB107, but the relative importance decreased with time [39]. The OH-PCB congener pattern has also been shown to vary to some extent between individuals. For human males, distinct differences in the OH-PCB congener pattern was observed in relation to the consumption of fatty fish from the Baltic [189]. In particular, the relative amounts of 4-OH-CB107 and 4-OH-CB187 were variable. The differences in the relative amounts of retained OH-PCBs in different species and individuals has yet to be explained. TTR is a highly conserved protein and present in all species studied, but differences in type and affinity of ligand binding can not be excluded. Other factors that may influence the OH-PCB metabolite pattern are species differences in metabolic capacity and metabolism kinetics.

5.3.3
Biological and Toxicological Effects

The potential toxicity of OH-PCBs was first investigated in the 1970s. Hydroxylated metabolites were reported having higher potencies for cell toxicity than their parent PCBs [64, 193]. Phenolic PCB metabolites were also found to affect mitochondrial respiration and the permeability of the inner membrane. The nature of the effect depended on the structure and pK_a of the phenolic metabolite [110, 194]. Low binding potencies toward the aryl hydrocarbon receptor (AhR), and low induction capacity of ethoxyresorufin-O-deethylase (EROD), respectively, have been reported for phenolic PCB metabolites of non-*ortho* CB-77 and mono-*ortho* CB-105 [34, 195]. Dihydroxylated PCBs may be oxidized to quinones that in turn may react with macromolecules to form adducts, and cause oxidative stress leading to cell death [116].

The thyroidogenic effects and corresponding biochemical mechanisms of PCBs and other OHS were recently reviewed by Brouwer et al. [44]. The selective retention of certain OH-PCB congeners in blood (Sect. 5.2.2 and 5.3.2) is concomitant with effects observed on the plasma levels of thyroid hormones. Thyroxine is transported in plasma by a protein complex consisting of TTR and retinol binding protein (RBP). Rats administered CB-77 were shown to have reduced plasma levels of both thyroxine and retinol [196]. A major metabolite of CB-77, 4-OH-3,3′,4′,5-tetrachlorobiphenyl, was identified as the active compound [40]. The same hydroxy-PCB metabolite was found to be retained in mouse fetal soft tissue [191, 197].

Administration of 4-OH-3,3′,4′,5-tetrachlorobiphenyl and 4-OH-CB107 to pregnant mice reduced the total thyroxine levels in both maternal and fetal plasma [34]. However, the decrease was less dramatic when the CB-77 was administered alone, and the metabolite formed *in vivo* [197]. Administration of Aroclor 1254 to adult rats resulted in high OH-PCB levels in plasma and a concomitant reduction in thyroxine [198]. In fetal and weanling rats, similar changes in OH-PCB and thyroxine levels were observed after *in utero* and lactational exposure to Aroclor 1254 [192]. A negative correlation between plasma levels of organohalogens and thyroid hormones was observed in pregnant Dutch women, probably due to dietary exposure of OHS. Changes in plasma thyroid hormones were observed in their newborn babies, however OH-PCBs were not analyzed within this study [199]. Changes in human plasma levels of thyroid hormones has also been observed following accidental or occupational exposure to PCBs or polybrominated biphenyls [200–203].

OH-PCBs can also influence thyroxine metabolism. Some of the OH-PCBs that are retained in blood were shown to strongly inhibit sulfation of thyroxine *in vitro* [204–206]. As sulfation is a major regulation pathway of thyroxine in the fetus, the OH-PCBs may negatively influence the development of the fetus, and in particular fetal brain development [44]. Diodinase mediation is another pathway for thyroxine metabolism; *e.g.*, to the active hormone triiodothyronine. Hydroxylated metabolites of CB-77 were shown to inhibit triiodothyronine formation in an *in vitro* assay using rat hepatic microsomes [207].

The endocrine disrupting effects of pollutants have lately received intense attention. Many persistent organic pollutants require screening to assess their potential for interacting with the estrogen receptor. OH-PCBs known to be retained in human blood were shown to be weakly anti-estrogenic in the MCF-7 human breast cancer cell line and HeLa cells [209, 210]. OH-PCB congeners without chlorine atoms in the hydroxylated phenyl ring and at least one *ortho*-chlorine in the other phenyl ring have been reported to have estrogenic activity in *in vitro* tests, and in a few cases also *in vivo* [211]. In the latter tests, the OH-PCB congeners under study were not environmentally relevant, since their parent PCBs are not present in technical PCB mixtures [152]. Considering that the OH-PCBs present in human blood are only weakly anti-estrogenic, the results from studies so far indicate that the environmentally relevant OH-PCB congeners are not significantly active as endocrine disrupters from the perspective of estrogen receptor interaction.

6
Concluding Remarks

The major pathway for distribution of xenobiotics in the environment is through, i) intentional application of pesticides, ii) the use of chemicals in the manufacturing chemical business for technical applications or for improving the quality of a product or of goods, and iii) formation and distribution of substances formed during combustion or as byproducts. In the present chapter we have shown that chemicals formed as metabolites of persistent OHS, in this case PCBs, may form metabolites that are *not* readily excreted, but instead are retained and accumulated in different compartments of an organism. Exceptionally high concentrations of these metabolites may be due to localization in certain cells or in organs/body fluids. Hence, this is also a pathway for the generation of environmental contaminants.

The number of species which have been analyzed for OH- and $MeSO_2$-PCBs remains limited mostly to seals, humans, polar bear and some bird species

(Table 3 and 5). With respect to the $MeSO_2$-PCBs, the growing body of studies in biota indicate that the formation of $MeSO_2$-PCBs is a common phenomena across various types of organisms. In cetaceans for example, $MeSO_2$-PCB formation seemed unlikely since previous catalytic and immunochemical studies indicated a particularly low capacity for CYP2B1/2-like metabolic activity. However, the *meta-para* PCB substrates of these enzymes are metabolized, although more slowly than in seals. Moreover, very recent studies on cetaceans show that persistent $MeSO_2$-PCBs are formed in baleen as well as toothed whales. An as yet unidentified CYP(s) may be operative in cetaceans in the mediation of *meta-para* PCB arene oxide formation, and subsequent $MeSO_2$-PCB formation. This is further supported by the occurrence of 1,2-shifted OH-PCB metabolites of both non-planar and co-planar PCB congeners in various species. Thus, *meta-para* PCBs are substrates for different CYP isozymes. Further research is necessary on different species and populations, temporal effects, and lactational and gestational transfer of both metabolite groups.

$MeSO_2$-PCB congener patterns in biota indicate that the biological half-life of a $MeSO_2$-PCB can vary depending on the structure. In species exposed to high OHS levels for example, $MeSO_2$-PCBs with 4 or less chlorines and/or adjacent *ortho-meta*, and especially *meta-para* chlorine unsubstituted carbons (*e.g.*, $MeSO_2$-CB52 and -CB95 metabolites) are apparently more susceptible to further metabolic degradation. Further research to clarify the $MeSO_2$-PCB biological half-life among $MeSO_2$-PCBs may involve *in vitro* metabolism studies with enzymatically viable microsomes of marine mammals such as seal.

A greater understanding is required regarding the specific retention properties, and the potential range of biological effects and toxicological activities of both OH-PCBs and $MeSO_2$-PCBs. $MeSO_2$-PCBs may have potential for perturbing other endocrine processes *in vivo* such as those operative via the estrogen receptor [212]. The formation and clearance of persistent $MeSO_2$-PCBs, and the formation of plasma retained OH-PCB clearly occurs in a variety of species and humans, but is species variable. The implications of exposure to $MeSO_2$-PCBs and OH-PCBs clearly needs to be included in the risk assessment of PCBs.

Persistent aryl methyl sulfones may be formed from several xenobiotics other than PCBs. $MeSO_2$-DDE metabolites also form in biota, and bind to *zona fasiculata* in the adrenals of mice and other species [25, 26, 119, 187, 213, 214], $MeSO_2$-polychlorobenzene have been shown to bind covalently and possess potent toxicity in the nasal mucosa [215]. Other OHS may be capable of forming persistent $MeSO_2$-metabolites as well. Ten tetrachloro- to hexachloro-$MeSO_2$-polychlorinated terphenyls were detected in the blood of Yusho patients [95]. OHS including polybrominated biphenyl, tetrachlorinated benzyltoluenes, and polychlorinated and polybrominated diphenyl ethers all could potentially form OH-metabolites and persistent $MeSO_2$-metabolites, which can be retained in an organism.

The presence of high levels of pentachlorophenol in mammalian blood is well-known. The fact that as many as 120 or more phenolic OHS are present in both human and fish (salmon) blood demonstrates, in addition to lipid rich tissues, the potential importance of blood for the accumulation of environmental contaminants or their metabolites. Phenolic OHS are not strongly accumulated in lipids due to their physico-chemical characteristics, but are instead retained due to protein binding. The identification of a number of selectively retained OH-PCBs indicates their importance as potential endocrine disrupters. The retention of OH-PCBs may influence also other adverse health outcomes, such as cancer, via secondary formation of catachol/quinone type substances. The latter are known as strong electrophiles that may react with biomacromolecules, including DNA.

In conclusion, PCBs, a representative class of OHS, have been shown to be metabolically transformed to products that are retained in the body or body fluids via protein binding or accumulation in lipid-containing tissues. These PCB metabolites may possess adverse health effects, but more research is required to define the appropriate toxicological endpoints and to better understand the mechanisms of action. Hopefully human beings have learned that great care must be taken in the future to avoid the use of substances with similar properties as the PCBs.

References

1. Bakke JE, Feil VJ, Price CE (1976) Biomed Mass Spectrom 3:226
2. Kuchar EJ, Geenty FO, Griffith WP, Thomas RJ (1969) J Agric Food Chem 17:1237
3. Jensen S (1972) Ambio 1:45
4. Jensen S, Jansson B (1976) Ambio 5:257
5. Sundström G, Hutzinger O, Safe S (1976) Chemosphere 5:267
6. Mio T, Sumino K, Mitzutani T (1976) Chem Pharm Bull 24:1958
7. Brandt I, Bergman Å (1987) Chemosphere 16:1671
8. Bergman Å, Haraguchi K, Athanasiadou M, Larsson C (1993) Organohalogen Compounds 14:199
9. Brandt I (1977) Acta Pharm (Toxicol Suppl) 48:1
10. Letcher RJ, Norstrom RJ, Bergman Å (1995) Sci Total Environ 160/161:409
11. Letcher RJ, Norstrom RJ, Muir DCG (1998) Environ Sci Technol 32:1656
12. Weistrand C, Norén K (1997) Environ Health Perspect 105:644
13. Olsson M, Andersson Ö, Bergman Å, Blomkvist G, Frank A, Rappe C (1992) Ambio 21:561
14. Weistrand C, Norén K (1998) J Tox Environ Health Part A 53:293
15. Haraguchi K, Athanasiadou M, Bergman Å, Hovander L, Jensen S (1992) Ambio 21:546
16. Norén K, Lundén Å, Pettersson E, Bergman Å (1996) Environ Health Perspect 104:766
17. Letcher RJ (1996) PhD thesis, Carleton University, Ottawa, Canada
18. Gillner M, Lund J, Cambillau C, Alexandersson M, Hurtig U, Bergman Å, Klasson-Wehler E, Gustafsson J-Å (1988) J Steroid Biochem 31:27
19. Larsen GL, Bergman Å, Klasson-Wehler E, Bass NM (1991) Chem-Biol Interact 77:315
20. Larsen GL, Bergman Å, Klasson-Wehler E (1990) Xenobiotica 20:1343
21. Lund J, Nordland L, Devereux T, Glaumann H, Gustafsson J-Å (1987) Chemosphere 16:1677
22. Bergman Å, Norstrom RJ, Haraguchi K, Kuroki H, Béland P (1994) Environ Toxicol Chem 13:121
23. Letcher RJ, Norstrom R, J, Muir DCG, Sandau C, Koczanski K, Michaud R, De Guise S, Béland P (1999) Environ Toxicol Chem. In press
24. Lund B-O, Bergman Å, Brandt I (1988) Chem-Biol Interact 65:25
25. Brandt I, Jönsson C-J, Lund B-O (1992) Ambio 21:602
26. Jönsson C-J, Lund B-O (1994) Toxicol Lett 71:169
27. Lund B-O, Lund J (1995) J Biol Chem 270:20895
28. Jansson B, Bergman Å (1978) Chemosphere 3:257

29. Koss G, Koransky W, Steinbach K (1979) Arch Toxicol 42:19
30. Kuroki H, Hattori R, Haraguchi K, Masuda Y (1989) Chemosphere 19:803
31. Kuroki H, Hattori R, Haraguchi K, Masuda Y (1987) Chemosphere 16:1641
32. Sundström G, Jansson B (1975) Chemosphere 6:361
33. Moir D, Viau A, Chu I, Klasson-Wehler E, Mörck A, Bergman Å (1996) Toxicol & Industr Health 12:105
34. Sinjari T, Klasson-Wehler E, Hovander L, Darnerud PO (1998) Xenobiotica 1:31
35. Jansson B, Jensen S, Olsson M, Renberg L, Sundström G, Vaz R (1975) Ambio 4:93
36. Brandt I, Bergman Å, Darnerud PO, Larsson Y (1982) Chem-Biol Interact 40:45
37. Asplund L, Athanasiadou M, Sjödin A, Bergman Å, Börjesson H (1999) Ambio 28:67
38. Klasson-Wehler E, Hovander L, Bergman Å (1997) Organohalogen Compounds 33:420
39. Bergman Å, Klasson-Wehler E, Kuroki H (1994) Environ Health Perspect 102:464
40. Brouwer A, Klasson-Wehler E, Bokdam M, Morse DC, Traag WA (1990) Chemosphere 20:1257
41. Klasson-Wehler E (1994) Organohalogen Compounds 20:437
42. Klasson-Wehler E, Bergman Å, Athanasiadou M, Ludwig JP, Auman HJ, Kannan K, van den Berg M, Murk AJ, Feyk LA, Giesy JP (1998) Environ Toxicol Chem 17:1620
43. Sandau CD, Norstrom RJ (1996) Organhalogen Compounds 29:412
44. Brouwer A, Morse DC, Lans MC, Schuur G, Murk AJ, Klasson-Wehler E, Bergman Å, Visser TJ (1998) Toxicol & Industr Health 14:59
45. Ballschmiter K, Mennel A, Butyen J (1993) Fresenius J Anal Chem 346:396
46. Sjödin A, Tullsten AK, Klasson-Wehler E (1998) Organohalogen Compounds 37:365
47. Bergman Å, Klasson-Wehler E, Kuroki H, Nilsson A (1995) Chemosphere 30:1921
48. Schnellman RJ, Vickers AEM, Sipes IG (1985) Rev Biochem Toxicol 7:247
49. Safe S (1980) Metabolism, uptake, storage and bioaccumulation. In: Kimborough RD (ed) Halogenated Biphenyls, Terphenyls, Naphthalenes, Dibenzodioxins and Related Products. Elsevier/North Holland Biomedical Press, Amsterdam, p 81
50. Bakke J, Gustafsson J-Å (1984) Trends Pharmac Sci 5:517
51. Norstrom RJ, Letcher RJ (1997) Role of biotransformation in bioconcentration and bio-accumulation. In: Sijm D, de Bruijn J, de Voogt P, de Wolf W (eds) Biotransformation in Environmental Risk Assessment. SETAC-Europe Publications, Brussels, p 103
52. Borlakoglu JT, Wilkins JPG (1993) Comp Biochem Physiol 105C:113
53. Bühler F, Schmid P, Schlatter C (1988) Chemosphere 17:1717
54. Safe SH (1989) Polychlorinated aromatics: uptake, disposition and metabolism. In: Kimborough RD, Jensen AA (eds) Halogenated Biphenyls, Terphenyls, Naphthalenes, Dibenzodioxins and Related Products. Elsevier, Amsterdam, p 131
55. Bergman Å, Athanasiadou M, Bergek S, Haraguchi K, Jensen S, Klasson-Wehler E (1992) Ambio 21:570
56. Brunström B (1992) Ambio 21:585
57. McFarland VA, Clarke JU (1989) Environ Health Perspect 81:225
58. Lewis DFV (1998) Xenobiotica 28:617
59. Boon JP, Van Arnhem E, Jansen S, Kannan N, Petrick G, Schulz D, Duinker JC, Reijnders PJH, Goksøyr A (1992) The toxicokinetics of PCBs in marine mammals with special reference to possible interactions of individual congeners with the cytochrome P450-dependent monooxygenase system: an overview. In: Walker CH, Livingstone DR (eds) Persistent Pollutants in Marine Ecosystems. Pergamon Press, Oxford, p 119
60. Goksøyr A (1995) Cytochrome P450 in marine mammals: isozyme forms, catalytic functions, and physiological regulations. In: Blix AS, Walløe L, Ulltang Ø (eds) Whales, Seal, Fish and Man. Elsevier Science, BV, Amsterdam, p 629
61. Haraguchi K, Kuroki H, Masuda Y, Koga N, Kuroki J, Hokama Y, Yoshimura H (1985) Chemosphere 14:1755
62. Preston BD, Allen JR (1980) Drug Metab Dispos 8:197
63. Koga N, Kikuichi-Nishimura N, Hara T, Harade N, Ishii Y, Yamada H, Oguri K, Yoshimura H (1995) Arch Biochem Biophys 317:464
64. Stadnicki SS, Allen JR (1979) Bull Environ Contam Toxicol 23:788

65. Reich HJ, Reich IL, Wollowitz S (1978) J Am Chem Soc 100:5981
66. Koga N, Kikuichi-Nishimura N, Yoshimura H (1995) Biol Pharm Bull 18:705
67. Ishida C, Koga N, Hanioka N, Saeki HK, Yoshimura H (1991) J Pharmacobio-Dyn 14:276
68. Ariyoshi N, Koga N, Yoshimura H, Oguri K (1997) Xenobiotica 27:973
69. Jerina DM, Daly JW (1974) Science 185:573
70. Klasson-Wehler E, Hovander L, Lund B-O (1996) Chem Res Toxicol 9:1340
71. Klasson-Wehler E, Lindberg L, Jönsson C-J, Bergman Å (1993) Chemosphere 27:2397
72. Yoshimura H, Yonemoto Y, Yamada H, Koga N, Oguri K, Saeki S (1987) Xenobiotica 17:897
73. Klasson-Wehler E (1989) PhD thesis, Stockhom University, Stockholm, Sweden
74. Ariyoshi N, Tanaka M, Ishii Y, Oguri K (1994) J Biochem 115:985
75. Norback DH, Seymore JL, Knieriem KM, Peterson RE, Allen JR (1976) Res Comm Chem Pathol Pharmacol 14:527
76. Bakke JE, Bergman ÅL, Larsen GL (1982) Science 217:645
77. Bergman Å, Larsen GL, Bakke JE (1982) Chemosphere 11:249
78. Bakke JE (1989) Metabolites derived from glutathione conjugation. In: Hutson DH, Caldwell J, Paulson GD (eds) Intermediary Xenobiotic Metabolism in Animals. Taylor & Francis, London, p 205
79. Vermeulen NPE (1996) Role of metabolism in chemical toxicity. In: Ioannides C (ed) Cytochromes P450: metabolic and toxicological aspects. CRC Press, New York, p 29
80. Bakke JE, Gustafsson J-Å (1986) Role of intestinal microflora in metabolism of polychlorinated biphenyls. In: Hallgren B (ed) Diet and Prevention of Coronary Heart Disease and Cancer. Raven Press, New York, p 47
81. Larsen GL (1985) Xenobiotica 15:199
82. Bakke JE (1990) Drug Metab Rev 22:637
83. Stevens JL, Bakke JE (1990) S-methylation. In: Mulder GJ (ed) Conjugation reactions in drug metabolism: an integrated approach. Taylor and Francis, London, p 251
84. Brandt I, Klasson-Wehler E, Rafter J, Bergman Å (1982) Toxicol Lett 12:273
85. Forrester LM, Henderson CJ, Glancey MJ, Back DJ, Park BK, Ball SE, Kitteringham NR, McLaren AW, Miles JS, Skett P, Wolf CR (1992) Biochem J 281:359
86. Brandt I, Lund J, Bergman Å, Klasson-Wehler E, Poellinger L, Gustafsson J-Å (1985) Drug Metab Dispos 13:490
87. Haraguchi K, Kato Y, Masuda Y, Kimura R (1997) Xenobiotica 27:831
88. Haraguchi K, Kato Y, Kimura R, Masuda Y (1998) Chem Res Toxicol 11:1508
89. Haraguchi K, Kuroki H, Masuda Y (1987) Chemosphere 16:2033
90. Haraguchi K, Kato Y, Kimura R, Masuda Y (1997) Drug Metab Dispos 25:845
91. Bergman Å, Wachtmeister CA (1978) Chemosphere 12:949
92. Haraguchi K, Kuroki H, Masuda Y (1987) J Agric Fd Chem 35:178
93. Mortimer RD, Newsome WH (1996) Chemosphere 32:935
94. Olsson A, Asplund L, Helander B, Bergman Å, Kylin H (1993) Organohalogen Compounds 14:113
95. Haraguchi K, Kuroki H, Masuda Y (1989) Chemosphere 19:487
96. Letcher RJ, Norstrom RJ, Bergman Å (1995) Anal Chem 67:4155
97. Haraguchi K, Kuroki H, Masuda Y (1987) Chemosphere 16:2299
98. Wiberg K, Letcher RJ, Sandau C, Duffe J, Norstrom RJ, Haglund P, Bidleman T (1998) Anal Chem 70:3840
99. Ellerichmann T, Bergman Å, Franke S, Hühnerfuss H, Jakobsson E, König WA, Larsson C (1998) Fresenius Envir Bull 7:244
100. Bergman Å, Jansson B, Bamford I (1980) Biomed Mass Spectrom 7:20
101. Buser HR (1985) Anal Chem 57:2801
102. Buser H-R, Zook DR, Rappe C (1992) Anal Chem 64:1176
103. Haraguchi K, Bergman Å (1991) Chemosphere 23:1837
104. Bergman Å, Wachtmeister CA (1987) J Labelled Comp Radiopharm 14:925
105. Baarschers WH, Krupay BW (1973) Can J Chem 51:156
106. Haraguchi K, Bergman Å, Jakobsson E, Masuda Y (1993) Fresenius J Anal Chem 347:441

107. Letcher RJ, Norstrom RJ (1997) J Mass Spectrom 32:232
108. Janák K, Becker G, Colmsjö A, Östman C, Athanasiadou M, Valters K, Bergman Å (1998) Environ Toxicol Chem 17:1046
109. Janák K, Grimvall E, Östman C, Colmsjö A, Athanasiadou M, Bergman Å (1994) J Microcol Sep 6:605
110. Ebner KV, Braselton WEJ (1987) Chem-Biol Interact 63:139
111. Hardt IH, Wolf C, Gehrcke B, Hochmuth DH, Pfaffenberger B, Hühnerfuss H, König WA (1994) J High Res Chromatog 17:859
112. Tulp MTM, Olie K, Hutzinger O (1977) Biomed Mass Spectrom 4:310
113. Jansson B, Sundström G (1974) Biomed Mass Spectrom 1:386
114. McOmie JF, Watts FL, West DE (1968) Tetrahedron 24:2289
115. Bauer U, Amaro AR, Robertson LW (1995) Chem Res Toxicol 8:92
116. Amaro AR, Oakley GG, Bauer U, Spielmann HG, Robertson LW (1996) Chem Res Toxicol 9:623
117. Haraguchi K, Kuroki H, Masuda Y (1989) Chemosphere 18:477
118. Troisi GM, Haraguchi K, Simmonds MP, Mason CF (1998) Arch Environ Contam Toxicol 35:121
119. Olsson A, Bergman Å (1995) Ambio 24:119
120. Mizutani T, Yamamoto K, Tajima K (1978) J Agric Food Chem 26:862
121. Haraguchi H, Kuroki H, Masuda Y, Shigematsu N (1984) Fd Chem Toxic 22:283
122. Haraguchi K, Kuroki H, Masuda Y (1986) Chemosphere 15:2027
123. Haraguchi K, Kuroki H, Masuda Y (1984) J Anal Toxicol 8:177
124. Haraguchi K, Kuroki H, Masuda Y (1986) J Chromatog 361:239
125. Karlson K, Ishaq R, Zebühr Y, Berggren P, Broman D (1998) Organohalogen Compounds 39:277
126. Weistrand C, Norén K, Nilsson A (1997) Environ Sci Pollut Res 4:2
127. de Voogt P, van Raat P, Rozemeijer M, Green N (1996) Organohalogen Compounds 28:517
128. de Voogt P, van Velzen MJM, Leonards PEG (1993) Organohalogen Compounds 14:105
129. de Voogt P, Häggberg L (1993) Chemosphere 27:271
130. Zook DR, Buser H-R, Bergqvist P-A, Rappe C, Olsson M (1992) Ambio 21:557
131. Yoshida S, Nakamura A (1979) Bull Environm Contam Toxicol 21:111
132. Simpson ER, Mahendroo MS, Means GD, Kilgore MW, Hinshelwood MM, Graham-Lawrence S, Amarneh B, Ito Y, Fisher CR, Michael MD, Mendelson CR, Bulun SE (1994) Endocrine Rev 15:342
133. Klasson-Wehler E, Bergman Å, Kowalski B, Brandt I (1987) Xenobiotica 17:477
134. Lans MC, Klasson-Wehler E, Willemsen M, Meussen E, Safe S, Brouwer A (1993) Chem-Biol Interact 88:7
135. Hovander L, Athanasiadou M, Asplund L, Jensen S, Klasson-Wehler E (1999). In preparation
136. Hovander L, Athanasiadou M, Asplund L, Klasson-Wehler E (1998) Organohalogen Compounds 35:115
137. Bakke JE, Feil VJ, Bergman Å (1983) Xenobiotica 13:555
138. Murk A, Morse D, Boon J, Brouwer A (1994) Eur J Pharm 270:253
139. Rozemeijer MJC, Olie K, de Voogt P (1997) J Chromatogr A 761:219
140. Hopper MI (1987) J Agric Fd Chem 35:265
141. Bergman Å, Wachtmeister CA (1977) Acta Chem Scand B31:90
142. Yoshida S, Nakamura A (1977) J Food Hyg Soc 18:387
143. Kuratsune M, Yoshimura H, Hori Y, Okumura M, Masuda Y (1996) Yusho – A Human Disaster Caused By PCBs and Related Compounds, Kyushu University Press, Kyushu
144. Haraguchi K, Bergman Å, Athanasiadou M, Jakobsson E, Olsson M, Masuda Y (1990) Organohalogen Compounds 1:415
145. Hansler R, Moisey J, Monte E, Norstrom RJ, Boon JP, van den Berg M, Seinen W, Letcher RJ (1999) Arch Environ Contam Toxicol. In preparation
146. Hansler RJ, Moisey J, Montie E, Norstrom RJ, Boon JP, van den Berg M, Seinen W, Letcher RJ (1999) Organohalogen Compounds 42:197

147. Green N, van Raat P, Jones K, de Voogt P (1996) Organhalogen Compounds 29:453
148. Letcher RJ, Moisey J, Hansler R, Norstrom RJ, (1999). In preparation
149. Tanabe S, Watanabe S, Kan H, Tatsukawa R (1988) Mar Mam Sci 4:103
150. Stegeman JJ, Brouwer M, Di Guilio R, Förlin L, Fowler BA, Sanders BM, van Veld PA (1992) Molecular responses to environmental contamination: enzyme and protein systems as indicators of chemical exposure and effect. In: Huggett RJ, Kimerle RA, Mehrle PMJ, Bergman HL (eds) Biomarkers: Biochemical, Physiological and Histological Markers of Anthropogenic Stress. Lewis, Chelsea, p 235
151. Lin JH, Lu AYH (1998) Clin Pharmacokinet 85:361
152. Schultz DE, Petrick G, Duinker JC (1989) Environ Sci Technol 23:852
153. Frame GM, Cochran JW, Bowadt SS (1997) J High Resol Chromatogr 19:657
154. Norstrom RJ, Muir DCG (1994) Sci Total Environ 154:107
155. Letcher RJ, Norstrom RJ, Lin S, Ramsay MA, Bandiera SM (1996) Toxicol Appl Pharmacol 137:127
156. Haraguchi K, Bergman Å, Masuda Y (1994) Organohalogen Compounds 20:501
157. Lund B-O, Örberg J, Bergman Å, Larsson C, Bergman A, Bäcklin B-M, Håkansson H, Madej A, Brouwer A, Brunström B (1999) Environ Toxicol Chem 18:292
158. Larsson C, Ellerichmann T, Franke S, Athanasiadou M, Hühnerfuss H, Bergman Å (1999) Organohalogen Compounds 40:427
159. Bergman Å, Brandt I, Jansson B (1979) Toxicol Appl Pharmacol 48:213
160. Bergman Å, Brandt I, Larsson Y, Wachtmeister CA (1980) Chem-Biol Interact 31:65
161. Brandt I, Bergman Å, Wachtmeister CA (1976) Experimentia 32:497
162. Larsen GL, Huwe JK, Bergman Å, Klasson-Wehler E, Hargis P (1992) Chemosphere 25:1189
163. Bakke JE, Bergmann ÅL, Brandt I, Darnerud P, Struble C (1983) Xenobiotica 13:597
164. Haraguchi K, Kuroki H, Masuda Y (1990) Chemosphere 20:1229
165. Bergman Å, Brandt I, Darnerud PO, Wachtmeister CA (1982) Xenobiotica 12:1
166. Bergman Å, Biessmann A, Brandt I, Rafter J (1982) Chem-Biol Interact 40:123
167. Brandt I, Bergman Å (1981) Chem -Biol Interact 34:47
168. Lund J, Brandt I, Poellinger L, Bergman Å, Klasson-Wehler E, Gustafsson J-Å (1985) Molecular Pharmacol 27:314
169. Andersson O, Lund J, Ripe E (1987) Chemosphere 16:1667
170. Härd T, Barnes HJ, Larsson C, Gustafsson J-Å, Lund J (1995) Nature Struct Biol 2:983
171. Johansson M, Nilsson S, Lund B-O (1998) Environ Health Perspect 106:769
172. Shigematsu N, Ishimaru S, Saito R, Ikedu T, Matsuba K, Suginama K, Masuda Y (1978) Environ Res 16:92
173. Nagayama J, Kiyohara C, Mohri N, Hirohata T, Haraguchi K, Masuda Y (1989) Chemosphere 18:701
174. Kiyohara C, Hirohata T (1994) Toxic in Vitro 8:1185
175. Kiyohara C, Mohri N, Hirohata T, Haraguchi K, Masuda Y (1990) Pharmacol Toxicol 66:273
176. Kiyohara C, Hirohata T, Mohri N, Masuda Y (1990) Toxic in Vitro 4:103
177. Haraguchi K, Kawashima M, Yamada S, Isogai M, Masuda Y, Kimura R (1995) Chem-Biol Interact 95:269
178. Kato Y, Haraguchi K, Kawashima M, Yamada S, Masuda Y, Kimura R (1995) Chem-Biol Interact 95:257
179. Kato Y, Haraguchi K, Kawashima M, Yamada S, Isogai M, Masuda Y, Kimura R (1995) Chem-Biol Interact 95:269
180. Kato Y, Haraguchi K, Tomiyasu K, Saito H, Isogay M, Masuda Y, Kimura R (1997) Environ Toxicol Pharmacol 3:137
181. Lund B-O, Bergman Å, Brandt I (1986) Toxicol Lett 32:261
182. Bannister R, Biegel L, Davis D, Astroff B, Safe S (1989) Toxicology 54:139
183. Kato Y, Haraguchi K, Shibahara T, Masuda Y, Kimura R (1998) Arch Toxicol 72:541
184. Letcher RJ, van Holsteijn I, Drenth, H-J, Safe S, Bergman Å, Norstrom RJ, Pieters R, van den Berg M (1999) Toxicol Appl Pharmacol 160:10

185. Letcher R, van Holsteijn I, Pieters R, Norstrom R, Bergman Å, van den Berg M (1998) Organohalogen Compounds 37:49
186. Johansson M, Larsson C, Bergman Å, Lund B-O (1998) Pharmacol Toxicol 83:225
187. Lund B-O (1994) Environ Toxicol Chem 13:911
188. Bergman Å, Olsson M (1986) Finn Game Res 44:47
189. Sjödin A, Hagmar L, Klasson-Wehler E, Björk J, Bergman Å (1999). In preparation
190. Olsson A, Ceder K, Helander B (1999). In preparation
191. Darnerud PO, Brandt I, Klasson-Wehler E, Bergman Å, D'Argy R, Dencker L, Sperber GO (1986) Xenobiotica 16:295
192. Morse D, Klasson-Wehler E, Wesseling W, Koeman J, Brouwer A (1996) Toxicol Appl Pharmacol 136:269
193. Yamamoto H, Yoshimura H (1973) Chem Pharma Bull 21:2237
194. Narasimhan TR, Kim HL, Safe SH (1991) Biochem Toxicol 6:229
195. Klasson-Wehler E, Brunström B, Rannug U, Bergman Å (1990) Chem-Biol Interact 73:121
196. Brouwer A, van den Berg M (1986) Toxicol Appl Pharmacol 85:301
197. Darnerud PO, Sinjari T, Jönsson C-J (1996) Pharmacol Toxicol 78:187
198. Lans MC (1995) PhD thesis, Wageningen Agricultural University, Wageningen, The Netherlands
199. Koopman-Esseboom C, Morse DC, Weiglas-Kuperus N, Lutke-Schipholt I, van der Pauw CG, Tuinstra LGMT, Brouwer A, Sauer PJJ (1994) Pediatr Res 36:468
200. Bahn AK, Mills JL, Snyder PJ, Gann PH, Bialik O, Hollman L, Utiger RD (1980) New Engl J Med 302:31
201. Emmett EA, Maroni M, Jefferys J, Schmith J, Levin BK, Lavares A (1988) Am J Ind Med 14:47
202. Kreis K, Roberts C, Humphrey HEB (1982) Arch Environ Health 37:141
203. Murai K, Okamura K, Tsuji H, Kajiwara H, Watanabe H, Akagi K, Fujishima M (1987) Environ Res 44:179
204. Schuur AG, van Meeteren ME, Jong WMC, Bergman Å, Visser TJ, Brouwer A (1996) Organohalogen Compounds 28:509
205. Schuur AG, Bergman Å, Brouwer A, Visser TJ (1999) Toxic In Vitro 13:417
206. Schuur AG, Legger FF, van Meeteren ME, Moonen MJH, van Leeuwen-Bol I, Bergman Å, Visser TJ (1998) Chem Res Toxicol 11:1075
207. Adams C, Lans MC, Klasson-Wehler E, van Engelen JGM, Visser TJ, Brouwer A (1990) Organohalogen Compounds 1:51
208. Rickenbacher U, McKinney JD, Oatly SJ, Blake CF (1986) J Med Chem 29:641
209. Kramer VJ, Helferich WG, Bergman Å, Klasson-Wehler E, Giesy JP (1997) Toxicol Appl Pharmacol 144:363
210. Moore M, Mustain M, Daniel K, Chen I, Safe S, Zacharewski T, Gillesby B, Joyeux A, Balaguer P (1997) Toxicol Appl Pharmacol 142:160
211. Connor K, Ramamoorthy K, Moore M, Nustain M, Chen I, Safe S, Zacharewski T, Gillesby B, Joyeux A, Balaguer P (1997) Toxicol Appl Pharmacol 145:111
212. Letcher RJ, van den Berg M, Brouwer A (1999) Organohalogen Compounds 42:39
213. Jönsson C-J, Lund B-O, Brunström B, Brandt I (1994) Environ Toxicol Chem 13:1303
214. Jönsson C-J, Lund B-O, Brandt I (1993) Ecotoxicol 2:41
215. Bahrami F, Brittebo EB, Bergman Å, Larsson C, Brandt I (1999) Toxicol Sci 49:1
216. Asplund L, Olsson A, Häggberg M, Athanasiadou M, Olsson M, Bergman Å (1995) Copenhagen, congress proceedings, SETAC-Europe
217. Kato Y, Kenne K, Haraguchi K, Masuda Y, Kimura R, Wärngård L (1998) Arch Toxicol 72:178
218. Andersson O, Nordlund-Möller L, Barnes HJ, Lund J (1994) J Biol Chem 269:19081
219. Sandau C, Ayotte P, Dewailly É, Bergman Å, Klasson-Wehler E, Norstrom R (1998) Organhalogen Compounds 38:29
220. Sundström G, Hutzinger O, Safe S (1976) Chemosphere 4:249

Subject Index

Printing (Computer to Film): Saladruck, Berlin
Binding: Stürtz AG, Würzburg

Printed in the United Kingdom
by Lightning Source UK Ltd.
127600UK00001B/145/A